W9-AAL-213

To all radiography students past, present, and future.

About the author

Robert DeAngelis is currently the president of the Radiography Program Consultants based out of Rutland Vermont.

Mr. DeAngelis has been an instructor in Radiography for over a quarter of a century and continued to provide educational services for the field.

His previous publications include the Radiologic Science Workbook by Praeger Scientific, The Radiography workbooks I & II, The Integrated Radiography Workbook, and a wide array of continuing education and simulated registry review products for the imaging fields.

Introduction

The new edition of The Integrated Radiography Workbook has undergone an extensive modification to help insure that it conforms with the most recent content specifications of the registry exam.

The reader will find that a new format has been used to improve the usefulness of this edition.

Hundreds of new questions and the inclusion of new and expanded explanations should provide the radiography students the resource they will need to complete their studies for the certification exam in radiography.

Cover design by Ken Tate and Mary DeAngelis
Section artwork compliments of Paula Nicolosi
Section 4 compliments of Nuclear Associates
ISBN 0943589-23-1
2004 by the Health and Allied Science Publishers, Rutland, VT 05702

New Table of Contents

Section III Image Production and Evaluation
Radiographic Graffiti II

Section IV   Anatomy and Physiology
Radiographic Graffiti III

Section V   Radiographic Procedures
Radiographic Graffiti IV

Section VI Patient Care and Management

Section VII Basic Mathematical Principles

Eyes

Thyroid

Gonads

Section I

Radiation Protection and Biology

1. The reduction in the size of a tissue or organ that occurs from an exposure to radiation, chemical or physical injury is called:

   A. Cellulitis
   B. Anemia
   C. Atrophy
   D. Desquamation

2. The ability of ionizing radiation to produce a more intense biologic response in a tissue is related to all of the following EXCEPT:

   A. An increase in the quality factor of the radiation
   B. An increase in the oxygenation of the tissue
   C. A decrease in the linear energy transfer
   D. An increase in the proliferation rate

3. At the present diagnostic exposure levels, which of the following early effects are likely to be seen after a patient has undergone a GI series with 4 minutes of fluoroscopy?

   A. A reduction in the number of red blood cells
   B. A reduction in the sperm count
   C. The opacification of the lens of the eye
   D. No early or late effects are likely

4. Which of the following cells is most likely to demonstrate the greatest amount of radiation damage from a 25,000 mrem exposure to x-rays?

   A. Oocyte
   B. Chondrocyte
   C. Neuron
   D. Adipose cell

5. The relative radiosensitivity of an individual is highest:

   A. In the neonatal period
   B. In early childhood
   C. During adolescence
   D. In the geriatric period

6. A total dose equivalent of 20,000 mrem during a one year period, is most likely to increase the possibility of developing which of the following disorders?

   A. Leukemia
   B. Cataracts
   C. Hepatitis
   D. Pyloric stenosis

7. The damage to the somatic tissues associated with sublethal exposure to radiation is considered:

   A. Non-repairable or permanent in nature
   B. Additive becoming more severe with each exposure
   C. Relatively nondestructive resulting in little or no damage
   D. Largely repairable through the regeneration of the tissues

8. The normal process by which somatic cells proliferate is termed:

   A. Metabolic synthesis
   B. Osmosis
   C. Enuresis
   D. Mitosis

9. Which of the following is one of the probable causes of death from the effects of the central nervous system syndrome?

   1. Vasculitis   2. Meningitis   3. High intracranial pressure

   A. 1 only
   B. 2 only
   C. 3 only
   D. 1, 2 & 3

10. The symptoms of fever and malaise associated with a reduced number of leukocytes is one of the major characteristics of the:

    A. Prodromal syndrome
    B. Hematologic syndrome
    C. Gastrointestinal syndrome
    D. Central nervous system syndrome

11. A threshold dose-response relationship is normally characterized by a:

   A. Minimal dose above which the effects of an exposure are detectable
   B. Small but noticeable effect to an exposure at any dose level
   C. Failure to produce any detectable effect even at high dose levels
   D. Progressive worsening of the effects over time

12. A somatic effect called erythema, which occurs from a large exposure of radiation, is most like what other condition?

   A. Premature hair loss   C. Diabetes
   B. Anemia   D. Sunburn

13. Which of the following is considered a late effect following a large single exposure to radiation?

   A. Chromosomal aberration   C. Temporary sterility
   B. Desquamation   D. Carcinogenesis

14. The most recent evidence seems to support the following statement concerning the damage to the genetic material of a human or animal cell:

   A. The damage occurring to the genetic material does not demonstrate any capacity for repair
   B. The damage to the genetic material appears to be repaired over time
   C. The damage to the genetic material is often manifested in beneficial traits
   D. The damage to genetic material is apparent at extremely small exposure levels

15. The period of major organogenesis normally occurs during the _____ of fetal development.

   A. First two weeks   C. Twelfth-twentieth week
   B. Second-eighth week   D. Last six weeks

16. Any response occurring in a tissue or organism in the absence of an identifiable stimulus is called a/an:

   A. Causative response   C. Manifest response
   B. Indirect response   D. Ambient response

17. The deaths related to large exposure to radiation appear to be related to damage to all of the following EXCEPT:

   A. Central nervous system   C. Hematologic system
   B. Gastrointestinal system   D. Reproductive system

18. The most sensitive organs to the effects of a radiation exposure of 200 rem, appears to be the:

   A. Thyroid gland and the kidney   C. Liver and the lungs
   B. Spinal cord and the brain   D. Testes and the ovaries

19. The dose to the gonads that is expected to produce a measurable genetic change in the entire population is termed:

   A. Mean marrow exposure   C. Reproductive limiting exposure
   B. Inherited radiation dose   D. Genetically significant dose

20. Cells such as the erythrocytes and spermatozoon that have a specialized function and are in the final stage of their maturation process are termed:

   A. Proliferation cells   C. Differentiated cells
   B. Ground cells   D. Precursor cells

21. The reddening of the skin that can follow radiation dose equivalent of more than 150 rem is termed:

   A. Hemoderma   C. Erythema
   B. Leukogenesis   D. Spondylitis

22. The most sensitive phase of the normal cell cycle occurs during the:

A. Beginning of the G1 phase
B. Middle of the G2 phase
C. Middle of the S-phase
D. Middle of the M-phase

23. The laws of Bergonié and Tribondeau state that radiation damage to a cell is related to all of the following EXCEPT:

A. Size and weight
B. Differentiation
C. Reproductive activity
D. Metabolic rate

24. The human fetus is most sensitive to radiation in the:

A. First trimester
B. Second trimester
C. Third trimester
D. Equally in all trimesters

25. An exposure to between 50-200 rem to the embryo/fetus in utero appears to be related to the development of:

A. Urinary tract defects
B. Nervous system defects
C. Circulatory system defects
D. Respiratory system defects

26. The first sign of the acute radiation syndrome following a large exposure of radiation is usually:

A. Diarrhea
B. Incontinence
C. Nausea
D. Excessive bleeding

27. Which of the following cells has been found to have the highest radiosensitivity?

A. Neuron
B. Chondrocyte
C. Lymphocyte
D. Osteocyte

28. A change in the normal proliferation or reproductive rate of a group of cells is called the:

A. Mitotic delay
B. Reproductive boost
C. Latent division
D. Prodromal syndrome

29. The property of x-rays that account for their effect on biological systems is in radiation's ability to:

A. Make the cells radioactive
B. Neutralize electric charges
C. Cause ionization in cells
D. Disrupt reproductive functions

30. Trisomy is any class of genetic disorders resulting from a cell with:

A. Extra chromosomes
B. Too few chromosomes
C. Twisted chromosomes
D. Mutated chromosomes

31. A direct effect usually refers to a _____ caused by x-ray photon or other form of high-energy radiation.

A. Point mutation on a DNA molecule
B. Rapid division of the cells
C. Free radical
D. Spontaneous mutation

32. Acute radiation exposure to the skin may result in erythema or an ulcerating condition called:

A. Desquamation
B. Vasculitis
C. Epilation
D. Isolysis

33. Which of the following tissue groups possesses the lowest radiosensitivity?

A. Reproductive and hemopoietic
B. Muscular and nervous
C. Gastrointestinal
D. Endocrine and dermis

34. A radiation exposure to which of the following systems will most likely result in damage to the blood forming organs?

A. Central nervous system
B. Urinary system
C. Biliary system
D. Skeletal system

35. The linear (non-threshold) curve of radiation dose-response relationships states that radiation damage is directly proportional to radiation dose, and that:

A. Only high exposures are associated with radiation's harmful effects
B. Any dose regardless of the magnitude can be associated with harmful effects
C. Radiation damage is apparent only after a minimum exposure is received
D. Somatic effects can occur at low exposures

36. The most common symptoms accompanying the gastrointestinal syndrome are:

A. Hypotension and apnea
B. Dehydration and vomiting
C. Uremia and hemorrhage
D. Tetanus and bradycardia

37. The normal human blood cell that is associated with the lowest radiosensitivity is the:

A. Mature erythrocyte
B. Mature lymphocyte
C. Immature leukocyte
D. Immature thrombocyte

38. The time following a large irradiation in which the signs of the exposure are absent is termed the:

A. Manifest period
B. Latent period
C. Conception period
D. Inherited period

39. Which of the following macromolecules has the greatest sensitivity to radiation?

A. Proteins
B. Mitochondria
C. Deoxyribonucleic acid
D. Adenosine triphosphate

40. The degree to which recovery is possible after exposure to radiation is related to all of the following EXCEPT:

A. Total dose received
B. LET of the radiation
C. Degree of protraction
D. Degree of reflection

41. A common by-product resulting from the irradiation of water in a living cell is a bleach called:

A. Hydrogen peroxide
B. Deoxyribose
C. Quinine
D. Ozone

42. The process by which energy from ionizing radiation is transferred from a directly ionized molecule to another molecule that is not directly ionized is called the:

A. Direct hit theory
B. Random interaction theory
C. Potential free radical theory
D. Indirect hit theory

43. The developing human is most sensitive to radiation during the _____ development stage.

A. Embryonic
B. Fetal
C. Neo-natal
D. Post-natal

44. The central nervous system syndrome that occurs following exposures of at least 2000 rem will normally cause death in approximately:

A. 30-60 days
B. 12-15 days
C. 3-5 days
D. 1-2 days

45. The study of a response to radiation or other stimulus occurring to tissues outside of the body (e.g. test tube) is termed an _____ response.

A. In vino
B. In vento
C. In vitro
D. In vivo

46. A biologic tissue is least sensitive to radiation damage:

A. During it's reproductive activities
B. When it is in an hypoxic state
C. During periods of rapid cell growth (cell division)
D. During periods of metabolic activity

47. Which of the following are the common symptoms associated with a large radiation exposure to the hemopoietic system?

   1. Reduced number of circulating blood cells
   2. Increased susceptibility to infection
   3. Longer coagulation time

A. 1 only
B. 2 only
C. 3 only
D. 1, 2 & 3

48. As a general rule, somatic damage to radiation is approximately _____ reparable when small exposures are received in fractionated doses.

A. 90%
B. 60%
C. 30%
D. 10%

49. The time between a lethal radiation exposure and death is termed:

A. Mean survival time
B. Mean threshold time
C. Mean recovery ratio
D. Latent-manifest ratio

50. Which of the following is the most sensitive cell of the body to radiation exposure?

A. Chondrocytes
B. Mature red blood cells
C. Spermatogonia
D. Nerve cells

Referring to the diagram of the relative radiosensitivity graph by age, answer questions 51 & 52.

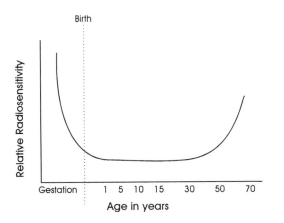

51. According to the diagram, the greatest sensitivity to radiation after a person is born is at the approximate age of:

A. 1 year
B. 10 years
C. 30 years
D. 70 years

52. In general, from ages 5-50 it can be said that radiosensitivity:

A. Increases with age
B. Decreases with age

C. Remains relatively constant with age
D. Varies greatly with age

53. Death resulting from radiation damage to the stem cells of the gastrointestinal tract is most likely to occur within:

A. 5 hours
B. 5 days

C. 4 weeks
D. 3 months

54. Organogenesis is the embryonic development stage that is associated with the:

A. Development of the major organs
B. Final maturation of the tissues

C. Formation of the skeletal system
D. Final development of genetic tissues

55. A large (5 gray) exposure to the hematologic system will result in the reduction of _____ in the circulating blood.

1. Leukocytes   2. Thrombocytes   3. Erythrocytes

A. 1 only
B. 2 only

C. 3 only
D. 1, 2 & 3

56. The manifest period of an acute exposure to radiation refers to the time during which (the):

A. Exposure causes mutations
B. Effects of the exposure are hidden

C. Effects of the exposure are apparent
D. Genetic effects occur

Referring to the survival curve, answer questions 57 & 58.

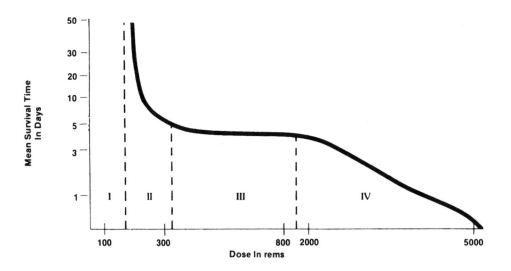

57. In the dose survival curve, the Roman numeral, which corresponds to death from the gastrointestinal tract, is:

A. I
B. II

C. III
D. IV

58. According to the diagram, at dose levels above 2000 rem, death is most likely to occur from damage to the:

A. Epithelial tissue
B. Urinary tract

C. Alimentary tract
D. Central nervous system

59. Which of the following types of radiation-induced conditions is classified as a late somatic effect?

    1. Cataract formation   2. Cancer   3. Leukemia

    A. 1 only   C. 3 only
    B. 2 only   D. 1, 2 & 3

60. A genetic mutation is defined as one that appears in the:

    A. Soft tissue of the body   C. Offspring of the exposed individual
    B. Blood forming organs   D. Cytoplasm of the exposed individual

61. Approximately _____ of the genetic mutations that occur are considered harmful to the individual.

    A. 99%   C. 75%
    B. 87%   D. 37%

62. The most radiosensitive part of a normal human cell is the:

    A. Cell membrane   C. Golgi Body
    B. Centriole   D. Chromosome

63. A large radiation exposure to the lens of the eye will most likely result in the clouding of the lens of the eye called:

    A. Glaucoma   C. Retinitis
    B. Scleroma   D. Cataracts

64. Death that follows an acute exposure to radiation of about 500 rem will generally occur in about:

    A. 3-5 hours   C. 3-5 weeks
    B. 3-5 days   D. 3-5 years

65. The majority of damage to the body from radiation exposure results from the:

    A. Direct effect   C. Target effect
    B. Indirect effect   D. Threshold effect

66. A general class of genetic disorders resulting from a deficient number of chromosomes is termed:

    A. Monosomy   C. Trisomy
    B. Oconsomy   D. Mutosomy

67. All of the following conditions can be classified as an early somatic effect from a radiation exposure EXCEPT:

    A. Nausea   C. Leukopenia
    B. Epilation   D. Life span shortening

68 The threshold dose refers the amount of exposure at which:

    A. All cells are damaged   C. A chromosomal defect is manifested
    B. No more damage can occur   D. A biologic response is detected

69. The loss of hair after exposure to large amounts of radiation is termed:

    A. Epistaxis   C. Bromidrosis
    B. Epilation   D. Erythema

70. Because of its high reproductive activity, which of the following tissues will show the greatest sensitivity to radiation:

    A. Central nervous tissue   C. Muscular tissue
    B. Alimentary tract tissue   D. Cardiovascular tissue

71. The immediate symptoms that appear after an acute radiation exposure are termed the:

A. Latent syndrome
B. Proportional syndrome

C. Chronic syndrome
D. Prodromal syndrome

Referring to the diagram of the dose-response curves, answer questions 72 & 73.

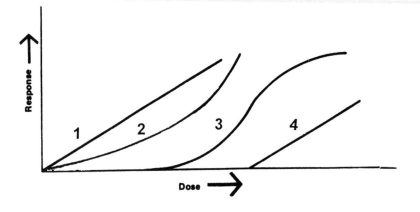

72. In the diagram, the linear non-threshold dose-response curve is represented by number:

A. 1
B. 2

C. 3
D. 4

73. The curve represented by number 3 is a:

A. Linear, threshold curve
B. Nonlinear, threshold curve

C. Linear, non-threshold curve
D. Nonlinear, non-threshold curve

74. The acute radiation syndrome associated with damage to the blood and blood-forming organs is called the:

A. Prodromal syndrome
B. Central nervous system syndrome

C. Gastrointestinal syndrome
D. Hematologic syndrome

75. A radiation exposure of more than 150 rem to the developing fetus is often associated with:

A. Spontaneous abortion
B. Cardiac abnormalities

C. Loss of hearing
D. Premature blindness

Explanations

1. C An exposure to radiation or other type of stimulus can result in a number of biologic effects. One of the more common is a reduction in the size of an irradiated tissue called atrophy.

2. C The classic laws of cellular sensitivity as proposed by Bergonié and Tribondeau relate cellular response to a chemical stimulus of radiation exposure to the cellular maturity (age), metabolic activity (O_2 consumption) and proliferation rate and the degree of tissue differentiation. Research has shown that cells that create new cells rapidly and have a short life span termed young cells are much more sensitive the radiation than older cells. Basel cells of the epidermis crypt cells of the intestines and erythroblasts are classified as young cells. The metabolism or the rate at which food is converted into energy will also effect the sensitivity of the cell. Tissues with a high metabolic rate (neoplasms) are generally more sensitive than less active tissues. Cells that have a higher proliferation rate will also be more radiosensitive. Cells that are undifferentiated will also be more radiosensitive than differentiated tissues.

3. D In a modern fluorographic system, the amount of exposure received during a GI series with 4 minutes of fluoro time is about 6-10 rem. At this level of exposure, no visible signs of radiation damage will be apparent.

4. A Of the cells listed, the female reproductive cell, the oocyte, is the most sensitive to the effects of an irradiation.

5. D The relative radiosensitivity in man is relatively constant from early infancy to late adulthood. See the graph for questions 51-52. The radiosensitivity increases dramatically in the more easily compromised geriatric patient.

6. A At a dose of 20,00 mrem (20 rem) there is no evidence to support the development of any type of somatic effects in an exposed individual. Evidence from the survivors of the atomic bombings in Japan, does indicate that an increase in the number of cases of leukemia has been seen following this and higher levels of exposure.

7. D Even following relatively large exposures as long as the stromal or germ tissues survive, the cells that are regenerated are likely to be normal.

8. D The usual process by which all somatic tissues reproduce or proliferate is called mitosis. During mitosis, the chromosomes are replicated to enable the formation of two daughter cells containing the same number of chromosomes at the original parent cell.

9. D The acute radiation syndrome associated with the central nervous system is only apparent at extremely high radiation dose levels. At these levels, the majority of the damage is due to the destruction of the cell membranes of the nerve cells and blood vessels. This can lead to vasculitis, meningitis and an increase in the intracranial pressure.

10. B The acute radiation syndrome associated with the blood and blood forming tissues is called the hematologic syndrome. Following an exposure of over 300 rem, a reduction in the number of all types of blood cells is seen. This will lead to anemia, clotting problems and depleted immune system. The fever and malaise are just two of many symptoms related to these effects.

11. A A threshold effect by definition is an effect that is only apparent after some amount of stimulus has been received. For most somatic conditions, the radiation threshold is usually above 150 rem.

12. D Erythema is the term used to describe the reddening of the skin following an exposure to ionizing radiation. It is similar in appearance to a sunburn.

13. D A late effect is defined as one that occurs only after a few months or years have passed following an irradiation. It has been found that carcinogenesis will normally not occur for 2-30 years following an exposure and is therefore classified as a long-term effect.

14. B   The ability to damage genetic tissues was clearly demonstrated in radiation experiments in the mid 1900's. Since then, it has been discovered that DNA has the capacity to repair most defects in the genetic material over a relatively short period of time following an irradiation.

15. B   The most important time in the development of the embryo/fetus is the period of major organogenesis. Beginning around the second week of gestation, the individual organs begin to develop. During this six-week process, the embryo is extremely sensitive to the effects of chemical or radiation stimuli.

16. D   A response that occurs in the absence of an identifiable stimulus is called an ambient effect. These can be caused by background radiation or any number of spontaneous genetic or chromosomal abnormalities.

17. D   Any exposure to radiation above the 300 rem level will cause a number of recognizable symptoms that are related to damage to organs that are sensitive to radiation. Depending on the amount of radiation that is absorbed, three systems have been identified with the death of the individual. These are the hematologic or blood forming tissues, the tissues of the gastrointestinal tract, and with massive doses the tissues of the central nervous system.

18. D   The testes and ovaries that contain the immature sperm and developing ova consist largely of undifferentiated tissues that are acutely sensitive to the effects of a radiation exposure.

19. D   In order to help estimate the genetic impact of low-dose radiation on large population group, an estimation of the actual value of the gonadal dose received by this population must be determined. This value is termed the genetically significant dose (GSD).

20. C   In biologic terms, cells or tissues that have a specialized function and are at the end of the cellular maturation process, and do not undergo any further mitotic activity, are called differentiated tissues. The erythrocyte and spermatozoon are two example of differentiated cells. These cells tend to have a low sensitivity to the effects of chemical stimuli or radiation. Cells are considered undifferentiated when they are not specialized and are involved in reproductive processes. Immature or undifferentiated cells, such as erythroblasts and spermatogonia, are much more sensitive to the effects of chemicals of radiation.

21. C   Following an exposure to large amounts of ionizing radiation (>150 rem) the skin can react with a sunburn type condition called erythema.

22. D   The normal cell cycle is divided into two phases and two periods between the phases called the G1 and G2 gaps. The first, called the M-phase or mitosis, is the part of the cycle concerned with the actual separation of a parent single cell into two smaller daughter cells. The cell appears to be more sensitive to the effects of an irradiation in the M-phase than at most other times in the cell cycle. The S-phase or DNA Synthesis is the portion of the cell cycle associated with replication of the genetic material in preparation for mitosis. The cells appear to be least sensitive to the effects of an irradiation during the S-phase. In the G1 gap between the mitosis and DNA synthesis, the relative sensitivity tends to rise steadily. Following the DNA synthesis but before the next mitotic cycle G2 gap, the relative sensitivity of the cell remains low.

23. A   The classic laws of cellular sensitivity, as proposed by Bergonié and Tribondeau, relate cellular response to a chemical stimulus of radiation exposure to the cellular maturity (age), metabolic activity (O_2 consumption) and proliferation rate and the degree of tissue differentiation. Research has shown that cells that create new cells rapidly and have a short life span are termed young cells. Young cells are much more sensitive to radiation than older cells. Basal cells of the epidermis crypt cells of the intestines and erythroblasts are classified as young cells. The metabolism or the rate at which food is converted into energy will also effect the sensitivity of the cell. Tissues with a high metabolic rate (neoplasms) are generally more sensitive to a stimulus than less active tissues. Cells that have a higher proliferation rate will also be more radiosensitive. Cells that are undifferentiated will also be more radiosensitive than differentiated tissues.

24. A   In utero irradiations have been found to be most harmful in the 2 week - 2 month development stage of the first trimester. It has been found that the relative risk in the period is about 5-6 times that of the last two trimesters.

25. B   Based in large part on the survivors of the atomic bombings in Japan it appears that large in utero irradiations of between 50-200 rem are associated with an increased incidence of spontaneous abortions, early neonatal death, future malignancies and severe developmental abnormalities of the central nervous and skeletal systems. When the estimated dose to the embryo-fetus was less than 15 rem, the data failed to show any type of abnormalities or future malignancies.

26. C   The high sensitivity of the crypt cells in the walls of the intestinal tract appears to be the main reason why the first signs of an acute radiation injury are nausea and vomiting.

27. C   The short life span of most lymphocytes makes them more sensitive to the harmful effects of irradiation than the longer lived thrombocytes or erythrocytes.

28. A   A large irradiation (100-900 rem) to a colony of cells is often associated with one of three responses called the mitotic delay, interphase death or reproductive failure. The first of these responses occurring at doses between 100-300 rem is a transient effect called the mitotic delay that alters the rate at which cells enter mitosis. The length and magnitude of the response is more pronounced as the level and dose rate increase. When the equivalent dose increases to more than 500 rem, the direct damage to the genetic material can prevent the future mitosis of the cells. Depending strongly on the cellular sensitivity, doses of between 25-900 rem, some colonies of cells will experience interphase death resulting from the disruption of the plasma membrane and the resulting mixing of the intra and extracellular environments.

29. C   The principal means by which radiation transfers energy to the biologic system is called ionization. Ionization is defined as the disassociation of compounds into their associated ions. Because ionization often leads to the modification of the chemical nature of the compounds. This is the principal mechanism by which radiation injuries occur.

30. A   Mongolism (Down's Syndrome) also called trisomy 21 is one of a number of hereditary disorders involving extra chromosomes which may be linked to a large radiation exposure to the reproductive cells.

31. A   Because of its sensitive molecular structure, deoxyribonucleic acid (DNA) is prone to damage from a direct interaction with ionizing radiation. Most changes that occur in this molecule are described as direct effects.

32. A   The denuding of the skin surface in response to large irradiation may result in the uncomfortable dry desquamation or more serious burn-like moist desquamation of the skin. These conditions that occur following massive exposures are sometimes seen during a course of external beam radiation therapies.

33. B   The high degree of specialization (differentiation) of nerve and muscle tissue tends to make them more resistant than most other cells of the body to the effects of radiation or chemical stimuli.

34. D   The red bone marrow which is found in the ends of all developing long bones and nearly all adult flat bones is responsible for the production of red blood cells, white blood cells, and platelets. This undifferentiated tissue is acutely sensitive to the effects of chemical stimuli or radiation.

35. B   A curve which originates at the dose-response junction (non- threshold) indicates that any exposure, no matter how small, may cause a potential response in the irradiated individual. The linear shape of the curve indicates the response will tend to occur in direct proportion to the radiation dose.

36. B   Because the stem (crypt) cells that line the tissues of the gastrointestinal tract that are responsible for the uptake of fluids, their destruction will lead to rapid loss of fluid volume and dehydration within the cells of the body.

37. A   From the most to least sensitive are the immature or precursor cells, the mature lymphocytes, mature platelets and the mature erythrocytes.

38. B   Most individuals receiving large doses >300 rem of radiation will have a short period of time during which the effects are not apparent. This transient period is called the latent period.

39. C   All of the macromolecules involved with the duplication of genetic material are sensitive to radiation. The relative sensitivity of these macromolecules in decreasing order are DNA, RNA, proteins, and ATP.

40. D   The amount of damage and ability of an organism to recover from a chemical stimulus or radiation exposure is related to a number of physical and biologic factors (explanation 23). Among the more important physical factors effecting the radiation response are the total dose received, the time over which the exposure occurs (protraction), the volume of the tissue exposed and the biologic effectiveness (quality factor, LET) of the radiation. An increase in the total dose, the size of the volume of tissue exposed, and the quality factor or LET will increase the response. An increase in the degree of protraction will lessen the effects of an exposure.

41. A   The irradiation of water will result in the removal of electrons and a chemical recombination of the hydrogen and oxygen ions that remain. This process, called the hydrolysis of water, is responsible for the formation of many different ions and free radicals. Some of the most important of these are hydrogen radicals, hydroxyl radicals, and hydrogen peroxide.

42. D   The majority of the radiation damage that occurs in the body is a result of the radiolysis of water. Since this chemical change is not triggered directly by ionization of water, but occurs from its byproducts, this event is classified as an indirect effect.

43. A   As a fetus develops, it goes through a number of developmental stages. By far the most sensitive period is the embryonic stage (2-8 weeks) and in order of decreasing sensitivity the ovum (0-2 weeks), fetal (8-40 weeks) and neonatal period after birth.

44. D   Because of the relatively low sensitivity of the central nervous system, a severe lethal effect such as excessive intracranial pressure will only appear at dose equivalents exceeding 2,000 rem (20 Sv). At these high dose levels the effects are seen within a few hours of the exposure and death is assured within 48 hours.

45. C   Two classic types of experimental testing methods are available for use with living organisms, in vivo and in vitro testing. In vivo testing accomplished within a living organism, though preferred, is often more difficult to accomplish because a test subject may be subjected to potentially dangerous chemical or radiation stimuli. In vitro testing involves the removal of the tissue from the body and its reaction to a stimulus. Though this type of testing is less dangerous it may not reflect the actual effect that will occur in organism as a whole.

46. B   It is possible to reduce the radiosensitivity of a cell by the lowering of the cell's oxygen content (hypoxia) through the use of antioxidants or radioprotectors. Some of the sulfhydryl compounds appear to have this effect in some tissues.

47. D   All types of blood cells can be reduced following an irradiation of more than 100 rem. Based on their function, the loss of the erythrocytes (red blood cells) is likely to be associated with anemia and the reduced capacity of the blood to carry oxygen. A reduced leukocyte (white blood cell) count may negatively impact the immune function while a deficiency in the thrombocyte (platelets) count will lengthen the coagulation or clotting times.

48. A   In an otherwise healthy individual, sublethal exposures are estimated to be about 90% repairable within just a few days following an exposure.

49. A   The survival time following a supra-lethal exposure to radiation (>800 rem) is primarily dependent upon the total dose, and time over which the exposure is received. The larger the dose, the shorter the mean survival time will be for the exposed individual.

50. C   The undifferentiated stem cells of the male reproductive system, the spermatogonia, the female egg cells and the short lived lymphocytes are considered the three most sensitive cells of the body to either chemical stimuli or radiation exposures.

51. D   In a graph of the relative sensitivity of an individual over their life, two periods of increased sensitivity are noted. The first occurs before birth during the embryonic development stage (2-8 weeks), and the second in the more easily compromised elderly patient.

52. C   Following the birth of an individual, the relative radiosensitivity remains nearly constant throughout childhood, adolescence and adulthood.

53. B   The death caused by the destruction of the crypt cells of the intestinal tract (GI). The symptoms of the gastrointestinal tract syndrome which include excessive fluid loss, electrolytic imbalance and secondary infections normally occurs within 3-5 days following an exposure to between 800-5000 rem exposure.

54. A   The development of the different organ systems from the undifferentiated cells occurs during the most sensitive portion of the development process called organogenesis. Organogenesis normally occurs in the embryonic stage between 2-8 weeks of intrauterine development.

55. D   All types of blood cells are dramatically reduced following an irradiation of more than 100 rem. Based on their function, the loss of the erythrocytes (red blood cells) is likely associated with anemia and the reduced capacity of the blood to carry oxygen. A reduced leukocyte (white blood cell) count may negatively impact the immune function while a deficiency in the thrombocyte (platelets) count will lengthen the coagulation or clotting times.

56. C   Radiation exposures and their visible effects on an irradiated individual can be divided into two categories; manifest effects, which are readily apparent in the person or latent (hidden) effects, which may remain dormant for a few hours, days or years.

57. C   Serious malabsorption and digestive problems result from dose equivalents in the 800-5000 rem, (850 Sv) range. The gastrointestinal symptoms including nausea, vomiting and diarrhea are often severe enough to cause death in a period of just 3-5 days.

58. D   A large dose equivalent in the range of 2,000-10,000 rem (20-100 Sv) will cause significant damage to the brain and spinal cord (central nervous system) causing high intracranial pressure, blindness, coma and death within 1-2 days.

59. D   Those effects which take many months or years to manifest themselves are called long term or late somatic effects. Most forms of leukemia, cancer and cataracts that take between 4-30 years to develop are classified as late effects.

60. C   Radiation induced mutations resulting from damage to the DNA molecules, genes, or chromosomes in the germ cells (sperm or egg) and appear in the offspring are called genetic effects. Radiation damage effects which occur in an exposed individual is classified as somatic effect.

61. A   Nearly all of the radiation or chemically induced genetic mutations are characterized by their negative impact on the viability of the exposed organism. It has been estimated that over 99% of the mutations that occur would have a negative impact on the developing fetus.

62. D   The genetic information of the cell is carried on a long chain of deoxyribonucleic acid (DNA) which is coiled into the subunits called chromosomes. DNA appears to be the most sensitive and important molecule to the action of chemical toxins or radiation.

63. D   Radiation exposures exceeding 150 R (150 cGy) to eye will substantially increase the rate of cataractogeneses (formation of cataracts) in the exposed individual. Cataracts refer to the opacification of the lens of the eye or its capsule.

64. C   An acute dose equivalent of 500 rem (5 Sv) is not sufficient to reach the threshold to produce the symptoms associated with gastrointestinal or central nervous system syndromes. Therefore the exposed individuals are most likely to die of damage to the hematologic organs which can take between 20-60 days.

65. B   Because it comprises nearly 80% of the body's mass, water is the most common molecule effected by radiation. The harmful byproducts, produced by its disassociation when struck by radiation such as hydrogen peroxide and hydroxyl ions, constitute most of the harmful chemicals associated with the indirect effect (see explanation 42).

66. A   Any condition resulting from a reduction in the normal number (46) of chromosomes is termed monosomy. This type of defect is normally associated with a number of severe mental and/or physical deformities.

67. D   Among the first somatic effects which appear in an irradiated individual are nausea, vomiting, erythema, diarrhea, epilation and a reduction in all types of blood cells which can cause anemia, bleeding disorders and low immunocompetence. Life span shortening has been demonstrated as one of the possible long term effects to large chronic exposures to radiation.

68. D   The visible manifestation of a biologic response is called the threshold or minimum dose threshold. For somatic effects, exposures under this level will not produce detectable effects.

69. B   At dose equivalents above 200 rem (2 Sv) to the skin, a common response is a temporary or permanent loss of hair called epilation. The condition becomes more severe with increasing dose, therefore is classified as a type of deterministic effect.

70. B   The rapid proliferation of the cells in the gastrointestinal tract make them highly sensitive to the effects of irradiation. Muscle tissue, cardiovascular tissue and nervous tissues are all differentiated and have a low sensitivity to radiation.

71. D   The immediate symptoms of nausea, vomiting, diarrhea and leukopenia that occur at acute doses above 100 rem (1 Sv) are often called prodromal syndrome. These symptoms may occur in few hours at higher doses or in a couple of days at lower doses.

72. A   Linear curves distinguished by their straight line appearance are used when the biologic response is directly proportional to the dose. All non-threshold curves originate at the junction of the dose response axis and indicate that any dose regardless of its magnitude may be harmful.

73. B   Nonlinear or sigmoid curves are characterized by a response that is not directly proportional to dose. All threshold curves require a minimal dose to cause a response and originate at some point on the dose axis away from the dose-response junction.

74. D   The blood and blood forming tissue are included as part of the hematologic or hematopoietic system which include spleen and bone marrow tissues.

75. A   The effects of a large irradiation to developing fetus are highly variable depending upon the total dose and the developmental stage. A radiation exposure above 50 rem(.5 Sv) to the embryo-fetus has been implicated in a number of bone, central nervous system disorders, as well as leukemia and many other types of cancers. Larger doses may lead to a spontaneous abortion or any number of serious growth defects to the fetus.

1. A monitoring device which is worn on the outside of protective apron at the level of the collar is primarily intended to estimate the exposure received by the:

   A. Auditory ossicles
   B. Thyroid gland
   C. Mandibular rami
   D. Parotid gland

2. A patient receives an estimated exposure of 4000 millirad of x-rays during tomographic images of the abdomen. What is this person's dose equivalent?

   A. 2.5 millirem
   B. 2.5 centigray
   C. 4000 millirem
   D. 4000 centigray

3. The concept known as the air kerma is employed to equate the exposure to radiation traditionally measured by the:

   A. Roentgen and expressed in the units of gray
   B. Rad and expressed in the units of curie
   C. Rem and expressed in units of becquerel
   D. RBE and expressed in units of the LET

4. Which of the following relationships relates the exposure to the exposure rate and time?

   A. $Exposure = \dfrac{Time}{Exposure\ rate}$
   B. Exposure = Exposure rate x Time
   C. $Exposure = \dfrac{Exposure\ rate}{Time}$
   D. Exposure = Exposure rate^2 x Time

5. Which of the following radiation detection devices is most commonly used to determine the amount of scatter produced in a room during the irradiation of a patient?

   A. Pocket dosimeter
   B. Ionization-type survey meter
   C. Film badge monitor
   D. Scintillation counter

6. In order to properly calibrate an ionization type meter, the radiation reading must be compared to a:

   A. Beta emitting source
   B. Short-lived gamma source
   C. Long-lived gamma source
   D. Alpha emitting source

7. The most sensitive device for the detection of gamma radiation is a special type of radiation device called a/an:

   A. Scintillation counter
   B. Ionization chamber
   C. Film badge
   D. Pocket dosimeter

8. The amount of x-ray or gamma exposure resulting in the production of 2.58 x 10^{-4} coulombs in 1 Kg of air defines the:

   A. Rad
   B. Sievert
   C. LET
   D. Roentgen

9. A person is exposed to 100 rad of beta having a quality factor (QF) of 1. This person's dose equivalency is:

   A. 100 Curie
   B. 100 Becquerel
   C. 100 Rem
   D. 10 Roentgen

10. All of the following are the normal components of a film badge EXCEPT:

    A. Plastic holder
    B. Various metal filters
    C. Light-tight pocket
    D. Self-reading meter

11. A patient is exposed to 70 rad of 1 MeV neutrons having a quality factor (QF) of 5. This person's dose equivalency is:

    A. 35 rem
    B. 70 rad
    C. 350 rem
    D. 700 rad

12. The primary portion of most types of survey instruments is a gas filled chamber called a/an:

A. Cloud chamber
B. Cathode ray chamber

C. Ionization chamber
D. Valve chamber

13. The unit of absorbed dose, the rad (gray), can be employed in the measurement of:

1. Alpha   2. Neutrons   3. Gamma rays

A. 1 only
B. 2 only

C. 3 only
D. 1, 2 & 3

14. The primary difference between a Geiger-Muller detector and cutie pie detector is:

A. The way in which the image is displayed
B. The speed at which the information can be acquired
C. The types of radiation that can be detected
D. The voltage impressed on the ionization chamber

15. Because of environmental factors, the manufacturers recommend that a film badge personal monitor should NOT be worn for more than:

A. Two weeks
B. One month

C. 3 months
D. 6 months

16. Recently a new class of reusable radiation monitoring devices have begun to replace film badge monitoring. The new (Luxel) badges detect radiation through a process called:

A. Bioluminescence
B. Liquid crystal stimulation

C. Optically stimulated luminosity
D. Multi-photon absorption

17. The location of a lost radioactive source is most easily accomplished by using a:

1. Survey meter   2. Film badge   3. Scintillation counter

A. 1 only
B. 2 only

C. 3 only
D. 1, 2 & 3

18. When lead aprons are employed, a radiation-monitoring device, should be worn at the level of the:

A. Waist outside of the apron
B. Waist under the apron

C. Collar outside of the apron
D. Collar under the apron

19. A standard film badge monitor can be used for the measurement of:

1. X-ray exposures   2. Beta exposures   3. Gamma exposures

A. 1 & 2 only
B. 1 & 3 only

C. 2 & 3 only
D. 1, 2 & 3

20. During pregnancy a second monitoring device is usually worn at the level of the waist to help determine the exposure to the:

A. Embryo/fetus
B. Reproductive organs

C. Fallopian tubes
D. Seminal vesicles

21. Which of the following is the only device that can be used in a radiation safety program to make absolute exposure dose measurements?

A. Baldwin-Farmer ionization chamber
B. Victoreen condenser chamber

C. Geiger-Muller counter
D. Luxel (OSL) badge

22. A person is exposed to 700 mR of x-rays during a series of radiographic studies in a two-week period of time. This individual's dose equivalent is approximately:

A. .7 rem (.07 Sv)
B. 14 (.14 mCi)

C. 35 rad (.35 Gy)
D. 1.4 rem (.14 Sv)

23. The smallest exposure that is detectable through the use of a standard film type personnel monitoring device is approximately:

A. 1.0 mrem

C. 50 mrem

B. 10 mrem

D. 100 mrem

24. The only personnel monitoring device which will enable a determination of the amount of an exposure immediately following an irradiation is a:

A. TLD

C. Film badge

B. G-M Counter

D. Pocket dosimeter

25. Personnel monitoring program is used in a radiation facility to help insure that:

A. Radiation sources are properly evaluated for their intensity
B. Radiation exposures to the personnel are kept below effective dose limits
C. Radiation exposures are maintained at the effective dose limits
D. All potential sources of radiation exposures have been eliminated

26. The smallest radiation exposure that can be detected by a biological method is approximately:

A. 5 mrem

C. 1,000 mrem

B. 100 mrem

D. 25,000 mrem

27. The radiation unit derived when a rad is multiplied by the quality factor results in the unit of dose equivalency termed the:

A. Roentgen

C. Rem

B. LET

D. Rep

28. The principal active component of a thermoluminescent dosimeter is a special crystal called:

A. Silver bromide

C. Lithium fluoride

B. Sodium iodide

D. Calcium tungstate

29. Which of the following types of detection device is normally classified as a type of rate meter?

A. Geiger-Muller counter

C. Pocket dosimeter

B. Thermoluminescent dosimeter

D. Film badge

30. The roentgen is most closely related to which other type of exposure?

A. Primary radiation

C. Remnant radiation

B. Secondary radiation

D. Secondary scatter radiation

31. A patient receives a dose of 2 gray of alpha having a quality factor (QF) of 20. This patient's dose equivalency is:

A. 10 rem

C. 40 sievert

B. 20 sievert

D. 40 rem

32. The amount of energy that is transferred to a material per unit length of travel in soft tissue is termed:

A. RBE

C. Rem

B. LET

D. Rad

33. The short interval of time during which a radiation counter is unable to respond to radiation exposure is called:

A. Interval time

C. Dead time

B. Spot time

D. Recycle time

34. A patient receives a dose of 3 rad of fast moving neutrons with a quality factor (QF) of 10. The total dose equivalency of this patient is:

A. 30 rem
B. 30 sievert

C. 300 rem
D. 300 sievert

35. Pocket ionization chamber is rarely used in a radiology personnel monitoring program because it:

A. Cannot provide accurate readings
B. Cannot be moved from location to location

C. Requires a long warm-up time
D. Requires daily attention to obtain readings

36. When film badges are employed for personnel radiation monitoring program, which of the following factors are likely to effect the badge reading?

1. Exposure of the badge to excessive heat
2. Submersion of the badge in fluids
3. Damage to the wrapper surrounding the film

A. 1 only
B. 2 only

C. 3 only
D. 1, 2 & 3

37. The sensitivity of G-M counter is most closely directly related to its:

A. Starting potential
B. Operating potential

C. Threshold potential
D. Discharge potential

38. A quenching agent is used in a G-M counter to improve the:

A. Resolving time
B. Hermetic seal integrity

C. Spectroscopic sensitivity
D. Pulse height discrimination

39. A 300 rem exposure is equivalent to a:

A. 0.3 sievert
B. 3 sievert

C. 30 sievert
D. 300 sievert

40. The linear energy transfer (LET) of an x-ray or gamma photon is low because of its:

A. Low energy
B. High penetration

C. Negative charge
D. Positive charge

41. An exposure to which of the following radiation will result in the person receiving the highest dose equivalency?

A. 4 gray of x-ray
B. 4 gray of beta

C. 4 gray of gamma
D. 4 gray of alpha

42. Which of the following nuclear emissions possesses the greatest threat to the internal organs from an external exposure?

1. Alpha   2. Beta   3. Gamma

A. 1 only
B. 2 only

C. 3 only
D. 1, 2 & 3

43. The traditional unit of absorbed dose is equivalent to an energy transfer of 100 ergs per gram of any matter and is termed the:

A. Rem
B. Rad

C. Roentgen
D. Sievert

44. The standard international unit for measurement of absorbed dose for all types of ionizing radiation is the:

A. Gray
B. Sievert

C. Becquerel
D. Curie

45. The curie is the traditional unit used to express the:

A. Activity of a radioactive material
B. Ionization potential of a tissue

C. Absorbed dose in a tissue
D. Exposure dose in air

Referring to the ionization chamber voltage curve, answer questions 46-49.

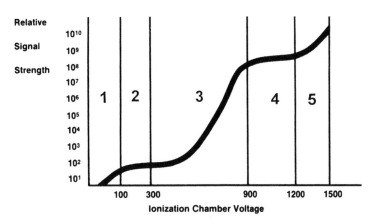

46. The area of the chamber voltage curve, letter number 4, represents the primary operating region for the:

A. Proportional counter
B. Geiger-Muller counter

C. Ionization chamber
D. Cloud chamber

47. Ionization chambers require a voltage in which no recombination of ion pairs takes place. This region is identified by:

A. Numbers 1 & 2
B. Number 3

C. Number 4
D. Number 5

48. In number 3 of the chamber voltage curve, the number of secondary ion pairs is proportional to the primary number of ion pairs produced and is termed the:

A. Threshold region
B. Recombination region

C. Proportional region
D. Glow region

49. In region number 5, the voltage is sufficient to cause spontaneous ionization of the chamber gases and is known as the region of:

A. Partial recombination
B. Maximum sensitivity

C. Minimum sensitivity
D. Continuous discharge

50. The rate of decay or half-life of a radionuclide is effected by changes in the:

A. Chemical bonding of the molecules
B. Temperature of the isotope

C. Atmospheric pressure
D. Nature of the nucleus

51. Which type of radiation detection devices will require the use of a photo-multiplier tube to detect the light emitted from the crystal source?

1. Scintillation counter   2. Thermoluminescent dosimeter   3. Pocket dosimeter

A. 1 & 2 only
B. 1 & 3 only

C. 2 & 3 only
D. 1, 2 & 3

52. An ionization-type survey meter can be used in the detection of all of the following EXCEPT:

A. Ultraviolet radiation
B. Primary radiation
C. Leakage radiation
D. Secondary scattered radiation

53. The most commonly employed inorganic material used in the crystal of a scintillation type counter is:

A. Calcium tungstate
B. Sodium iodide
C. Silver bromide
D. Calcium fluoride

54. Which type of radiation survey instrument is best suited for the detection of a high intensity radiation field?

1. G-M counter   2. Scintillation counter   3. Cutie pie

A. 1 only
B. 2 only
C. 3 only
D. 1, 2 & 3

55. In an Luxel (optically stimulated luminosity) crystal, the amount of light obtained when the crystal is exposed, is most closely related to the _____ by the crystal.

A. Amount of radiation absorbed
B. Mass of the particles absorbed
C. Wavelengths of light absorbed
D. Frequency of the x-rays absorbed

Explanations

1. B   A personnel monitoring device is normally intended to obtain the highest dose that would be received by the user for a particular anatomic structure. For this reason the monitor is usually placed on the outside of any protective apparel that is employed. The location of the monitoring device at the collar is intended to estimate the exposure to the thyroid gland. A monitoring device worn at the level of the waist will is used to determine the exposure to the gonads.

2. C   In the past, most measurements that were used to estimate the dose to an individual were expressed in the units of the rad or millirad. It is now more appropriate to express the absorbed dose in the units of the rem or millirem, the unit of dose equivalent. The conversion of rad (millirad) to rem (millirem) normal is considered to be 1 to 1. Therefore 4000 mrad = 4000 mrem.

3. A   It has been the usual practice to express radiation exposures in the units called the roentgen (R). The unit of radiation intensity or exposure, the roentgen (R), is equal to the amount of radiation that will produce 2.58×10^{-4} coulombs of charge in 1 kilogram of air. In order to facilitate the use of more appropriate units the concept of air kerma was developed. This unit provides a way to express the exposure in terms of an absorbed dose. The air kerma expresses the amount of exposure in roentgens in the units of gray. Through this conversion an exposure of 1R is estimated at an absorbed dose of 8.7 mGy.

4. B   The amount of an exposure is the product of the exposure rate and the time the individual or object is exposed to a source.

5. B   In order to determine the amount of scatter produced in an area or to locate a lost radioactive source will require the use of a portable device that can determine the exposure rate. The device most often used for this purpose is an ionization-type survey meter such as a Geiger-Muller counter.

6. C   The accurate calibration of a survey or rate meter is normally accomplished through the use of a long half lived radioactive source. By using a long-lived calibrated gamma source it is possible to check the accuracy of many different types of radiation detectors. To insure consistency, the test source must be precisely prepared and should meet the National Bureau of Standards (NBS) standards.

7. A   When precise measurements of photon radiation is required a scintillation counter is often employed. This type of device, which can actually record a single photon event, is most often used in nuclear medical imaging devices and laboratory analyzers. Scintillation type devices can be designed to detect both the quantity (counters) and quality of the radiation (analyzers).

8. D   The unit of radiation intensity or exposure, the roentgen (R), is equal to the amount of radiation that will produce 2.58×10^{-4} coulombs of charge in 1 kilogram of air. In more recent years it has been recommended the roentgen be replaced by an SI unit, the Coulomb per kilogram (C/Kg).

9. C   The unit of dose equivalence, the rem, is derived from the formula:
Dose equivalent (100 rem) = absorbed dose (100 rad) x quality factor (QF of 1).

10. D   The plastic film badge holder is designed with a series of filters and openings to selectively enable the detection of high and low energy x-rays and gamma rays, as well as beta particles.

11. C   The unit of dose equivalence, the rem, is derived from the formula:
Dose equivalent (350 rem) = absorbed dose (70 rad) x quality factor (QF of 5).

12. C   A gas filled chamber containing two electrodes, called an ionization chamber, works by recording the change voltage that occurs within a closed chamber when the gas is exposed to radiation. The majority of the survey or rate meters use some form of ionization chamber to detect radiation.

13. D   The unit of radiation absorbed dose, the rad, is often preferred over the roentgen because it can be used in the measurement of all types of radiation, including alpha, beta, gamma, x rays and neutrons.

14. D   The main difference in the types of ionization chamber type detectors is their sensitivity, which is directly related to the operating voltage or potential difference that exists between the two electrodes of the ionization chamber.

15. B   If film badges are worn for more than a month, they may show abnormally high dose readings caused by the film's increased sensitivity following its initial exposure to radiation and the accumulation of background radiation.

16. C   In the past, most radiation detection devices used photographic means (film badge) to detect radiation. More recently, film badges have been replaced by a new monitoring badge that uses the principle of optically stimulated luminosity (Luxel). These badges have a greater sensitivity to radiation (1 millirem) and can be used dozens of times. The most commonly used material in these badges, are crystals of aluminum oxide.

17. A   The constant readout and audible signal produced by a rate or survey meter is useful in the location of lost radioactive sources. The cutie pie and Geiger-Muller counters are the most well known of the rate meters.

18. C   Most experts in radiation safety favor the placement of a radiation monitor badge at the collar level outside the protective apron. This position enables the recording of the maximum estimated skin exposure to the unprotected surfaces of the body. The lower reading obtained by placing it behind the apron may give an underestimation of the actual dose received by both shielded and unshielded areas.

19. D   The plastic film badge holder is designed with a series of filters and openings to selectively enable the detection of high and low energy x-rays and gamma rays, as well as beta particles. Alpha particles due to their high LET and low penetrability cannot pass through the paper surrounding the film.

20. A   To help estimate the dose to the fetus, a pregnant worker is normally instructed to wear a second monitoring badge at the level of the waist.

21. D   The measurement of an individual's absolute dose is most commonly determined with a device called a dosimeter. Four common types of dosimeters that are currently employed are the pocket dosimeter, thermoluminescent dosimeter, and the film badge and optically stimulated luminosity dosimeter (OSL).

22. A   Patient exposures are most often expressed in the unit of the roentgen (R) while the dose equivalent is what accounts for biological effectiveness of an exposure is expressed in rem or sievert. Because 1 R is nearly equal to 1 rem an exposure of 700 mR would nearly equal to 700 mrem, .7 rem, or .07 sievert.

23. B   Photographic film can indicate exposures as low as about 5 - 10 mR or dose equivalent levels of 5 - 10 mrem. Exposures below this level are normally reported as M or minimal.

24. D   A pocket dosimeter is the only type of radiation detection device that permits the evaluation of an exposure after the completion of a study. This is made possible by the self-reading meter incorporated into the design of this device.

25. B   Radiation monitoring devices are employed by all radiology departments to help insure proper radiation protection practices are followed and to insure all workers remain within established annual effective dose equivalent limits.

26. D   Blood testing is no longer considered acceptable as a means of determining radiation exposure because of its high threshold of about 25,000 mrem (.25 Sv).

27. C   The unit of dose equivalent, the rem, can be obtained from the formula:
Dose equivalent (rem) = absorbed dose (rad) x QF.

28. C The active component of a thermoluminescent dosimeter (TLD) is a small chip of the crystalline material, lithium fluoride. This material has the unusual property of emitting a different intensity of light when heated, depending on the amount of ionizing radiation to which it has been exposed.

29. A In order to determine the amount of scatter produced in an area or locate a lost radioactive source requires the use of a portable device that can determine the exposure rate. The device most often used for this purpose is an ionization-type survey meter such as a Geiger-Muller counter.

30. A The output of an x-ray tube (primary radiation), because it is measured in air, is normally stated in the unit of roentgen (R).

31. C In the international system of units (SI), the gray (Gy) is the unit of absorbed dose and the Sievert (Sv) is the unit of dose equivalence. They are related by the formula: Sv = Gy x QF. Dose equivalent (40 Sv) = 2 Gy x QF of 20

32. B The ability of radiation to cause ionization in living tissues is in large part related to a physical factor called the linear energy transfer(LET). This unit helps to express the relative biologic effect of different radiations based on the amount of ionization occurring along the path of the radiation in a tissue. Highly penetrating photons such as x-rays and gamma rays are associated with a low LET and produce less ionization than alpha and neutrons that have a high LET.

33. C The time during which a detection device is unable to respond to an additional exposure of radiation is called its dead time. Detection devices with long dead times cannot be used in the measurement of short-lived radiation events such as those associated with pulsed x-ray sources.

34. A In the traditional system of units (SI), the rad is the unit of absorbed dose and the rem is the unit of dose equivalence. They are related by the formula: rem = rad x QF. Dose equivalent (30 rem) = 3 rad x QF of 10

35. D Pocket dosimeters, though sensitive, are relatively expensive, fragile, require daily monitoring and a manual recording of all readings by a qualified radiation safety officer (RSO).

36. D Among the many drawbacks to the use of film badge monitoring are the erroneous readings that can occur when film is exposed to excessive temperatures, liquids or visible light. The newer optically stimulated luminosity badges are much less likely to be effected by these external factors.

37. B The sensitivity of any type of ionization type chamber detector is related to its operating potential. The higher operating voltage of a Geiger-Muller counter is the main reason for this device's higher sensitivity to radiation.

38. A A quenching agent such as chlorine gas is added to the ionization chamber to absorb some of the positive ions produced. This can substantially improve the time it takes for the device to be able to recover between counts (resolving time.

39. B The conversion from rem to Sievert is based on relationship that 1 Sievert is equivalent to 100 rem. Therefore 300 rem is equal to 3 Sievert.

40. B X-ray and gamma photons have a high degree of penetrability and therefore deposit only small amounts of energy per length of travel. These radiations are said to have a low linear energy transfer (LET) and low ionization potential.

41. D The high quality factor of alpha emissions (QF of 20) will result in the highest dose equivalency of these particles.

42. C Because gamma rays are highly penetrating, they are most likely to contribute to a high degree of damage from an external exposure. When taken internally an alpha source, because of its high LET, will pose a much greater hazard than a low LET emission.

43. B By definition, the energy transfer of 100 erg per gram of any matter by any type of radiation defines 1 rad.

44. A   In an attempt to standardize all units of radiation, the international system of units has selected the gray (Gy) as the unit of absorbed dose.

45. A   The curie is a unit of radiation activity and is equal to 3.7×10^{10} disintegrations per second of a radionuclide. This unit is normally used to express the activity of the radio nuclides used in nuclear medicine or radiation therapy.

46. B   In the ionization chamber voltage curve, the region between 900-1200 volts, is called the Geiger region. Any counter operating in this voltage range is termed a Geiger-Muller (GM) counter.

47. A   In the ionization chamber voltage curve, the regions under 300 volts (numbers 1 & 2) are characterized by a lack of ion pair recombination and the failure to produce a measurable signal making this region unsuitable for the operation of a detection device.

48. C   The slope of region number 3 indicates that the number of primary and secondary ion pairs produced are proportional to each other, hence the name the proportional region.

49. D   At ionization chamber voltages above 1500 volts, a single electron creates a cascade or continuous discharge within the chamber and therefore is useless from a detection standpoint. Region number 5 is called the continuous discharge region.

50. D   The rate at which a radionuclide decays is always constant and is not effected by changes in temperature, pressure or the age of the radioactive material.

51. A   The small amount of light emitted by a scintillation crystal and the thermally stimulated emission of a thermoluminescent dosimeter must first pass through a photo-multiplier tube. This device uses a photocathode to convert the visible light into an electronic signal and a series of charged plates (dynodes) to convert the electrons into a detectable electrical signal.

52. A   Ionization type detectors can be used to measure most high-energy photon radiation (primary, secondary, and exit) as well as most particulate radiations (alpha, beta, neutrons) through the attachment of specialized probes.

53. B   The principal scintillator used in most scintillation detectors is hermetically sealed thallium activated sodium or cesium iodide crystal.

54. C   The lower operating potential of the Cutie pie type ionization detection makes it more suitable for its use with high intensity radiation fields.

55. A   In an optically stimulated luminosity crystal (aluminum chloride), a physical change occurs when it is exposed to radiation. When exposed to a special light source, the energy absorbed from its radiation exposure is detected as a change in the light intensity that it emits. A careful analysis of this light indicates the amount of radiation the crystal has absorbed

Chapter 3

1. The protective aprons and gloves that are used to reduce the exposure to occupational personnel should be checked periodically for cracks by performing a/an:

   A. Visual inspection
   B. Tactile inspection

   C. Radiographic inspection
   D. Electronic inspection

2. During a mobile radiographic procedure the radiographer can reduce their exposure by all of the following EXCEPT:

   A. Wearing a protective apron
   B. Standing at the maximum distance from the tube and patient
   C. Using a high speed image receptor
   D. Removing the added filtration from the tube

Referring to the exposure contour of this fluoroscopic unit, answer questions 3 & 4.

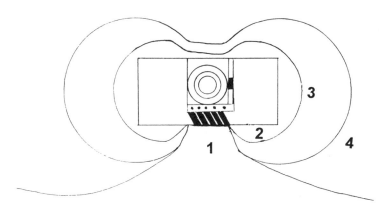

3. Which number corresponds to the region receiving the highest exposure?

   A. Number 1
   B. Number 2

   C. Number 3
   D. Number 4

4. The reduction of the exposure level to the operator standing at location number 1 has been accomplished through the use of:

   A. Protective aprons and gloves
   B. Fluoroscopic drape and a Bucky slot cover
   C. Radiation detection devices and personnel monitors
   D. Compensating filters and the air gap

5. Which of the following procedures is normally associated with the highest exposures to a radiographer?

   A. Tomographic procedures
   B. Routine radiographic procedures

   C. Fluoroscopic procedures
   D. Routine computed radiographic procedures

6. Which of the following factors need be considered in the determination of the thickness for a radiation barrier?

   1. Weekly workload   2. Use factors   3. Occupancy factors

   A. 1 only
   B. 2 only

   C. 3 only
   D. 1, 2 & 3

7. The principal source of scatter when performing a radiographic or fluoroscopic examination is the:

A. Primary shutters of the collimator
B. Patient being imaged
C. Floor of the radiographic room
D. Radiographic tabletop

8. Primary barriers are those located perpendicular to the line of travel of the primary beam, and in the normal diagnostic range should have a thickness of at least:

A. 0.5mm lead (1/40")
B. 0.75mm lead (1/32")
C. 1.5mm lead (1/16")
D. 2.5mm lead (1/10")

9. Which of the following technical factors or equipment should be employed to reduce the gonadal exposure of the radiographer during a fluorographic procedure? The use of a:

1. Low fluoroscopic mA technique   *2. Radiographic grid*   *3. Smaller exposure field*

A. 1 & 2 only
B. 1 & 3 only
C. 2 & 3 only
D. 1, 2 & 3

10. During a radiographic procedure if the patient dose is increased by a factor of two, the dose to the radiographer is likely to:

A. Increase by 2 times
B. Increase by 3 times
C. Decrease by 2 times
D. Not to be effected

11. The protective drape or sliding panel that is used during a fluoroscopic examination to intercept the scattered radiation from the patient must have a lead equivalency of at least:

A. .25mm
B. .50mm
C. .75mm
D. 1.5mm

12. Any device that reduces the amount of scatter produced in a patient will also serve to:

A. Reduce the occupational exposure
B. Reduce the amount of tube filtration
C. Reduce the radiographic contrast of the image
D. Increase the optical density of the image

13. The primary purpose of the Bucky slot cover or the protective drape on a fluoroscopic unit is the reduction of:

1. Back scatter radiation   *2. Operator's exposure dose*   *3. Dose to the patient*

A. 1 only
B. 2 only
C. 3 only
D. 1, 2 & 3

14. Which of the following radiation sources would require the greatest amount of shielding to reduce the radiation leakage to acceptable limits?

A. 50 KeV x-ray
B. 250 KeV x-ray
C. $Tc^{99}m$ gamma (150 KeV)
D. Co^{60} gamma (1.5 MeV)

15. An exposure switch on a mobile radiographic unit should be long enough to allow the operator to remain:

A. 2 feet from the image receptor
B. 100 cm from the patient
C. 200 cm from the x-ray tube or the patient
D. 400 cm from the image receptor

16. The length of time it takes for a radionuclide to reduce its intensity to one-half of its original intensity by a normal decay process is called the:

A. Half-value layer
B. Effective half-life
C. Biological half-life
D. Physical half-life

17. A secondary protective barrier must have a lead equivalency of at least:

A. .75mm lead (1/32")
B. 1.5mm lead (1/16")
C. 2.5mm lead (1/8")
D. 5mm lead (1/4")

Chapter 3   Protection of Personnel

18. During a fluoroscopic procedure that requires the radiographer to be in the room, a protective apron needs to be worn that has an equivalency of at least:

A. .25 mm lead
B. .5 mm lead
C. 1.5 mm aluminum
D. 2.5 mm aluminum

19. The half-life of a radioactive substance:

1. Varies with time   2. Is characteristic of the isotope   3. Is constant

A. 1 & 2 only
B. 1 & 3 only
C. 2 & 3 only
D. 1, 2 & 3

20. Which of the following represents the amount of protective lead required around the average x-ray tube?

A. 1.5 mm
B. 2.0 mm
C. 2.5 mm
D. 3.5 mm

21. In order to minimize the radiographer's exposure while operating fixed radiographic unit, the beam is required to scatter at least _____ before the remaining radiation can enter the shielded booth where the operator stands.

A. Two times
B. Three times
C. Five times
D. Ten times

22. Which of the following would lower a radiographer's dose more during a fluoroscopic study in which their presence in the room is required?

A. Reducing the fluoroscopic kilovoltage peak setting by 15%
B. Reducing the fluoroscopic mA output by 30%
C. Doubling the distance between the source and the radiographer
D. Reducing the fluoroscopic timer setting by 15%

23. In order to prevent an occupational worker from potentially exceeding the annual effective dose limit, the shielding for a controlled area must reduce the exposure to less than:

A. 5mR/week
B. 10mR/week
C. 50mR/week
D. 100mR/week

24. Which of the following is TRUE concerning the recommendations required for the holding of a patient during a radiographic procedure?

A. One radiographer should be assigned as the usual patient holder
B. Those holding a patient should always be shielded
C. Shielding is only required for radiation workers holding patients
D. No patient can be held by a radiation worker

25. When exposed to a high energy gamma source of radiation, which of the following would lower the radiation exposure by the greatest amount?

A. Wearing a standard protective apron
B. Wearing a rubber glove when handling the source
C. Doubling the distance from the source
D. Venting of the room through the use of a positive pressure system

26. The percentage of scattered radiation emanating from a patient during fluoroscopy is generally considered to be_____ of the primary beam when measured at one meter from the patient.

A. 0.1%
B. 0.5%
C. 1.0%
D. 10.0%

27. A diagnostic-type protection housing should reduce leakage radiation to an air kerma level of less than _____ when a tube is operated at its leakage technique.

A. 50 rad (50cGy)/hr.
B. 10 rad (10cGy)/hr.

C. 5 rad (5cGy)/hr.
D. .1 rad (.1cGy)/hr.

Referring to the diagram of the half-life of this radioactive isotope, answer question 28.

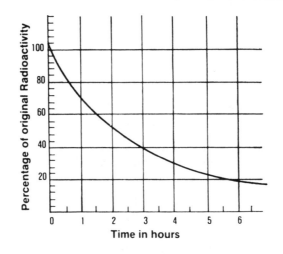

28. In the diagram, the amount of time it takes the radionuclide to decay to 50% of its original value is approximately:

A. 30 minutes
B. 1 hour

C. 2 hours
D. 4 hours

29. A scattered photon at 90 degrees and one meter from an object reduces intensity by about 1000 times. A beam that undergoes two scatterings will reduce the original intensity by about:

A. 10^{-2} times
B. 10^{-3} times

C. 10^{-4} times
D. 10^{-6} times

30. A primary protective barrier must extend upwards from the floor to a height of:

A. 3' (1 meter)
B. 5' (1.5 meters)

C. 7' (2.1 meters)
D. 10' (3 meters)

31. A radioactive source has a half-life of six hours. How long would it take to reduce a 200 millicurie/hr. source to a level of 12.5 millicurie/hr.?

A. 24 hours
B. 18 hours

C. 12 hours
D. 6 hours

32. After a radioisotope is injected into the body, the rate at which it is expelled by the body is termed the:

A. Physiologic half-life
B. Biologic half-life

C. Effective half-life
D. Affected half-life

33. A radioactive source has a half-life of eight hours. How long would it take to reduce a 400 millicurie/hr. source to a level of 50 millicurie/hr.?

A. 8 hours
B. 16 hours

C. 24 hours
D. 32 hours

34. In regard to the three important factors for radiation protection, the lowest exposure is possible if the:

1. Distance from a source is as short as possible
2. Time of exposure is as short as possible
3. Shielding is as thin as possible

A. 1 only
B. 2 only
C. 3 only
D. 1, 2 & 3

35. For the purpose of radiation protection, a building can be divided into controlled and uncontrolled areas. Which area is generally considered an uncontrolled area?

1. Radiology waiting room   **2. The darkroom**   **3. File room**

A. 1 & 2 only
B. 1 & 3 only
C. 2 & 3 only
D. 1, 2 & 3

36. If it is necessary to restrain a patient during a radiographic examination, the LEAST acceptable person to do this is a/an:

A. 50-year-old female radiology secretary
B. 25-year-old male technologist
C. 28-year-old pregnant friend of the patient
D. 18-year-old orderly

37. As the atomic number of a substance increases, the half-value layer thickness:

A. Increases
B. Decreases
C. Remains the same

38. The intensity of a beam is measured at 72 mR for a 200 mA, 150 ms, 77 kVp exposure. If all other factors remain unchanged, the approximate intensity at 300 mA, 200 ms, and 77 kVp will be:

A. 199 mR
B. 168 mR
C. 144 mR
D. 126 mR

39. Lead is usually used in the walls of diagnostic x-ray rooms instead of concrete because:

A. A smaller total mass of lead is required
B. Mass for mass, lead walls are cheaper than concrete
C. Lead is a good structural material
D. The protection calculations are easier

40. The switch that activates the radiographic exposure in most modern radiographic units is permanently attached to the control console. This helps insure that the operator will:

1. Not use excessive prep times
2. Use the proper technical settings
3. Remain behind the shielded booth during the exposure

A. 1 only
B. 2 only
C. 3 only
D. 1, 2 & 3

41. The use of finger badge monitoring is of particular importance to those persons involved with the:

A. Holding of infants
B. Positioning of patients
C. Injections of contrast media
D. Injections of radionuclides

42. In order to maintain the radiation exposure of a non-occupational worker to within effective dose limit the walls, floors etc. for an uncontrolled area must reduce the air kerma to less than:

A. .01 cGy (.01 rad)/week
B. .1 cGy (.1 rad)/week
C. .5 cGy (.5 rad)/week
D. 1 cGy (1 rad)/week

43. If the intensity of a radioactive source is 400 mR/hr. and has a half-life of 6 hours, what will the intensity of this source be in one day?

A. 133 mR/hr.
B. 25 mR/hr.

C. 12.5 mR/hr.
D. 6.25 mR/hr.

44. One-tenth value layer thickness will reduce the intensity of the beam by as much as:

A. 1.5 HVL
B. 2.5 HVL

C. 3.3 HVL
D. 5 HVL

45. The fraction of time that a radiation beam is directed at a specific barrier or area is termed the:

A. Use factor
B. Workload factor

C. Occupancy factor
D. Attenuation factor

In the following place an "A" next to a factor which is principally designed to protect a patient, a "B" next to a factor which is principally used to protect the radiographer or "C" next to a factor which is applicable to both persons.

46. _____ The use of diagnostic filtration
47. _____ The use of collimation during a fluoroscopic procedure
48. _____ The use of primary protection barriers
49. _____ The use of a protective apron by a patient
50. _____ The use of a protective curtain during fluoroscopy

51. If the exposure rate of an x-ray beam is initially 50 R/min., and it requires 2 mm aluminum filtration to decrease the exposure rate to 25% of the original value, the half-value thickness of this beam is:

A. 1 mm aluminum
B. 2 mm aluminum

C. 4 mm aluminum
D. 8 mm aluminum

52. A fixed cone or spacer of NOT less than 30 cm (12") must be provided on _____ to insure this minimum source-to-skin distance.

1. Stationary radiographic units
2. Mobile radiographic units
3. Stationary fluoroscopic units

A. 1 only
B. 2 only

C. 3 only
D. 1, 2 & 3

Referring to the diagram of the effects of filtration on a quantity of x-rays, answer questions 53-55.

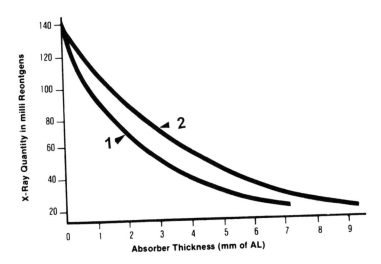

53. The half-value layer thickness for curve I is approximately:

    A. 2 mm aluminum
    B. 3 mm aluminum

    C. 4 mm aluminum
    D. 6 mm aluminum

54. In order to reduce the intensity of curve II to 25% of its original value will require about:

    A. 2 mm aluminum filtration
    B. 4 mm aluminum filtration

    C. 6 mm aluminum filtration
    D. 10 mm aluminum filtration

55. Comparing the curves I and II, which would represent the beam with the highest intensity?

    A. Curve I
    B. Curve II

    C. Unable to determine

Explanations

1. C   To insure that the protective apparel that is worn by a radiographer is able to properly protect the individual it should be checked for cracks or other defects at least once each year. This is best accomplished by the radiographic evaluation of the apparel.

2. D   During mobile radiographic procedures it is possible to reduce the exposure to the operator by requiring the radiographer to: wear protective apparel (aprons), use high speed imaging systems to reduce the overall exposure employed, and stand as far from the radiation source and the patient as possible. The removal of the added filtration will increase the skin exposure to the patient but will have little if any effect on the radiographer.

3. B   The isodose curves for this fluoroscopic unit indicate that the highest exposure would occur within the curve designated as number 2, closest to the table but beyond the protection of the protective drapes.

4. B   The isodose curves for this fluoroscopic unit indicate that the lowest exposure would occur at the location designated as number 1. In this region the operator is protected by the Bucky slot cover and the fluoroscopic drape that are required to have a lead equivalency value of at least .25 mm of lead.

5. C   The relative exposure received by a patient or a radiographer during a procedure is based on the total amount of radiation delivered during an imaging procedure. Assuming the same region of the body is imaged, the amount of exposure received will be the lowest during routine radiographic procedures followed by an increase for CR procedures and Tomography and a significant increase for fluoroscopic procedures that are performed in real time.

6. D   In order to determine the thickness required for a radiation barrier to reduce the intensity of radiation to allowable limits a number of different factors must be evaluated These include:

    1. Energy of the beam
    2. Use factors (time the beam is directed at the barrier)
    3. Occupancy factors (the amount of time the area being protected is likely to be occupied with people)
    4. Weekly work load (the average exposure for the unit multiplied by the maximum time the unit is likely to be used during the week)
    5. The distance of the source from the region of interest

7. B   The principal scattering object during the radiographic or fluoroscopic procedures is the Compton's scattering that occurs in a patient undergoing an examination.

8. C   In order to provide the required radiation protection to persons working in or around a radiation area, all diagnostic x-ray rooms must have a primary barrier with a lead equivalent thickness of about 1.5 mm or 1/16". To insure that even the tallest of individuals are shielded during radiographic procedure, the shielding must extend up from the floor to a height of 7 feet or 2.1 meters.

9. B   The dose to the gonads can be reduced by a combination of gonadal shielding, high speed imaging systems, tight collimation and low mAs techniques. Though a radiographic grid will improve the image quality, it is associated with higher patient exposures.

10. A   The operator and patient dose can be shown to be directly proportional. In theory, due to the design of the room and barriers, the operator's exposure will always be approximately 1 million times less than that of the patient.

11. A   The majority of all temporary protective drapes and panels will require a minimum lead equivalency of not less than .25 mm.

12. A   All beam limiting devices such as diaphragms, cones, cylinders, and collimators are effective at reducing the scattered radiation produced in the patient, thus lowering the patient and operator doses, as well as improving the contrast of the radiographic image.

13. B   The Bucky slot cover that automatically covers the opening between the tabletop and the table support, and the protective drape on a fluoroscope, both serve to reduce the dose to the operator of the equipment and any other personnel that must remain in the room during a procedure. These devices have little effect on the dose that is received by the patient.

14. D   The photon source with the highest energy, the 1.5 MeV gamma of Co60, also has the greatest penetration; therefore, the shielding material needed for this source would require the greatest thickness.

15. C   According to the NCRP Report 105 the cord for a mobile unit must be of sufficient length to permit the operator to be at least 2 meters (200 cm) or 6 feet from the x-ray tube or the patient during an exposure.

16. D   The half-life of a radionuclide that involves the normal decay process is termed its physical half-life.

17. A   Secondary protective barriers or those which are only exposed to leakage or scattered radiation must have a lead equivalency of .75 mm (1/32").

18. B   Protective aprons which are designed to have a minimum of a .5 mm lead equivalence will normally attenuate between 80-90% radiation hitting the aprons. According to current regulations, all personnel required to be present during fluoroscopy shall wear a protective apron of this thickness during the procedure.

19. C   The physical half-life of a radionuclide is unique for a specific isotope and is unaffected by temperature, pressure, time or chemical bonding.

20. A   Since the tube housing is exposed directly to primary radiation, it is classified as a primary barrier and therefore will require at least 1.5 mm of lead equivalency.

21. A   Each scattering event lowers the radiation intensity by about 1000 times as compared to that in the directly exposed tissue. By locating the exposure booths in a position that requires at least two scattering events can reduce the operator's dose by at least 1 million times compared to that of the patient.

22. C   The greatest reduction in dose would result from a doubling of the distance that lowers the radiographers dose by about four times. A 15% reduction in the kVp will result in an increase of the mA and exposure by approximately 50%.

23. D   The shielding for a controlled area must be thick enough to reduce the exposure to less than 100 mR/week or 5000 mR per year, or an air kerma of .1 rad (.1 cGy) per week.

24. B   The following regulations regarding the holding of patients are outlined in the NCRP Report 105. Patients should only be held when patient restraining devices have proven inadequate. No medical personnel should be routinely assigned to hold patients. All persons holding patients are required to wear protective apparel and should not be within the primary beam. Pregnant women, children under 18, are not allowed to hold patients at any time.

25. C   The use of standard protective apparel is not recommended for high energy gamma sources because the shielding is insufficient to stop a large portion beam and may even increase the dose by producing secondary emissions of ionized electrons. With any gamma source the reduction of the time the person is exposed and the use of the principles of the inverse square law provide the best means for reducing exposures.

26. A   The percentage of scattered radiation emitted from the patient is estimated to be 1000 times lower or .1% of the primary beam when measured at a distance of 1 meter from the patient.

27. D   The maximum allowable limit for leakage of radiation from the protective housing surrounding the x-ray tube is less than 100 mR/hr. or an air kerma of .1 cGy in one hour when operated at its leakage technique.

28. C   The half-life is found by following the curve to the 50% line at about 2 hours.

29. D   The intensity of scatter radiation 1 meter from a patient is approximately 1000 times (10^3) less than the intensity of the useful beam at the surface of the patient. If the beam is scattered twice the original intensity, the level of the beam is reduced by 1 million (10^6) times.

30. C   In order to protect the average to above average person, a protective barrier must extend to a height of at least 2.1 meters or 7 feet from the floor.

31. A   Since the intensity of a 200 mCi source is reduced by one-half every 6 hours, the following values can be obtained: 100 mCi after 6 hours; 50 mCi after 12 hours; 25 mCi after 18 hours; and 12.5 mCi after 24 hours.

32. B   The rate at which a radioactive material decays by true radioactive means is termed the physical half-life. The rate at which a substance is eliminated from the body due to a biologic process is the biologic half-life.

33. C   Since the intensity of a 400 mR/hr source is reduced by one-half every 8 hours, the following values can be obtained: 200 mR/hr after 8 hours; 100 mR/hr after 16 hours; 50 mR/hr after 24 hours; and 25 mR/hr after 32 hours.

34. B   In order to keep one's exposure to radiation as low as possible, it is important to keep as much distance between the individual and the source of radiation as possible, reduce the time the individual is exposed, and place as much shielding between the individual and source as possible.

35. D   Areas assigned the controlled designation are occupied by monitored occupational workers. These will normally include the x-ray rooms, NM scanning areas, hot labs, and CT scanning areas. Non-controlled areas are frequented by non-occupational workers, and include such areas as offices, waiting rooms, darkrooms, file rooms, and corridors.

36. C   Though the holding of patients should always be avoided when possible, when necessary, it should not be accomplished by pregnant women or persons under 18 years of age. All persons used for this purpose shall be shielded for primary radiation.

37. B   The higher the atomic number of the material, the more radiation it attenuates, thereby reducing its half-value layer thickness.

38. C   The intensity of the beam is directly related to the mAs when the kVp remains constant. Therefore if a 200 mA, 150 mA (30 mAs) provides and an exposure of 72 mR at 300 mA, 200 mS, (60 mAs) exposure will result in a new intensity of about 144 mR.

39. A   Because lead has many times the stopping power of concrete, less total mass or thickness to obtain the same protective equivalency is required.

40. C   In all newer stationary units, the exposure switch is attached to the control console to prevent its activation while the operator is standing inside the x-ray room.

41. D   The nuclear medical technologist or physician responsible for radionuclide injections, or the person responsible for handling injections during special procedures, should be assigned ring badge monitors in addition to the collar and waist badges normally recommended.

42. A   The shielding of a non-controlled area must reduce the exposure to an individual standing behind the barrier to an air kerma level of less than .01 rad (.01 cGy) per week.

43. B   Since the intensity of a radioactive source is 400 mR/hr. is reduced by one-half every 6 hours, the following values can be obtained: 200 mR/hr. after 6 hours; 100 mR/hr. after 12 hours; 50 mR/hr. after 18 hours; and 25 mR/hr. after 24 hours.

44. C   One-tenth value layer thickness is defined as the amount of material that will reduce the radiation intensity by 1/10th of its original value. This is equal to the 3.3 half-value layer thickness.

45. A   The specific term that is used to estimate the time a beam is directed at a protective barrier is called the use factor. In most radiographic rooms the beam is directed at the floor for most studies and is given a use factor of 1. The walls of the room which are only occasionally the focus of the primary beam will have a use factor of ¼.

46. A   The use of diagnostic filtration is primarily intended to reduce the exposure to the skin of the patient.

47. C   Collimation reduces the amount of exposed tissue in the patient and lowers the amount of scattered radiation. These factors combine to reduce the exposure to both the patient and the operator of the radiographic unit.

48. B   Primary barriers are intended to reduce the exposure to the operator of the equipment, the radiographer.

49. A   Lead aprons placed on a patient are for their protection and do not reduce the exposure to the radiographer.

50. B   Protective drapes and Bucky slot covers are primarily designed to protect the operator of the equipment and the radiographer.

51. A   If the initial exposure rate is 50 R/min. and 2 mmAl filtration reduces this rate by 25%, then 1 mmAl filtration would reduce the rate by 50% for 1 half-value layer thickness.

52. D   The regulations concerning stationary radiographic or fluoroscopic units, require that the minimum source-to-skin distance (SSD) shall not be less than 30 cm (12 in.) and should not be less than 38cm (15 in.). All mobile and fixed radiographic and fluorographic units must be provided with a fixed cone collimator or spaced to insure a minimum SSD of not less than 30 cm.

53. A   The half-value layer thickness (HVL) for curve I which starts at an intensity of 140 mR is about 2 mmAl.

54. C   The half-value layer thickness (HVL) for curve I is about 3 mmAl, therefore it would take twice this amount (6mmAl) to reduce the intensity to 25% of the original value.

55. B   The beam with the greatest half-value layer thickness (HVL) (curve II) will have a higher penetrability and provide a higher intensity for a given exposure.

1. During a right PA oblique (RAO) position of the stomach the area of the body receiving the greatest exposure is the:

    A. Skin on the left dorsal surface of the abdomen
    B. Internal organs of the right abdomen
    C. Skin on the right ventral surface of the abdomen
    D. Internal organs of the left abdomen

2. A radiographic exposure of 200 mA at 100ms, 70 kVp is made registering 50 mR per exposure. Assuming no other changes are made, what will exposure be using a technique of 600 mA, 100 ms, 70 kVp?

    A. 12.5 mR   C. 150 mR
    B. 25 mR   D. 300 mR

3. Which of the following would help in reducing the radiation exposure to a child during a radiographic study of the abdomen?

    1. The use of gonadal shielding
    2. The use of high speed imaging system
    3. The use of a tightly collimated beam

    A. 1 only   C. 3 only
    B. 2 only   D. 1, 2 & 3

Referring to the diagram image, answer questions 4 & 5.

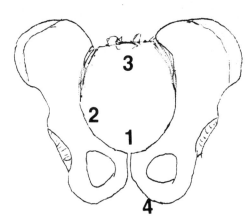

4. During the radiographic examination of the pelvis of a male patient, choose the number that most closely approximates the location of the testes.

    A. Number 1   C. Number 3
    B. Number 2   D. Number 4

5. During the radiographic examination of the female pelvis, choose the number that most closely approximates the location of the ovary.

    A. Number 1   C. Number 3
    B. Number 2   D. Number 4

6. If the filtration for a radiographic unit were inadvertently omitted for a series of chest exposures, the greatest increase in exposures would occur in the:

A. Diaphragm
B. Skin of the chest

C. Mediastinal structures
D. Pulmonary vessels

7. The principal reason that diagnostic filtration should be employed for all radiographic studies is to:

A. Decrease the ionization potential of the beam
B. Reduce the entrance skin exposure of the patient
C. Reduce the amount scatter produced in the patient
D. Decrease the penetrability of the beam

8. Which of the following exposure and technical factors will result in a patient receiving the lowest exposure?

A. 300 mA, 30 ms, 90 kVp 400 speed screen
B. 200 mA, 30 ms, 95 kVp 200 speed screen

C. 400 mA, 30 ms, 85 kVp 50 speed screen
D. 100 mA, 30 ms, 85 kVp 400 speed screen

9. All of the following devices can be used to reduce the amount of scattered radiation and reduce the exposure to the patient EXCEPT:

A. An aperture diaphragm
B. A variable aperture collimator

C. A radiographic grid
D. An extension cone

10. Whenever possible a PA projection of the thorax should be used during imaging studies instead of an AP projection to:

A. Reduce involuntary motion caused by the heart
B. Decrease skin-to-image receptor distance
C. Reduce the superimposition of the heart, lungs and ribs
D. Reduce exposures to the breast

11. Which of the following is a type of shielding device currently available for the reduction of dose to the gonads?

1. Flat contact shield   **2. Shadow shield**   **3. Shaped contact shield**

A. 1 only
B. 2 only

C. 3 only
D. 1, 2 & 3

12. To protect the unborn fetus of a pregnant technologist in a modern radiology department, the following recommendation is generally made:

A. Quit work immediately
B. Work in the low radiation areas only

C. Do not work in the x-ray department
D. No special precautions are required

13. All of the following can be used to reduce the repeat rate for radiographic images EXCEPT:

A. Effective communication skills
B. Careful selection of exposure factors

C. Proper immobilization techniques
D. High ratio radiographic grids

14. Which of the following film-holder combinations will best reduce exposure during a radiographic examination?

A. 50 speed film-screen system
B. 200 speed film-screen system

C. 400 speed film-screen system
D. Non screen system

15. An x-ray machine operating at potentials above 70 kVp requires at least:

A. 1.5mm aluminum of the tube filtration
B. 2.5mm aluminum of the tube filtration

C. 1.5mm lead of the tube filtration
D. 2.5mm lead of the tube filtration

16. The principal advantage of a rare earth imaging system as compared to a calcium tungstate imaging system is:

A. A reduction in the exposure to the patient
B. An increase in the low contrast resolution

C. A decrease in the amount of quantum mottle
D. A reduction in the amount of size distortion

17. The use of image intensified fluoroscopy increases the brightness of an image with a substantial reduction in _____ as compared to conventional fluorography.

    A. Spatial resolution
    B. Patient exposure

    C. Visual acuity
    D. Patient motion

18. The total amount of radiation received by a patient during a radiographic procedure is effected by the:

    1. Size of the exposure field
    2. Total amount of filtration in the tube
    3. Milliampere-seconds used during the exposure

    A. 1 & 2 only
    B. 1 & 3 only

    C. 2 & 3 only
    D. 1, 2 & 3

19. Which of the following techniques should be utilized if a reduction of patient exposure is desirable?

    A. 100 mA, 50ms, 80 kVp
    B. 400 mA, 25ms, 80 kVp

    C. 400 mA, 8ms, 90 kVp
    D. 200 mA, 24ms, 90 kVp

Referring to the radiographic image, answer question 20.

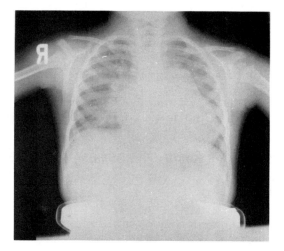

20. In the radiograph, the low optical density area in the bottom of the image indicates the use of:

    A. Leaded waist apron
    B. Shaped contact shield

    C. Compensating bolus
    D. Excessive collimation

21. Which of the following types of spot films will result in the lowest patient dose for a single exposure?

    A. 70 mm photo spot film
    B. 105 mm photo spot film

    C. 8-10" cassette type spot film
    D. 9x9" cassette type spot film

22. Which of the following combination exposure and technical factors will result in a patient receiving the lowest entrance skin exposure?

	mA	Time	kVp	Screen Speed	Grid
A.	100	100ms	87	250	12:1
B.	200	80ms	88	200	12:1
C.	100	150ms	84	50	6:1
D.	200	20ms	86	400	5:1

23. A diagnostic fluorographic unit operating at 2 mA, 80 kVp with 2.5mm aluminum filtration has an output of 150 milliroentgens per minute of operation at the tabletop. How long would it take this unit to deliver an exposure of 600 milliroentgens?

A. 1.5 minutes
B. 2 minutes
C. 3 minutes
D. 4 minutes

24. In the following radiographic examinations, list, from greatest to least, the expected fetal dose for a pregnant female.

1. AP hip image   *2. Kidney, ureter, bladder (KUB)*   *3. AP thoracic spine image*

A. 1, 2, 3
B. 2, 1, 3
C. 3, 2, 1
D. 1, 3, 2

25. Patient dose in fluoroscopy may be minimized by:

A. Leaving the shutters as wide open as possible
B. Increasing the mA used during the exposure
C. Restricting the size of the field of view
D. Reducing the source-to-image receptor distance

26. When gonad shielding is employed during fluoroscopy, the radiographer should place the shield:

A. On the lateral surface of the table
B. Over the pelvis of the patient
C. Under the pelvis of the patient
D. No shielding can be used during fluoroscopy

Referring to the radiographic image, answer question 27.

27. In the radiograph, what technical error caused the unnecessary radiation exposure to the patient? The failure to:

A. Select the proper focal spot
B. Remove inherent filtration
C. Properly orientate the collimator
D. Collimate the beam

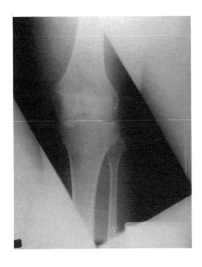

28. Gonadal shielding should be employed during a radiographic procedure whenever the:

1. Person has a reasonable reproductive potential
2. Gonads lie within or close to the exposure field
3. Shielding will not compromise the objectives of the study

A. 1 only
B. 2 only
C. 3 only
D. 1, 2 & 3

29. Whenever possible, women of childbearing ages should have radiographic examinations scheduled during the _____ of the menstrual cycle:

A. First 10 days
B. 10th - 20th day
C. Last 10 days
D. The manifest period

30. Which of the following factors will have the least effect on the radiation exposure to the patient if the mAs and kVp remain constant?

A. The total amount of tube filtration   C. The size of the exposure field
B. The source-to-image receptor distance   D. The size of the focal spot

Explanations

Chapter 4

1. A When a patient is in the right PA oblique position the central ray enters the body on the dorsal surface of the body. This will be the region that is associated with the highest entrance skin exposure.

2. C The amount of exposure the patient receives is directly related to amount of mAs employed when the kVp remains constant. Therefore if an exposure of 200 mA and 100 ms (20 mAs) results in an exposure of 50 mR a technique of 600 mA and 100 ms (60 mAs) will result in a tripling of the exposure or a exposure of about 150 mR.

3. D A radiographer can help reduce patient exposures in a number of ways, including the use of higher speed image receptors, proper shielding, tight collimation, and the selection of the appropriate technical factors.

4. D The primary reproductive tissues of the male are the testes. These are located in the scrotal sac just below the bony pelvis. The number that most closely represents this region is number 4.

5. B The female reproductive tissue is found in the ovaries. These are located on either side of the pelvis just above the mid pelvis. The number that most closely corresponds to these structures is region number 2.

6. B The principal function of diagnostic filtration is the reduction of low energy photons that do not contribute to the image but add to the entrance skin exposure of the chest of this patient. The use of diagnostic filtration can reduce the exposure to the skin by approximately 50%. Two other positive effects will occur from the use of diagnostic filtration; namely, an increase in quality of the beam and an increase in the homogeneity of the beam.

7. B See the previous explanation.

8. D Any factor which enables the use of lower mAs and kVp values will enable a reduction in the patient exposure. The use of screens with high speed number requires less radiation to provide the same optical density and therefore can substantially lower exposure to the patient.

9. C Any device that reduces scatter radiation before it reaches the patient such as beam limiting devices; i.e., cone, cylinder, aperture diagram, and collimator, will reduce patient exposures. Grids that eliminate scatter after it leaves the patient require an increase in the incident beam intensity to maintain the desired optical density.

10. D During a radiographic procedure, the tissues of the body will act like a filter, significantly lowering the dose on the side of the body on which the beam exits. Because the breasts are be more sensitive than the skin of the back, a PA projection can reduce the exposure to the breast by as much as 60%.

11. D When patients have a substantial reproductive potential, gonadal shielding should be used whenever the gonads are within 5 centimeters (2") of the collimated beam unless the placement of the shield will compromise the objectives of the study . In a modern department three types of gonad shields are used. The first which is often made of an old lead apron is called a flat contact shield. The second is a shaped contact shield that can accommodate the testes of the male patient individual. That last is a leaded shadow shield that hangs from the bottom of the collimator. This shield is most useful in the OR where the other types of devices can contaminate the surgical field

12. D The most recent regulations concerning pregnant radiation workers are designed to avoid possible discrimination of the woman. Because it is extremely unlikely that any trained worker is likely to exceed either their own or the embryo-fetus effective dose equivalent limits, other than providing additional monitoring devices, no special precautions are advised or permitted.

13. D   One of the largest contributing factors to the patient exposure dose is the number of repeat images that may be required. The repeat rate can be reduced by the proper maintenance of the equipment, use of immobilization techniques, careful positioning, careful selection of the technical factors and the use of appropriate communication skills.

14. C   The (400) speed rare earth screens that are commonly employed in nearly all modern radiology departments must be used in conjunction with a specifically matched screen type film will provide the optimum speed. In this question the 400 speed imaging system would provide the lowest patient exposure when a similar radiographic contrast and optical density are obtained.

15. B   Because diagnostic filtration can reduce the entrance skin exposure by as mush as 50% with no negative effect on the quality to the image, the government has mandated that all radiographic or fluoroscopic tubes operating above 70 kVp must be filtered by less than 2.5 mm of aluminum equivalent. This approximates 1 half value layer of the beam at 80 kVp.

16. A   The main advantages of the rare earth imaging systems are an increased speed enabling the use of less radiation thus reducing the patient's exposure. The spatial resolution and radiographic contrast of 400 speed rare earth and 100 speed calcium tungstate screen system is nearly identical.

17. B   There are a number of things a radiographer or physician can do to lower the patient's exposure during fluoroscopy. These include reduction of the field size, the use of a lower mA value, the use of a higher kilovoltage setting, a reduction in the time that the fluoroscope is activated and the application of protective shielding when possible. But the most important is the use of the image intensifier system. This electronic device that brightens the image is also associated with significant reduction in amount of radiation exposure received by the patient.

18. D   Among the numerous factors that effect the dose a patient receives are mAs, screen speed, grid ratio, SID, field size or field of view, filtration, and kVp. A reduction in the exposure to a patient is accomplished by the use of low mAs values, the use of a high speed imaging system, reduced field size (FOV) and the use of diagnostic filtration. The use of lower kVp for a given optical density and the use of radiographic grids are normally associated with higher patient exposures.

19. C   Any factor which enables the use of lower mAs and kVp values will enable a reduction in the patient exposure. In this question the least mAs and patient exposure would be obtained from the use of the technique in answer letter C. These values require less radiation to provide the same optical density and therefore can substantially lower exposure to the patient.

20. A   The use of protective (lead) aprons to limit gonadal exposures during chest radiography is common in modern radiography. The low optical density artifact seen in the lower portion of the image is the plastic waistband of one of these aprons.

21. A   A photo spot film will require about 1/10th the exposure required to produce a conventional screen type spot film. A further reduction will occur when a smaller format (70 mm) film system is employed.

22. D   Any factor which enables the use of lower mAs and kVp values will enable a reduction in the patient exposure. The use of screens with high speed number and the radiographic grid with the lowest ratio require less radiation to provide the same optical density at the lowest entrance skin exposure to the patient.

23. D   The time required to deliver a given dose is found by dividing the dose to be delivered by the dose rate. In this example the unit has an output of 150 mR /min. Therefore, it would take four minutes to deliver a dose of 600 mR.

24. B   The estimated entrance-skin exposure (mR) and embryo-fetal dose (mrad) for the three examinations in the question is based on the use of a 400 speed imaging system. The estimated entrance-skin exposure is 110 mR fetal dose and 35 mrad for a KUB, 110 mR and 25 mrads for a hip and 90 mR and 1 mrad for the thoracic spine.

25. C There are a number of things a radiographer or physician can do to lower the patient's exposure during fluoroscopy. These include reduction of the field size (field of view), the use of a lower mA value, the use of a higher kilovoltage setting, a reduction in the time that the fluoroscope is activated and the application of protective shielding when possible.

26. C Since the fluoroscopic tube is normally located under the tabletop, in order to be effective any shielding must be placed on top of the table between the patient and the tube.

27. C The image was obtained with improper orientation of the collimator to the body part.

28. D In order to protect the individual from the possibility of damage to the genetic tissue the following guidelines have been made. Gonadal shielding should be employed when ever a person has a reasonable reproductive potential. The gonads lie within 5 centimeters (2") of the collimated beam and the placement of the shield does not compromise the objectives of the study.

29. A Because a pregnancy is highly unlikely during menstruation and the few days following, radiographic examinations, when possible, should be performed during this 10 day period of time.

30. D Nearly all the technical and equipment factors will have an effect on the optical density and patient dose. One exception to this rule is the size of the focal spot that controls the focal spot blur and the spatial resolution of the resulting image. A change in the size of the focal spot should not effect the optical density or exposure received by the patient.

1. A pregnant female has undergone a series of radiographic studies resulting in an estimated fetal dose equivalent of 1300 mrem (13 mSv). A reasonable recommendation by the radiation safety officer would be to:

   A. Strongly suggest the abortion of the fetus
   B. Start the female on drugs to prevent the spontaneous abortion of the fetus
   C. Closely monitor the fetus for any signs of radiation-induced malformations
   D. Indicate to the female that no significant risk to the fetus has occurred

Referring to the following diagram, answer questions 2.

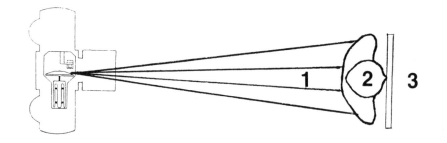

2. Which of the locations could be used to make measurements of the radiation absorbed dose?

   A. Number 1   C. Number 3
   B. Number 2

3. Because of its ability to support the weight of heavy patients while permitting a high transmission of the primary beam, most x-ray tabletops are made of a material such as?

   A. Leaded glass   C. Calcium tungstate
   B. Carbon-fiber   D. Lucite

4. A dead man type switch is employed during a fluoroscopic study to:

   A. Activate the leaves of the automatic collimation system
   B. Place the image intensifier into the high dose mode
   C. Initiate the start of the fluoroscopic exposure
   D. Activate the automatic brightness control system

5. According to the current recommendations, which of the following would be allowed the highest annual effective dose limit?

   A. The lens of the eye of a member of the public   C. The hands of an occupational worker
   B. The hands of a member of the public   D. The skin of the fetus/embryo

6. The annual effective dose limits for the stochastic effects are based on the assumption that there is:

   A. Only a temporary risk for these exposures
   B. A nominal lifetime risk for these exposures
   C. A high probability of cure for any conditions associated with these exposures
   D. Only short-term effects are considered dangerous

7. As the distance from a radiation source increases, the radiation exposure to an individual will:

   A. Decrease inversely with the square of the distance
   B. Decrease directly with the distance
   C. Increase inversely with the square of the distance
   D. Increase directly with the distance

8. In which of the following areas of a radiation facility is a member of the public likely to receive an exposure exceeding 10 mR/week?

   A. In the patient area
   B. In the radiographic darkroom
   C. In the main control area
   D. In the radiographic imaging area

9. The permanent dose equivalent for an occupational worker is .85 rem. What is this person's dose equivalent in mrem?

   A. 85,000 mrem
   B. 8,500 mrem
   C. 850 mrem
   D. 85 mrem

10. A pregnant radiographer receives a monthly badge report showing dose equivalent of 90 mrem for the waist monitor. The approximate dose equivalent to the fetus is:

    A. 270 mrem
    B. 180 mrem
    C. 90 mrem
    D. 30 mrem

11. According to the ALARA concept the dose an occupational worker receives should be:

    A. Maintained at the levels which are as low as possible
    B. Too low to measure by standard monitoring methods
    C. No greater than amount of exposure received from background sources
    D. At or just below effective dose limits

12. Though the earth is struck by large amounts of solar and other forms of radiation, relatively little reaches the surface because:

    A. The clouds absorb most of the harmful rays
    B. The earth's atmosphere filters most of the harmful rays
    C. The radiation is repelled by the earth's magnetic field
    D. Most of the photons are absorbed by the oceans

13. A total dose equivalent limit of .05 rem (.5 mSv) per month will apply to a/an:

    1. Embryo/fetus   *2. Pregnant radiographer*   *3. Uranium mineworker*

    A. 1 only
    B. 2 only
    C. 3 only
    D. 1, 2 & 3

14. A radiographer receives an exposure of 4 mR/hr during a 40-hour work week in a controlled area. What was the total dose received by this individual for the week?

    A. 10 mR
    B. 80 mR
    C. 160 mR
    D. 640 mR

15. A gonadal shield with a .5 mm lead equivalent is able to attenuate approximately _____ of the incident x-rays for a 75 kVp beam.

    A. 33%
    B. 50%
    C. 88%
    D. 99%

16. The recommended annual effective dose limit for deterministic effects for the skin of the extremities and all other organs for an occupational worker should NOT exceed:

    A. 500 mSv (50 rem)
    B. 300 mSv (30 rem)
    C. 150 mSv (15 rem)
    D. 50 mSv (5 rem)

17. A reasonable effective dose limit for a radiographer under ALARA might by:

A. 150 mSv (15 rem) per year
B. 80 mSv (8 rem) per year
C. 50 mSv (5 rem) per year
D. 5 mSv (.5 rem) per year

18. A somatic effect which increases in severity with an increase in the radiation dose above the threshold is called a:

A. Deterministic effect
B. Committed effect
C. Optimization effect
D. Reference effect

19. A student radiographer is admitted into training six months before his/her 18th birthday. This person's effective dose limit until the age of 18 is:

A. 1 mSv (.1 rem)
B. 3 mSv (.3 rem)
C. 5 mSv (.5 rem)
D. 10 mSv (1 rem)

Refer to the statement below in answering questions 20-21.

A radiation worker received an occupational dose of 3 mSv, a natural background dose of 2 mSv, and had 50 mSv of medical exposures during the year.

20. The total occupational dose of radiation this person received during the year is:

A. 2 mSv
B. 3 mSv
C. 52 mSv
D. 53 mSv

21. The total dose of radiation this person received during the year is:

A. 2 mSv
B. 3 mSv
C. 52 mSv
D. 55 mSv

22. Which of the following amount of shielding will attenuate the greatest number of high energy (100 keV) x-ray photons from a beam?

A. 2.5 mm of aluminum
B. .5 mm of lead
C. 1.5 mm of Lucite
D. .25 mm of copper

23. A thirty-year-old occupational worker has received a 6.5 mSv whole body exposure in the first month of employment. What is the maximum effective dose limit this person can receive in the next 11 months?

A. 3.5 mSv
B. 15.5 mSv
C. 23.5 mSv
D. 43.5 mSv

24. A pregnant radiographer receives a monthly badge report showing 70 mrem for the badge worn at the waist level. The approximate dose equivalent to the fetus is:

A. 170 mrem
B. 70 mrem
C. 62 mrem
D. 23 mrem

25. A high radiation area sign should be posted in an area where the exposure rate exceeds:

A. 10 mR/hr.
B. 100 mR/hr.
C. 100 mR/day
D. 100 mR/week

26. According to the NCRP, the annual effective dose limit for deterministic (somatic) effects for the lens of the eye is:

A. 10 mSv (1 rem)
B. 30 mSv (3 rem)
C. 50 mSv (5 rem)
D. 150 mSv (15 rem)

27. Once a pregnancy becomes known, the exposure to an embryo-fetus shall be no greater than _____ per month.

A. .5 mSv (.05 rem)   C. 18 mSv (1.8 rem)
B. 5 mSv (.5 rem)   D. 36 mSv (3.6 rem)

28. A member of the public will have a maximum annual effective dose limit of _____ for infrequent annual exposures.

A. 50 mSv (5 rem)   C. 10 mSv (1 rem)
B. 30 mSv (3 rem)   D. 5 mSv (.5 rem)

29. The guidance level for cumulative exposures of a 36-year-old occupational worker with 10 years of experience is:

A. 130 mSv (13 rem)   C. 460 mSv (46 rem)
B. 360 mSv (36 rem)   D. 900 mSv (90 rem)

30. For a fixed fluoroscopic unit minimum, the source-to-skin distance shall NOT be less than:

A. 12" (30 cm)   C. 20" (52 cm)
B. 15" (38 cm)   D. 25" (64 cm)

31. The average annual dose equivalent for all monitored radiation workers was found to be about:

A. 46 mSv (4.6 rem)   C. 19 mSv (1.9 rem)
B. 32 mSv (3.2 rem)   D. 2.1 mSv (.21 rem)

32. A member of the public is accidentally exposed to 4 mSv (.4 rem). This person has exceeded the annual effective dose limit.

A. True   B. False

33. An occupational radiation worker is exposed to an average 675 mrem per month. How many months can this individual work without exceeding the annual effective dose limit?

A. 2 months   C. 6 months
B. 4 months   D. 7 months

34. In order to remain within the permitted effective dose equivalent limits for an occupational worker, an individual should NOT receive more than:

A. 1 mSv/week   C. 10 mSv/week
B. 5 mSv/week   D. 20 mSv/week

35. Which of the following is normally considered a type of stochastic effect:

1. Cataract formation   *2. Cancer*   *3. Genetic conditions*

A. 1 & 2 only   C. 2 & 3 only
B. 1 & 3 only   D. 1, 2 & 3

Refer to the statement below in answering questions 36-40.

For each of the following workers place an A if they are classified as an occupational, or a B if they are classified as a non-occupational.

36. _____ Ultra sonographer with no general radiographic responsibilities.
37. _____ Darkroom technologist in the radiology department.
38. _____ CT technologist with no general radiographic responsibilities.
39. _____ Staff radiographer with general radiographic responsibilities.
40. _____ Mammographer with no general radiographic responsibilities

41. According to the NCRP recommendations, the guidance level for cumulative exposures should be calculated using the formula:

A. Age x 50 mSv
B. 5 mSv (age - 18)

C. 10 mSv x age
D. (Age x 2 mSv)

42. The effective dose limits are based on this type of dose relationship.

A. Linear, non-threshold
B. Non-linear, threshold

C. Linear, energy transfer
D. Non-linear, energy transfer

43. When calculating the annual effective dose for a technologist, which of the following exposures should be considered? The exposure received:

A. From your own bone scan study
B. From your own exposure while helping in a fluoroscopic room
C. From nuclear fallout
D. From your own dental radiographs

44. Which of the following is an example of a deterministic effect attributable to the exposure to relatively high levels of ionizing radiation?

1. Lens opacification   *2. Decreased leukocyte count*   *3. Decreased sperm count*

A. 1 only
B. 2 only

C. 3 only
D. 1, 2 & 3

45. A pregnant woman is scheduled for an intravenous urogram to rule out serous renal failure. What principle is used to justify the possible injury to the fetus?

A. The risk of not doing the procedure outweighs the potential risks to the mother or fetus
B. Medical risks always outweigh the radiation risks of a procedure
C. Radiation is less risky for the fetus than the mother
D. No justification is possible for this dangerous medical procedure

46. The NCRP recommends that the effective dose limit to individuals in the general population be limited to _____ the occupational limit.

A. 1/100 of
B. 1/10 of

C. 2 times
D. 10 times

47. The smallest amount of artificial exposure to the general population is derived from:

A. Nuclear fallout
B. Medical x-rays

C. Cosmic radiation
D. Terrestrial gamma sources

48. It is recommended that the fetus should NOT exceed a dose equivalent limit of _____ for the entire gestation period.

A. 1 mSv (.1 rem)
B. 5 mSv (.5 rem)

C. 15 mSv (1.5 rem)
D. 30 mSv (3 rem)

49. The effective dose limit for the stochastic effects for an occupational radiation worker in a one-year period is:

A. 5 mSv (.05 rem)
B. 10 mSv (1 rem)

C. 50 mSv (5 rem)
D. 250 mSv (25 rem)

50. According to the NCRP, the air kerma at the level of the patient during fluoroscopy, shall NOT exceed:

A. 1 cGy (1 rad) per min
B. 3 cGy (3 rad) per min

C. 5 cGy (5 rad) per min
D. 10 cGy (10 rad) per min

51. On rare occasions when essential radiation related tasks must be performed by an occupational worker, a single emergency occupational exposure limit of _____ may be permitted.

A. .5 Sv (50 rem)
B. 1.5 Sv (150 rem)
C. 35 Sv (300 rem)
D. 55 Sv (500 rem)

52. An occupational worker receives 80 mSv/yr. while working in a controlled area, 30 mSv from yearly x-rays, and 110 mSv from all other environmental sources. This person's occupational dose is:

A. 30 mSv
B. 80 mSv
C. 140 mSv
D. 170 mSv

53. The largest estimated average annual whole body radiation exposure from natural sources, excluding radon, are from:

A. Cosmic rays from the sun
B. Internal terrestrial sources
C. External terrestrial sources
D. Atmospheric ultraviolet rays

54. In the calculation of an occupational worker's effective dose limit, consideration shall be given to:

A. Internal exposures
B. External exposures
C. Both of the above
D. Neither of the above

55. According to the NCRP recommendations, the guidance level for cumulative effective dose for a 28 year-old occupational worker with three years of experience is:

A. 28 mSv
B. 76 mSv
C. 280 mSv
D. 350 mSv

1. D   Since no evidence exists that shows any increased risk to the embryo/fetus during a pregnancy for exposures under 25,000 millirem (250 mSv), it would be appropriate to tell the patient that no significant risk has occurred.

2. B   The measurement of the exposure will be made at location number 1, the measurement of the radiation-absorbed dose would be made at location number 2.

3. B   In order to limit the amount of absorption by the x-ray table, it should be made of a material that has aluminum equivalent of less than 1 mm. The carbon-fiber tabletop, due to its exceptional strength and low atomic number, is one of the few materials that will meet both of these requirements..

4. C   All type of permanent and mobile radiographic and fluoroscopic units must be equipped with a type of exposure switch will start and maintain an exposure only while pressure is applied to the switch. The name of the class of switch that requires constant pressure to keep the circuit open is a "dead man" switch. Taking the pressure off this type of switch will automatically terminate the exposure.

5. C   According to present regulations, the annual dose equivalent limit for an occupational worker for all other organs including the skin and the extremities is 50 rem (500 mSv). This far exceeds the annual dose limit for a member of the public which is 1.5 rem (15 mSv) for the eyes, 5 rem (50 mSv) for the hands, and .5 rem (5 mSv) for the fetal limit.

6. B   All effective dose limits are based on the principle that exposures at or below the recommended limits will be associated with a nominal lifetime risk.

7. A   The three cardinal principles of radiation protection that help to insure the radiation exposures to persons working around radiation are time, distance and shielding. The distance from a radiation source should always be as long as possible to ensure the lowest dose. Because the amount of radiation exposure will decrease inversely with the square of the distance, the use of a long distance is often the best way to reduce radiation exposures.

8. D   For the purposes of radiation safety, a facility is often divided into controlled and uncontrolled areas. In most facilities, an uncontrolled area is one that is occupied by anyone including the members of the public. Since an uncontrolled area is only allowed, a maximum exposure rate of 10 mR (2.6 micro C/Kg) per week the thickness of any barrier must be considerable. Under normal circumstances, waiting rooms, darkrooms, main control areas and offices are classified as uncontrolled areas. Controlled areas such as radiographic imaging areas, and hot labs, will require barriers that reduce the exposure rate to less than 100 mR (2.6 micro C/Kg) per week.

9. C   Because the equivalent dose levels can be reported in units of the rem or millirem (mrem)it is important to be able to convert these units. The number of rem can be converted into mrem by moving the decimal three places to the right. Millirem can be converted to rem by moving the decimal three places the to the left.

10. D   Because of the fluids and overlying tissues of the mother, the dose to the fetus is normally estimated to be 1/3rd of that received by the mother.

11. A   Though the effective dose limits serve as a guideline for radiation workers it is always desirable to keep radiation exposures to a level that is as low as is reasonably achievable (ALARA).

12. B   Solar radiation is responsible for an annual exposure of about 40 mrem per year. This many times lower than the actual amount that reaches the surface of the planet. The most important factor related to this reduction is the thick layer of atmosphere surrounding the surface of the earth that is absorbs much of this radiation.

13. A Since a pregnant radiographer and a uranium mineworker are both considered occupational workers, they will have an annual effective dose equivalent limit of 5 rem or 50 mSv or approximately .4 rem or 4 mSv per month. The embryo-fetus a non-occupational worker has a dose equivalent limit of .5 rem or 5 mSv during gestation, a .05 rem, or .5 mSv limit per month.

14. C The total dose from radiation is calculated by the formula:
Total Dose (160 mR) = Dose rate (4 mR/hr) x time (40 hours)

15. C According to published tables a .5 mm lead equivalent barrier (gonadal shield) will attenuate approximately 99% of a 50 kVp beam, 88% of a 75 kVp beam and 75% of a 100 kVp beam.

16. D The annual effective dose limit for the deterministic (somatic) effects for all other organs including the skin, extremities, gonads, lungs, breasts, and the red bone marrow , is 500 mSv (50 rem) per year.

17. D All radiation exposures should be kept to levels that are as low as reasonably achievable (ALARA). Unlike effective dose limits that are mandated absolute limits, ALARA will also consider economic and social factors. The ALARA limits enable a degree of flexibility that may differ significantly from the ideal. In general, a reasonable recommendation is to try to maintain a radiation worker's exposure to that of a non-occupational worker or 5 mSv (.5 rem) annually whenever possible.

18. A A deterministic effect, formerly called a non-stochastic effect, is defined as a somatic effect that increases in severity as a radiation dose increases above the dose threshold. Deterministic effects may be early, i.e., erythema, or late as with lens opacification, decreased fertility or organ atrophy.

19. A Any student under 18 years of age is allowed an annual effective dose equivalent limit of 1 mSv (.1 mrem) until his/her 18th birthday.

20. B The total occupational dose does not include background or medical exposures, so it is simply 3 mSv for this worker.

21. D The total exposure received by this individual is the sum of all of the individual exposures or 55 mSv.

22. B The amount of radiation attenuated is related to the thickness of the barrier and the atomic number of the material. Because of lead's high atomic number .5 mm thickness of this material will attenuate considerably more than 100 keV photons than any of the other listed materials.

23. D Since an occupational worker has a yearly effective dose limit of 50 mSv, the 6.5 mSv dose is subtracted, leaving a 43.5 mSv balance for the next 11 months.

24. D According to Bushong, it is possible to estimate fetal exposures based on monitor badge readings at the collar or waist. For a collar badge the fetal dose is estimated at about 10 % of the collar reading. The fetal dose because of the attenuation of maternal tissues is about 30% of the waist badge reading.

25. B Two main radiation area signs are commonly employed in or around a controlled area. A radiation area sign is used when the rate is between 5-100 mR/hr and a high radiation area sign is used when the exposure rate exceeds 100 mR/hr.

26. D According to the NCRP Report 116 specific annual effective dose limit for deterministic effects to the lens of the eye is 150 mSv (15 rem).

27. A According to current recommendations of the NCRP the highest annual effective dose equivalent limit is permitted for an adult occupational worker (5 rem or 50 mSv). This limit is the same during pregnancy though special limits for the embryo-fetus must also be observed.

28. D All members of the public (non-occupational) workers have an effective dose limit of .1 mSv (10 mrem) per week, or 5 mSv (500 mrem) per year.

29. B According NCRP Report 116 the individual worker's lifetime effective dose can be calculated by the formula 10 mSv x age in years providing the answer 10 mSv x 36 = 360 mSv (36 rem).

30. B   The regulations concerning stationary radiographic or fluoroscopic units, requires that the minimum source-to-skin distance (SSD) shall not be less than 30 cm (12 in.) and should not be less than 38 cm (15 in.).

31. D   Though all radiation workers may receive up to 50 mSv (5000 mrem) per year, the average annual dose is normally many times lower. It has been estimated that the dose equivalent for all radiation workers is about 2.1 mSv (210 mrem) per year.

32. B   Since a member of the public (non-occupational worker) has a yearly dose equivalent of 5 mSv (500 mrem), a 4 mSv or .4 rem exposure will not have exceeded this individual's annual effective dose equivalent limit.

33. D   This problem is solved by dividing annual limit 5000 mrem by the monthly dose (675 mrem) to give the number of months that this person can work without exceeding the permitted annual effective dose equivalent limit. 5000 mrem / 675 mrem = 7.4 months.

34. A   Because the year consists of 50 work weeks an occupational worker can receive up to 1 mSv per week without exceeding the 50 mSv permitted each year.

35. C   A stochastic effect is a radiation-induced injury that is assumed to have no dose threshold. Stochastic effects are normally limited to cancer and genetic effects that are considered all or nothing responses.

36. B   Occupational workers are classified through their work with ionizing radiation. Radiographers, CT technologists, NM technologists, and mammographers, are all classified as occupational workers. Ultrasonographers, MRI technologists, darkroom technologists are classified as non-occupational workers because they are not normally exposed to diagnostic x-rays.

37. B   See explanation number 36 of this section.

38. A   See explanation number 36 of this section.

39. A   See explanation number 36 of this section.

40. A   See explanation number 36 of this section.

41. C   The NCRP handbook No. 116 has recommended the elimination of the old 5(n-18) formula in favor of the age x 10 mSv (age x 1 rem) formula for the cumulative effective dose limit.

42. A   Annual effective dose limit, formerly the maximum permitted dose (MPD) is based on a linear non-threshold relationship at low or nominal risk levels.

43. B   Occupational exposures are intended to reflect only those exposures received as part of one's employment or participation in work related activities in a controlled area. Exposures from your own thyroid study, nuclear fallout and your own dental x-rays are classified as non-occupational.

44. D   Among the deterministic effects seen following a large irradiation are the formation of radiation cataracts (lens opacification), decreased blood cells counts and decreased sperm counts.

45. C   The vast majority of the recommendations concerning radiation and its control are provided by the National Council Radiation Protection and Measurement (NCRP).

46. B   Untrained individuals working in or around a radiation area are not considered different from any other member of the general public. These individuals can receive an annual effective dose equivalent limit of 5 mSv (500 mrem) or 1/10th of that of an occupational worker.

47. A   The following are the estimated annual whole body dose equivalents received by an individual from artificial radiation sources. Medical x-rays .39 mSv or 39 mrem , radiopharmaceuticals .14 mSv or 14 mrem, nuclear fallout .02 mSv or 2 mrem . The total from all artificial sources is estimated at .65 mSv or 65 mrem

48. B   The effective dose equivalent limit for the gestation period is 5 mSv (.5 rem) per year or .05 rem (.5 mSv) per month.

49. C   An occupational worker has the following effective dose limit for the stochastic effects of 1 mSv (100 mrem) per week or 50 mSv (5 rem) per year.

50. D   Under normal operating conditions, the kerma rate at the center of the point where the beam enters the patient for a fluoroscopic unit should not exceed 5 cGy (5 rad) per min. A rate of up to 10 cGy (10 rad) min is permitted when a high level control including a continuous audible signal is provided with the fluoroscopic unit.

51. A   When life saving action justifies acute exposures, they may be permitted to exceed annual effect dose limitations using the following guidelines. Older workers that volunteer with low lifetime accumulated effective doses should be selected first. The maximum recommended effective dose be limited to .5 Sv (50 rem) or an equivalent dose of 5 Sv (500 rem) to the skin.

52. B   Only the doses received in a controlled area should be considered in the calculation of the occupational dose. Medical x-rays, therapeutic x-rays and all other environmental sources are non-occupational in nature.

53. A   Natural sources of radiation include cosmic radiation, internal and external gamma sources and radon gas. Collectively these give an annual exposure of about 295 mrem.

   Though a number of environmental factors may effect these values, the average whole body dose for natural sources is .30 mSv or 30 mrem for cosmic radiation, .39 mSv or 30 mrem for internal gamma and .30 mSv or 30 mrem for external gamma sources and 2 mSv or 200 mrem for radon.

54. C   All occupational exposures, whether internal or external, should be considered in the calculation of the effective dose equivalent limit.

55. C   According to the current age formula 10 mSv x age of the individual, the answer is 280 mSv (28 rem).

Inverse Square Law

1. A radiographer receives an exposure of 100 mR/week at a 200 cm source-to-skin distance (SSD) . What would this person's exposure have been at a 100 cm source-to-skin distance (SSD)?

2.-10. State the new exposure that would be received by a radiographer for the following source-skin distances (SSD) and original exposure factors.

	Original SSD	New SSD	Original Exposure	New Exposure
2.	40 inches	80 inches	100 mR/week	_____
3.	72 inches	36 inches	50 mR/week	_____
4.	40 inches	60 inches	10 mR/week	_____
5.	72 inches	60 inches	100 mR/week	_____
6.	100 cm	200 cm	50 mR/week	_____
7.	180 cm	90 cm	10 mR/week	_____
8.	200 cm	50 cm	100 mR/week	_____
9.	100 cm	140 cm	50 mR/week	_____
10.	60 cm	180 cm	90 mR/week	_____

11. A radiographer receives an exposure of 5 mR/hr. standing at a 200 cm source-to-skin distance (SSD). In order to reduce this exposure to 1.25 mR/hr., the radiographer should move to a distance of _____.

12.-16. State the new source-to-skin distance (SSD) which will result from the use of the following source-to-skin distance (SSD)s and exposures.

	Original Exposure	New Exposure	Original SSD	New SSD
12.	40 mR/hr.	160 mR/hr.	72 inches	_____
13.	20 mR/hr.	5 mR/hr.	60 inches	_____
14.	5 mR/hr.	45 mR/hr.	100 cm	_____
15.	2 mR/hr.	18 mR/hr.	300 cm	_____
16.	100 mR/hr.	11 mR/hr.	180 cm	_____

17. The time it takes a patient to receive a dose of 1 gray is 10 minutes at a 50 cm source-to-skin distance (SSD). How long would it take to receive a dose of 2 gray at a 100 cm source-to-skin distance (SSD)?

18. If the time to deliver a dose of 100 cGy at a 100 cm source-to-skin distance (SSD) is 5 minutes, what is the dose rate per minute at this distance?

19. If the time it takes to deliver a dose of 200 cGy is 1 minute at 100 cm source-to-skin distance (SSD) , what dose would be received in 3 minutes at a 200 cm source-to-skin distance (SSD)?

20. A patient receives an exposure of 100 mR at a 180 cm source-to-skin distance (SSD). If the source-to-skin distance (SSD) is doubled with no other changes made, the exposure will be reduced by:

A. 25%
B. 50%

C. 200%
D. 400%

The following formulas have been used to solve these problems.

$$\text{New Exposure} = \frac{\text{Original Exposure x Original Distance (SSD)}^2}{\text{New Distance (SSD)}^2}$$

$$\text{New Distance}^2 = \frac{\text{Original Distance (SSD)}^2 \text{ x New Exposure}}{\text{Original Exposure}}$$

Inverse Square Law

1. A radiographer receives an exposure of 200 mR/week at a 300 cm source-to-skin distance (SSD). What would this person's exposure have been at a 100 cm source-to-skin distance (SSD)?

2.-10. State the new exposure that would be received by a radiographer for the following source-skin distances and original exposure factors.

	Original SSD	New SSD	Original Exposure	New Exposure
2.	30 inches	90 inches	100 mR/week	_____
3.	72 inches	60 inches	50 mR/week	_____
4.	40 inches	80 inches	20 mR/week	_____
5.	36 inches	72 inches	40 mR/week	_____
6.	200 cm	100 cm	50 mR/week	_____
7.	180 cm	90 cm	30 mR/week	_____
8.	200 cm	50 cm	70 mR/week	_____
9.	90 cm	180 cm	60 mR/week	_____
10.	80 cm	160 cm	20 mR/week	_____

11. A radiographer receives an exposure of 10 mR/hr. standing at a 200 cm source-to-skin distance (SSD). In order to reduce this exposure to 2.5 mR/hr., the radiographer should move to a source-to-skin distance (SSD) of _____ .

12.-16. State the new source-to-skin distance (SSD) that will result from the use of the following source-to-skin distance (SSD) and exposure.

	Original Exposure	New Exposure	Original SSD	New SSD
12.	30 mR/hr.	120 mR/hr.	72 inches	_____
13.	60 mR/hr.	15 mR/hr.	36 inches	_____
14.	50 mR/hr.	450 mR/hr.	300 cm	_____
15.	3 mR/hr.	27 mR/hr.	240 cm	_____
16.	200 mR/hr.	22 mR/hr.	60 cm	_____

17. The time it takes a patient to receive a dose of 2 gray is 1 minute at a 60 cm source-to-skin distance (SSD). How long would it take to receive a dose of 4 gray at a 120 cm source-to-skin distance (SSD)?

18. If the time to deliver a dose of 10 cGy at a 200 cm source-to-skin distance (SSD) is 10 minutes, what is the exposure rate per minute at this distance?

19. If the time it takes to deliver a dose of 300 cGy is 3 minutes at 200 cm source-to-skin distance (SSD), what dose would be received in 6 minutes at a 100 cm source-to-skin distance (SSD)?

20. A patient receives an exposure of 40 mR at a 90 cm source-to-skin distance (SSD). If the source-to-skin distance (SSD) is doubled with no other changes made, the exposure will be reduced by:

A. 25%
B. 50%

C. 200%
D. 400%

The following formulas have been used to solve these problems.

$$\text{New Exposure} = \frac{\text{Original Exposure x Original Distance (SSD)}^2}{\text{New Distance (SSD)}^2}$$

$$\text{New Distance}^2 = \frac{\text{Original Distance (SSD)}^2 \text{ x New Exposure}}{\text{Original Exposure}}$$

Answers

Inverse Square Law
Exercise 1

1. 400 mR/week
2. 25 mR/week
3. 200 mR/week
4. 4.4 mR/week
5. 144 mR/week
6. 12.5 mR/week
7. 40 mR/week
8. 1600 mR/week
9. 25.5 mR/week
10. 10 mR/week
11. 400 cm
12. 36 inches
13. 120 inches
14. 33 cm
15. 100 cm
16. 60 cm
17. 80 minutes
18. 20 cGy
19. 150 cGy
20. C

Inverse Square Law
Exercise 2

1. 1800 mR/week
2. 11.1 mR/week
3. 86.4 mR/week
4. 5 mR/week
5. 10 mR/week
6. 200 mR/week
7. 120 mR/week
8. 1200 mR/week
9. 15 mR/week
10. 5 mR/week
11. 400 cm
12. 36 inches
13. 72 inches
14. 100 cm
15. 80 cm
16. 180 cm
17. 8 minutes
18. 2 cGy
19. 2400 cGy
20. C

Section II

Equipment Maintenance and Operation

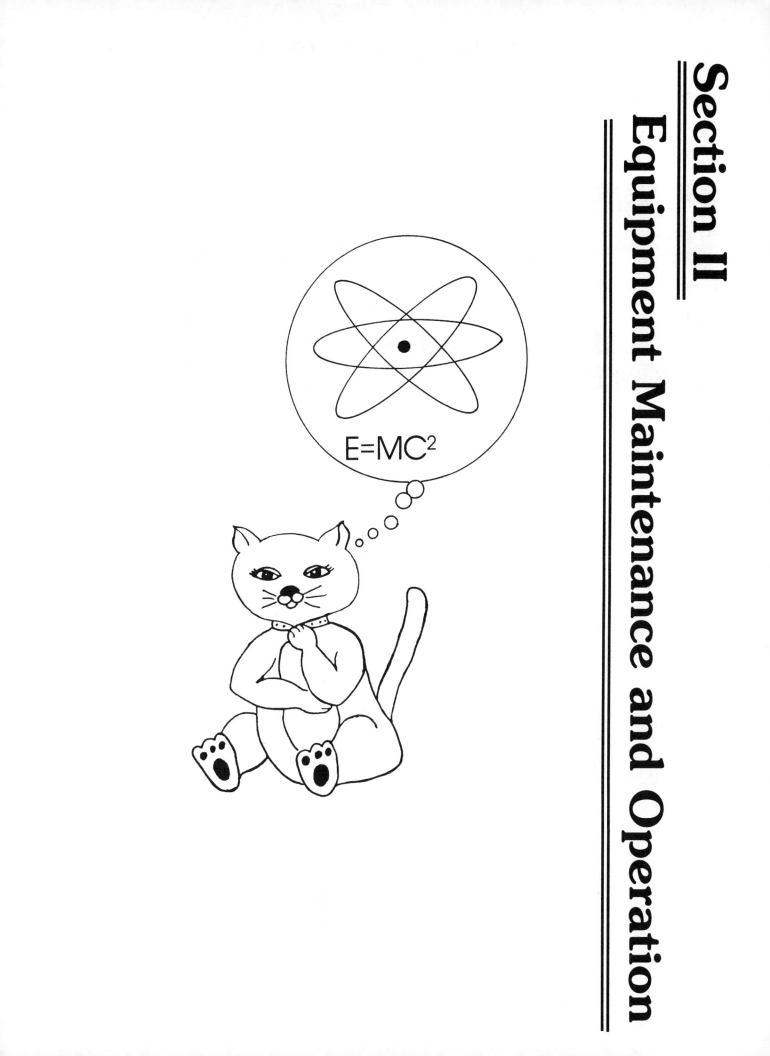

Radiographic Graffiti I

Can you identify these common patient related artifacts? (Answers appear at the bottom of the page.)

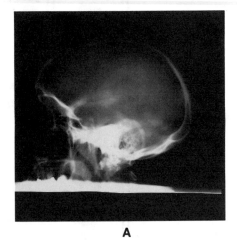

A

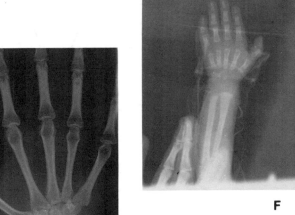

B

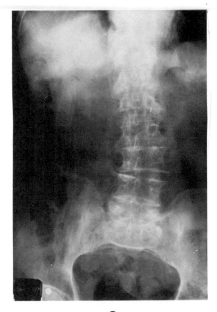

C

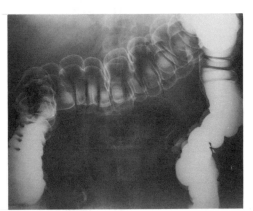

D

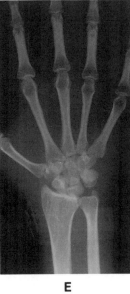

E

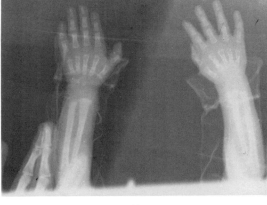

F

Atomic Structure

1. The force of attraction between the nucleus and an orbital electron is called the electron's_____ .

   A. Valence
   B. Transition number
   C. Binding energy
   D. Valence number

2. The schematic display most often employed to represent the structure of the atom was developed by:

   A. Charles Dalton
   B. Neils Bohr
   C. Michael Crookes
   D. Charles Darwin

3. The chemical properties of an atom are primarily controlled by its:

   A. Number of protons
   B. Number of neutrons
   C. Atomic weight
   D. Valence number

4. The orderly arrangement of the element by their atomic number and chemical properties is found in a/an:

   A. Emission spectrum
   B. Table of contents
   C. Auger table
   D. Periodic table

5. The principal unit of energy measurement on the atomic scale is the:

   A. Electron volt
   B. Neutron
   C. Coulomb
   D. Farad

6. The atomic number of the atom is determined by those particles that have an atomic mass unit of 1 and a single positive charge called:

   A. Electrons
   B. Positrons
   C. Protons
   D. Neutrons

7. The majority of the mass of the nucleus is derived from:

   A. Protons and electrons only
   B. Protons and neutrons only
   C. Protons, neutrons and electrons only
   D. Alpha, beta, gamma only

8. An atom that loses an orbital electron from the electrical influence of the nucleus has undergone:

   A. Radioactive decay
   B. Subluxation
   C. Ionization
   D. Covalent bonding

9. The number of outer shell electrons is known as the:

   A. Electron quota
   B. Orbital energy
   C. Valence
   D. Quantum mottle

10. In the symbol $^{12}_{6}C$ the number in the lower left-hand corner indicates the:

    A. Mass number
    B. Atomic number
    C. Isotope number
    D. Ionization level

11. In the previous question, the number in the upper left-hand corner indicates the:

    A. Mass number
    B. Atomic number
    C. Isotope number
    D. Ionization level

12. A neutral atom that loses an electron by ionization is termed a:

    A. Negative atom
    B. Positive atom
    C. Negative ion
    D. Positive ion

13. The smallest division of a substance possessing the same chemical and physical properties of the substance as a whole is called a/an:

   A. Atom
   B. Element

   C. Molecule
   D. Isotope

14. The tiny particles possessing a negative charge of 1.6×10^{-19} Coulombs that orbit the nucleus of an atom are termed:

   A. Electrons
   B. Protons

   C. Neutrons
   D. Photons

15. The centralized portion of an atom, the nucleus, is primarily composed of particles called the:

   1. Neutrons   *2. Electrons*   *3. Protons*

   A. 1 & 2 only
   B. 1 & 3 only

   C. 2 & 3 only
   D. 1, 2 & 3

16. The photons produced during the orbital transitions of electrons from a higher to lower energy are termed:

   A. Characteristic radiations
   B. Corpuscular radiations

   C. Vacancy radiations
   D. Field radiations

17. The formula Zn^2 is used to calculate:

   A. Minimum number of electrons in an orbital shell
   B. Maximum number of electrons in an orbital shell
   C. Total number of electrons in the atom
   D. Number of outer shell electrons in an atom

18. The atomic mass of an atom can be estimated by adding the atomic masses of the:

   1. Protons   *2. Neutrons*   *3. Electrons*

   A. 1 & 2 only
   B. 1 & 3 only

   C. 2 & 3 only
   D. 1, 2 & 3

19. The highest bonding energy of an atom is found in the electron's _____ level.

   1. First energy orbital   *2. Second energy orbital*   *3. Third energy orbital*

   A. 1 only
   B. 2 only

   C. 3 only
   D. 1, 2 & 3

20. A given atom has the following binding energies: K-shell - 1000 eV, L-shell - 70 eV, M-shell - 10 eV. During a M-L transition, the photon energy emitted will be:

   A. 930 eV
   B. 80 eV

   C. 60 eV
   D. Unable to determine

21. The chemical bond formed when two or more atoms share electrons is termed:

   A. Ionic bonding
   B. Polar bonding

   C. Covalent bonding
   D. Valence bonding

22. The interconvertability of mass and energy first proposed by Einstein is expressed by the equation:

   A. $E = \frac{1}{2} mv^2$
   B. $E = hu$

   C. $E = mc^2$
   D. $E = 2v^2$

23. The octet rule states that the most chemically stable atoms have an outer shell configuration with:

   A. Two electrons
   B. Four electrons

   C. Seven electrons
   D. Eight electrons

24. The natural state of matter with the highest energy due to the movement of atoms is the:

   A. Gaseous state
   B. Liquid state
   C. Plastic state
   D. Solid state

25. About 99% of the atom's mass is located within the:

   A. Orbital shell
   B. Valence shell
   C. Nucleus
   D. Alpha particle

26. Ionization of an atom may occur from exposing the atom to:

   1. Heat   2. Electric current   3. Radiation

   A. 1 & 2 only
   B. 1 & 3 only
   C. 2 & 3 only
   D. 1, 2 & 3

27. The electron binding energy is dependent upon the:

   1. Valence number   2. Amount of protons   3. Electron to nucleus distance

   A. 1 & 2 only
   B. 1 & 3 only
   C. 2 & 3 only
   D. 1, 2 & 3

28. Which of the following atoms would possess the highest binding energy for its K shell electron?

   A. $^{16}_{8}O$
   B. $^{23}_{11}Na$
   C. $^{63}_{29}Cu$
   D. $^{126}_{53}I$

29. An atom that has seven outer shell orbital electrons belongs to a group of elements known as:

   A. Inert gasses
   B. Noble gasses
   C. Halogens
   D. Rare earth elements

30. Two or more chemicals with the same chemical formula, but having different chemical properties, are termed:

   A. Isomers
   B. Isotopes
   C. Isobars
   D. Isotones

31. The formation of molecules can be accomplished through the interaction of valence electrons by a process termed:

   A. Ionic bonding
   B. Covalent bonding
   C. Both of the above
   D. Neither of the above

32. The mass of an orbital electron is about:

   A. 1/2000 the mass of a proton
   B. 1/10 the mass of a neutron
   C. 20 times the mass of a neutron
   D. 2000 times the mass of a proton

33. Which of the following is TRUE concerning the K-shell electrons of two different atoms?

   1. The electron binding energy is the same for both atoms
   2. The electron binding energy is characteristic of each atom
   3. The heavier element will have more K-shell electrons

   A. 1 only
   B. 2 only
   C. 3 only
   D. 1, 2 & 3

34. In a neutral atom the electrical neutrality is achieved by maintaining a balance between the:

   A. Protons and electrons
   B. Neutrons and positrons
   C. Electrons and positrons
   D. Protons and neutrons

35. The atomic particle which has a weight of 1 amu and carries no charge is called a/an:

A. Proton
B. Electron
C. Neutron
D. Negatron

36. The vertical columns of the periodic table contain groups of elements related by the same:

A. Atomic mass
B. Atomic number
C. Nuclear energy
D. Valence number

37. The physical state of matter is related to the degree of molecular attraction. Which state corresponds to the highest degree of molecular attraction?

A. Solid
B. Liquid
C. Plastic
D. Gas

38. A neutral atom that gains an electron will have a:

A. Neutral charge
B. Negative charge
C. Positive charge

39. The chemical bonding of two atoms by the transfer of an electron is a form of:

A. Ionic bonding
B. Covalent bonding
C. Mutual bonding
D. Polar bonding

40. Characteristic radiation emission occurs when:

A. Electrons move from lower to higher orbital shells
B. Electrons move from higher to lower orbital shells
C. The atom has an unstable nucleus
D. The atom undergoes nuclear fission

41. The mass energy equivalent value for an electron is:

A. .60 MeV
B. .147 MeV
C. .391 MeV
D. .511 MeV

42. An electron with a lowest binding energy is most likely located in the:

A. K shell
B. L shell
C. M shell
D. It is the same for all electrons

43. Which of the three fundamental particles has the lowest atomic mass?

A. Electron
B. Photon
C. Proton
D. Neutron

44. The most chemically stable elements having a filled outer shell configuration are termed:

A. Alkali metals
B. Alkaline metals
C. Nobel gases
D. Halogens

45. The physical characteristics of the atom are controlled by the atom's:

A. Atomic number
B. Atomic mass
C. Valence number
D. Electron binding energy

46. The stability of a nucleus is highly dependent upon the:

A. R/N ratio of the nucleus
B. e^- / e^+ ratio of the nucleus
C. N/Z ratio of the nucleus
D. e^-/P ratio of the nucleus

47. The nuclear particle possessing a single positive charge with a mass of 1.67×10^{-24} gm is termed the:

A. Neutron
B. Positron
C. Proton
D. Electron

48. The term applied to the raising of an electron to a higher energy orbit within a given atom is called:

A. Excitation
B. Ionization
C. Stabilization
D. Neutralization

49. The maximum number of inner or K-shell electrons in any atom is:

A. 1
B. 2
C. 4
D. 8

50. The smallest part of an element that retains all the characteristics properties of that element in bulk is called a/an:

A. Molecule
B. Atom
C. Coulomb
D. Nuclide

51. The following atoms are: $^{12}_{6}C$ $^{13}_{6}C$ $^{14}_{6}C$

A. Isotopes
B. Isobars
C. Isotones
D. Isomers

Fill in the appropriate value for the following neutral atoms, for questions 52-69.

	$^{23}_{11}Na$	$^{7}_{3}Li$	$^{35}_{17}Cl$
State their atomic number	52 _____	58 _____	64 _____
State their atomic mass	53 _____	59 _____	65 _____
State their valence	54 _____	60 _____	66 _____
State their number of protons	55 _____	61 _____	67 _____
State their number of neutrons	56 _____	62 _____	68 _____
State their number of electrons	57 _____	63 _____	69 _____

70. In the three elements above, which will have similar chemical properties?

1. Lithium   2. Sodium   3. Chlorine

A. 1 & 2 only
B. 1 & 3 only
C. 2 & 3 only
D. 1, 2 & 3

71. The principal force involved with nuclear binding is the:

A. Electromagnetic force
B. Weak force
C. Gravitational force
D. Strong force

72. Which of the following is a decay process by which an unstable nucleus tries to become stable?

1. Alpha decay   2. Beta decay   3. Positron decay

A. 1 only
B. 2 only
C. 3 only
D. 1, 2 & 3

73. Atoms with different types of nuclear configurations are termed:

A. Elements
B. Nuclides
C. Molecules
D. Mixtures

74. The term used to describe the process in which light nuclei combine to form a heavier nucleus is:

A. Fusion
B. Fission
C. Nuclear capture
D. De-excitation

75. Nuclides having the same atomic number but different atomic masses are termed:

   A. Isotopes
   B. Isotones

   C. Isobars
   D. Neutrinos

76. The rate of radioactive decay is greatly influenced by:

   A. Chemical bonding
   B. Pressure changes

   C. Temperature changes

77. After alpha decay, the atomic number (Z) of the parent nuclide will:

   A. Increase by 2
   B. Increase by 4

   C. Decrease by 1
   D. Decrease by 2

78. The atomic number of a daughter nuclide following a beta- decay will:

   A. Increase by 1
   B. Decrease by 1

   C. Increase by 2
   D. Decrease by 4

79. The fission process is best described as the:

   A. Combining of two small nuclides
   B. Combining of two large nuclides

   C. Splitting of a large nucleus
   D. Photo disintegration of a small nuclide

80. The most likely radionuclides undergoing spontaneous beta- decay will possess an unfavorably high N/Z ratio with an excess of:

   A. Neutrons
   B. Protons

   C. Positrons
   D. Electrons

Explanations

1. C The binding energy is the force by which the positively charged nucleus holds onto the negatively charged orbital electrons.

2. B An atom is often said to resemble a miniature solar system with the nucleus of the atom representing the sun and the electrons appearing as the planets. This model which is still used today was first proposed by a Danish physicist Neils Bohr in 1913.

3. D The chemical properties of an atom are a direct result of rearrangements in the number of the outer shell or valence electrons.

4. D In the 1850's a Russian scholar Dimitri Mendeleev described an arrangement of the known elements by their atomic mass and known chemical properties (valence) into what is now called the periodic table.

5. A The basic unit of measurement of the atomic scale is the electron volt (eV).

6. C The subatomic particle possessing a mass of one atomic mass unit (amu) and a single positive charge is the proton.

7. B The largest portion of the atom's mass (atomic mass) is found in the nucleus of the atom. The nucleus is formed by two subatomic particles, the neutrally charged neutron and positively charged proton.

8. C Following an atom's exposure to some forms of electromagnetic radiation, the orbital electrons can absorb enough energy to jump to higher orbital shells (excitation) or be completely removed from the atom's sphere of influence in a process called ionization.

9. C The number of outer shell orbital electrons is called the valence and is directly related to the chemical properties of the atom.

10. B The two numbers seen next to the atomic symbol represent the atomic mass and the atomic number for the atom. The top number corresponds to the atomic mass (A) which indicates the total number of neutrons and protons and the lower number corresponds to the atomic number (Z) and is equal to the number of protons in the nucleus.

11. A The two numbers seen next to the atomic symbol represent the atomic mass and the atomic number for the atom. The top number corresponds to the atomic mass (A) which indicates the total number of neutrons and protons and the lower number corresponds to the atomic number (Z) and is equal to the number of protons in the nucleus.

12. D Since an electron carries a negative charge, its loss by ionization will leave a positively charged ion.

13. C After a chemical bonding has occurred, a substance or compound is formed. The smallest subdivision of this compound is called a molecule.

14. A The only particles which commonly exist in the orbital shells surrounding the nucleus are the negatively charged electrons.

15. B The largest portion of the atom's mass (atomic mass) is found in the nucleus of the atom. The nucleus is formed by two subatomic particles, the neutrally charged neutron and positively charged proton.

16. A The binding energies of each atom are characteristic for a given element, when electrons move from higher to lower energy states or drop from higher to lower orbital shells. The photon energy created is identical to the difference between binding energies of the orbital shells.

17. B   The maximum number of orbital electrons which can occupy a given orbital shell can be calculated by taking the squared value of the orbital shell number and multiplying that value by 2.

18. A   The atomic mass (A) is found by adding the number of protons and neutrons.

19. A   Because of their proximity to the nucleus, inner shell electrons possess the highest force of attraction or binding energy.

20. C   The amount of energy of the photons emitted during an M-L transition is equal to the differences in binding energies of the electrons. 70eV - 10eV = 60eV.

21. C   The two main types of chemical bonds are the ionic bond, in which an outer shell electron is passed between the atoms, and the covalent bond in which outer shell electrons are shared between two or more atoms.

22. C   Perhaps the most well known and significant equation of Albert Einstein was E = mc2 in which mass, energy and speed of a particle or photon were found to be interrelated.

23. D   The major driving force behind chemical interactions is the desire of an atom to have a filled valance (outer) shell consisting of 8 electrons often called the octet rule.

24. A   Matter normally occupies one of the three normal states of matter as a solid, liquid or gas. The principal difference between these is in the energy of the atoms. In the lowest energy state or solid state, little atomic energy is found. By increasing the energy or heating the material it can pass into the liquid higher energy gaseous stages. Some material such as glass and most plastics reside in so-called quasi liquid or plastic state.

25. C   Over 99% of the atom's mass resides in the nucleus with the protons and neutrons.

26. D   The movement of electrons from lower to higher energy orbital shells (excitation) or the complete loss of the electrons (ionization) can be a result of an atom's exposure to the various forms of electromagnetic radiation including infrared (heat), visible light, ultraviolet, x-rays and gamma radiation.

27. C   The force of attraction or binding energy between the nucleus and the orbital electrons is dependent upon the number of protons and the distance between the nucleus and the electrons.

28. D   The highest binding energy for a given orbital shell is found in the atom with the highest atomic number.

29. C   Among the most chemically reactive elements are those found in the seventh group of the periodic table called the halogens. These include the elements fluorine, chlorine, bromine and iodine.

30. A   The arrangements of atoms into compounds does not always result in identical compounds. Isomers are chemicals that have the same chemical formula but combine in different patterns to form compounds with different chemical natures. The sugars fructose and dextrose share the same C6, H12, O6 formula but are chemically different.

31. C   The two main types of chemical bonds are the ionic bond in which an outer shell electron is passed between the atoms and the covalent bond in which outer shell electrons are shared between two or more atoms.

32. A   The mass of an electron is about 2000 times smaller than that of a proton or neutron.

33. B   All atoms which share the same atomic number are called elements and will have binding energies which are characteristic of that element.

34. A   In order for an atom to remain neutral, it must have a balance between the number of negatively charged electrons and positively charged protons.

35. C   The principal uncharged nuclear particle is the neutron which has a mass of 1amu.

36. D   The groups of the periodic table share similar chemical properties because they posses the same number of outer shell or valence electrons.

37. A   Matter normally occupies one of the three normal states of matter as a solid, liquid or gas. The principal difference between these is in the energy of the atoms. In the lowest energy state or solid state, little atomic energy is found. By increasing the energy or heating the material it can pass into the liquid higher energy gaseous stages. Some material such as glass and most plastics reside in so-called quasi liquid or plastic state.

38. B   The addition of a negatively charged electron to a neutral atom will result in an excess charge and the renaming of this particle into a negative ion.

39. A   The specific type of chemical bond which involves transfer of an electron is called ionic bonding.

40. B   The binding energies of each atom are characteristic for a given element, when electrons move from higher to lower energy states or drop from higher to lower orbital shells. The photon energy created is identical to the difference of the shells binding energies.

41. D   All mass bearing particles according to Einstein's theory have an energy equivalence. The value for an electron is .511meV and about 1GeV for a proton.

42. C   The highest binding energy is found in an inner shell (K-shell) electron and the lowest in the loosely bound outer shell electrons.

43. A   Electrons are about 2000 times less massive than protons or neutrons.

44. C   Though no elements are entirely chemically inert, the noble gasses such as helium, neon, argon, krypton, and xenon are all extremely unreactive.

45. A   The physical properties such as appearance, color, texture, smell and taste are most closely related to the atomic number of the element.

46. C   The degree of nuclear stability is closely related to the neutron-proton (N/P) ratio. Elements with favorable ratios are found to be more stable than those with unfavorable N/P ratios.

47. C   The proton is the largest mass bearing particle carrying a positive charge.

48. A   Following an atom's exposure to some forms of electromagnetic radiation, the orbital electrons can absorb enough energy to jump to higher orbital shells (excitation) or be completely removed from the atom's sphere of influence in a process called ionization.

49. B   All atoms regardless of their atomic number can have a maximum of two inner shell electrons.

50. B   The fundamental form of matter that has all the characteristics of that matter is the atom.

51. A   By definition, any atoms which have the same atomic number and different atomic mass are termed isotopes.

52. 11	53. 23	54. 1	55. 11	56. 12	57. 11
58. 3	59. 7	60. 1	61. 3	62. 4	63. 3
64. 17	65. 35	66. 7	67. 17	68. 18	69. 17

70. A   Because sodium (Na) and lithium (Li) have the same number of valence electrons they will have similar chemical properties.

71. D   Perhaps the strongest force in nature is that which binds the nucleus together called the strong force or nuclear glue.

72. D   In an attempt to become stable or non-radioactive, a nucleus may emit a number of decay particles, among these are: alpha particles, beta particles, neutrons and positrons.

73. B The term nuclide is used to describe the nuclear configuration or the relative number of nucleons, protons and neutrons, that are found in the nucleus of the atom. The term nuclide is used to express any of the different nuclear configurations that may exist.

74. A The process by which our sun and most other stars emit energy is due to the fusion of hydrogen into the heavier element helium. This is the same process that enabled the development of the hydrogen bomb.

75. A By definition, any atoms which have the same atomic number and different atomic mass are termed isotopes.

76. D The rate of radioactive decay occurs independently of all outside forces.

77. D Because the alpha particle consists of 2 protons and 2 neutrons, its emission is associated with a decrease of 2 in the atomic number of the parent nuclide.

78. A During a beta decay a neutron is converted to a proton resulting in an increase of the atomic number of 1.

79. C Most nuclear reactors and the first atomic bombs utilized the fission process in which a nucleus of uranium or plutonium is split into two or more smaller atoms.

80. A Beta decay is most likely in nuclides possessing too few protons or an excess of neutrons.

1. The two types of electrical charges are derived from their corresponding charged atomic particles called the:

   A. Electrons and protons
   B. Photons and neutrons
   C. Quarks and Lepton
   D. Nucleons and quantum

2. The rubbing of fur and amber or silk with glass result in electrification by a process called:

   A. Hydrolysis
   B. Induction
   C. Friction
   D. Transference

3. Any materials which tend to oppose the flow of electrons are termed:

   1. Non-conductors   2. Conductors   3. Insulators

   A. 1 & 2 only
   B. 1 & 3 only
   C. 2 & 3 only
   D. 1, 2 & 3

4. The transfer of same charge to an uncharged object by placing the two objects in contact with each other is termed:

   A. Friction
   B. Conduction
   C. Impedence
   D. Induction

5. According to the first law of electrostatics, two objects with opposite charges will tend to:

   A. Repel each other
   B. Attract each other
   C. Both of the above
   D. Neither of the above

6. An object with a net positive charge can be described as having a deficiency of:

   A. Protons
   B. Photons
   C. Neutrons
   D. Electrons

7. The electrical discharge that occurs between two oppositely charged objects, when the insulating properties of the transfer medium is exceeded, is called:

   A. Crosstalk
   B. Static
   C. Characteristic discharge
   D. Grounding

8. Electrification may occur from which of the following processes?

   1. Spontaneity   2. Conduction   3. Friction

   A. 1 & 2 only
   B. 1 & 3 only
   C. 2 & 3 only
   D. 1, 2 & 3

9. In a solid conductor, current consists of a flow of moving:

   A. Photons
   B. Neutrons
   C. Protons
   D. Electrons

10. The electrostatic force between two charged objects is _____ proportional to the product of their magnitudes.

    A. Directly
    B. Inversely
    C. Indirectly
    D. Conversely

11. Which of the following are most often classified electrical conductors?

    1. Copper   2. Rubber   3. Water

    A. 1 & 2 only
    B. 1 & 3 only
    C. 2 & 3 only
    D. 1, 2 & 3

12. According to the principle stated in Coulomb's law, the electrostatic force between two charged objects is _____ proportional to the_____of the distance between them.

A. Inversely, prime
B. Directly, prime

C. Inversely, square
D. Directly, square

13. The electrical resistance which occurs in a conductor is most often manifested as the amount of _____ created.

A. Heat
B. Electrical noise

C. X-rays
D. Magnetism

14. The force within an electrical circuit most responsible for the movement of current is termed the:

A. Resistance
B. Wattage

C. Amperage
D. Potential difference

15. A circuit which has 6.3×10^{18} electrons flowing per second is said to have a current of:

A. One ampere
B. One ohm

C. One volt
D. One watt

16. The type of electricity associated with the flow of current in a single direction is termed:

A. Alternating current
B. Direct current

C. 3-phase current
D. Repulsive current

17. A major source of electrical current which is produced by a chemical process is a:

A. Battery
B. Generator

C. Transformer
D. Capacitor

18. Which of the following class of materials have a high electrical resistance?

1. Metals   2. Rubber   3. Wood

A. 1 & 2 only
B. 1 & 3 only

C. 2 & 3 only
D. 1, 2 & 3

19. The law which relates to the relative amperage, voltage and resistance in a direct current circuit is:

A. Coulomb's law
B. Joule's law

C. Boyle's law
D. Ohm's law

20. The resistance of a wire is related to its:

1. Length   2. Temperature   3. Cross-sectional area

A. 1 & 2 only
B. 1 & 3 only

C. 2 & 3 only
D. 1, 2 & 3

21. An electromagnetic device which converts mechanical into electrical energy is termed:

A. Transformer
B. Generator

C. Electroscope
D. Autotransformer

22. Which of the following is a type of device used for storage of electricity?

1. Capacitor   2. Condenser   3. Battery

A. 1 & 2 only
B. 1 & 3 only

C. 2 & 3 only
D. 1, 2 & 3

23. Select the copper wire with the least electrical resistance.

A. 100 cm 20 gauge
B. 200 cm 20 gauge

C. 100 cm 10 gauge
D. 200 cm 10 gauge

24. The physical contact of two or more non-insulated wires carrying current will result in a/an:

A. Open circuit
B. Short circuit
C. Closed circuit
D. Relay circuit

25. The power loss as heat that occurs in a conductor is expressed by the formula P=:

A. IR
B. IV
C. VR2
D. I2R

26. The amount of electrical potential, or potential difference in an electrical circuit, is measured by a unit called the:

A. Volt
B. Amp
C. Ohm
D. Forad

27. The power of a circuit is equal to the product of current and voltage and is measured by the unit:

A. Joule
B. Erg
C. Watt
D. Newton

28. The resistance of a wire will increase as its:

1. Length increases   *2. Temperature increases*   *3. Diameter decreases*

A. 1 & 2 only
B. 1 & 3 only
C. 2 & 3 only
D. 1, 2 & 3

29. An electromagnetic device consisting of a single iron core surrounded by a wire carrying current is termed a/an:

A. Voltaic pile
B. Electroscope
C. Ammeter
D. Electromagnet

30. How many wet cell batteries must be connected in series to provide the 220 volt power supply of a battery powered mobile unit?

A. 55
B. 110
C. 220
D. 440

31. The production of electricity by the rotation of a coil of wire within a stationary magnetic field is termed:

A. Electromagnetic irradiation
B. Electric impedance
C. Magnetic fluctuation
D. Electromagnetic induction

32. If an x-ray unit has a maximum output of 120 kilovolts and 600 milliamperes, the generator must have a power rating of at least:

A. 63 kilowatts
B. 72 kilowatts
C. 120 kilowatts
D. 600 kilowatts

33. Which of the following is classified as a type of natural magnet:

1. Alnico   *2. Lodestone*   *3. The earth*

A. 1 & 2 only
B. 1 & 3 only
C. 2 & 3 only
D. 1, 2 & 3

34. The strong magnetic dipole of a hydrogen atom is due in large part to the:

1. Unpaired orbital electron
2. Imbalance of the protons and neutrons
3. Alternating nuclear spin

A. 1 only
B. 2 only
C. 3 only
D. 1, 2 & 3

35. The magnetic dipole resulting from the proton spin in a hydrogen nucleus is termed the:

A. Magnetic moment
B. Random domain

C. Unifield force
D. Attractive force

36. In a magnet the individual magnetic domains are aligned in:

A. Many different directions
B. Single direction

37. The magnetic force lines or lines of flux in a magnetic field flow:

A. Within the magnet itself
B. Outside the magnet

C. Both of the above
D. Neither of the above

38. Which of the following are units used in the measure of magnetic flux density?

A. Watt or ohm
B. Erg or Joule

C. Newton or compass
D. Tesla or gauss

39. A commonly used device used in the detection of a magnetic field is a:

A. Monopole
B. Compass

C. Solenoid
D. Spectroscope

40. Magnetic fields are a phenomenon associated with:

A. Moving neutrons
B. Moving charges

C. Stationary monopoles
D. Stationary photons

41. The laws of magnetism are similar to the laws of electrostatic because they both:

1. *Cause photons to deflect*
2. *Have the same atomic configuration*
3. *Are produced by the same physical process*

A. 1 only
B. 2 only

C. 3 only
D. 1, 2 & 3

42. Which of the following will have an associated magnetic moment (field)?

1. *A Hydrogen atom*   2. *An orbital electron*   3. *Iron atom*

A. 1 & 2 only
B. 1 & 3 only

C. 2 & 3 only
D. 1, 2 & 3

43. The ability of a magnetic material to resist demagnetization is termed:

A. Retentivity
B. Grounding

C. Permeability
D. Excitation

44. Which of the following is an appropriate form of Ohm's law?

1. *V = IR*   2. *R = V/I*   3. *I = V/R*

A. 1 only
B. 2 only

C. 3 only
D. 1, 2 & 3

Referring to the following series circuit, answer questions 45-50.

A series circuit with a current of 10 amperes contains three separate resistive elements with values of

R1 = 5 ohms, R2 = 10 ohms, and R3 = 20 ohms.

45. What is the total resistance for this circuit?

    A. 3 ohms   C. 15 ohms
    B. 5 ohms   D. 35 ohms

46. The total voltage for this circuit is:

    A. 350 volts   C. 110 volts
    B. 220 volts   D. 100 volts

47. What will the voltage measure across R1?

    A. 20 volts   C. 100 volts
    B. 50 volts   D. 200 volts

48. The voltage across R2 will measure:

    A. 50 volts   C. 200 volts
    B. 100 volts   D. 350 volts

49. The voltage across R3 will measure:

    A. 20 volts   C. 200 volts
    B. 110 volts   D. 220 volts

50. The total current through each resistor is the same for all parts of the series circuit.

    A. True
    B. False

Referring to the following parallel circuit, answer questions 51-55.

A parallel circuit carrying a 120 voltage contains four separate resistive elements with values of
R1 = 30 ohms, R2 = 40 ohms, R3 = 20 ohms and R4 = 30 ohms.

51. What is the total resistance for this circuit?

    A. 111 ohms   C. 17 ohms
    B. 100 ohms   D. 7 ohms

52. The total amperage found in this circuit is:

    A. 17 amp   C. 43 amp
    B. 21 amp   D. 57 amp

53. What will the current measure through R1?

    A. 1.5 amp   C. 3.6 amp
    B. 2.5 amp   D. 4.0 amp

54. What will the current measure through R2?

    A. 1.2 amp   C. 3 amp
    B. 1.6 amp   D. 4 amp

55. What will the current measure through R3?

    A. 1 amp   C. 4 amp
    B. 3 amp   D. 6 amp

Explanations

1. A The two basic charges of all matter are the positive charge derived from the proton and negative charge obtained from electrons.

2. C The movement of certain materials across each other may cause the generation of charges during a process called friction.

3. B In order to protect an individual from electrical shocks most electrical components are surrounded by materials which oppose the flow of electricity called non-conductors or insulators.

4. B Conduction is defined as the transfer of a like charge by the physical contact of two bodies.

5. B The first law of electrostatics states that two charged objects with opposite charges will be attracted to each other and the objects with like charges will be repelled.

6. D In electrical terms a positive charge is normally caused by having a shortage of negatively charged electrons.

7. B A static discharge such as lightning occurs as electrons overcome the insulating properties of the surrounding air and travel from a negatively charged cloud to positively charged object.

8. C The three processes by which electrification can occur are conduction, induction, and friction.

9. D Though both positive ions and electrons can move in liquids only electrons can move in a solid conductor.

10. A According to Coulomb's law, the magnitude of the electrostatic force is directly proportional to the product of the charges and inversely proportional to the square of the distance between the charged objects.

11. B Electrical conductors which have a low resistance to the flow of electrons are usually made of copper or other metals. Tap water is also a relatively good conductor of electricity.

12. C According to Coulomb's law, the magnitude of the electrostatic force is directly proportional to the product of the charges and inversely proportional to the square of the distance between the charged objects.

13. A The opposition to the flow of electrons (resistance) is most often manifested in the form of heat.

14. D The driving force for the movement of electrical current is an electrical potential or potential difference that is caused by a difference in charge between the various components in a circuit.

15. A By definition, 1 amp of current is equal to 1 coulomb of electrons flowing in 1 second.

16. B The current from a battery, capacitor and DC generator all travel in a single direction and are referred to as types of direct current.

17. A The major source of chemically produced electricity is a battery. At the present time a number of batteries are available which are classified as either wet or dry cell types.

18. C The best insulations are usually made of rubber or plastic which have an extremely high electrical resistance. Certain types of oils are also very good insulators.

19. D The behavior of electric currents in a circuit are best described by Ohm's law which is stated as $V = RI$ where V=voltage, I=ampere, R=resistance.

20. D In addition to the material from which the wire is made, the electrical resistance can be minimized by employing a larger cross sectional area (smaller gauge), lower conductor temperature, and reduced length of the wire.

21. B   One of the most important devices in the production of current electricity is an electromagnetic device called a generator which can convert mechanical into electrical energy.

22. D   Electricity can be stored for short periods of time in a capacitor or condenser and for longer periods of time in wet or dry cell batteries.

23. C   In addition to the material from which the wire is made, the electrical resistance can be minimized by employing a larger cross sectional area (smaller gauge), lower conductor temperature, and reduced length of the wire.

24. B   Short circuits can be avoided by the proper insulation of the wires carrying electricity.

25. D   The heat loss in a conductor can be calculated using the power loss formula (P=I2R) where P is the power loss as heat, I2 is ampere squared, and R is the resistance of the wire.

26. A   There are 4 main units of electricity: the volt which is the unit of potential difference; the ampere, the unit of electrical quantity; the ohm, the unit of electrical resistance; and the watt or unit of power.

27. C   The unit of power, the watt, is obtained by multiplying the number of Amp x volts; i.e., P=IxV.

28. D   In addition to the material from which the wire is made, the electrical resistance can be minimized by employing a larger cross sectional area (smaller gauge), lower conductor temperature, and reduced length of the wire.

29. D   An electromagnet is a simple device consisting of an iron core surrounded by a coil of wire that is carrying current.

30. B   Since voltage is an additive, it would take 110 separate 2 volt wet cell batteries to provide 220 volts of power supply.

31. D   A generator which consists of a coil of wire which is rotated within a magnetic field, produces electricity by a process called electromagnetic induction.

32. B   According to the power formula (P=Ix V), the minimum rating for this device is 120 kVp x 600 mA = 72,000 watts.

33. C   The earth and lodestone are the two common types of natural magnets. Alnico is a very strong type of artificial magnet.

34. A   Because magnetism is always associated with movement of electrons, the unpaired orbital electron of hydrogen will make this matter magnetic.

35. A   The induced magnetic field due to the nuclear spin of a hydrogen atom creates a magnetic field known as the magnetic movement.

36. B   In a strong permanent magnet all the individual magnetic domains will be aligned in a single direction.

37. C   The magnetic lines of force or magnetic flux move from the south magnetic pole to the north magnetic pole within the magnet and from north to south poles in the space surrounding the magnet.

38. D   The Tesla and gauss are the two main units for measuring magnetism. The earth's magnetic field measures about 1 gauss or .0001 Tesla.

39. B   The simplest device for the detection of a magnetic field is a compass which, due to its magnetic properties, aligns itself with any other external magnetic field.

40. B   A magnetic field is associated with the movement of a charged particle; i.e., electron and proton or atoms with an imbalance in the number of orbital electrons, i.e,: hydrogen and iron.

41. C   The laws of magnetism and the law of electrostatics are similar because they are created simultaneously when charged particles are in motion.

42. D See explanation number 40 of this section.

43. A Permanent magnets are described as having two main properties; retentivity, or the ability of a material to resist demagnetization, and permeability, the ability of a material to become magnetized.

44. D All are correct forms of Ohm's law.

45. D All are solved using the versions of Ohm's law in previous question and the formula $R_T = R_1 + R_2 + R_3$.

46. A See explanation number 45 of this section.

47. B See explanation number 45 of this section.

48. B See explanation number 45 of this section.

49. C See explanation number 45 of this section.

50. A See explanation number 45 of this section.

51. D All are solved using the versions of Ohm's law in question 44 and the formula
$$\frac{1}{R_T} = \frac{1}{R_1} + \frac{1}{R_2} + \frac{1}{R_3}$$

52. A See explanation number 51 of this section.

53. D See explanation number 51 of this section.

54. C See explanation number 51 of this section.

55. D See explanation number 51 of this section.

1. A line voltage compensator is incorporated into the primary circuit of a modern radiographic unit to:

    A. Compensate for variations of the timer
    B. Control the desired speed of the anode
    C. Provide the desired filament charge
    D. Maintain a constant kVp level

2. The type of meter that is used to measure the current passing between the electrodes of an x-ray tube is the:

    A. Pre-reading kilovolt meter
    B. Multiphasic ohm meter
    C. Milliammeter-seconds meter
    D. Synchronous ammeter

3. A transformer has 60 primary and 12,000 secondary turns of wire. How much voltage will be induced on to the secondary side if a 400-volt primary power supply is impressed on the device?

    A. 18 kVp
    B. 80 kVp
    C. 120 kVp
    D. 160 kVp

4. The voltage in the secondary circuit of a modern x-ray circuit is normally about _____ than the voltage in the primary circuit.

    A. 20 times lower
    B. 500 times lower
    C. 100 times higher
    D. 1000 times higher

5. The high-tension portion of a modern x-ray circuit contains all of the following components EXCEPT:

    A. The x-ray tube cables
    B. The filament selector
    C. The solid state diodes
    D. The milliammeter

6. The use of the oil within the sealed housing of the x-ray tube is provided to:

    1. Increase the thermal cooling of the tube envelope
    2. Insure that the motor bearing of the induction motor remain lubricated
    3. Provide electrical insulation for the tube

    A. 1 & 2 only
    B. 1 & 3 only
    C. 2 & 3 only
    D. 1, 2 & 3

7. Which of the following changes would be associated with a higher amount of current passing through the filament of the x-ray tube?

    A. The selection of a longer timer setting
    B. The selection of a higher mA station
    C. The application of a shorter source-to-image distance
    D. The use of a kilovoltage compensation circuit

8. The high tension (step-up) transformer of a modern x-ray circuit is most closely related to the:

    A. Production of the high amperage needed to develop the space charge
    B. High potential required to accelerate the electrons across the tube
    C. Maximum speed achieved by the rotating anode
    D. Rate at which the x-ray photons are formed in the tube

9. Which of the following components are likely to be found in the primary circuit of a modern x-ray unit?

    1. Autotransformer   **2. Timer switch**   **3. The anode of the x-ray tube**

    A. 1 & 2 only
    B. 1 & 3 only
    C. 2 & 3 only
    D. 1, 2 & 3

10. The high temperatures that are needed to produce the thermionic emission in the filament of an x-ray tube require a current of about:

A. 1-2 milliamps
B. 3-6 milliamps
C. .5-1 amp
D. 4-6 amps

11. The majority of components in a modern x-ray circuit are found in the primary circuit. This is done to:

A. Reduce the risk from electrical shock
B. Decrease the size of the transformer
C. Reduce the size of the unit
D. Decrease the size of the tube housing

12. The current passing between the secondary of the high tension transformer and the rectifiers is best described as a type of:

A. High voltage (kV) AC
B. Low voltage AC
C. High voltage (kV) DC
D. Low voltage DC

13. In nearly all modern units the use of the large focal is initiated when the an exposure setting above _____ is employed.

1. 1 second   2. 300 mA   3. 100 kVp

A. 1 only
B. 2 only
C. 3 only
D. 1, 2 & 3

14. Most x-ray machines used in the United States are designed to operate on a 220 volt or 440 volt:

A. 60 hertz, alternating current power supply
B. 60 hertz direct current power supply
C. 120 hertz, alternating current power supply
D. 50 hertz, direct current power supply

15. Which of the following current regulating devices are required in the production of the high-voltage direct current needed in the x-ray production process?

1. Rectifiers   2. High-tension transformer   3. Filament transformer

A. 1 & 2 only
B. 1 & 3 only
C. 2 & 3 only
D. 1, 2 & 3

16. The control for the amount of filament current in most modern radiographic units is directly related to the:

A. Timer setting selection
B. Kilovoltage selection
C. Milliampere selection
D. Focal spot size selection

17. In nearly all-radiographic units, the high-tension transformer is located in an oil-filled tank that also houses the:

1. Rectifiers   2. Filament transformer   3. Autotransformer

A. 1 & 2 only
B. 1 & 3 only
C. 2 & 3 only
D. 1, 2 & 3

18. In order to maintain a constant filament current as changes are made in the kilovoltage requires the use of a:

A. Space charge compensator
B. Ballistics stabilizer
C. Filament boost circuit
D. Line voltage compensator

19. The milliamperage that is impressed across the x-ray tube will have a substantial effect on the:

1. Beam penetration   2. Filament heat   3. Rate of photon production

A. 1 & 2 only
B. 1 & 3 only
C. 2 & 3 only
D. 1, 2 & 3

20. In the filament branch of the primary circuit, the type of meter employed to measure this current is a/an:

A. Micro-ammeter
B. Milliammeter

C. Ammeter
D. Volt meter

21. The purpose of a ballistic-type milliampere-second meter is to measure current flow when:

A. High kilovoltage settings are employed
B. Low milliamperage settings are employed

C. Rapid exposure time settings are employed
D. Long exposure time settings are employed

22. The main component of a line voltage compensator is a device that is capable of storing small amounts of electrical charge called a:

A. Capacitor /condenser
B. Ballistic collector

C. Wavetail reserve
D. Phase converter

23. The selection of voltage to the primary of the high-tension transformer in a 3-phase unit is made by (a):

A. Single variac
B. Twin rectifier

C. Three autotransformers
D. Three circuit breakers

24. In order to operate properly x-ray tube requires direct electrical connections to the:

1. Filament circuit   **2. Primary circuit**   **3. Secondary circuit**

A. 1 & 2 only
B. 1 & 3 only

C. 2 & 3 only
D. 1, 2 & 3

25. The midpoint of the high-tension transformer in a modern x-ray circuit is grounded. This serves to reduce the:

A. Number of rectifiers that are required
B. Storage capacity of the anode

C. Amount of cable insulation
D. Amount of insulating oil

26. The selection of voltage to the high-tension transformer is most commonly made by adjusting the settings on the:

A. Exposure timer
B. Line voltage compensator

C. Filament transformer
D. Autotransformer

27. The pre-reading kV meter is incorporated into the x-ray control panel to measure the potential difference on the:

A. Primary side of the autotransformer
B. Secondary side of the high tension transformer

C. Secondary side of the autotransformer
D. Filament side of the x-ray tube

28. Which component is NOT generally located in the high voltage portion of a modern x-ray circuit?

A. Rectifiers
B. X-ray tube

C. Filament transformer
D. Milliampere meter

29. Circuit overloads occurring in sensitive components of an electrical circuit can be prevented by the use of which of the following devices?

1. Thermionic diodes   **2. Fuses**   **3. Circuit breakers**

A. 1 & 2 only
B. 1 & 3 only

C. 2 & 3 only
D. 1, 2 & 3

30. Which of the following is generally found on the operating console of a x-ray machine?

1. kV control switch   **2. mA control switch**   **3. Timer control switch**

A. 1 only
B. 2 only

C. 3 only
D. 1, 2 & 3

31. Which of the following devices can be used for the storage of electrical charge?

A. Rectifier
B. Transformer

C. Battery
D. Variac

32. The most common device used to create the high potential difference (kilovoltage) across the x-ray tube is the high tension:

A. Autotransformer
B. Variable resistor

C. Saturable reactor
D. Transformer

33. The reduction of voltage and development of high amperage by a step-down transformer occurs in the:

A. Primary portion of the circuit
B. Secondary portion of the circuit

C. Filament portion of the circuit
D. High tension portion of the circuit

34. In a radiographic unit, the milliammeter will register only during the activation of the:

A. Boost (prep) stage of the exposure switch
B. Second (exposure) stage of the exposure switch

C. Both of the above
D. Neither of the above

35. A transformer has 2000 primary and 200 secondary turns of wire. How much voltage will be induced onto the secondary side if a 200 volt primary power supply is impressed on the device?

A. 20 volts
B. 200 volts

C. 1000 volts
D. 2000 volts

Match the following for questions 36-38.

36. _____ Primary circuit
37. _____ Filament circuit
38. _____ Secondary circuit

A. 25,000-150,000 volts
B. 110, 220 or 440 volts
C. 10-12 volts

Referring to the diagram of the basic x-ray circuit, answer question 39-50.

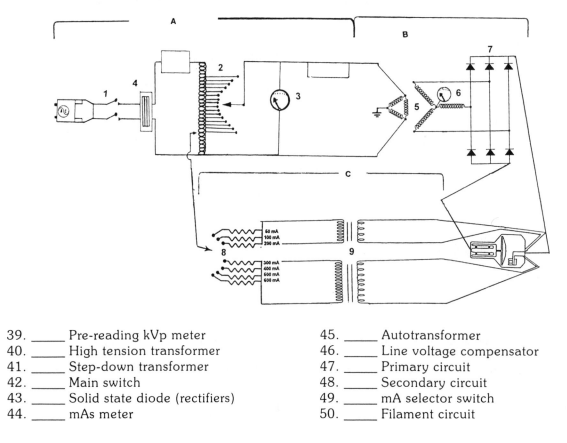

39. _____ Pre-reading kVp meter
40. _____ High tension transformer
41. _____ Step-down transformer
42. _____ Main switch
43. _____ Solid state diode (rectifiers)
44. _____ mAs meter

45. _____ Autotransformer
46. _____ Line voltage compensator
47. _____ Primary circuit
48. _____ Secondary circuit
49. _____ mA selector switch
50. _____ Filament circuit

Explanations

1. D   In order to insure that there is a consistent voltage supply to the high voltage components in the radiographic unit, a capacitor, called a line voltage compensator, is employed to adjust the incoming line voltage. This will help to maintain consistent kVp levels during an exposure.

2. C   During the activation of the exposure switch, the amount of electrons passing between the electrodes of the x-ray tube the cathode, and anode, during the actual radiographic exposure constitutes the tube current. This is measured by a device called the milliampere-seconds (mAs) meter located in the secondary circuit.

3. B   The amount of voltage change occurring in a transformer is based on the ratio of the number of primary to secondary turns. Using the transformer law:

   Secondary voltage (80,000 volts/ 80 kVp) = Primary voltage (400 volts) x Number of secondary turns (12,000)

   Number of primary turns (60)

4. D   In most modern high voltage transformers the voltage from the autotransformer is increased by about 1 thousand times before it passes into the rectification system.

5. B   The high tension or high voltage circuit consists of five major components; the transformer, the milliampere-seconds (mAs) meter, the rectifiers, the high tension cables and the anode of the x-ray tube.

6 B   In order to increase the amount of heat that can be dissipated away from the x-ray tube, the tube housing surrounding the tube envelope, is normally filled with a layer of oil. Because this oil is also a poor conductor of electricity it also helps to electrically insulate the x-ray tube. This will help prevent any possible shock hazards.

7. B   The amount of current impressed on the filament is proportional to amount of heat (thermionic emission) or the temperature of the filament. In nearly all modern systems the control of the filament temperature is selected by changes in the milliampere (mA) setting. A higher mA setting will increase the amount of heat in the filament

8. B   The high tension or step-up transformer creates the high potential difference that is necessary to move the tube electrons from the cathode to the anode during the x-ray production process.

9. A   The majority of components in an x-ray are located in the primary circuit. Among these are the main switch, autotransformer, pre-reading kVp meter, timer switches and the line voltage compensator.

10. D   The high heat necessary to boil off the electrons of the cathode or filament requires a temperature of about 2200°C. This is usually accomplished by the application of a current between 4-6 amperes of current onto the filament. In most modern systems this is determined by selection of the desired mA station.

11. A   The primary circuit of the radiographic unit carries current which is similar in form and voltage to normal household alternating current. Because this type of current is relatively safe to work with, the majority of accessory devices (motors, conveyers, lights) are placed into the primary circuit.

12. A   The high voltage (kilovoltage) alternating current (AC) which is developed on the secondary side of the high tension transformer needs to be converted into a direct current form before passing through the x-ray tube. The use of solid rectifiers accomplishes this conversion in a modern x-ray unit.

13. B   In addition to the selection of the amount of current passing into the filament the selection of the mA station is also related to the selection of the size of the focal spot. In most units, the selection of a 25 mA, 50 mA 100 mA, and 200 mA station will provide a small focal spot. The large focal spot is activated when the mA setting is 300 mA or greater.

14. A   All of the power supplied to the electrical devices used in this country are in the form of a 60 cycle per second alternating current at voltages of 110, 220 or 440 volts depending on the power needs and nature of the equipment.

15. A   The production of high voltage direct current used during the x-ray production process, requires the use of the high-tension transformer to provide the high kVp and the rectifiers to convert the alternating current into direct current.

16. C   In nearly all modern equipment, the selection of the filament temperature is tied into one of the preset mA selector buttons (stations). The higher the mA station selected, the higher the current and temperature achieved in the filament.

17. A   For safety reasons, any components which carry extremely large amounts of voltage or current when possible are placed in a highly insulated oil filled tank. The main components placed in this tank are the high tension transformer, the filament transformer and the rectifiers.

18. A   Any increase in the kilovoltage peak impressed on the electrodes of the x-ray tube will increase the percentage of the electron cloud (space charge) moved across the tube. In order to provide a consistent exposure (mAs) at various kVp levels, the use of a space charge compensator is employed in a modern filament circuit.

19. C   The milliampere (mA) setting and the change in the filament temperature it causes (explanation 16) is directly related to the exposure rate of photon production and the resulting optical density on the radiographic image. The amount of penetration of the beam is controlled by the kVp setting.

20. C   In many older radiographic units, the selection of the filament temperature was a function of a filament control knob. An ammeter provided a measure of the actual current and means for estimating the filament heat employed. In most modern units, this meter has been replaced by preset mA stations.

21. C   In a radiographic unit which uses a meter to display the tube current or milliampere - seconds values, a rapid exposure time of less than 1/20th (.05) seconds would be imperceptible to the human eye. A ballistic type of meter which lengthens or slows the swing of the indicator enables the operator to see the mAs values for the exposure.

22. A   A line (voltage) compensator is a type of electrical storage device called a capacitor which is able to store and release small amounts of current to compensate for small drops in the incoming voltage supply. Surge protectors often recommended for computers are similar in function to the line compensators employed in the x-ray circuits.

23. C   In a 3-phase circuit, each of the three voltage phases must be selected individually by its own autotransformer. Therefore, 3-phase circuits require the use of 3 separate autotransformers.

24. B   The cathode or filament of a modern radiographic tube has a dual electrical connection. The first is the connection to the filament circuit which is activated during the boost or prep stage to pre-heat the filament (space charge). The second connection from the high-tension circuit provides the potential difference that is required to move the electrons from the cathode to anode in the x-ray tube.

25. C   By grounding the mid-point of the secondary coil of the transformer, half of the selected voltage is below zero and one half above zero. In effect, each cable needs to be insulated for only ½ of the total amount of voltage, saving both cost and weight of the x-ray cables.

26. D   The incoming voltage for the x-ray unit can be modified into single volt increments by an electromagnetic device called an autotransformer. The autotransformer used in the x-ray circuit to select the desired voltage to the primary side of the high tension transformer. Because it is not safe to operate an autotransformer in a high voltage circuit it is located within the primary circuit.

27. B   To ensure the appropriate voltage has been provided by the autotransformer, a voltmeter called pre-reading kVp meter, is placed in the circuit on the secondary side of the autotransformer.

28. C The inherent danger associated with the high voltage section can be minimized by placing only essential devices into this circuit; namely, the high tension transformer, rectifiers, insulated cables, milliampere-seconds meter and the x-ray tube.

29. C Of the two types of overload protective devices, the fuse and circuit breaker, the most commonly employed device in a modern circuit is the electromagnetic circuit breaker which has the distinct advantage of being resettable.

30. D The operating console of a modern radiographic unit will normally provide the monitoring and control of the mA, kVp, timer settings as well as any number of accessory systems such as AEC, tomographic, and fluoroscopic controls.

31. C Two main types of devices are available for the storage of electrical charges. Relatively small amounts of electricity can be stored in dry or wet cell-type batteries and a device called a capacitor or condenser. In addition to the storage functions of a battery these are also able to generate limited amounts of electrical power resulting from the chemical breakdown of its components.

32. D A high tension, (step-up) transformer is the primary type of electrical device that is used to increase the amount of voltage in any electrical circuit. A transformer with more secondary than primary turns is used to increase the voltage across the electrodes of the x-ray tube.

33. C The production of a high ampere, low voltage current is possible through the use of a so called step-down transformer. The high amperage current is needed to pre-heat the cathode. This is accomplished by the use of a step-down transformer which is placed in the filament circuit.

34. B The milliampere-seconds (mAs) meter measures the amount of current passing across the electrodes of the x-ray called the tube current. Since the tube current is only generated during the second phase of x-ray production process it is only possible to obtain readings during or immediately after a radiographic exposure.

35 A The amount of voltage change occurring in a transformer is based on the ratio of the number of primary to secondary turns. Using the transformer law:

$$\text{Secondary voltage (20 volts)} = \frac{\text{Primary voltage (200 volts) x Number of secondary turns (200)}}{\text{Number of primary turns (2000)}}$$

36. B The main difference in the three portions of the x-ray circuit is related to amount of voltage found in each part of the circuit. The voltage in the filament circuit is normally only about 10-12 volts. In the primary circuit the voltage is normally about 110-440 volts. In the secondary circuit the voltage is between 25-150 kilovolts.

37. C See explanation 36.

38. A See explanation 36.

| 39. | 3 | 40. | 5 | 41. | 9 | 42. | 1 | 43. | 7 | 44. | 6 |
| 45. | 2 | 46. | 4 | 47. | A | 48. | B | 49. | 8 | 50. | C |

Transformers

1. A transformer has 100 primary and 400 secondary turns of wire. What secondary voltage will be induced if 200 volts are placed on the primary coil (assuming no power losses occur)?

2.-10. Using a transformer, calculate the secondary voltage that will be induced based on the following number of primary and secondary turns and primary voltages, assuming no power losses occur.

	Primary Voltage	# Primary Turns	# Secondary Turns	Secondary Voltage
2.	220	100	1,000	_____
3.	110	200	20,000	_____
4.	220	1,000	10,000	_____
5.	110	10,000	100	_____
6.	440	5,000	500	_____
7.	200	2,000	20,000	_____
8.	90	300	30,000	_____
9.	380	30,000	300	_____
10.	80	880	8,800	_____

11. A transformer has 100 primary and 400 secondary turns of wire. What secondary current (amperage) will be induced if 100 amps are placed on the primary coil (assuming no power losses occur)?

12.-17. Using a transformer, calculate the secondary current that will be induced based on the following number of primary and secondary turns and primary currents.

	Primary Current	# Primary Turns	# Secondary Turns	Secondary Current
12.	100 amps	66	660	_____
13.	80 amps	120	12,000	_____
14.	50 amps	80	800	_____
15.	1 amp	10,000	100	_____
16.	10 amps	12,000	1,200	_____
17.	1 kilo amp	100	10,000	_____

18. If 220 volts are impressed on the primary side of an autotransformer that consists of 110 total turns, what secondary voltage will be induced if the 60th tapped turn is selected?

19.-25. Using an autotransformer, calculate the secondary voltage that will be induced for the following primary voltages, total number and tapped turns for the autotransformer, assuming no power losses occur.

	Primary Voltage	Total # of Turns	# of Tapped Turns	Secondary Voltage
19.	100	100	50	_____
20.	200	400	100	_____
21.	110	220	110	_____
22.	220	110	55	_____
23.	440	880	110	_____
24.	80	500	250	_____
25.	140	500	300	_____

The following formula has been used in calculating these problems.

Transformer:

$$\text{Secondary Voltage} = \frac{\text{Primary voltage x \# of Secondary Turns}}{\text{\# of Primary Turns}}$$ $$\text{Secondary Amperage} = \frac{\text{Primary amperage x \# of Primary Turns}}{\text{\# of Secondary Turns}}$$

$$\text{Autotransformers: Secondary Voltage} = \frac{\text{Primary voltage x \# of Tapped Turns}}{\text{Total \# of Turns}}$$

Transformers

1. A transformer has 600 primary and 1800 secondary turns of wire. What secondary voltage will be induced if 220 volts are placed on the primary coil (assuming no power losses occur)?

2.-10. Using a transformer, calculate the secondary voltage that will be induced based on the following number of primary and secondary turns and primary voltages, assuming no power losses occur.

	Primary Voltage	# Primary Turns#	Secondary Turns	Secondary Voltage
2.	220	200	20,000	_____
3.	110	400	20,000	_____
4.	220	2,000	10,000	_____
5.	110	20,000	80	_____
6.	440	6,000	60	_____
7.	220	2,200	22,000	_____
8.	90	330	330,000	_____
9.	380	15,000	300	_____
10.	80	20	8,000	_____

11. A transformer has 100 primary and 1,000 secondary turns of wire. What secondary current (amperage) will be induced if 100 amps are placed on the primary coil (assuming no power losses occur)?

12.-17. Using a transformer, calculate the secondary current that will occur for the following number of primary and secondary turns and primary currents, assuming no power losses occur.

	Primary Current	# Primary Turns	# Secondary Turns	Secondary Current
12.	100 amps	160	3,200	_____
13.	80 amps	100	24,000	_____
14.	50 amps	80	1,800	_____
15.	1 kilo amp	14,000	200	_____
16.	10 amps	12,800	3,200	_____
17.	1 kilo amp	200	14,000	_____

18. If 440 volts are impressed on the primary side of an autotransformer that consists of 110 total turns, what is the primary voltage that will be induced if the 100th tapped turn is selected?

19.-25. Using an autotransformer, calculate the secondary voltage that will be induced based on the following primary voltages, total number and tapped turns for the autotransformer, assuming no power losses occur.

	Primary Voltage	Total # of Turns	# of Tapped Turns	Secondary Voltage
19.	220	400	50	_____
20.	110	1,000	10	_____
21.	800	880	110	_____
22.	220	220	55	_____
23.	440	880	220	_____
24.	80	750	125	_____
25.	200	450	150	_____

The following formulas have been used in solving these problems:

$$\text{Secondary Voltage} = \frac{\text{Primary voltage x \# of Secondary Turns}}{\text{\# of Primary Turns}}$$

$$\text{Secondary Amperage} = \frac{\text{Primary amperage x \# of Primary Turns}}{\text{\# of Secondary Turns}}$$

$$\text{Autotransformers: Secondary Voltage} = \frac{\text{Primary voltage x \# of Tapped Turns}}{\text{Total \# of Turns}}$$

Answers

Transformers
Exercise 1

1. 880 V
2. 2200, 2.2 kVp
3. 11,000 V, 11 kVp
4. 2200 V, 2.2 kVp
5. 1.1 V
6. 44 V
7. 2000 V, 2 kVp
8. 9000 V, 9 kVp
9. 3.8 V
10. 800 V
11. 25 amps
12. 10 amps
13. .8 amp, 800 mA
14. 5 amps
15. 100 amps
16. 100 amps
17. 10 amps
18. 120 V
19. 50 V
20. 50 V
21. 55 V
22. 110 V
23. 55 V
24. 36 V
25. 84 V

Transformers
Exercise 2

1. 660 V
2. 2200, 22 kVp
3. 5500 V, 5.5 kVp
4. 1100 V, 1.1 kVp
5. 4.4 V
6. 44 V
7. 2000 V, 2.2 kVp
8. 90,000 V, 90 kVp
9. 7.6 V
10. 32,000 V, 32 kVp
11. 10 amps
12. 5 amps
13. .33 amps
14. 2.2 amps
15. 70 amps
16. 40 amps
17. 14.3 amps
18. 400 V
19. 27.5 V
20. 11 V
21. 100 V
22. 55 V
23. 110 V
24. 13.3 V
25. 66.6 V

Rectification

1. In a modern x-ray circuit, the rectification of high voltage alternating current into pulsating direct current occurs between the secondary of the high-tension transformer and the:

    A. X-ray tube
    B. Solid state diodes
    C. Autotransformer
    D. Inverter circuit

2. In a 3-phase, 6-pulse rectified unit, how many voltage peaks are produced for each cycle of incoming line voltage?

    A. 3 peaks
    B. 6 peaks
    C. 9 peaks
    D. 12 peaks

3. All of the following are advantages of a 3-phase power supply in a modern radiographic unit EXCEPT:

    A. Shorter permitted exposures
    B. Higher beam quality
    C. Reduced scatter production
    D. Higher beam intensity

4. The principal material that is used in the formulation of a solid state diode is:

    A. Tungsten
    B. Lanthanum
    C. Gadolinium
    D. Silicon

5. In a 3-phase radiographic unit, if one of the rectifiers is damaged prior to a radiographic exposure, which of the following is likely to occur?

    A. An image with an increased amount of focal spot blur
    B. An increased amount of prep time to start the exposure
    C. A reduced amount of optical density due to a decreased radiation output
    D. A increased number of anode heat units

6. A radiographic exposure is taken on a 3-phase 6-pulse unit. In order to maintain the same optical density using a full wave unit would require:

    A. An increase of about 15% of the mAs value
    B. A decrease of about 30% of the kVp value
    C. A decrease of about 15% of the kVp value
    D. A doubling of the mAs value

7. The process by which alternating current is converted into pulsating direct current is termed:

    A. Electrification
    B. Polarization
    C. Depolarization
    D. Rectification

8. In a 3-phase, 12-pulse unit, rectification is accomplished by employing:

    A. 4 thermionic diodes
    B. 6 solid state diodes
    C. 12 solid state diodes
    D. Delta and star windings

Referring to the voltage waveforms, answer questions 9-12.

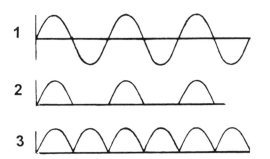

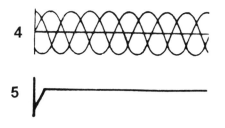

9. A full-rectified waveform is represented by number?

A. 1
B. 2
C. 3
D. 4

10. A 3-phase, alternating current waveform, is represented by number?

A. 1
B. 2
C. 3
D. 4

11. The curve represented by number 5 is best described as a/an:

A. ½ wave rectified curve
B. Alternating current curve
C. 3-phase curve
D. Steady direct current curve

12. Which of the following represent direct current waveforms:

A. 1, 2, & 3 only
B. 1 & 4 only
C. 2, 3, & 5 only
D. 1, 2, 3, 4, & 5

13. The main advantage of the 3-phase, over single-phase (full wave) current, is that in 3-phase current, the voltage:

A. Never drops below the peak voltage
B. Never drops to a zero voltage
C. Is self-rectified
D. Is inverted

14. The type of unit most likely to possess the smallest voltage ripple is a:

A. Full wave, single-phase unit
B. 3-phase, 6-pulse unit
C. Self-rectified unit
D. High frequency unit

15. The most common type of rectifier for use in 3-phase radiographic equipment is the:

A. Thermionic diode
B. Solid state diode
C. Rheostat
D. Valve tube

16. Rectification is used to suppress the inverse voltage associated with alternating current before it reaches the x-ray tube. This helps to:

A. Prevent the flow of electrons towards the filament
B. Focus the electron stream onto the anode
C. Reduce the formation of eddy currents
D. Increase the charge on the tube envelope

17. An exposure is made on a full-wave rectified unit. In order to maintain the same optical density for an exposure made with a 3-phase, 6-pulse radiographic unit, the mAs value employed should be decreased by about:

A. 5-10%
B. 15-25%
C. 25-30%
D. 50-55%

18. With an alternating current source the electron flow will alternate:

A. 12 times a second
B. 60 times a second
C. 120 times a second
D. 1000 times a second

19. A 3-phase, 6-pulse unit has an effective kV of about _____ of the peak kVp.

A. 87%
B. 64%
C. 57%
D. 37%

20. The ripple value is the variation in voltage across the x-ray tube as expressed as a percentage of the maximum value. With a single-phase system, a ripple value of _____ is considered normal.

A. 3%
B. 13%
C. 79%
D. 100%

21. A satisfactory radiograph is produced using a 3-phase, 6-pulse unit. If the same technical factors are used on a single-phase unit, the radiograph would show a:

A. Higher optical density
B. Less radiographic contrast

C. Lower optical density
D. Less focal spot blur

22. During the operation of an x-ray tube being supplied by full-wave rectified current during 50 millisecond exposure, the tube electrodes will change (reverse) polarity:

A. 24 times
B. 12 times

C. 6 times
D. No change in polarity occurs

23. The most common method for obtaining a fully rectified tube current involves the use of:

A. Two diodes (valve tubes)
B. Four diodes (valve tubes)

C. Six diodes (valve tubes)
D. Twelve diodes (valve tubes)

24. All of the following are advantages of a high frequency generator EXCEPT:

A. A lower radiation exposure to the patient
B. Higher tube rating for short exposures

C. An increased quality of the beam
D. An increased intensity of the beam

25. Compared to a single-phase x-ray unit, a 3-phase unit using the same technical factors will create:

1. More x-ray photons   2. More anode heat units   3. Higher energy photons

A. 1 only
B. 2 only

C. 3 only
D. 1, 2 & 3

Referring to the synchronous spin top film, answer question 26.

26. The synchronous spin top diagram represented by this film is an example of a 100 millisecond exposure taken on a:

A. Self rectified unit
B. High frequency unit
C. Full wave rectified unit
D. Half wave rectified unit

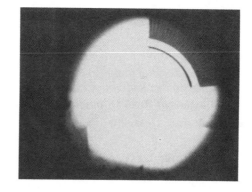

27. The main advantages of the use of a high-frequency generating system includes all of the following EXCEPT:

A. Reduced size
B. Higher efficiency

C. Reduced cost
D. Reduced radiation output

28. The use of 3-phase, radiographic units, has the advantage of _____ when compared to single-phase units.

1. Shorter exposure timer settings
2. Higher milliamperage settings
3. More of the beam penetrability

A. 1 only
B. 2 only

C. 3 only
D. 1, 2 & 3

29. The current produced in a high-frequency generator is accomplished be a series of high-speed switches called the:

A. Inverter module
B. Capacitor bank

C. Thyristors
D. Extinction circuit

30. What is the power rating for a unit energized at 800 mA, 100 millisecond and 100 kVp?

A. 8 kilowatts
B. 80 kilowatts

C. 800 kilowatts
D. 8000 kilowatts

31. 3-phase current consists of three single-phase voltage waveforms separated by:

A. 30 degrees
B. 60 degrees

C. 120 degrees
D. 180 degrees

32. Solid state rectifiers as compared to thermionic diode have/are:

1. More durable   **2. More efficient**   **3. A longer life**

A. 1 only
B. 2 only

C. 3 only
D. 1, 2 & 3

33. In modern x-ray equipment, the device most often used for current rectification is a/an:

A. Modified x-ray tube
B. Thermionic valve tube

C. Solid state diode
D. Ionization chamber

34. Which of the following types of single-phase radiographic units will permit the greatest instantaneous loading capacity?

A. Self-rectified current
B. Half-wave rectified current

C. Full wave rectified current
D. Rectification is not required

35. The small variation of the voltage waveform of 3-phase rectified current is termed the:

A. Constant potential
B. Phase curve

C. Helix effect
D. Voltage ripple

State the number of dots that should appear on a spin top test image for a single-phase full wave rectified unit for questions 36-38.

	Timer Setting	**Expected Number of Dots**
36.	250 milliseconds	_____
37.	150 milliseconds	_____
38.	16 milliseconds	_____

Referring to the radiographic images, answer questions 39-40.

A

B

39. If the manual spin top test film represented by letter A was taken on a properly functioning single-phase full wave rectified unit, what timer setting was employed?

   A. 200 milliseconds

   B. 100 milliseconds

   C. 50 milliseconds

   D. 10 milliseconds

40. If the manual spin top test represented by film B was taken at 33ms (1/30) second on a single-phase full wave rectified unit, which statement best describes the unit?

   A. Timer is inaccurate

   B. Faulty valve tube

   C. Pitted anode

   D. Nothing is wrong

Explanations

1. A   The process by which alternating current is converted into pulsating direct current is called rectification. This process is accomplished through the use of a number of thermionic diodes (valve tubes) or solid-state (silicon) diodes.

2. B   In a 3-phase circuit the incoming (single-phase) alternating current waveform is separated into 3 separate waveforms by the special delta and star winding of the high tension transformer. During this process each phase of current is offset by 120 degrees. When the 3-phase waveform is rectified, the original AC waveform will now appear to have 6 separate overlapping impulses and 6 separate voltage peaks.

3. C   Among the main advantages of a 3-phase power supply compared to a single-phase unit is the higher average energy (quality) of the photons in the beam. Because of the overlapping of the waveforms the electrons move across the tube in a more or less constant stream thus eliminating the pulsations of the x-rays that are characteristic of single-phase beams. A 3-phase unit will enable the use of timer settings that are shorter than with single-phase systems. A 3-phase system will increase the intensity of the beam which will increase the radiation on output by nearly 40%. The amount of scatter produced in either unit is virtually the same.

4. D   The majority of solid state diodes and semiconductors are made of crystalline silicon which has been modified (doped) with arsenic or gallium. These impurities are added to the silicon to adjust forward or reverse bias to the flow of electrons.

5. C   The malfunction of one of the six rectifiers in a 3-phase radiographic system will reduce the radiation output by about 20%. This will reduce the optical density seen on the radiographic image.

6. D   The use of a 3-phase rectification unit is associated with about a 5% increase in the effective kVp and about a 40% increase in the intensity of the beam compared to a single-phase unit. When the same imaging system is employed, a given 3-phase technique will provide nearly twice the radiation equivalent for a given set of exposure factors. In order to maintain the same optical density on a single-phase unit will therefore require a doubling of the mAs value or an increase of about 15% in the kVp.

7. D   The process by which alternating current is converted into pulsating direct current is termed rectification. In most modern systems this is accomplished by a series of solid state (silicon) diodes in the secondary circuit.

8. C   Rectification is accomplished by the desired arrangement of a number of thermionic or solid state diodes. The type of rectification and the number of diodes required are as follows: self rectification - no diodes; half wave rectification - 2 diodes; full wave rectification - 4 diodes; 3-phase 6-pulse - 6 diodes; 3 phase 12-pulse - 12 diodes.

9. C   See corresponding curve.

10. D   See corresponding curve.

11. D   See corresponding curve.

12. C   Wave forms which alternate above and below the zero voltage lines are forms of alternating currents (graphs 1 & 4); direct current waveforms, which stay above line in a pulsed form (graphs 2 & 3), are types of pulsating direct currents. The graph 5 represents a steady direct current waveform associated with the output of a battery or capacitor.

13. B   The electrons in the charged cloud around the filament under the influence of a fully rectified electrical current are accelerated toward the anode as the voltage begins to build toward its maximum. When the voltage falls back to zero, the electron movement is slowed or stopped, causing the pulsing of the x-ray beam. In a 3-phase system, the electron stream is more constant providing higher beam intensity and quality.

14. B   The voltage waveform is found by plotting the relative voltage verse time (see diagrams in questions 9-12). In a fully rectified single-phase unit (curve 3) the voltage begins at zero volts and builds to a maximum voltage before returning to zero creating what is termed a 100% voltage ripple. In 3-phase 6-pulse units, the voltage never drops to zero and varies between 87% and 100% producing a 13% ripple and from 97-100% for a 3% ripple. In a 3-phase 12-pulse type system the voltage varies between 97-100% providing a 3% voltage ripple. The smaller the voltage ripple the greater the consistency of the radiation out put. In recent years to development of high frequency units has further improved the voltage ripple to about 1%.

15. B   The majority of solid state diodes and semiconductors are made of crystalline silicon which has been modified (doped) with arsenic or gallium. These impurities are added to the silicon to adjust forward or reverse bias to the flow of electrons.

16. A   The suppression of the inverse voltage with direct current enables x-ray tube electrons to pass in a single direction from cathode to the anode during the exposure. If this current were permitted to flow in both directions, the high-energy electron stream would reverse its direction causing serious damage to the filament.

17. D   For a given set of technical factors, the output of a full wave rectified unit is lower in both quantity and quality than that of a 3-phase unit. In order to provide a similar optical density, during a 3-phase technique a decrease of 50-55% of the mAs is required.

18. C   With an alternating current at 60 cycles per second the electron flow is reversed every 1/120th of a second or 120 times per second.

19. A   A 3-phase 6-pulse unit will have an effective kilovoltage that is about 87% of the initial peak kVp.

20. D   In a fully rectified single-phase unit the voltage begins at zero volts and builds to a maximum voltage before returning to zero creating what is termed a 100% voltage ripple.

21. C   Because a 3-phase unit provides a higher quality and quantity of radiation, an identical set of exposure factors used on a single and 3-phase unit will result in a lower optical density and a slightly higher contrast on images obtained on the single-phase unit.

22. D   In direct current, no reversal or change in the electrical potentials of the cathode and anode occur.

23. B   Rectification is accomplished by the desired arrangement of a number of thermionic or solid state diodes. The type of rectification and the number of diodes required are as follows: self rectification - no diodes; half wave rectification - 2 diodes; full wave rectification - 4 diodes; 3-phase 6-pulse - 6 diodes; 3-phase 12-pulse - 12 diodes.

24. A   A high frequency unit is capable of producing a beam that has an increased quality and beam intensity and allows for higher tube rating. Patient dose being more a function of imaging system employed and the optical density is not affected when all other factors are the same.

25. D   The higher efficiency of a 3-phase radiographic unit provides a tube current with more electrons per unit time and greater heat formation in the anode. In a 3-phase unit the mA x time x kVp heat unit formula requires a correction value to obtain the actual number of heat units. The correction for a 3-phase 6-pulse unit is 1.35 and 1.41 for a 3-phase 12-pulse unit.

26. B   A synchronous spin top tester is motor driven to spin at a fixed rate of one revolution per second. Because of this fixed rotation, it can be used to test the timer settings of both single and 3-phase radiographic units. A high frequency or 3-phase unit will not produce individual dots and will appear as a solid line.

27. D   There are many advantages of high frequency generating system, most noticeable are the reduced size and cost of the system. When combined with the higher efficiency an increase in the use of these units in the future can be anticipated.

28. B   The over-all power and improved beam quality of a 3-phase unit will permit shorter exposure timer setting, higher mA settings and a slightly improved beam penetrability.

29. B   The nearly constant potential waveform with a high frequency unit is in large part due to the increased frequency of the current. A special circuit called an inverter circuit employs a series of high speed electrical switches called an invertor module. This cuts a direct current wave into an alternating current-like (500 Hz) that can be placed onto the small high tension transformer. A series of high voltage capacitors complete the conversion of the AC-like wave back into a nearly smooth steady DC wave.

30. B   The power rating of a unit is based on the formula:

$$\text{power rating (80kW)} = \frac{\text{mA (800 mA) x kVp (100 kVp)}}{1000}$$

31. C   3-phase current consists of 3 single-phase current wave forms which are shifted (phases) through an angle of 120 degrees with each other.

32. D   The solid state silicon diodes used in most modern units have no filament to burn out, produce no low energy x rays, are more efficient and durable than the valve tubes they replace.

33. C   Until the 1970's, thermionic diodes or valve tubes were the principal type of rectifying devices. The advent of semiconductor technology allowed their replacement with less expensive, more durable, and longer lasting, solid state diodes.

34. C   Since rectification improves the instantaneous loading capacity for a radiographic unit, a full wave rectified unit would provide the highest instantaneous loading capacity for the units listed.

35. D   The small variation of the voltage waveform in any type of unit is termed the voltage ripple.

36. C   The number of dots on a spin top test is found as the product of the timer setting in seconds times the number of pulses per second. In a full wave rectified system the:

Number of expected dots (30) = Timer setting (.250 s) x Pulses per second (120).

37. D   Number of expected dots (18) = Timer setting (.50 s) x Pulses per second (120).

38. D   Number of expected dots (2) = Timer setting (.017 s) x Pulses per second (120).

39. C   In fully rectified single-phase radiograph unit the 60 cycle per second alternating current is converted by a 4 diodes arrangement to provide a 120 impulse per second direct current. When a high voltage direct current is passed into an x-ray tube, a short burst of x-rays is given off every 1/120 sec. (120 dots per second). When a metallic spin top with a single cutout is spun over radiographic film during an exposure, a series of dots representing the number of impulses seen. Because U.S. power supplies are strictly maintained at 60 cps, the likely cause of a change in the number of expected dots is either a faulty x-ray timer or inoperative valve tube (diode). Image A shows 6 dots on a properly functioning unit. The timer setting is found by dividing: number of dots on the image by the number of dots per second (120) this will provide a setting of .05 seconds or 50 milliseconds. Since Image B was taken at a 1/30 second, 4 dots should be visualized. Therefore this unit is operating within acceptable limits.

40. D   Since this unit imaged the desired amount of dots, it can be assumed that nothing is wrong with the rectification system.

1. Which of the following types of radiographic units will permit the lowest minimum exposure timer setting?

   A. Half wave, single-phase
   B. Full wave, single-phase
   C. 3-phase (with AEC)
   D. 3-phase (without AEC)

2. Using an automatic exposure controlled (AEC) unit, 50 speed image receptor (cassette) is mistakenly employed instead of a 400 speed image receptor. Which of the following is likely to occur?

   A. The exposure time will be longer than expected
   C. The exposure time will be shorter than expected
   B. The radiograph will have an excessive optical density
   D. The radiograph will have insufficient optical density

3. The accuracy of a modern electronic exposure timer of a 3-phase radiographic unit which permits exposures as low as 4 milliseconds is best evaluated using a/an:

   A. Oscilloscope or digital timing meter
   B. Thyratron or manual timing meter
   C. Ballistics meter or analog timing meter
   D. Scintillation counter or quantum timing meter

4. The sensors used in the majority of modern automatic exposure controlled units for the detection of the remnant beam, consist of three or more thin:

   A. Photomultipliers
   B. Ionization chambers
   C. Laser scanners
   D. Optical readers

5. Which type of timer is most commonly employed with most modern 3-phase radiographic units?

   A. Silicon controlled rectifier
   B. Impulse timer
   C. Synchronous timer
   D. Thyratron

6. An automatic exposure controlled (AEC) exposure is attempted using 100 kVp on a patient with intestinal gas from an obstruction. The most likely result will be a/an:

   A. Overpenetrated radiograph showing a low optical density
   B. Underpenetrated radiograph showing an excessive optical density
   C. Radiograph with excessive optical density
   D. Radiograph with insufficient optical density

7. With a grid controlled x-ray tube, the radiographic exposure is terminated when a preselected:

   A. Anode voltage is reached
   B. Cathode voltage is reached
   C. Grid voltage is reached
   D. Filament amperage is reached

8. During a procedure using an automatic exposure controlled device, which of the following factors will normally increase the optical density of the image?

   A. An increase in the mA setting
   B. An increase in the kVp setting
   C. A decrease in the source-to-image receptor distance
   D. An increase in the optical density setting

9. The use of an impulse or synchronous timer in a single-phase x-ray unit will permit accurate exposure times as short as:

   A. 1 ms (1/1000)
   B. 4 ms (1/240)
   C. 17 ms (1/60)
   D. 34 ms (1/30)

10. A radiographic exposure requires 24 mAs. What timer setting will result in the appropriate values using the 200 mA station?

    A. 20 millisecond
    B. 120 millisecond
    C. 150 millisecond
    D. 250 millisecond

11. Which of the following can be used to reduce involuntary motion during an abdominal exposure if an automatic exposure controlled technique is attempted?

    1. The application of a shorter backup time
    2. An increased source-to-image receptor distance
    3. The use of a higher kVp setting

    A. 1 only   C. 3 only
    B. 2 only   D. 1, 2 & 3

12. During an automatic exposure controlled (AEC) technique a cassette with a 400 speed image receptor is mistakenly employed instead of a 100 speed screen image receptor. Which of the following is likely to occur?

    A. A radiographic image having an excessive optical density
    B. A longer than expected exposure timer value
    C. A radiographic image having an insufficient optical density
    D. A shorter than expected exposure timer value

13. For safety reasons, the timer circuits for most modern radiographic units are located in the:

    A. Filament circuit   C. Secondary circuit
    B. Primary circuit   D. High voltage circuit

14. The radiation sensors that are employed in most automatic exposure controlled (AEC) units are located between the _____ and the _____.

    A. Tube/collimator   C. Tabletop/grid
    B. Collimator/tabletop   D. Grid/Bucky tray

15. What is the likely cause of an excessively long automatic exposure controlled (AEC) exposure?

    1. The failure to properly center the x-ray tube
    2. The selection of an excessive backup time
    3. The selection of an excessive kVp setting

    A. 1 only   C. 3 only
    B. 2 only   D. 1, 2 & 3

16. A manual radiographic exposure utilizes a 300 millisecond exposure timer setting. What exposure setting should be employed it a reduction of 40% of this value is desired?

    A. 120 millisecond   C. 240 millisecond
    B. 180 millisecond   D. 280 millisecond

17. The termination of the radiographic exposure during an automatic exposure controlled (AEC) technique is accomplished by a device called a/an:

    A. LCD   C. Thyristor
    B. Photomultiplier tube   D. Cumulative timer

18. A manual exposure requiring 40 mAs is desired for an image of the shoulder. What timer setting can be used to obtain this mAs value if an 800 mA station is employed?

    A. 50 milliseconds   C. 500 milliseconds
    B. 100 milliseconds   D. 750 milliseconds

19. Which of the following can be used to increase the optical density on a radiographic image during a normal automatic exposure controlled (AEC) exposure?

   1. *An increase in the backup timer*
   2. *An increase in the density control setting*
   3. *A Decreasing in the source-to-image receptor distance*

   A. 1 only   C. 3 only
   B. 2 only   D. 1, 2 & 3

20. If a 200 mA, 240 milliseconds and 84 kVp exposure was used to produce an image with the desired optical density, what new time factor would produce a similar optical density using a 400 mA station and 84 kVp?

   A. 30 milliseconds   C. 120 milliseconds
   B. 60 milliseconds   D. 480 milliseconds

21. An automatic exposure controlled (AEC) exposure of an abdomen is taken. Which of the following is likely to occur during a second exposure after the administration of two cups of barium ?

   A. The optical density of the image is likely to increase
   B. The time of the exposure is likely to increase
   C. The optical density of the image is likely to decrease
   D. The kVp setting is likely to decrease

22. A radiographic exposure is taken using the 200 mA station and a 50 millisecond timer setting at 80 kVp. What amount of mAs was used during this exposure?

   A. 4 mAs   C. 16 mAs
   B. 10 mAs   D. 40 mAs

23. An automatic exposure controlled (AEC) unit is being used to produce a radiographic PA image of the chest. Which of the following changes would be likely to increase the optical density of the image?

   A. Use the center AEC pickup or sensor   C. Use a lower density control setting
   B. Increase the kilovoltage setting   D. Use a longer backup timer setting

24. The silicon controlled rectifier (SCR) that is often used in most modern 3-phase or high frequency units, will normally permit exposure times as short as 1/1000 second or:

   A. 1 millisecond   C. 8 milliseconds
   B. 5 milliseconds   D. 10 milliseconds

25. The use of a backup timer, required in an automatic exposure controlled (AEC) unit, is primarily intended to:

   A. Improve the quality of the radiation exposing the patient
   B. Protect the unit and patient for inadvertent overexposures
   C. Enable the unit to permit multiple exposures
   D. Provide the means for checking the exposure accuracy

26. What is the smallest mAs value obtainable for an automatic exposure controlled (AEC) exposure using 300 mA 30 milliseconds if the unit has a minimum response time of 20 milliseconds?

   A. 2 mAs   C. 6 mAs
   B. 3 mAs   D. 9 mAs

27. A patient is to be imaged using an automatic exposure controlled (AEC) exposure at 600 mA and 80 kVp. If the minimum response time is 20 milliseconds, with a backup timer setting of 200 milliseconds, which of the following mAs values is possible?

   1. *12 mAs*   2. *48 mAs*   3. *100 mAs*

   A. 1 only   C. 3 only
   B. 2 only   D. 1, 2 & 3

28. In order to help prevent unwanted exposures, the timer on a modern diagnostic unit shall provide a/an _____ indication when a radiographic exposure is attempted.

   A. Thermal and tactile   C. Audible and visible
   B. Gustatory and motor

29. Which of the following is a likely cause of an overload or backup time light to appear during an automatic exposure controlled (AEC) exposure?

   A. Failure to use the appropriate image receptor
   B. Failure to use the appropriated amount of collimation
   C. Failure to properly center the tube to the center of the Bucky tray
   D. Failure to restrain the patient during the procedure

30. An automatic exposure controlled (AEC) exposure is attempted using a 400 mA station at 80 kVp. If a 4 mAs exposure is desired but the minimum response time for this unit is 20 milliseconds, what timer setting will be indicated when the exposure is attempted?

   A. 10 milliseconds   C. 100 milliseconds
   B. 20 milliseconds   D. 200 milliseconds

31. During an automatic exposure controlled exposure of the pelvis, the primary sensor is placed over a large pocket of gas. The resulting radiographic image is likely to:

   A. Have an excessive optical density   C. Have a great deal of involuntary motion
   B. Have an insufficient optical density   D. An increased amount of focal spot blur

32. An automatic exposure controlled (AEC) exposure is attempted at 400 mA and 88 kVp. If the unit has a minimum response time of 25 milliseconds and a back tier setting of 100 milliseconds which of the following exposures will be permitted?

   1. 6 mAs   *2. 12 mAs*   *3. 40 mAs*

   A. 1 & 2 only   C. 2 & 3 only
   B. 1 & 3 only   D. 1, 2 & 3

33. What is the shortest permitted timer setting for a full wave rectified unit using an impulse timer?

   A. 8 milliseconds   C. 16 milliseconds
   B. 12 milliseconds   D. 24 milliseconds

34. During an automatic exposure controlled (AEC) exposure what is the likely result if the cassette is placed upside down in the Bucky tray?

   A. The resulting image will have a deficient optical density
   B. The resulting image will have an excessive optical density
   C. The resulting image will have the desired optical density
   D. The resulting image will possess a higher contrast

35. All of the following changes can be made during an automatic exposure controlled (AEC) exposure without causing a noticeable change in the optical density EXCEPT:

   A. The removal of the added filtration in the tube
   B. Decreasing the source-to-image receptor distance by 20 centimeters
   C. Removing the radiographic grid from the Bucky tray
   D. Increasing the kVp by 15% for the exposure

36. During an automatic exposure controlled (AEC) exposure an increase in the density control setting of minus 2 is likely to:

   A. Reduce the back timer setting by 70%
   B. Increase the minimum response time by 50%
   C. Decrease the optical density by the equivalent of 50% of the mAs
   D. Increase the optical density by the equivalent of 15% of the kVp

Select the appropriate sensor pattern for the following anatomic regions for an automatic exposure controlled image for questions 37-40.

37. PA lungs ☐ ☐ ☐

38. AP thoracic spine ☐ ☐ ☐

39. Lateral chest ☐ ☐ ☐

40. AP pelvis ☐ ☐ ☐

Explanations

1. D The shortest permissible timer setting allowed by a radiographic unit is related to its type of rectification employed and the type of exposure timer used. The impulse timer that is used with nearly all fully rectified single-phase units requires a minimum of one impulse or 1/120 second(8 ms)to terminate an exposure. 3-phase units employ modern electronic (silicon rectifier) timers which enable exposures as short as 1 ms (1/1000 second). The automatic exposure controlled (AEC) circuits have timer systems that can permit exposures as low as 10 ms (1/100sec.).

2. D The operation of an automatic exposure control (AEC) system is based on the assumption that an appropriate image receptor (cassette), film, grid, degree of collimation, patient centering points and density control settings have been employed. Any failure to utilize the appropriate factors will result in a radiograph possessing an excessive or deficient optical density. The use of a slower speed image receptor which is not recognized by the AEC system will result in an image possessing an insufficient optical density.

3. A The accuracy of any radiographic timing system can be tested by the use of the highly accurate oscilloscope or solid state digital timing device. In the past, determinations of the timer accuracy were made on single-phase systems by the use of a manual or synchronous spin top tester.

4. B The sensors or radiation detectors of nearly all modern automatic exposure control (AEC) units consist of three 2"-4" rectangular or circular ionization chambers which are incorporated into the Bucky tray apparatus. The familiar □ □ pattern enables the selection of any number of single or   □

    multiple sensors pattern based on the nature of the anatomical region. Most manufacturers recommend a single-center sensor for non-contrast abdominal studies, skull positions, spine and skeletal structures and the lateral chest projections. All three sensors are selected for the majority of positive contrast studies of the abdomen and routine views of the pelvis. The two outer sensors are required to obtain the appropriate optical density of the lung during PA projections of the chest.

5. A The most common type of device used in a modern 3-phase timer circuit is a silicon controlled rectifier (SCR) which is capable of exposures as short as 1 millisecond (.001 sec). Impulse, synchronous mechanical, and AEC timers are not normally capable of accurate exposures under 10 milliseconds or .01 seconds.

6. D Whenever one of the selected sensors is inadvertently placed under a gas filled loop in the intestine, radiation floods the sensor causing a premature termination of the exposure. This is one of the more common causes of AEC images providing an insufficient optical density. This is most frequently seen with obstructed patients and when double contrast techniques are employed.

7. C In some mobile and cine radiographic systems, a special triode or grid controlled x-ray tube is used to terminate the exposure by the placement of a secondary charge on the grid. Compared to a more conventional system, these tubes can be rapidly turned on and off or pulsed to meet this requirement during cine radiography.

8. D With a properly functioning automatic exposure control device, changes in the kVp, mA and source-to-image receptor distance should not normally effect the optical density because of the compensation in timer setting which will be made by the unit to adjust the exposure. Optical density modifications can best be accomplished through the adjustment of the density setting controls located on the control console of an AEC unit.

9. C In older full wave rectified units timer selections were based on the number of impulses of current. In these units, though maximum settings of 1/120 sec. or 8 ms was possible, in practice, 1/60 sec. or 17 ms setting was the shortest setting which provided consistent exposures.

10. B The timer setting can be found by the application of the mAs formula where Time (.12) = mAs (24)/mA (200). A time of .12 seconds is equal to 120 milliseconds.

11. C During an automatic exposure controlled (AEC) technique, the main variable is the timer setting. In order to eliminate motion, high quality and intensity radiation is required to reduce the time required for the exposure. This can be accomplished by the use of a higher mA setting, higher kVp setting, or shorter source-to-image receptor distance.

12. A All automatic exposure control (AEC) units are adjusted for a specific image receptor combination for a given exposure. The use of an inappropriate image receptor will not be recognized by the AEC system, resulting in a loss of optical density when slower speed combinations are employed. An overexposed radiographic image if a higher speed imaging system is used.

13. B The timing devices which are incorporated into the radiographic unit to terminate the exposure are located in the low voltage or primary portion of the x-ray circuit.

14. C The ionization type sensors or detectors used in nearly all modern automatic exposure control units are housed in a thin metal or plastic frame which sits directly above the radiographic grid in the Bucky tray, just below the top of the x-ray table.

15. A An excessive exposure time with an automatic exposure control (AEC) unit indicates that the intensity or exposure rate is low and that the sensors have not received adequate amounts of radiation to terminate the exposure. The main causes of long exposure times are insufficient kVp, low mA settings, excessive SID's and improper alignment of the tube to the primary sensor.

16. B In order to obtain the proper timer setting, the original setting is multiplied by 40% to obtain the required amount of correction (120 ms). The answer is found by subtracting this value from the original timer setting.   *300 ms-120 ms = 240 ms*

17. C The physical device, which triggers the termination of the x-ray exposure when the appropriate amount of radiation has been detected by the ionization chambers, in older AEC systems, was a triode tube called a thyratron. In more modern units, a solid state switch called a thyristor is employed.

18. A The timer setting can be found by the application of the mAs formula where Time (.05) = mAs (40)/mA (800). A time of .05 seconds is equal to 50 milliseconds.

19. B The optical density of an AEC exposure is not normally effected by changes in the mA setting, kVp setting or backup timer setting. The AEC unit will have a density control setting which can be adjusted to increase (plus setting) or decrease (negative setting) to change the optical density of the image.

20. C The new mAs setting can be found by the application of the mAs formula where mAs (48) = mA (200)x time (.24). When a new mA of 400 is employed the new Time (.12) = mAs (48)/mA (400).

21. C During an automatic exposure controlled (AEC) technique an increase in the thickness or physical density of the patient will require a greater amount of incident radiation and a longer exposure time to obtain the appropriate optical density. Since barium adds to the physical density of the patient, the exposure will normally be longer.

22. B The mAs can be determined by the application of the mAs formula:
mAs (10) = mA (200) x time (.05).

23. A The optical density of an AEC exposure is not normally effected by changes in the mA setting, kVp setting or backup timer settings. The AEC unit will have a density control setting which can be adjusted to increase (plus setting) or decrease (negative setting) to change the optical density of the image. During a PA chest image, using an AEC unit, the optical density of the image can be increased by an increase in the density control setting or by using the center sensor that sets over the spine rather than the two lateral sensors that are normally placed over the lung fields.

24. A The most common type of device used in a modern 3-phase timer circuit is a silicon controlled rectifier (SCR) which is capable of exposures as short as 1 millisecond (.001 sec).

25. B   In theory an automatic exposure control device has no upper limit for the duration of exposure. In practice, this limit is determined by the backup timer. During an exposure the backup timer is only activated when the selected maximum timer value is exceeded. The timer's principal function is to protect the unit or patients from excessively long exposures.

26. C   In an automatic exposure controlled (AEC) unit, the permitted range of exposures is found between the values obtained from multiplying the selected mA value (300) x the minimum response time (20 ms) = 6 mAs, the selected mA value (300), and the backup timer setting (30 ms) = 9 mAs. The shortest permissible exposure with an AEC unit is called the minimum response time. The longest permissible exposure value is related to the backup timer setting.

27. D   In an automatic exposure controlled (AEC) unit, the shortest permissible exposure is called the minimum response time. This can be found from multiplying the selected mA value (600) x the minimum response time (20 ms) = 12 mAs. The maximum exposure permitted by this unit is 600 mA x 200 ms or 120 mAs. Therefore, all exposures are permitted.

28. C   To help insure that no accidental patient exposure occurs the activation of the exposure switch must provide an audible and visual indication that this exposure is taking place.

29. C   Though different manufacturers may employ different indications, a light and/or audible sound signal will be activated when the maximum or backup time has been exceeded. Most units will display an overload, maximum output, or exposure reset light. These overloads are most frequently seen in radiography of obese patients and when the x-ray tube is improperly centered to the radiation sensors or during decubitus images when the sensors are not in line with the x-ray source.

30. B   Since this AEC nit is incapable of exposures below the minimum response time, the smallest obtainable is the product of the mA (400) and minimum response time (20 ms) = 8 mAs. This will be the time indicated for the exposure.

31. B   Whenever one of the selected sensors is inadvertently placed under a gas filled structure, radiation floods the primary sensor causing a premature termination of the exposure. This is one of the more common causes of AEC images providing an insufficient optical density.

32. C   In an automatic exposure controlled (AEC) unit, the permitted range of exposures is found between the values obtained from multiplying the selected mA value (400) x the minimum response time (25 ms) = 10 mAs and the selected mA value (400), and the backup timer setting (100 ms) = 40 mAs. Therefore, only exposures between 10-40 mAs will be permitted.

33. D   The impulse timer that is used with nearly all fully rectified single-phase units requires a minimum of one impulse or 1/120 second (8 ms)to terminate an exposure. In practice, 1/60 sec. or 17 ms setting was the shortest setting which provided consistent exposures.

34. B   During an AEC exposure the lead that is placed in the back lid of most cassettes to reduce backscatter will absorb sufficient amounts of primary radiation to increase the length of the radiographic exposure and increase the optical density of the resulting image.

35. C   The AEC unit will compensate for all of the changes except the removal of the radiographic grid. If the system expects a grid to be used and it is eliminated, a considerable overexposure is the result.

36. C   A minus 2 setting in the density control setting will normally reduce the exposure y an amount that will be nearly equal to a 50% reduction in the mAs value.

37. ■ ■   The sensors that should be activated for a routine PA chest are the two lateral sensors located
     □   over the lung fields.

38. □ □   The sensor that should be activated for an AP thoracic spine is the center sensor located over the
     ■   thoracic vertebrae.

39. □ □   The sensor that should be activated for a lateral projection of the chest is the center sensor
     ■   located over the lung field.

40. ■ ■   All three sensors should be activated for an AP to average of the value for this large region.
     ■

Chapter 11

1. In a modern diagnostic x-ray tube the vast majority of the energy of the electron stream striking the target of the anode is converted into:

   A. Thermal energy (heat) at the surface of the anode
   B. X-radiation just below the surface of the anode
   C. Compression waves just above the surface of the anode
   D. High frequency sound waves at the surface of the focus cup

2. In order to prevent movement of electrons from the anode to the cathode during the exposure cycle, the type of current employed for an x-ray tube is:

   A. Alternating current
   B. Direct current
   C. Inversion current
   D. Multiphase current

3. The main function of the Pyrex glass that forms the protective envelop of the x-ray tube is:

   1. The containment of the electron stream near the filament
   2. The containment of the vacuum within the x-ray tube
   3. The production of the static electrical field around the filament

   A. 1 only
   B. 2 only
   C. 3 only
   D. 1, 2 & 3

4. All of the following are advantages of a rotating type anode tube EXCEPT:

   A. An increased rate of heat dissipation from the surface of the anode
   B. An increase in the allowable exposures (tube ratings) for the unit
   C. A reduction in the amount of roughening that will occur on the focal tract
   D. A reduction in the amount of low energy photons produced during an exposure

5. The principal disadvantage of the use of a smaller focal spot during a radiographic exposure is:

   A. An increase in the amount of focal spot blur
   B. The limitation of the kilovoltage peak to diagnostic levels
   C. The reduction in the speed of the anode rotation
   D. The limitation of the amount of mAs that can be applied to the anode

6. The rotation speed of an anode in most single-phase x-ray units is about:

   A. 200 rpm
   B. 1000 rpm
   C. 3000 rpm
   D. 10,000 rpm

7. The temperature of the filament in a modern x-ray tube is most closely related to:

   A. The time selected for a radiographic exposure
   B. The amount of current impressed on the filament
   C. The speed of the electron stream across the tube
   D. Time required to accelerate the anode to its maximum speed

8. Tungsten is the principal material used in the formation of the:

   1. Surface of the target of a rotating anode
   2. Surface of the target of a stationary anode
   3. Wire used in the helix of the x-ray tube filament

   A. 1 only
   B. 2 only
   C. 3 only
   D. 1, 2 & 3

9. The reduction of effective focal spot size from about 2 mm to 1 mm can be accomplished by:

A. Decreasing the voltage impressed on the focus cup
B. Selecting the smaller of the two filaments
C. Increasing mA and kVp values
D. Increasing speed at which the anode rotates

10. The thin, flattened surface of the x-ray tube envelope which allows for the minimum absorption of x-rays is termed the:

A. X-ray window   C. Primary shutter
B. Primary diaphragm   D. Focus cup

11. A modern dual focus x-ray tube contains:

1. One filament   2. Two filaments   3. One anode

A. 1 & 2 only   C. 2 & 3 only
B. 1 & 3 only   D. 1, 2 & 3

12. In a modern diagnostic x-ray tube, the majority of the heat that is produced during the x-ray production process is created and concentrated on the:

A. Surface of the focal tract   C. Bearing of the induction motor
B. Surface of the filament support D. Inner surface of the tube envelope

13. The amount of space charge created at the cathode is primarily controlled by the:

A. Impressed amperage   C. Tube current
B. Impressed kilovoltage   D. Hysteresis

14. In a stationary anode tube, the target is imbedded in an anode made of:

A. Tungsten   C. Copper
B. Molybdenum   D. Rhenium

15. The speed and energy of the electron stream as it passes across an x-ray tube is primarily controlled by the:

A. Applied kilovoltage   C. Temperature of the filament
B. Vacuum material   D. Focus cup voltage

16. The process by which heating of the filament results in the liberation of electrons in the x-ray tube is called:

A. Isotropic propagation   C. Thermoluminescence
B. Thermionic emission   D. Phosphorescence

17. In a modern x-ray tube, the positively-charged electrode that serves as the target for the electron stream is the:

A. Primary filament   C. Anode disk
B. Induction coil   D. Electron gun

18. The unit of power, the watt, is derived from the product of:

A. Amperage x time   C. Resistance x amperage
B. Voltage x resistance   D. Amperage x voltage

19. The measurement of the focal spot size in most diagnostic x-ray tubes is based on the computation of the:

1. Actual focal spot   2. Effective focal spot   3. Electronic focal spot

A. 1 only   C. 3 only
B. 2 only   D. 1, 2 & 3

20. The first modern radiographic tube having a heated filament was originally developed in the early 1900's by:

A. Wilhelm Roentgen
B. William Coolidge
C. Neils Bohr
D. Madame Curie

21. The angle on the anode of a x-ray tube helps to insure that more of x-ray photons produced at the target are:

A. Directed toward the patient
B. Transmitted through the glass housing
C. Scattered in all directions equally
D. Absorbed in the surface of the envelope

22. If the angle on a x-ray tube anode is decreased from 12 degrees to 7 degrees it will result in a decrease in:

1. Size of the focal spot   *2. Loading capacity of the tube*   *3. Field coverage*

A. 1 & 2 only
B. 1 & 3 only
C. 2 & 3 only
D. 1, 2 & 3

23. The majority of heat in a modern rotating anode tube is dissipated through the:

A. Conduction of heat through the stem of the anode
B. Transfer of the space charge to the anode disk
C. Passage of radiant heat to the tube envelope
D. Absorption of electrons by the ion sink

24. In a rotating anode, the molybdenum stem is a poor conductor of heat. This material helps in preventing:

A. Thermal expansion of the anode
B. Pitting of the anode
C. Bearing damage of the induction motor
D. Excessive prep time

25. Which is the proper sequence of materials traversed by the x-ray beam after leaving the anode of the x-ray tube?

A. Oil, aluminum, Pyrex glass
B. Pyrex glass, oil, aluminum
C. Aluminum, Pyrex glass, oil
D. Oil, Pyrex glass, aluminum

26. The amount of heat storage capacity of a rotating anode is dependent upon all of the following EXCEPT:

A. Diameter of the anode disk
B. Speed of the anode rotation
C. Prep (boost) time
D. Size of the focal spot

27. The x-ray tube envelope is composed of a heat resistant material called:

A. Bakelite
B. Pyrex glass
C. Lead glass
D. Wratten 6B glass

28. Which of the following factors will effect size of the focal spot?

1. Angle of the anode   *2. Speed of the rotation*   *3. Size of the filament*

A. 1 & 2 only
B. 1 & 3 only
C. 2 & 3 only
D. 1, 2 & 3

29. As mA increases, blooming of the focal tract results in a/an:

A. Increase in the size of the effective focal spot
B. Decrease in the optical density of the image
C. Reduction in the amount of focal spot blur
D. Increase in the radiographic contrast of the image

30. The material that is often added to a rotating anode disk to reduce surface defects and cracks is called:

A. Silver
B. Cesium

C. Lead
D. Rhenium

31. The main reason tungsten is employed as filament material in an x-ray tube is its:

A. Low phosphorescence
B. High degree of flexibility

C. High atomic number
D. High melting point

32. The discovery of x-rays was made by Wilhelm Roentgen in 1895 while experimenting with a partially evacuated glass tube called a/an:

A. Crooke's tube
B. Coolidge tube

C. Edison tube
D. Compton tube

33. The target of a modern rotating anode is bevelled or inclined to an angle of between:

A. 3-7 degrees
B. 10-12 degrees

C. 15-25 degrees
D. 25-37 degrees

34. The durability and efficiency of the tungsten filament is often improved by the addition of the metal:

A. Copper
B. Silver

C. Lead
D. Thorium

35. The anode of a modern rotating anode tube is generally composed of a combination of tungsten and:

1. Copper   *2. Rhenium*   *3. Molybdenum*

A. 1 & 2 only
B. 1 & 3 only

C. 2 & 3 only
D. 1, 2 & 3

36. The line-focus principle concerns the effect of target angulation and focal spot size. Which is a TRUE statement of this principle?

A. The actual focal spot is larger than the effective focal spot
B. As tube angulation decreases, the effective focal spot increases
C. Radiation intensity is greatest on the cathode side of the tube
D. More than one but not all the above

37. The collective repulsion of the electron cloud around the cathode that creates an equilibrium with the rate of electron emission by the filament is termed the:

A. Space charge effect
B. Anode heel effect

C. Thermionic emission
D. Edison effect

38. The major reason that tungsten is the material chosen for the electrodes of a modern x-ray tube is its high melting point of about:

A. 1000 degrees C
B. 1300 degrees C

C. 2300 degrees C
D. 3400 degrees C

39. The maximum anode heat storage capacity of a radiographic tube is related to the type and size of the:

1. Anode disk   *2. Thermal cushion*   *3. External housing*

A. 1 only
B. 2 only

C. 3 only
D. 1, 2 & 3

40. When a radiographer changes from a small to a large focal spot for an exposure, he/she is actually:

A. Selecting a larger filament
B. Increasing the filament-anode distance

C. Increasing the diameter of the anode disk
D. Increasing the angle of the anode disk

41. Radiographic examinations requiring large tube currents or serial exposures will often employ high speed rotating anodes with rotational speeds of about:

A. 400 rpm
B. 1,000 rpm

C. 3,000 rpm
D. 10,000 rpm

Referring to the diagram of the X-ray tube, answer questions 42-47.

In the diagram of the x-ray tube place the number which corresponds to the following structures.

42. _____ Induction motor ball bearings
43. _____ Anode Stem
44. _____ Stator windings
45. _____ Anode disc
46. _____ Tube envelope
47. _____ Filament

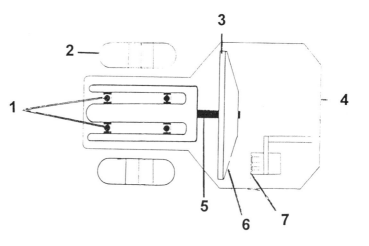

48. In a modern rotating anode, the motion of the rotating disk is accomplished by the use of a/an:

A. Mechanical motor
B. Induction motor

C. Synchronous motor
D. Solar motor

49. The steepness of the anode angle will effect the amount of the:

1. Anode heel effect   2. Field coverage   3. Thermionic emission

A. 1 & 2 only
B. 1 & 3 only

C. 2 & 3 only
D. 1, 2 & 3

50. Vibration of the x-ray tube during a radiographic exposure will most likely result in:

1. An increase in optical density
2. A decrease in radiographic contrast
3. An increase in the amount of focal spot blur

A. 1 & 2 only
B. 1 & 3 only

C. 2 & 3 only
D. 1, 2 & 3

51. The ball bearings which aid in the rotation of the induction motor require a special low vaporizing lubricating material called:

A. Graphite
B. PCB's

C. Powdered silver
D. Grease

52. The general formula that is used to determine the anode heat units created with a fully rectified x-ray tube is:

A. HU = mA x kV x SID
B. HU = mA x Time x FSS

C. HU = mA x kV x Time
D. HU = mA x SID x FSS

53. A reduction in the kilovoltage setting during the operation of an x-ray tube will most likely:

> *1. Reduce the speed of the electron stream*
> *2. Increase the filament temperature*
> *3. Decrease the space charge*

    A. 1 only   C. 3 only
    B. 2 only   D. 1, 2 & 3

54. All of the following portions of the induction motor and anode disc assembly of a rotating anode are normally contained within the vacuum of the tube EXCEPT:

    A. Stem of the anode   C. Rotor
    B. Ball bearings   D. Stator windings

55. The principal method by which the space charge can be increased in a modern tube is by employing a higher:

    A. mA station   C. Timer setting
    B. kVp setting   D. Screen speed

56. Which of the following events occur within the x-ray tube during the exposure (second) stage of the timer mechanism?

> *1. A high potential difference is developed*
> *2. Filament heating begins*
> *3. The movement of Bucky grid begins*

    A. 1 only   C. 3 only
    B. 2 only   D. 1, 2 & 3

57. The formation of the space charge in the filament of the x-ray tube is accomplished during the _____ of the normal exposure cycle.

    A. Prep or boost   C. Exposure stage
    B. Rectification stage   D. Selection stage

58. The negative electrode of an x-ray tube is commonly called the:

    A. Filament   C. Target
    B. Anode   D. Diaphragm

59. The filaments in most modern x-ray tubes are surrounded by metal collars which serve to:

    A. Increase filament heat   C. Accelerate the electrons
    B. Compress the electron stream   D. Prevent shock hazards

60. The temperature at which thermionic emission occurs is approximately:

    A. 1000 degrees C   C. 1570 degrees C
    B. 1250 degrees C   D. 2200 degrees C

1. A   The electrons that are freed by the heating of the filament of the x-ray tube are moved toward the anode by a large difference in electrical potential. Upon striking the anode much of the energy of the electron stream is converted into electromagnetic photons. In the diagnostic range of energies the majority (99%) of this electron energy is converted into thermal energy or heat. Less than 1% of this energy is converted into x-radiation.

2. B   The movement of the electrons across the tube is accomplished through the creation of a high potential difference between the filamen and large positively charged anodes. If an alternating current were used in the tube, the polarity changes of the current would lead to a reverse of the charge on the anode and the possible return of the electron flow to the filament. This is avoided by maintaining a direct current across the x-ray tube.

3. B   The transparent x-ray tube envelope that serves to maintain the vacuum in the x-ray tube is composed of a special heat resistant glass made from the oxides of boron called Pyrex.

4. D   A rotating anode has been designed to increase the rate at which heat is taken away from or dissipated away from the anode during x-ray production process. The rotation of the anode will also reduce the damage to the surface of the anode target (surface roughening) and raise instantaneous loading and tube rating capacities for the tube.

5. D   The use of a smaller focal spot is associated with an improved spatial resolution and higher image quality. The main disadvantage of the use of the smaller focal spot is the limited output (lower mAs) that can be generated during the exposure.

6. C   The speed at which the anode rotates is normally related to the type of rectification system employed. In most single-phase units a rotation speed of the anode disk is about 3000-3600 revolutions per minute (rpm's). In 3-phase rectified systems, the rotation speed is three normally 3 times faster or about 10,000 rpm's.

7. B   The heat or temperature of the filament is directly related to the amount of current, amperage flowing through the filament. In most modern units the selection of the current to the filament is tied into the selection of the desired mA station. The higher mA station the higher the filament current, the greater its temperature and the larger the number of electrons boiled off and x-ray photons produced.

8. D   The high melting point of about 3400 degrees C of tungsten makes it highly desirable for its use in components that are exposed to large amounts of heat. In the x-ray production process both the filament and the surface of the target (anode) will be exposed to a great deal of heat stress. These components are normally made of tungsten.

9. B   In most modern radiographic units, the selection of smaller filament is associated with the choice of lower mA stations with currents of less than 200 mA. The larger filament and focal spot is normally activated at mA settings above 300 mA. The principal means for reducing the effective focal spot size is through the selection of a smaller filament.

10. A   The Pyrex glass envelope surrounding the x-ray tube always causes some attenuation of the x-ray beam (inherent filtration). Some manufactures have designed x-ray tubes that have a thinner flattened area (window) that minimizes the absorption of the x-rays by this portion of the tube envelope.

11. C   The vast majority of modern diagnostic type x-ray tubes are designed with two filaments, a (small and large) (dual focus) and a single rotating anode (target).

12. A   During the actual exposure the electron cloud surrounding the filament is drawn across the x-ray tube by the high potential difference. When the stream collides with the surface of the anode (focal tract) the majority of kinetic energy of this stream of electrons is converted into a low energy form of electromagnetic radiation called infrared rays or heat.

13. A The amount of free electrons produced at the cathode (space charge) is directly dependent upon the impressed amperage (mA station) selected for the exposure.

14. C A stationary anode is normally from a solid block of copper that is embedded with tungsten target.

15. A The speed developed by the electron stream as it passes from the cathode (filament) to the anode (target) side of the x-ray tube is directly proportional to the potential difference or the kilovoltage applied to the electrodes. The electrons in this 2-3 cm gap obtain speeds as high as 1.5×10^{10} cm/s or about ½ the speed of light.

16. B The current that is applied to the filament of the x-ray tube produces a heating which frees the tightly bound electrons. This process called thermionic emission forms an electron cloud or space charge around the filament. In tungsten this emission occurs when the temperature reaches a level of about 2200^0 C or 4000^0 F.

17. C The anode disk of a rotating anode x-ray tube is made of an alloy of tungsten and rhenium. The stem of the anode is made of a poor heat conducting material called molybdenum. The tungsten coated disk serves as the positively charged target that attracts the high speed electron stream originating at the heated filament.

18. D The watt is a unit used to express the amount of electrical power and is defined as the product of amperes and voltage; i.e., watt = amps x volts.

19. B The effective (projected) focal spot is directly related to the amount of focal spot blur which occurs in a radiographic image. Unlike the actual or electronic focal spot located on the surface of the anode the projected focal spot is located in space a few inches below the tube.

20. B The use of a heated filament was one of the first major advances in the development of the modern x-ray tube. This was recommended by William Coolidge. In order to accomplish this heating, the unit will require the use of a stepdown (filament) transformer

21. A Since x-rays are produced equally in all directions at the surface of the anode, many x-rays are emitted away from the patient. The tilting or angling of the anode helps direct a greater portion of unattenuated photons toward the image receptor by reducing the absorption of the photons in the heel of the anode.

22. D A reduction in the anode angle from the usual 12^0 to 7^0 will provide for a smaller focal spot due to the line focus principle. The steeper angle of the 7^0 anode will result in a smaller field coverage because of the increase in the anode heel effect which occurs when the anode is more vertical.

23. C In a modern rotating anode tube the vast majority of heat is dissipated by the radiation of heat from the anode to the tube envelope. Since the radiation of heat is a much more efficient more heat can be removed from the anode in a shorter period of time.

24. C The low heat conductivity of molybdenum makes this an ideal material for the stem of the anode disk. If the heat that builds up on the anode during the x-ray production process were to reach the ball bearing of the induction motor a great deal of damage would occur. The stem helps to prevent the heat from the anode from reaching the induction motor.

25. B After leaving the anode, the x-rays first encounter the Pyrex glass of the tube envelope. Since most tubes are surrounded by a layer of insulating oil, the beam may pass through some oil before hitting the aluminum (filtration) prior to its encounter with the collimator mirror and plastic collimator cover.

26. C The heat storage capacity of a rotating anode tube can be increased by increasing the cross-sectional area or diameter of the disk, using an anode with a high rotational speed, using a larger focal spot, increasing the size of the thermal cushion (external housing) and by the application of an external cooling (fan) device.

27. B The transparent x-ray tube envelope that serves to maintain the vacuum in the x-ray tube is composed of a special heat resistant glass made from the oxides of boron called Pyrex.

28. B   In most modern radiographic units, the selection of smaller filament is associated with the choice of mA stations with currents of less than 200 mA. During the manufacturing process, the use of a focus cup, steeper angled anode, and use of a smaller filament can be used to reduce the size of the effective focal spot.

29. A   As the mA station is increased, the heat of the filament and the amount of space charge expands. Since this can result in the electrons striking a larger area on the anode, the effective focal spot may be increased by a process called blooming. The amount of focal spot blur increases when a larger focal spot is employed.

30. D   Pure tungsten is an extremely brittle metal and is subject to surface cracks at high temperatures. The addition of rhenium increases both the thermal capacity and the resistance of the target to surface defects.

31. D   The main characteristic required in a filament material is its need to resist melting at high temperatures. Pure tungsten or a mixture of thorium and tungsten provide wire filaments with melting points of about 3400^0 C or 6000^0 F.

32. A   Sometimes called the first x-ray tube, the glass filled Crooke's-Hittorf tube provided Roentgen with the opportunity to make his important discovery of x-rays.

33. B   The tilting of the face of the anode is related to the length of the exposure field. Anodes with 12° angles are used for general purpose x-ray rooms. This anode will permit the coverage of a 17" image receptor at a 100 cm SID.

34. D   The addition of thorium to the brittle tungsten improves its ability to be coiled as well as improving its efficiency as an electron emitter. This material has been credited with increase in both durability, longevity and efficiency of the filament.

35. C   A modern rotating anode is composed of three different materials. The surface of the anode 00consists of a thin layer of tungsten that covers a solid rhenium disk. The molybdenum stem is attached to the disk and the induction motor that rotates the disk assembly.

36. D   Due to the line focus principle, the effective (projected) focal spot will appear smaller than the actual focal spot. As long as the anode angulation is less than 45 degrees the effective focal spot will be smaller than the actual focal spot. The reduction in the angle of the anode will decrease the effective focal spot for a given filament size. The steeper angle of the 7^0 anode will result in a smaller field coverage because of the increase in the anode heel effect which occurs when the anode is more vertical. According to the anode heel effect the radiation intensity will be greater on the cathode side of the tube than on the anode side of the tube.

37. A   When electrons are emitted by the filament in the absence of a large potential difference, they remain in the vicinity of the filament forming an electron cloud or space charge. Because of the repulsive force of the electrons the further emission of additional electrons is limited by the so-called space charge effect.

38. D   The high melting point of about 3400^0 C of tungsten makes it highly desirable for its use as filament material, as well as for the target (anode) of the electron stream.

39. D   The heat storage capacity of a rotating anode tube can be increased by increasing the cross-sectional area or diameter of the disk, using an anode with a high rotational speed, increasing the size of the thermal cushion (external housing) and by the application of an external cooling (fan) device.

40. A   During the manufacturing process, the use of a focus cup and steeper angled anode or the use of a smaller filament will provide an x-ray tube with smaller effective focal spot. In most modern radiographic units, the selection of smaller filament is associated with the choice of mA stations with currents of less than 200 mA.

41. D   The use of high speed anodes that rotate at speeds of 10,000 rpm will provide the higher tube ratings and heat storage capacity needed to produce serial exposures.

42. 1   43. 5   44. 2   45. 3   46. 4   47. 7

48. B The induction motor of a rotating anode x-ray tube consists of a movable coil of wire (rotor) which is attached to the anode stem and disk. Two sets of ball bearings reduce the friction enabling the rotor to spin within the vacuum of the tube envelope. The stator (electromagnets) windings that provide the driving force for the rotation of the anode disk are located out side the vacuum surrounding the anode end of the tube outside the envelope.

49. A The angulation of the anode is related to the size of the effective focal spot, the area that is covered by the beam (field coverage) and the amount of the anode heel effect.

50. C Tube vibrations, though relatively rare, are almost always caused by an imbalanced induction motor assembly or a warped anode disk. Both will lead to inadequate rotation speed and an increase in the amount of focal spot blur and loss of radiographic contrast on the radiographic images.

51. C Because the rotor of the induction motor is located inside the vacuum of the tube envelope, oil or graphite lubricants which vaporize at relatively low temperatures cannot be used. Powdered metals such as silver are successfully used as rotor lubricants of many modern x-ray tubes.

52. C In a single-phase fully rectified unit the number of anode heat units created for a given exposure is calculated using the formula: Anode heat units = mA x kVp x time. In a 3-phase 6-pulse unit a 1.35 correction factor is required. In a 3-phase 12-pulse system a 1.41 correction factor is required.

53. A A lower tube potential or lower kVp will decrease the speed and energy of the electron stream crossing the tube. This lower energy will decrease the quality of the beam by creating fewer photons with shorter wavelengths and higher frequencies.

54. D The induction motor of a rotating anode x-ray tube consists of a movable coil of wire (rotor) which is attached to the anode stem and disk. Two sets of ball bearings reduce the friction enabling the rotor to spin within the vacuum of the tube envelope. The stator (electromagnets) windings that provide the driving force for the rotation of the anode disk are located outside the vacuum surrounding the anode end of the tube outside the envelope.

55. A The heating of the filament which occurs in the first (prep or boost) stage of the exposure is directly related to the development of the space charge. A higher mA station is used to increase the amount of the space charge.

56. A In the first stage of an exposure, three events occur: the filament is heated, the Bucky begins to move and the anode rotation commences. During the second or exposure stage, high potential difference moves the electron cloud toward the anode (target).

57. A The heating of the filament which occurs in the first (prep or boost) stage of the exposure is directly related to the development of the space charge.

58. A An x-ray tube contains two electrodes, the negatively charged tungsten filament (cathode) and the positively charged tungsten target called the anode.

59. B In order to compress and focus the electron cloud created around the filament of an x-ray tube, it is surrounded by a negatively charged collar called a focus cup.

60. D The process by which metallic elements release free electrons when heated is called thermionic emission. In tungsten this emission occurs when the temperature reaches a level of about 2200° C or 4000° F.

Anode Storage Capacity

In the following exercise, all of the questions pertain to the anode cooling chart appearing below.

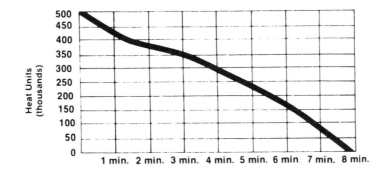

1. How many rapid full-wave rectified exposures at 100 mA, 250 ms, 80 kVp can be made before exceeding the anode storage capacity of this tube?

2-7. State the maximum number of full-wave rectified exposures at the following settings that can be made without exceeding the anode storage capacity of this tube.

	mA	Time	kVp	Rectification	Maximum # of Exposures
2.	300	250 ms	90	Full-wave	_____
3.	500	300 ms	80	Full-wave	_____
4.	200	1.20 sec.	90	Full-wave	_____
5.	400	600 ms	100	Full-wave	_____
6.	100	1.5 sec.	95	Full-wave	_____
7.	200	750 ms	105	Full-wave	_____

8. How many rapid 3-phase, 6-pulse exposures at 100 mA, 250 ms, 80 kVp can be made before exceeding the anode storage capacity of this tube?

9-14. State the maximum number of 3-phase, 6-pulse exposures at the following settings that can be made without exceeding the anode storage capacity of this tube.

	mA	Time	kVp	Rectification	Maximum # of Exposures
9.	300	250 ms	100	3-phase, 6-pulse	_____
10 .	500	300 ms	90	3-phase, 6-pulse	_____
11.	200	1.5 sec.	90	3-phase, 6-pulse	_____
12.	400	750 ms	100	3-phase, 6-pulse	_____
13.	800	50 ms	95	3-phase, 6-pulse	_____
14.	600	150 ms	105	3-phase, 6-pulse	_____

15. How many rapid 3-phase, 12-pulse exposures at 100 mA, 250 ms, 80 kVp can be made before exceeding the anode storage capacity of this tube?

16-20. State the maximum number of 3-phase, 12-pulse exposures at the following settings that can be made without exceeding the anode storage capacity of this tube.

	mA	Time	kVp	Rectification	Maximum # of Exposures
16.	400	70 ms	98	3-phase, 12-pulse	_____
17.	300	250 ms	90	3-phase, 12-pulse	_____
18.	600	250 ms	75	3-phase, 12-pulse	_____
19.	800	270 ms	90	3-phase, 12-pulse	_____
20.	300	330 ms	80	3-phase, 12-pulse	_____

Correction factors used in this exercise for the 3-phase, 6-pulse unit is 1.35 and for the 3-phase, 12-pulse unit is 1.41.

In the following exercise, all of the questions pertain to the anode cooling chart appearing below.

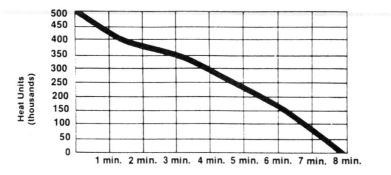

1. How many rapid full-wave rectified exposures at 200 mA, 250 ms, 80 kVp can be made before exceeding the anode storage capacity of this tube?

2.-7. State the maximum number of full-wave rectified exposures at the following settings that can be made without exceeding the anode storage capacity of this tube.

	mA	Time	kVp	Rectification	Maximum # of Exposures
2.	800	250 ms	80	Full-wave	
3.	300	300 ms	90	Full-wave	
4.	400	1.20 sec.	80	Full-wave	
5.	100	600 ms	110	Full-wave	
6.	200	1.5 sec.	95	Full-wave	
7.	400	750 ms	105	Full-wave	

8. How many rapid 3-phase, 6-pulse exposures at 200 mA, 250 ms, 80 kVp can be made before exceeding the anode storage capacity of this tube?

9.-14. State the maximum number of 3-phase, 6-pulse exposures at the following settings that can be made without exceeding the anode storage capacity of this tube.

	mA	Time	kVp	Rectification	Maximum # of Exposures
9.	800	250 ms	100	3-phase, 6-pulse	
10.	600	300 ms	80	3-phase, 6-pulse	
11.	300	1.5 sec.	90	3-phase, 6-pulse	
12.	100	750 ms	110	3-phase, 6-pulse	
13.	300	50 ms	95	3-phase, 6-pulse	
14.	400	150 ms	105	3-phase, 6-pulse	

15. How many rapid 3-phase, 12-pulse exposures at 200 mA, 250 ms, 80 kVp can be made before exceeding the anode storage capacity of this tube?

16.-20. State the maximum number of 3-phase, 12-pulse exposures at the following settings that can be made without exceeding the anode storage capacity of this tube.

	mA	Time	kVp	Rectification	Maximum # of Exposures
16.	200	70 ms	98	3-phase, 12-pulse	
17.	300	150 ms	60	3-phase, 12-pulse	
18.	400	250 ms	75	3-phase, 12-pulse	
19.	800	70 ms	90	3-phase, 12-pulse	
20.	600	330 ms	70	3-phase, 12-pulse	

Correction factors used in this exercise for the 3-phase, 6-pulse unit is 1.35 and for the 3-phase, 12-pulse unit is 1.41.

Tube Rating Chart

In the following exercise, all of the questions pertain to the tube rating chart appearing below.

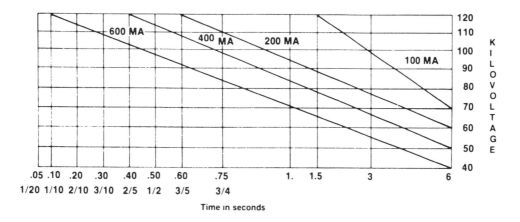

1. What is the highest exposure timer setting that can be used at 120 kVp for a 100 mA exposure?

2. What is the highest exposure timer setting that can be used at 120 kVp for a 400 mA exposure?

3. What is the highest exposure timer setting that can be used at 100 kVp for a 200 mA exposure?

4. What is the highest exposure timer setting that can be used at 100 kVp for a 600 mA exposure?

5. What is the highest mA station that can be used for a .4 second exposure at 90 kVp?

6. What is the highest mA station that can be used for a .75 second exposure at 120 kVp?

7. What is the highest mA station that can be employed for a .2 second exposure at 120 kVp?

8. What is the highest mA station that can be employed for a .05 second exposure at 120 kVp?

9. What is the highest mA station that can be employed for a 1.5 seconds exposure at 90 kVp?

10. What is the highest mA station that can be employed for a 3 seconds exposure at 60 kVp?

11. What is the highest kilovoltage setting that can be used at 400 mA, .2 second?

12. What is the highest kilovoltage setting that can be used at 600 mA, .5 second?

13. What is the highest kilovoltage setting that can be used at 100 mA, .3 second?

14. What is the highest kilovoltage setting that can be used at 200 mA, .75 second?

15. Will this unit permit a 400 mA, .5 second, 80 kVp exposure?

16. Will this unit permit a 600 mA, .3 second, 100 kVp exposure?

17. Will this unit permit a 100 mA, .5 second, 120 kVp exposure?

18. Will this unit permit a 200 mA, .05 second, 120 kVp exposure?

19. Will this unit permit a 400 mA, 1.0 second, 100 kVp exposure?

20. Will this unit permit a 600 mA, 3.0 seconds, 60 kVp exposure?

Tube Rating Chart

In the following exercise, all of the questions pertain to the tube rating chart appearing below.

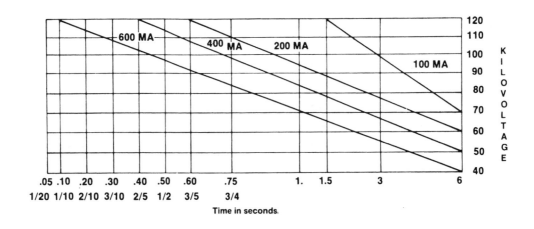

1. What is the highest exposure timer setting that can be used at 90 kVp for a 100 mA exposure?

2. What is the highest exposure timer setting that can be used at 100 kVp for a 400 mA exposure?

3. What is the highest exposure timer setting that can be used at 90 kVp for a 200 mA exposure?

4. What is the highest exposure timer setting that can be used at 120 kVp for a 600 mA exposure?

5. What is the highest mA station that can be used for a .4 second exposure at 120 kVp?

6. What is the highest mA station that can be used for a .75 second exposure at 110 kVp?

7. What is the highest mA station that can be employed for a .3 second exposure at 120 kVp?

8. What is the highest mA station that can be employed for a 1 second exposure at 100 kVp?

9. What is the highest mA station that can be employed for a 3 seconds exposure at 70 kVp?

10. What is the highest mA station that can be employed for a 1.5 seconds exposure at 70 kVp?

11. What is the highest kilovoltage setting that can be used at 600 mA, .6 sec.?

12. What is the highest kilovoltage setting that can be used at 400 mA, .75 second?

13. What is the highest kilovoltage setting that can be used at 100 mA, 3 seconds?

14. What is the highest kilovoltage setting that can be used at 200 mA, .1 second?

15. Will this unit permit a 600 mA, .5 second, 80 kVp exposure?

16. Will this unit permit a 200 mA, .3 second, 80 kVp exposure?

17. Will this unit permit a 100 mA, 1.5 second, 120 kVp exposure?

18. Will this unit permit a 200 mA, .10 second, 120 kVp exposure?

19. Will this unit permit a 600 mA, 1.0 second, 100 kVp exposure?

20. Will this unit permit a 400 mA, 1.5 seconds, 60 kVp exposure?

Answers

Anode Storage Capacity
Exercise 1

1. 250	11. 13
2. 74	12. 12
3. 41	13. 97
4. 23	14. 39
5. 20	15. 177
6. 35	16. 129
7. 31	17. 52
8. 185	18. 31
9. 49	19. 18
10. 27	20. 44

Anode Storage Capacity
Exercise 2

1. 125	11. 9
2. 31	12. 44
3. 61	13. 259
4. 13	14. 58
5. 75	15. 88
6. 17	16. 258
7. 15	17. 131
8. 92	18. 47
9. 18	19. 70
10. 25	20. 25

Tube Rating Chart
Exercise 1

1. 1.5 sec.	11. 120 kVp
2. .40 sec.	12. 90 kVp
3. .75 sec.	13. 120 kVp
4. .40 sec.	14. 110 kVp
5. 600 mA	15. Yes
6. 100 mA	16. Yes
7. 400 mA	17. Yes
8. 600 mA	18. Yes
9. 100 mA	19. No
10. 400 mA	20. No

Tube Rating Chart
Exercise 2

1. 3 sec	11. 90 kVp
2. .60 sec.	12. 90 kVp
3. 1 sec.	13. 90 kVp
4. .10 sec.	14. 90 kVp
5. 400 mA	15. Yes
6. 200 mA	16. Yes
7. 400 mA	17. Yes
8. 100 mA	18. Yes
9. 200 mA	19. No
10. 400 mA	20. Yes

1. The process by which instruments and equipment are maintained to operate within normal tolerance limits is called:

   A. Acceptance testing
   B. Error correction

   C. Quality control
   D. Practice standards

2. In order to insure the proper functioning of an automatic control system which of the following must be evaluated at least once each year?

   1. Backup timer   2. Minimum response time   3. Density control functions

   A. 1 only
   B. 2 only

   C. 3 only
   D. 1, 2 & 3

3. To insure that protective apparel is functioning properly protective aprons and gloves should be:

   A. Visually inspected at least once each month
   B. Sensitometrically inspected at least once each week
   C. Radiographically inspected at least once each year
   D. Ultrasonically inspected every 6 months

4. An exposure of 600 mA, 300 ms, and 89 kVp is used on a full wave rectified single-phase x-ray unit. How many heat units will this exposure produce?

   A. 1,800
   B. 6,720

   C. 16,020
   D. 160,020

5. A series of exposures are made using the following sets of exposure factors.

	mA	Time	kVp
1.	100	100 ms	75
2.	200	50 ms	75
3.	400	25 ms	75

   These are most commonly used in the evaluation of the x-ray units:

   A. Linearity
   B. Congruency

   C. Consistency
   D. Illumination potential

6. If a 180 cm (72") source-to-image receptor distance is employed for a chest unit, to be in compliance the light and radiation fields must coincide to within:

   A. 1.8 cm on each border of the image receptor
   B. 3.6 cm on each border of the image receptor

   C. 7.2 cm on each border of the image receptor
   D. 18 cm on each border of the image receptor

7. All of the following conditions will help to properly warm up the anode of a modern rotating anode EXCEPT:

   A. The use of a higher mA setting
   B. The use of a lower kVp setting

   C. The use of a long prep time
   D. The use of a long timer setting

8. Which of the following devices is used in a quality control program to check the exposure rate of a radiographic or fluoroscopic unit?

   A. Synchrotron
   B. mR/ mAs meter

   C. Photometer
   D. Flood field tester

9. The maximum heat storage capacity of the anode of a radiographic tube is related to the:

   1. *Diameter of the anode disk*
   2. *Design of the thermal envelope*
   3. *Rotational speed of the anode disk*

   A. 1 only   C. 3 only
   B. 2 only   D. 1, 2 & 3

10. In compliance with current regulations all stationary radiographic units shall:

   1. *Provide a means to align the center of the radiation bean with the center of the image receptor*
   2. *Be equipped with a light localizing adjustable collimator*
   3. *Provide an audible and visible indication for the production of x-rays*

   A. 1 only   C. 3 only
   B. 2 only   D. 1, 2 & 3

11. The radiology department plans on changing the film-screen imaging system it currently employs. Which of the following testing procedures would be an essential part in this selection process?

   A. Image uniformity   C. Chromatography
   B. Psychodometry   D. Sensitometry

12. The source-to-image receptor distance indicator on all new radiographic units must be accurate to within:

   A. 2% of the actual measurable source-to-image receptor distance
   B. 5% of the actual measurable source-to-image receptor distance
   C. 10% of the actual measurable source-to-image receptor distance
   D. 20 % of the actual measurable source-to-image receptor distance

13. The instantaneous loading capacity rating of the anode is primarily effected by the:

   A. Area of the focal tract   C. Size of the thermal envelop
   B. Atomic number of the target D. Size of the glass envelope

14. A full wave single-phase rectified radiographic exposure produces 900 heat units in the anode. Using the same exposure factor on a 3-phase 6-pulse unit would be associated with the production of:

   A. 993 heat units   C. 1137 heat units
   B. 1064 heat units   D. 1215 heat units

15. A routine exposure is attempted during a radiographic examination but no ready light appears. The radiographer should check to see if the:

   1. *Entrance door to the room is closed*
   2. *Proper ratio grid has been selected*
   3. *Proper film has been placed in the image receptor*

   A. 1 only   C. 3 only
   B. 2 only   D. 1, 2 & 3

16. Which type of equipment evaluation is required for all new pieces of radiographic units prior to being placed into service?

   A. Target composition testing C. Hypo retention testing
   B. Repeat analysis-testing   D. Acceptance testing

17. When a pinhole or resolution phantom test is performed for focal spot measurements, the hole or the phantom should be located:

    A. As close to the source as possible
    B. Midway between the source and the image receptor
    C. 1 meter behind the image receptor
    D. As close to the image receptor as possible

18. How many heat units would be produced in the anode of a 3-phase 12-pulse radiographic unit during an exposure of 300 mA 100 ms and 84 kVp?

    A. 2520
    B. 3042

    C. 3402
    D. 3553

19. A failure to obtain a ready light prior to a radiographic exposure may indicate:

    A. The patient has exceeded the weight limit for the table
    B. The x-ray tube is NOT located at the correct distance from the image receptor
    C. The sensors of the AEC unit are NOT under the thickest region of the body
    D. The speed of the Bucky grid is slower than the exposure timer setting

20. Surface irregularities on the face of the anode (pitting) are most often associated with a/an:

    A. Excessive amount of filtering by the vacuum in the tube
    B. Reduction in the radiation output during the exposure
    C. Increased output of radiation due to the blooming of the focal spot
    D. Reduction in the potential difference across the x-ray tube

21. The most common cause of a radiograph coming out of the processor with numerous regions possessing low optical densities is:

    A. Improper handling of the film in the darkroom
    B. Placing a film on a wet feed tray
    C. Failing to regulate the temperature of the developer solution
    D. Failing to clean the dust from the surface of the intensifying screen

22. Which of the following devices can be used to evaluate the accuracy of the timer of a 3-phase radiographic unit?

    A. Densitometer/sensitometer
    B. Oscilloscope/digital timing meter

    C. Flow meter/LCD
    D. Ionization chamber/thyratron

23. A loud, grinding noise is heard from an x-ray tube during the prep stage of the exposure. The most likely cause is:

    A. Gassy x-ray tube
    B. Loose filament collar

    C. Damaged induction motor bearing
    D. Cracked tube envelope

24. All of the following can be used to extending the life of an x-ray tube EXCEPT:

    A. Avoiding prolonged prep times
    B. Employing proper warm-up techniques

    C. Observing anode heat values
    D. Using higher anode rotation speeds

25. A synchronous spin top tester can been used in a quality control program to determine the:

    A. Accuracy of the timing devices in a single-phase rectified system
    B. Slice thickness of a tomographic unit
    C. Quality of the photons in the x-ray beam of a high frequency unit
    D. Resolution of the cathode ray tube in a digital image system

26. An overload light appears when an exposure is attempted on a radiographic unit using the 100 mA station. If no overload light is seen using the same technique on the 300 mA station, the most likely cause is:

A. Burnout of the small filament
B. Faulty exposure timer
C. Failure of the induction motor
D. A defective rectifier

27. The common indication of a gassy x-ray tube is _____ during the exposure.

A. Fluctuations in mA readings
B. Fluctuations in kVp readings
C. Vibrations of the tube housing
D. No mA readings

28. The life of the filament can be dramatically increased by employing:

A. Lower kilovoltage values
B. Lower mA settings
C. Smaller spot settings
D. Higher focus cup voltages

29. The radiation output of an x-ray unit can be accurately determined by the use of a:

A. Multichannel analyzer
B. Spectrometer
C. mAs/mR ionization type meter
D. Thermoluminescent dosimeter

30. If a radiographer does NOT employ the proper warm-up technique before making an exposure, which is most likely to occur?

A. Spark over
B. Cracking of the anode
C. Pitting of the anode
D. Melting of the anode

31. Smoke or fire is detected coming from the equipment cabinet. Your first action should be to:

A. Throw water on the cabinet
B. Turn off all power to the unit
C. Close the door to the room
D. Pull the fire alarm box

32. A white or clear area appears on a series of films taken on the same cassette and in the same location on the films. The most likely cause is (a):

A. Opaque contrast residue on the table
B. Defect in the manufacturing of the film
C. Defect in the automatic processor
D. Stain on the intensifying screen

33. Which of the following types of fire extinguishers are generally considered safe for use on electrical equipment?

1. Halon   2. Water   3. Carbon dioxide

A. 1 & 2 only
B. 1 & 3 only
C. 2 & 3 only
D. 1, 2 & 3

34. The output radiation intensity for a series of identical radiographic exposures (reproducibility) must NOT exceed:

A. 50%
B. 30%
C. 5%
D. .1%

35. According to national averages what is the most common cause for images that need to be repeated?

A. Positioning errors
B. Exposure selection errors
C. Film handling and processing errors
D. Failures in maintaining the equipment

36. The principal method of heat dissipation in a rotating anode x-ray tube is intended to be:

A. Conduction through the anode stem
B. Radiation to the tube envelope
C. Conduction through the anode bearings
D. Radiation to the focus cup

37. Which of the following is likely to result in the appearance of an overload light?

   1. The use of an excessive source-to-image receptor distance
   2. The use of a technique that exceeds the capacity of the anode
   3. The failure of the radiographer to push in the Bucky tray

   A. 1 only
   B. 2 only
   C. 3 only
   D. 1, 2 & 3

38. The accuracy of distance indicators and collimation devices is required to be within:

   A. 2% of the SID
   B. 2% of the SOD
   C. 10% of the SID
   D. 18% of the SID

39. A radiographic (tube) rating chart is used in the determination of the:

   A. Maximum exposure limits for a unit
   B. Maximum heat capacity for an anode
   C. Cooling characteristics of the anode
   D. Cooling characteristics of the housing

40. Which of the following is the best method to determine if the tube/collimator possesses an adequate amount of diagnostic filtration?

   A. The evaluation of the half-life values
   B. The evaluation of the spin top values
   C. The evaluation of the voltage waveform curves
   D. The evaluation of the half-value layer thickness

41. Nearly all causes of tube failure are related to the _____ of the x-ray tube.

   A. Mechanical characteristics of the housing
   B. Electrical characteristics of the power supply
   C. Thermal characteristics of the x-ray tube
   D. Movement of the induction motor

42. A line pair or star phantom is often employed in a quality control program for determination of the:

   A. Spatial resolution of the imaging system
   B. Contrast resolution of a digital image
   C. Speed of the imaging system
   D. Conversion efficiency of the sensors

43. In general, the proper method for warming up an x-ray anode may include exposures using:

   A. A 30-second filament prep time technique
   B. A short time, high kilovoltage technique
   C. A long-time, low kilovoltage technique
   D. A low mAs, long prep time technique

44. The determination of effective focal spot size can be accomplished by the use of a:

   A. Spinning top tester
   B. Wire mesh test
   C. Pinhole camera
   D. Penetrometer test

45. The consistency of the beam intensity that can be maintained by different combinations of mA and timer settings that provide the same mAs values is termed:

   A. Proportionality/equality
   B. Reciprocity/linearity
   C. Congruency/coherence
   D. Reproducibility/equivalence

46. All of the following are factors which will lead to the blurring of a radiographic image due to poor screen contact EXCEPT:

   A. An improperly mounted screes
   B. A broken hinge in the image receptor
   C. A warped front in the image receptor
   D. An upside down image receptor

47. During a tomographic exposure it is required that the device that indicates the level of the tomographic section be accurate to within:

   A. ± 2% mm
   B. ± 2% cm
   C. ± 2% m
   D. ± 2% km

48. Even when proper warm-up procedures are followed, the life of the x-ray tube is limited. A possible reason for this is:

   1. Spin top leakage   2. Filament burnout   3. Aperture diaphragm expansion

   A. 1 only
   B. 2 only
   C. 3 only
   D. 1, 2 & 3

49. Dry chemical and carbon dioxide type fire extinguishers can be used for fires involving:

   1. Solid combustibles   2. Flammable liquids   3. Electrical equipment

   A. 1 only
   B. 2 only
   C. 3 only
   D. 1, 2 & 3

50. Most older radiographic tubes will appear to have a thin coating plated on the inside of the tube envelope. This is usually formed from vaporized:

   A. Oil
   B. Silver
   C. Lead
   D. Tungsten

51. A wire mesh test is the most common method for determination of:

   A. Poor screen contact
   B. Excessive phosphorescence
   C. Opaque screen artifacts
   D. Conversion efficiency

52. The rotating anode of a modern x-ray unit requires about one second to accelerate to its designated speed. To avoid exposures during this time, rotation starts:

   A. When the machine is turned on
   B. When the technique is selected
   C. During the prep stage of the exposure
   D. During the exposure stage

53. The induction motor of a rotating anode is damaged and stops before a series of multiple exposures is made. This will most likely result in:

   A. Pitting of the anode
   B. Cracking of the anode
   C. Crazing of the anode
   D. Cracking of the tube envelope

54. A step wedge (penetrometer) is a device that can be used for:

   1. Development of a characteristic curve
   2. Calibrating x-ray equipment
   3. Determination of kVp

   A. 1 only
   B. 2 only
   C. 3 only
   D. 1, 2 & 3

55. Excessive heating of the tube may lead to vaporization of the filament or anode. Over a long period of time this may result in any of the following EXCEPT:

   A. Arcing within the tube
   B. Filtration of the beam
   C. The deterioration of the vacuum
   D. An increase in the beam intensity

56. A collimator test tool and the nine-penny test are normally used in a quality control program for the evaluation of the:

   A. Light field-radiation field congruence
   B. Overload protection
   C. Illuminator bulb brightness
   D. kVp accuracy

57. All modern units are equipped with heat sensors that will provide a warning whenever the anode heat reaches:

   A. 20% of the maximum value
   B. 40% of the maximum value
   C. 50% of the maximum value
   D. 75% of the maximum value

58. A photometer is used in a quality control program to:

A. Evaluate the brightness of a view box
B. Measure the portion of light passing through an x-ray film
C. Evaluate the intensity of a radiation beam
D. Measure the amount of luminescence produced during a film-screen exposure

59. The measurement of the half-value layer thickness is often employed to insure:

A. The proper amount of filtration is present in the x-ray tube
B. The anode is rotating at the desired speed
C. The current has been properly rectified
D. The proper exposure factor has been selected

60. The maximum permitted variance between the actual and the selected kVp values during a radiographic exposure is:

A. ± 5%   C. ± 45%
B. ± 20%   D. ± 60%

61. A mid-density or speed point that is below the lower control limit on a processor control chart is normally associated with a/an:

A. Depleted or exhausted developer solution
B. Increase in the time the film is immersed in the processing solutions
C. Increase in the activity of the developer solution
D. Over-replenished developer solution

62. According to the American National Standards Institute (ANSI) which of the following indicators should be evaluated to determine the sensitivity and contrast of images that have been obtained from an automatic processor?

1. Base-plus-fog levels of the film   *2. Speed point levels*   *3. Contrast index*

A. 1 only   C. 3 only
B. 2 only   D. 1, 2 & 3

63. All of the following are possible causes of a reduction of the contrast index on a processor control chart EXCEPT?

A. The presence of contaminated solutions   C. An insufficient developer solution temperature
B. A depleted or exhausted developer solution D. An excessive fixer solution temperature

64. Which of the following is likely to be associated with an increase in the base-plus fog level on a processor control chart?

A. A developer with a reduced chemical activity   C. A contaminated fixer solution
B. A contaminated developer solution   D. Improperly seated drying tubes

65. A filtered light source that is used to provide a film with a graduated scale of densities is called:

A. Densitometer   C. Penetrometer
B. Photometer   D. Sensitometer

Explanations

1. C   The process by which instruments and equipment are maintained within operable limits is called quality control. The three main steps in the quality control process related to all types of radiographic equipment includes acceptance testing, routine performance evaluation and error correction testing.

2. D   According to the NCRP Report 99 the proper functioning of an automatic exposure control system should evaluated at least once each year. Among some of the factors that must be checked include the minimum response time, backup time, density control function, AEC reproducibility, mA and kVp variability and density consistency.

3. C   In order to insure that protective apparel do not have cracks or gaps these devices need to be radiographically inspected at least once each year.

4. C   The number of heat units in a full wave rectified unit is based on the formula:
Heat units (16,020) = mA (600) x time in seconds (.3) x kVp (89).

5. A   The term reciprocity or exposure linearity refers to output of radiographic units when a series of exposures are made with different combinations of mA and timer values to obtain the same mAs. All radiographic units should be tested for exposure linearity at least once each year. The acceptable variation of the reciprocity or linearity is + 10%.

6. B   The light field of the collimator and the actual dimensions of the radiation field measured at the film must correspond to within ± 2% of the SID or 3.6 cm at a distance of 180 cm or 2 cm at a distance of 100 cm.

7. C   All x-ray tubes are subject to a number of thermal and mechanical stresses. In order to limit the effect of thermal stress on the anode all radiographic units should be warmed-up prior to the start of a radiographic procedure. The exposures used for this purpose will normally include 2-3 high mA, long time, low kVp techniques; i.e., 300 mA, .5 seconds and 70 kVp. Warm-up exposures will help prevent the cracking of the anode disk.

8. B   The exposure rate of any radiation type unit can be measured by a special type of ionization type detector called an mR/mAs meter.

9. D   The heat storage capacity of a rotating anode tube is related to the cross-sectional area or diameter of the disk, the rotational speed of the anode disk, size and design of the thermal cushion (external housing) and the use of external cooling (fan) devices.

10. D   According to the NCRP Report # 102 all stationary radiographic units must have a means for aligning the center of the beam to the center of the image receptor, have a collimator and provide an audible and visual indication for an exposure.

11. D   The most important test used in the evaluation of a film or film screen imaging system is called sensitometry. With this type of testing it is possible to compare the relative speed, contrast and exposure latitudes of the imaging systems.

12. A   The distance indicator on any fixed or mobile radiographic unit must be accurate to within + 2% if the actual source-to-image receptor distance.

13. A   The instantaneous loading capacity of an anode is related to the surface area of the focal tract and the type of generating system used to produce the radiation.

14. D   The number of heat units in a full wave rectified unit is based on the formula:
Heat units = mA x time in seconds x kVp.

    In a 3-phase unit a correction factor of 1.35 is required to obtain the appropriate number of heat units. 900 HU x 1.35 = 1215 HU in a 3-phase system.

15. A   Most modern radiographic units have a number of relays or interlocks that must be closed to permit the unit to start a radiographic exposure. Among the most common are the interlocks related to the entrance and/or bathroom doors, closure of the Bucky tray, placement of the fluoro carriage, and the use of the proper SID. The unit has no way of knowing if the proper grid ratio or film has been used in an image receptor.

16. D   The process by which instruments and equipment are maintained within operable limits is called quality control. The three main steps in the quality control process related to all types of radiographic equipment includes acceptance testing, routine performance evaluation and error correction testing. Acceptance testing helps insure that a unit will work as prescribed when it placed into service.

17. B   During a pin hole camera test to determine the size of the effective focal spot of an x-ray tube a piece of lead with a small pinhole must be placed at an equal distance from the focal spot of the tube and the imaging receptor. A radiographic exposure will enable only a portion of the x-rays to pass through the hole.

18. D   The number of heat units in a 3-phase 12-pulse rectified unit is based on the formula: Heat units = mA x time in seconds x kVp x 1.41 The correction factor of 1.41 is required to obtain the appropriate number of heat units. mA (300) x time (.1) x kVp (84) x 1.41 = 3553 HU

19. B   Most modern units have a number of relays or interlocks that must be closed to enable the radiographic unit to start the exposure. Among the most common are the interlocks related to the entrance and or bathroom doors, closure of the Bucky tray, placement of the fluoro carriage, and the use of the proper SID. The unit has no way of determining if the proper grid ratio or film has been used in an image receptor.

20. B   Any irregularity on the surface of the anode disc such as pitting or crazing results in an increased attenuation of x-rays by the disc and a reduction of the radiation output for a given tube current.

21. D   The only common artifacts that have a low optical density are those related to a reduced amount of light getting to the x-ray film. The must common of these are dust or dirt on the intensifying screens.

22. B   An evaluation of timer accuracy in a radiographic unit (single or 3-phase) can be made by a device called an oscilloscope or a digital timing meter.

23. C   The only moving component of a modern x-ray tube is the induction motor of the x-ray tube. The ball bearings incorporated into this motor helps reduce the friction and noise associated with its operation. Bearing damage is commonly associated with loud grinding (growl) noise heard during the prep or boost stage of the exposure.

24. D   Since the burn-out of the filament is one of the most likely reasons for shortened tube life, the use of a lower mA station and shortened boost time will extend the life of the filament. The use of proper warm-up times, observation of the heat units and tube rating limits will substantially reduce anode malfunctions.

25. A   A special motor driven spin top tester which spins at precisely one revolution per second can be used to test for valve tube competency in a single-phase unit and timer accuracy on both single-phase and 3-phase radiographic units.

26. A   The 25 mA, 50 mA and 100 mA stations are all related to the choice of the small filament and small focal spot. A faulty filament will negate all exposures attempted at any of these settings. Techniques selected on the large filament or high mA stations will be permitted if this filament is not damaged.

27. A   As a tube ages, the tungsten from the filament and anode may begin to vaporize, destroying the vacuum within the x-ray tube. Since some of the tube's electrons will have difficulty reaching the anode, it may cause fluctuations of the mA meter readings during the exposure.

28. B   Not unlike a light bulb, the life filament can be extended by limiting its use. The use of lower mA stations which require less space charge, lower filament temperatures, and the avoidance of the long boost or prep times will help extend life of the filament.

29. C The radiation output of an x-ray tube is measured in air by a milliroentgen/milliampere seconds meter. This modified ionization type meter is designed to accurately measure the short exposure bursts associated with diagnostic exposures.

30. B All x-ray tubes are subject to a number of thermal and mechanical stresses. In order to limit the effect of thermal stress on the anode all radiographic units should be warmed-up prior to the start of a radiographic procedure. The exposures used for this purpose will normally include 2-3 high mA, long time, low kVp techniques; i.e., 300 mA, .5 seconds and 70 kVp. Warm-up exposures will help prevent the cracking of the anode disk.

31. A The first and most important action of a radiographer faced with an electrical fire is the shut down of the main power to the unit. Secondly is the removal of the patient from the room. Third is to close all doors and lastly to pull the fire alarm box. Throwing water on an electrical fire can worsen the situation by increasing the potential for electrical shock.

32. D A low optical density artifact that appears on a series of images when using the same cassette is most commonly the result of a stain on the intensifying screen.

33. B Electric fires require the use of a C type extinguisher. These include the dry chemical type and gaseous type devices like Halon or carbon dioxide.

34. C The reproducibility of radiographic units is critical in maintaining a consistent density for a given exposure. This can be tested by making a series of identical radiographic exposures and measuring the intensity of each using an mR/mAs meter. When properly adjusted, a maximum variability of ± 5% in the reproducibility is allowed.

35. B A repeat analysis is likely to show the most common cause for a repeated image is improper exposure factor selection. This may be surprising in light of the widespread use of automatic exposure control devices which are intended to eliminate this type of error.

36. B In a rotating anode tube excessive heat conduction through the stem would cause damage to the induction motor. The poorly conductive molybdenum stem of the anode encourages radiation of heat off the anode while resisting the conduction of the heat to the induction motor.

37. B A tube rating chart provides a means for determining the maximum exposure limits for the unit. The rating chart is used to see if a given exposure can be safely permitted by the radiographic unit. In most modern units, these limits are programmed into the unit to protect the anode and cathode from excessive heat loads. The red overload load or maximum rating light will normally control console will appear when unacceptable exposure settings have been selected.

38. A The recent recommendations of the NCRP requires that all new radiographic units posses distance indicators and collimation devices. Each must be accurate to within 2% of a given source-to-image distance; i.e., at a 100 cm SID the collimator must be accurate to within 2 cm and 3.6 cm at a distance of 180 cm.

39. A A tube rating chart provides a means for determining the maximum exposure limits for the unit. The rating chart is used to see if a given exposure can be safely permitted by the radiographic unit. In most modern units, these limits are programmed into the unit to protect the anode and cathode from excessive heat loads. The red overload load or maximum rating light will normally control console will appear when unacceptable exposure settings have been selected.

40. D The best way to measure the amount of filtration in a radiographic system is through the determination of the half-value layer thickness of the beam. In most modern units the 2.5 mm aluminum filtration is equivalent to about 1HVL at an exposure of 80 kVp.

41. C The vast majority of the malfunctions associated with the operation of the x-ray tube are related to the high heat and thermal stresses associated with the x-ray production process.

42. A The resolution of an image receptor is most easily determined through the use of a resolution (phantom) consisting of precisely spaced lead strips and radiolucent spacers. The two types most commonly employed are called line pair and star phantoms. Examples of a star phantom and two line pair phantoms are found on page 43 at the beginning of section 2.

43. C   Most modern radiographic tubes require a series of warm-up techniques which serve to prevent uneven expansion of the anode and the development of heat stress cracks. All warm-up techniques share 3 common properties: the use of a low kVp setting (70), large focal spot, and a long timer setting (500 ms).

44. C   A pinhole camera is used to determine the size of the focal spot of an x-ray tube. This is accomplished by passing the beam through a piece of lead that has a small pinhole. To insure the proper geometry is maintained the pinhole must be placed at an equal distance from the focal spot of the tube and the imaging receptor. When performed properly this test will determine the size of the effective (projected) focal spot.

45. B   One of the most important of the required equipment tests is in the determination of the consistency or exposure linearity of a series of radiographic exposures. This test requires an exposure at each different mA station using the appropriate timer settings to provide the same mAs value. Using an mR/mAs ionization type meter the intensity of each exposure is recorded and compared to other values. The reciprocity between exposures is considered acceptable if the variation is within 10% for all exposures at different mA stations.

46. B   In the recent past, the most common cause of image blurring was poor screen-film contact due to loose or low tension cassette straps. Though rarely seen with the new wafer cassettes, it can result from improperly mounted screens, faulty latches or hinges and warping of the cassette.

47. C   All tomographic units should be evaluated at least once each year. This will help to insure that the tomographic level, ± 2%, and exposure angle, ± 2% remain within prescribed limits.

48. B   All x-ray tubes are subject to a number of thermal and mechanical stresses. Though tube warm ups are critical for extending the life of the anode, little can be done to increase the longevity of the filament which is one of the more frequent causes of tube failure.

49. D   Dry chemical and gaseous type devices like Halon or carbon dioxide are generally classified as ABC type extinguishers because they can be used on any type of fire.

50. D   The normal operation of the x-ray tube involves the production of extremely high temperatures of the filament and anode. Over time, this heat causes the vaporization of tungsten and the formation of a thin rust-colored layer which coats the inside of the Pyrex envelope called bronzing or tanning.

51. A   Poor film-screen contact results in a rapid diffusion of the light from the screens before it reaches the film which noticeably blurs the radiographic image. The film-screen contact of an imaging system is best evaluated through the placement of a number 40 or 60 wire mesh over the cassette. Any irregularities shown will indicate a failure of the system to maintain the desired contact between the intensifying screen and the radiographic film.

52. C   To avoid accidental exposures to an anode which is stopped or rotating at less than its maximum speed, a relay or interlock switch is incorporated into the timer circuitry. This device requires a brief .5 sec. delay between the activation of the prep stage and the actual start of x-ray production process.

53. A   If a radiographic exposure is made on a stationary or slowly rotating anode, the heat is concentrated on a small area causing localized melting called pitting of the anode.

54. D   The aluminum penetrometer (step wedge) is one of the oldest and widely used devices in the evaluation and testing of the imaging equipment. The step-like design enables the production of a graduated scale of densities that can be employed to form a characteristic curve for the imaging components. A step wedge can be used to compare and calibrate x-ray films, intensifying screens, exposure control settings, and automatic processing systems.

55. A   Any overheating of the x-ray tube can be associated with the vaporization of tungsten and the coating of the inside of the tube envelope. This layer can act as a third electrode, and additional filtration for the beam, and if it remains as a vapor, it can over time destroy the vacuum.

56. D   The alignment of the collimated field and the actual radiation field need to be evaluated at least once each year. A commercial collimator test tool or a simple nine-penny test can be used to evaluate this congruence.

57. B A heat sensor is now commonly incorporated in all modern units to warn when more than 75% of the maximum heat storage capacity has been exceeded.

58. D A photometer is an electro-optical devise that is used in the measurement of the amount of light emitted from a view box.

59. A All tubes operating above 70 kVp are required to have an amount of at least 2.5 mmAl equivalent filtration. At 80 kVp this constitutes about a one half-value layer thickness for the beam. An exposure made with and without the collimator can be used to estimate the half-value layer thickness for the beam.

60. B The variation between the stated and actual kVp should be measured annually to insure that the values are within ± 5%.

61. A In a processor control program sensitometry is the most commonly employed means for maintaining the consistency of the processor. This involves the production of a step pattern with a series of different optical densities (sensitometric strip). This strip can be used in the evaluation of the base-plus fog density, the speed point of mid-density and the contrast index for an image. A mid-density of speed point below the desired value is most often related to an exhausted developer solution.

62. D The x-ray images must be evaluated for three main properties; the base-plus fog density, the speed point of mid-density and the contrast index for an image.

63. A All of the following are possible causes for a reduction in the contrast index, a contaminated developer solution, exhausted developer solution, insufficient developer temperature and increase fog levels.

64. B An increase in the base-plus fog level is associated with white light leaks, unsafe safe lights or excessive developer solution temperatures, and contaminated developer solutions.

65. D The sensitometer consists of a light source which has a series of graduated optical filters to provide a graduated scale of optical densities on a film.

Filtration

Chapter 13

1. The use of diagnostic x-ray filtration is associated with all the following EXCEPT:

   A. A 10% increase in the quality of the beam
   B. A 50% decrease in the entrance skin exposure to the patient
   C. A decrease in the number of long wavelength photons in the beam
   D. A 45% increase in the intensity of the beam

2. The thickness of absorbing material necessary to reduce the x-ray intensity to half of its original value is called:

   A. Half-life
   B. Half-wave reduction
   C. Half-penetration value
   D. Half-value layer thickness

3. The inherent filtration of a radiographic machine is usually composed of the material:

   A. Pyrex glass
   B. Aluminum
   C. Bakelite
   D. Lead

4. The half-value layer thickness of most diagnostic x-ray beams operating at a level of 90 kVp is about:

   A. 3.1 mm aluminum
   B. 3.1 mm copper
   C. 3.1 mm tin
   D. 3.1 mm lead

Referring to the diagram of the x-ray energy, answer question 5.

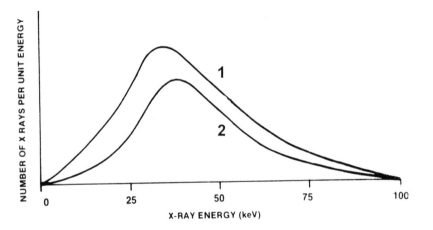

5. If the x-ray emission spectra above were produced using the same mAs and kVp which of the curve most likely represents a beam possessing the greatest amount of filtration?

   A. Curve 1
   B. Curve 2
   C. Unable to determine

6. Diagnostic filtration is principally employed to reduce the number of:

   1. Low energy photons reaching the skin surface
   2. Short wavelength photons reaching the skin surface
   3. Scattered photons reaching the image receptor

   A. 1 only
   B. 2 only
   C. 3 only
   D. 1, 2 & 3

7. An 80 kVp x-ray tube source is filtered by 4 mm aluminum which reduces the output to 25% of its original intensity. What is the half-layer thickness of this x-ray source?

A. 1 mm aluminum
B. 2 mm aluminum

C. 8 mm aluminum
D. 12 mm aluminum

8. During a radiographic examination a more uniform optical density of structures of unequal thickness can be accomplished through the use of a/an:

A. Thoraeus filter
B. Compensating filter

C. Compound filter
D. Inherent filter

Referring to diagram of thickness of aluminum absorber, answer question 9.

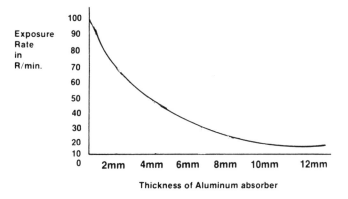

Thickness of Aluminum absorber

9. Approximately what is the half-value layer thickness for this beam?

A. 2 mm aluminum
B. 4 mm aluminum

C. 6 mm aluminum
D. 8 mm aluminum

10. All of the following are substantially altered by filtration of a heterogeneous diagnostic x-ray beam EXCEPT:

A. The speed of the photons
B. The quality of the photons in the beam

C. The number of low energy photons
D. The homogeneity of the beam

11. The normal diagnostic x-ray tube has a filtration equivalency of about .5 mm aluminum. This is called the:

A. Inherent filtration
B. Added filtration

C. Compensating filtration
D. Total filtration

12. A 100 mR/min. x-ray source has a HVL of 6 mm aluminum. How much filtration would be necessary to reduce the beam intensity to a level of 12.5 mR?

A. 9 mm aluminum
B. 12 mm aluminum

C. 15 mm aluminum
D. 18 mm aluminum

13. Which of the following are most common filtering materials used for the filtration of a diagnostic x-ray beam?

A. Thorium and tungsten
B. Aluminum and Lucite

C. Copper and tin
D. Lead and barium

14. The total amount of diagnostic filtration can be determined by finding the sum of:

A. Inherent & compensating filtration
B. Compensating & added filtration

C. Compression & compensating filtration
D. Inherent & added filtration

15. An x-ray beam has an intensity of 100 mR/min. unfiltered. If the addition of 10 mm aluminum reduces this level of 25 mR/min., the half-value layer thickness is about:

A. 2.5 mm aluminum
B. 5.0 mm aluminum

C. 7.5 mm aluminum
D. 15 mm aluminum

Referring to the diagram for beam filtration, answer questions 16-17.

16. Pertaining to the diagram above, each of the lines 1-4 represents the intensity change due to filtration at different kVp levels. According to the graph, the highest kVp is represented by:

A. Curve 1
B. Curve 2
C. Curve 3
D. Curve 4

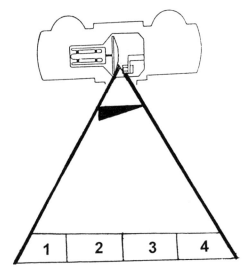

17. Pertaining to the diagram above, the greatest amount of attenuation is seen for values represented by:

A. Curve 1
B. Curve 2
C. Curve 3
D. Curve 4

18. A high energy x-ray source requires 4 mm of copper to reduce the original intensity by 50%. This source has a half-value layer of:

A. 22 mm copper
B. 15 mm copper

C. 8 mm copper
D. 4 mm copper

Referring to the diagram, answer question 19.

19. An exposure is taken using a wedge filter. Which area of the image will posses the least amount of optical density?

A. Area 1
B. Area 2
C. Area 3
D. Area 4

20. If it is desired to increase the amount of primary beam attenuation, the material used for the filter should have a _____ than the material it replaces.

A. Higher atomic number
B. Higher electron density

C. Lower atomic number
D. Lower electron density

21. Which of the following x-ray beams will possess the highest radiographic quality?

A. 80 kVp unfiltered

B. 90 kVp unfiltered

C. 80 kVp, 2.5 mm aluminum

D. 90 kVp, 2.5 mm aluminum

22. The amount of material required to reduce radiation intensity by one-tenth of the original value is termed:

A. Filtration value layer

B. Half-value layer

C. Tenth-value layer

D. Half-life layer

23. A beam that is filtered by two half-value layers of absorber is considered to be:

1. Polychromatic   **2. Monochromatic**   **3. Heterogeneous**

A. 1 only

B. 2 only

C. 3 only

D. 1, 2 & 3

24. A single tenth-value layer will have about the same protective value as:

A. 2.4 half-value layers

B. 3.3 half-value layers

C. 5.2 half-value layers

D. 7.3 half-value layers

25. A 100 mR/hr. x-ray source has a half-value layer of 1mm copper. How many half-value layers are required to reduce the beam intensity to 12.5 mR/hr.?

A. 1 HVL

B. 2 HVL

C. 3 HVL

D. 4 HVL

1. D   A minimum of 2.5 mm aluminum equivalent of diagnostic filtration is required for all radiographic units operating at above 70 kVp of energy. Though this amount of filtration will decrease the beam intensity by about ½, its primary purpose is the removal of the long wavelength (low energy) photons that increase the exposure to the skin of the patient without substantially effecting the radiographic image. Diagnostic filtration is estimated to reduce entrance skin exposures by about 50% compared to an unfiltered beam, while increasing the average energy (quality) of the beam by about 10%.

2. D   The amount or thickness of a material required to reduce the intensity of the heterogeneous x-radiation beam to one-half of its original value is called the half-value layer thickness.

3. A   The total filtration required for a radiographic unit consists of an inherent layer formed primarily by the Pyrex glass of the x-ray tube envelope and an added layer usually consisting of a thin sheet of aluminum or Lucite. A small amount of additional filtration is provided by the positioning mirror and plastic cover of the collimator.

4. A   The half-value layer thickness of a diagnostic x-ray beam is highly variable depending on the kilovoltage peak selected. According to the NCRP Report # 102 table 1.3, the HVL thickness at a 90 kVp is 2.6 mm aluminum for single-phase units and 3.1 mm aluminum for 3-phase radiographic units. At 110 kVp, the HVL thickness for single-phase and 3-phase units will be 3.1 mm aluminum and 3.6 mm aluminum respectively.

5. B   The beam with the greatest filtration curve 2 has the lower intensity but higher beam quality shown by the curve being skewed to the higher energy portion of the graph.

6. A   A minimum of 2.5 mm aluminum equivalent of diagnostic filtration is required for all radiographic units operating at above 70 kVp of energy. Though this amount of filtration will decrease the beam intensity by about ½, its primary purpose is the removal of the long wavelength (low energy) photons that increase the exposure to the skin of the patient without substantially effecting the radiographic image. Diagnostic filtration is estimated to reduce entrance skin exposures by about 50% compared to an unfiltered beam, while increasing the average energy (quality) of the beam by about 10%.

7. B   With a heterogenous x-ray source, if 4 mm aluminum reduces the intensity by 25% of its original value then its half-value layer thickness would be ½ that amount or 2 mm aluminum.

8. B   In some regions of the body; i.e., entire lower leg or entire spine, a single set of exposure factors will not provide the desired optical density in all areas. The use of specially designed (compensating) filters modifies the intensity of the beam thus compensating for these changes in tissue thickness or physical density, and therefore produces more uniform images of these structures.

9. B   The correct value is obtained by following the curve from the 100 mR line to the 50 mR point and reading the thickness (4 mm) at that point.

10. A   All x-ray beams produce a broad spectrum of photon energies and are referred to as being heterogeneous. The selective removal of the low energy photons during the filtering process will increase the average energy (quality) of the beam, reduce the intensity (quantity), and make the beam more homogeneous or monochromatic.

11. A   The Pyrex glass envelope which is designed to provide only a minimal resistance to the passage of the x-ray beam has an equivalent thickness of about only .5 mm aluminum.

12. D   If an x-ray beam with a 100 mR/min. intensity has a half-value layer thickness (HVL) of 6 mm aluminum, the following reduction in intensity is seen: 50 mR/min. with 6 mm aluminum (1HVL), 25 mR/min. with 12 mm aluminum (2 HVL) and 12.5 mR/min. with 18 mm aluminum of filtration (3 HVL).

13. B   The low cost of aluminum and Lucite help to make these the most commonly employed materials for the filtering x-rays in the diagnostic range. Special high density filters; i.e., gadolinium and holmium have been used to advantage with some types of contrast agents.

14. D   The total amount of diagnostic filtration is found by adding the amount of inherent filtration to the amount of added filtration.

15. B   If 10 mm aluminum reduces the original beam intensity to one quarter of its original intensity, then half that amount or 5 mm aluminum would be the half-value layer thickness for this beam.

16. A   The curve with the highest kVp will undergo the least loss of intensity. According to the diagram the uppermost curve (1) meets this criteria.

17. D   Attenuation will increase as the beam energy decreases. The highest attenuation occurs in curve 4 that was produced at the lowest kVp value.

18. D   By definition, the half-value layer is the amount of material which will reduce the original intensity of the beam by ½ or 50%.

19. A   The thickness portion of a wedge filter will cause a greater decrease in the intensity of the beam than the thinner portion of the wedge. Therefore less radiation will reach the image at region number 1 causing a reduction in the optical density.

20. A   If higher degree of attenuation is desired, the atomic number of the filter should be increased.

21. D   A beam with the highest quality can be produced by using a combination of the highest peak energy and the greatest amount of filtration.

22. C   A tenth-value layer is defined as the amount of material which will reduce the original intensity of the beam by 1/10th of its original value.

23. B   After a heterogeneous beam has been filtered by two half-value layer thicknesses of an absorber, most of the photons passing through the filter will have similar wavelengths. This is referred to as a homogeneous beam.

24. B   By definition, a tenth-value layer reduces the beam intensity to 10% or 1/10th its original value. This is nearly identical to the effect of 3.3 half-value layer thicknesses.

25. C   If a beam with a 100 mR/min. intensity has a half-value layer (HVL) of 1 mm copper, the following reduction in intensity is seen: 50 mR/min. with 1 mm copper (1HVL), 25 mR/min. with 2 mm copper (2 HVL) and 12.5 mR/min. with 3 mm copper of filtration (3 HVL).

Electromagnetic Radiation

1. The dual wave and particle natures of electromagnetic radiation was first proposed by (the):

   A. Coulomb's law
   B. Einstein's theory of relativity
   C. Ohm's law
   D. Quantum theory

2. The principal difference between an x-ray photon and a gamma ray is the:

   A. Shape of the electrical disturbance
   B. Magnitude of the electrical disturbance
   C. Region in the atom from where the photon originates
   D. Velocity at which the wave is propagated in matter

3. Which of the following relationships is TRUE concerning the energy of electromagnetic radiation?

   A. The energy of a wave will increase as its wavelength increases
   B. The energy of a wave will increase as its frequency increases
   C. The energy increases as the velocity of the wave increases
   D. The energy of the wave increases as the mass of the wave increases

4. The frequency (hertz) of an electromagnetic wave is best defined as the:

   A. Distance the wave travels per unit of time
   B. Number of oscillations or cycles per second
   C. Distance between the successive wave crests
   D. Distance between the maximum and minimum height of the wave

5. All electromagnetic radiations produce a type of variation in space known as a:

   A. Shear wave
   B. Step wave
   C. Skip wave
   D. Sine wave

6. All of the following electromagnetic waves are most likely to interact with matter as if they were waves EXCEPT?

   A. X-rays
   B. Visible light
   C. Microwaves
   D. Infrared rays

7. Diagnostic x-ray photons are normally produced at wavelengths of between:

   A. 1-5 kilometers
   B. 100-150 centimeters
   C. 1-5 decimeters
   D. 1-5 nanometers

8. Which members electromagnetic spectrum will normally have photon wavelengths longer than those of diagnostic x-ray photons?

   1. Gamma rays   2. Infrared rays   3. Microwaves

   A. 1 & 2 only
   B. 1 & 3 only
   C. 2 & 3 only
   D. 1, 2 & 3

9. A sunburn is a common reaction to overexposure to a normal component of sunlight called:

   A. Ultraviolet radiation
   B. Cosmic radiation
   C. Ultrasonic radiation
   D. Monochromatic radiation

10. The relationship of energy keV and photon wavelength is expressed by the formula:

    A. $E = keV/v$
    B. $E = 124/keV\ \lambda$
    C. $E = keV/\lambda$
    D. $E = keV \times c$

11. The partial absorption of energy by any form of matter is termed:

    A. Attenuation
    B. Transmission
    C. Reflection
    D. Diffraction

12. Which of the following forms of energy have shown the ability to deflect the path of an electromagnetic photon beam?

A. Strong magnetic field
B. Strong electrostatic field
C. Weak gravitational field
D. Strong gravitational field

13. The member of the electromagnetic spectrum possessing a wavelength of 700 nanometers is:

A. Green light
B. Blue light
C. Yellow light
D. Red light

Referring to the following diagram, answer questions 14 & 15.

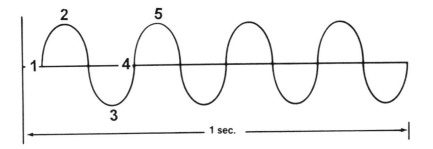

14. In the above diagram of an electromagnetic wave, which numbers best define wavelength?

A. 1-2
B. 3-4
C. 2-5
D. 3-5

15. What is the frequency of the electromagnetic wave represented in the diagram:

A. 2 cps
B. 3 cps
C. 4 cps
D. 8 cps

16. The energy of an electromagnetic radiation is directly proportional to the:

A. Wavelength of the radiation
B. Frequency of the radiation
C. Velocity of the photon
D. Shape of the electrical disturbance

17. All of the following are members of the electromagnetic spectrum EXCEPT:

A. Gamma rays
B. X-rays
C. Ultraviolet rays
D. Alpha rays

18. The velocity of all electromagnetic radiation in a vacuum is:

A. 3×10^{10} cm/sec.
B. 3×10^5 m/sec.
C. 3×10^3 km/sec.
D. Photons cannot travel in a vacuum

19. One of the principal dangers of exposure to radiation is the ability to cause ionization in the exposed tissue. Which type of radiation possesses sufficient energies to cause ionization?

1. Radio waves   2. Infrared rays   3. X-rays

A. 1 only
B. 2 only
C. 3 only
D. 1, 2 & 3

20. Heat is a form of electromagnetic radiation produced at relatively low energies called:

A. Ultraviolet radiation
B. Gamma rays
C. Microwaves
D. Infrared radiation

21. What is the wavelength of a 248 KeV gamma photon?

    A. .5 nanometers
    B. .05 millimeters

    C. .1 centimeter
    D. 2 kilometers

22. If the frequency of an electromagnetic photon is increased by a factor of 4, the wavelength will:

    A. Increase 16 times
    B. Increase 4 times

    C. Decrease 16 times
    D. Decrease 4 times

23. The principal unit used to measure the energy of a photon is the:

    A. Milliampere
    B. BTU

    C. Electron volt
    D. Neutron

24. The mass-energy equivalency of electromagnetic energy and matter was first predicted by:

    A. Albert Einstein
    B. Max Plank

    C. Neils Bohr
    D. Wilhelm Roentgen

25. Which formula is used in calculating the photon energy of electromagnetic radiations?

    A. $E = mc2$
    B. $E = hv$

    C. $E = 1/2\ mv2$
    D. $E = 12.4\ KeV$

26. The smallest subdivision of electromagnetic radiation is an individual packet of energy called a:

    A. Vector or pixel
    B. Photon or quantum

    C. Spiral or curve
    D. Bit or byte

27. Which of the following electromagnetic radiation is most likely to interact with matter as if it were a wave?

    1. Visible light   2. X-rays   3. Ultraviolet radiation

    A. 1 only
    B. 2 only

    C. 3 only
    D. 1, 2 & 3

28. The alteration in the speed of visible light when it encounters transparent matter of varying densities is termed:

    A. Reflection
    B. Attenuation

    C. Refraction
    D. Absorption

29. The energy of electromagnetic radiation is related to its wavelength by the relationship E=:

    A. hc
    B. hv

    C. hc/v
    D. ½ mv^2

30. The principal difference between electromagnetic radiation and sound waves is the:

    A. Mass of the waves
    B. Relative shape of the waves

    C. Electrical charge of the waves
    D. Velocity at which the waves travel

1. D One of the basic concepts of the quantum theory is the dual nature of the photons. This theory states that photons can either act as a wave or a particle. The penetration of light through glass or x-rays through low-density matter represents the photon's wave nature. The particle nature involves a substantial exchange in energy such as that between photons and electrons during excitation or ionization.

2. C If detected in free space, an x-ray photon and a gamma ray are identical in every measurable characteristic. Their classification is based entirely on their point of origin. X-rays are produced during from the exchange of energy that occurs in the electron (orbital) shells of an atom. Gamma rays are the result of changes that occur in the nucleus of the atom.

3. B All electromagnetic waves have a consistent velocity in a vacuum of 3×10^8 meters or 3×10^{10} centimeters per second called the speed of light. Because their speed is constant in a vacuum, the relative energy is directly proportional to frequency of the wave and inversely proportional to the wavelength of the photon. In other words, an increased frequency of the wave is associated with a higher energy while an increase in the wavelength of a photon is associated with a lower photon energy.

4. B All electromagnetic waves if visible would appear as an electrical magnetic disturbance that has the shape of a sine wave. Using this analogy, it is possible to define the wave in terms of the number of cycles passing by a given point each second (frequency). The frequency of a wave is directly related to the energy of the wave and is measured in the number of cycles per second called hertz (Hz).

5. D A sine wave is often conceptualized as an invisible piece of string of a fixed length that is being vibrated in an up and down fashion.

6. A The particle nature of electromagnetic radiation is most frequently seen in waves possessing a large enough amount of energy to cause ionization. The three principal waves with this amount of energy are ultraviolet radiations, gamma rays and x-rays.

7. B The normal range of energy for diagnostic x rays is from about 25 keV to 125 keV according to the formula: minimum wavelength in nanometers = 124/keV. The wavelength for this range of energies is found to be 1 nm - 5 nm.

8. C X-rays are a highly energetic form of electromagnetic radiation, possessing short wavelengths and high frequencies. Therefore, all of the low energy members of the spectrum including radio waves, microwaves, infrared radiation, visible light and ultraviolet radiation would all have wavelengths longer than x-rays.

9. A The ionization of the chemicals within the cells of the skin may cause a reddening that appears as the skin cells die. This damage when due to exposure to the sun (sunburn) is caused by one of the more energetic types of EMR called ultraviolet (UV) radiation.

10. B The normal range of energy for diagnostic x rays is from about 25 keV to 125 keV according to the formula: minimum wavelength (λ) = 124/keV. The wavelength for this range of energies is found to be 1 nm - 5 nm.

11. A Attenuation is defined as the partial absorption of energy.

12. D Because electromagnetic waves have no measurable charge or mass, they are unaffected by magnetic, electrostatic or weak gravitational fields. It has been demonstrated that extremely strong gravitational fields such as those around black holes, neutron stars and our own sun are capable of bending photons.

13. D The visible spectrum of light includes the various colors of light from the long wavelength red light (700 nm) to the shorter wavelength blue or violet (400 nm) light.

14. D The wavelength of a photon is defined as the linear distance between the crests of two adjacent waves. As the wavelengths of a photon increases or becomes longer, its energy is decreased. Therefore, the numbers 2-5 best defines the wavelength.

15. C The frequency is defined as the number of cycles or oscillations passing through a given point in a second. The frequency of the wave is determined by counting the number of peaks seen in the upper portion of the wave in one second or 4 cps.

16. B According to the formula E=h(λ) where h =planks constant and λ = frequency of the wave, the energy of the wave is found to be directly proportional to its frequency.

17. D The members of the electromagnetic spectrum include from lowest to highest energies: radio waves, microwaves, infrared radiation, visible light, ultraviolet radiation, x-rays, gamma rays, and cosmic radiation. Alpha rays are not part of this spectrum

18. A The velocity of all electromagnetic radiation in a vacuum is 3 x 10^{10}cm/sec. or 186,000 miles/sec. The velocity changes somewhat depending on the nature of the medium in which the waves travel.

19. C Many of the higher energy forms of electromagnetic radiation, including ultraviolet rays, x-rays and gamma rays, are capable of removing (ionizing) electrons from the atoms exposed to this radiation.

20. D Though heat or infrared radiation is invisible to the human eye, its production is often associated with the inadvertent production of some form of visible light. The red color seen with a heating element of an electric stove is an example of this inadvertent light production.

21. A The normal range of energy for diagnostic x rays is from about 25 keV to 125 keV according to the formula: minimum wavelength (λ) = 124/keV. The wavelength for this range of energies is found to be 1 nm - 5 nm. A 248 keV photon will have a wavelength of .5mm.

22. D Since the frequency and wavelength of a photon are inversely proportional to each other, a four-time increase in frequency results in a four time decrease in the photon wavelength.

23. C In radiology it is most convenient to measure the energy of a photon in terms of its electron voltage (eV). The Joule is a unit by which photon energy is expressed in the SI system of measurement.

24. A One of the first concepts in Albert Einstein's theory of relativity was the equivalency he proposed between energy and matter E = mc^2. In recent years, the vast majority of this theory has been tested and found to be accurate.

25. B According to the formula E=h(λ) where h =planks constant and λ = frequency of the wave, the energy of the wave is found to be directly proportional to its frequency.

26. B Not unlike the relationship of the atom as the smallest division of matter, the photon is the smallest unit of electromagnetic radiation.

27. A Most low energy electromagnetic radiation such as radio waves, microwaves, and infrared will interact with matter weakly as if they were waves. Higher energy waves that are responsible for ionizing effects of radiation act more like particles.

28. C The deviation of visible light when it encounters glass or other transparent matter is termed refraction.

29. C The energy of a photon can be found by the formula E=hc/λ where h is planks constant, c is the speed of light, and λ is wavelength of the photon.

30. D Sound differs from electromagnetic radiation in two major ways. The speed of sound is only 640 miles per hour in air and sound, unlike EMR, requires a propagation media and will not travel within a vacuum.

Nature of X-rays

Referring to the diagram, answer question 1.

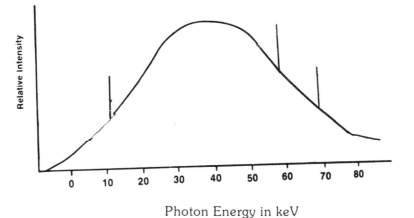

Photon Energy in keV

1. In the diagram of the x-ray spectral distribution curve, for a tungsten anode the three peaks seen 11, 59, 69 keV represent:

   A. K-edge absorption spectrum of the tungsten target
   B. Bremsstrahlung x-rays of the tungsten target
   C. Characteristic x-rays of the tungsten target
   D. Energy of the primary photons

2. The energy lost as the x-ray tube electrons pass through the outer orbital shells of tungsten target are commonly associated with the production of:

   A. High energy gamma rays
   B. Low energy beta particles
   C. Infrared radiation
   D. High energy x-ray photons

3. During the x-ray production process the photons are created at a broad spectrum of energies. The term that best describes this type of emission spectrum is:

   A. Monochromatic
   B. Heterogeneous
   C. Homogeneous
   D. Uniform

4. In the diagnostic range, the majority of the x-ray photons produced during the x-ray production process are termed:

   A. Bremsstrahlung radiation
   B. Compton radiation
   C. Photoelectric radiation
   D. Characteristic radiation

5. The amount of filtration and the kilovoltage peak used with a heterogeneous beam are the two main factors that will effect the:

   A. Field of view and the quantity of photons
   B. Direction and intensity of the beam
   C. Penetrability and the beam quality
   D. Consistency and the dispersion of the beam

6. The energy of the photons produced during the bremsstrahlung process are dependent upon the:

   1. Energy of the projectile electrons
   2. Atomic number of the target material
   3. The distance between electron and the nucleus

   A. 1 only
   B. 2 only

   C. 3 only
   D. 1, 2 & 3

Referring to the diagram, answer questions 7 & 8.

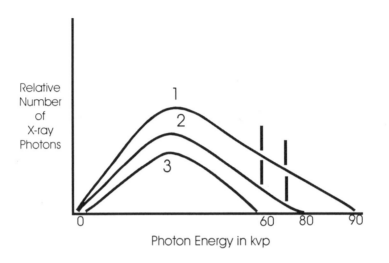

7. Which of the curves is associated with an exposure taken at the highest tube potential or kilovoltage peak?

   A. Curve 1
   B. Curve 2

   C. Curve 3
   D. All exposures were taken at the same kVp

8. The vertical spikes seen on curves 1 and 2 are absent in curve 3 because:

   A. The inherent filtration of the tube envelope reduced the spike intensity
   B. The energy was insufficient to produce these bremsstrahlung photons
   C. There was an insufficient milliampere setting used during the exposure
   D. The tube potential was below the threshold for these characteristic photons

9. The penetrability of an x-ray photon is effected by all the following EXCEPT:

   A. Frequency of the photon
   B. Mass of the photon

   C. Wavelength of the photon
   D. Energy of the photon

10. When a high speed stream of projectile electrons strikes the anode of an x-ray tube, the vast majority of the electron's kinetic energy is converted into:

   A. X-rays
   B. Infrared radiation

   C. Radio frequencies
   D. Ultrasonic waves

11. An increasing in the velocity of the projectile electron stream across an x-ray tube will result in production of x-ray:

   A. Photons with a lower frequency
   B. Photons with an increased amplitude

   C. Photons with shorter wavelengths
   D. Photons with a higher velocity

12. During bremsstrahlung x-ray production process, the x-rays that are produced will contain photons:

    A. With a small number of discrete energies
    B. Having only peak energy levels
    C. Having energies ranging for minimal to peak energies
    D. At a single photon energy level

13. Which of the following factors will have a significant effect on the energies produced during characteristic x-ray production process?

    A. The atomic number of the material in the target of the x-ray tube
    B. The type of gas used to fill the filament-anode gap
    C. The composition of the material used in the tube envelope
    D. Temperature to which the filament is heated

14. The frequency (number of cycles per second) of a diagnostic x-ray photon is directly proportional to all of the following EXCEPT:

    A. The quality of the beam   C. The ionization potential of the beam
    B. The penetrability of the beam   D. The intensity of the beam

15. All of the following factors are required for the production of x-ray EXCEPT:

    A. A source of high velocity projectile electrons   C. A target with a high atomic number
    B. A source of loosely bound (ionized) electrons   D. A strong magnetic field

16. The efficiency of x-ray production can be calculated with the formula:

    A. $E = K \times Z \times kVp$ C. $E = mc^2$
    B. $E = \frac{1}{2} mv^2$ D. $E = kV \times T \times mA$

17. A soft x-ray beam has a low penetrability. This beam is also said to possess a:

    A. High quality   C. High resolution
    B. Low quality   D. Low intensity

18. The majority of x-rays produced in an x-ray tube are the result of:

    A. A rapid deceleration of the projectile electrons as they pass through the target material
    B. A rapid deceleration of projectile neutrons as they pass through the target material
    C. A rapid acceleration of projectile protons as they pass through the target material
    D. A rapid acceleration of projectile positrons as they pass through the target material

19. It would be possible to produce a homogeneous x-ray beam if the kilovoltage were held constant and all off-target radiation were prevented at the anode.

    A. True
    B. False

20. During a normal radiographic exposure using 100 kVp, bremsstrahlung radiation accounts for about:

    A. 99% of the beam   C. 75% of the beam
    B. 95% of the beam   D. 50% of the beam

Referring to the x-ray intensity distribution curve, answer questions 21 & 22.

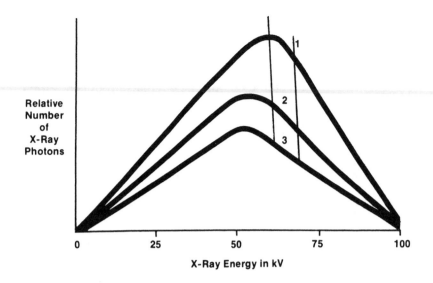

21. Which of the following curves resulted from the use of the lowest milliampere setting?

    A. Curve 1   C. Curve 3
    B. Curve 2

22. Curve 1 corresponds to the x-ray beam accomplished at a _____ than the other curves.

    1. Longer source-to-image receptor distance
    2. Milliampere setting
    3. Kilovoltage peak

    A. 1 & 2 only   C. 2 & 3 only
    B. 1 & 3 only   D. 1, 2 & 3

23. At a potential of 70 kVp the projectile electrons will reach a speed of approximately:

    A. 9.1×10^4 cm/s   C. 3.0×10^8 cm/s
    B. 1.9×10^3 cm/s   D. 1.5×10^{10} cm/s

24. The quality or penetrability of an x-ray photon is chiefly affected by changes in the:

    1. Kilovoltage peak setting   2. Milliampere setting   3. Timer setting

    A. 1 only   C. 3 only
    B. 2 only   D. 1, 2 & 3

25. In the diagnostic energy region, approximately what percentage of electron stream energy is converted into x-rays?

    A. .2%   C. 37%
    B. 7%   D. 86%

Referring to the diagrams, answer questions 26-28.

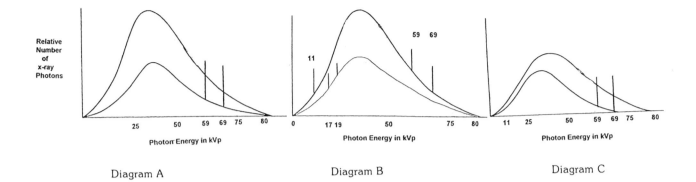

Diagram A

Diagram B

Diagram C

26. The differences resulting in the two curves in diagram A were caused by a change in the:

A. Atomic number of the target materials used during the exposures
B. Amount of filtering used during the exposures
C. Different milliampere-seconds values used during the exposures
D. Different source-to-image receptor distances used for the exposures

27. The differences resulting in the two curves in diagram B were caused by a change in the:

A. Change in the grid ratio used during the exposures
B. Atomic number of the target materials used during the exposures
C. Different kilovoltage peak value used during the exposure
D. Different image receptor speed for the exposures

28. The differences resulting in the two curves in diagram C were caused by a change in the:

A. Different milliampere-seconds values used during the exposures
B. Different kilovoltage peak value used during the exposure
C. Different image receptor speed for the exposures
D. Different source-to-image receptor distances used for the exposures

29. The loss of energy by a projectile electron as it passes by a tungsten nucleus in the anode results in a photon being produced by an event termed:

A. Characteristic radiation
B. Compton interaction
C. Bremsstrahlung radiation
D. Pair production

30. When the stream of fast-moving electrons interacts with the target of the anode, x-rays are generated by two different processes called:

A. Bremsstrahlung and characteristic radiation
B. Primary and secondary radiation
C. Leakage and scattered radiation
D. Compton and photoelectric radiation

Explanations

1. C   A graph of the x-ray distribution curve has a smooth bell curve appearance with three higher intensity spikes at the energies of 11, 59 and 69 keV. These represent the greater number of photons produced from the characteristic x-rays of the tungsten target. The majority of the bell curve is formed by those x-rays formed during the Bremsstrahlung production process.

2. C   The vast majority of high-speed tube electrons which pass from the filament and interact with electrons in the outer orbital shells in the anode produce low energy infrared (heat) radiation.

3. B   During the x-ray production process, the incident energy of the electron beam is converted into broad spectrums of low, medium and high energy photons. Because of the great variance in the nature of the beam, it is described as being heterogeneous (polychromatic) or consisting of many different photon wavelengths.

4. A   The collision of a high energy electron stream with the target material of the x-ray tube will result in the production of x-rays by two different processes. Bremsstrahlung or Brems radiation results from the slowing down of the incident electron as it travels past the nucleus of the tungsten atom in the anode. This accounts for about 70-85% of all the diagnostic quality photons in a beam. The remaining 15-30% of the photons are formed by a process called characteristic x-ray production which occurs when an incident electron hits and ejects on of the inner shell electrons of the tungsten target. The x-rays produced result from the characteristic photons produced during the orbital filling of the vacant K shells.

5. C   The beam quality refers to the average energy of the x-ray photons in the beam existing from the tube. An increase in the potential difference (kVp) across the x-ray tube will produce a greater number of photons with a higher energy. The addition of 2-3 mm aluminum diagnostic filtration will selectively remove the majority of the low energy photons while increasing the average energy or quality of the beam.

6. D   Bremsstrahlung (Brems) radiation is formed from the energy lost by a high speed electron as it passes near the positively charged tungsten nucleus. The loss of energy involved in this process appears in the form of an electromagnetic photon. The energy of these photons is dependent upon three of the original energy or velocity of the incident (tube) electrons, the atomic number of the target and the distance between projectile electrons and the nucleus of the target.

7. A   The diagram shows three spectral distribution curves produced with the same x-ray tube at different kVp values. The highest of the three curves (number 1) was taken at a kilovoltage peak of 90 kVp, number 2 at a kVp of 84, and number 3 at a kVp of 58.

8. D   The spikes in curve number 3 taken at 58 kVp are missing because the minimum energy required to product the characteristic photons of tungsten has not been reached. Therefore, eliminating the characteristic spikes at 59 and 69 keV emission of tungsten target are not produced.

9. B   The ability of an x-ray photon to penetrate through matter is dependent on both the nature of the matter and energy of the photon. The thicker or more dense the material the lower the degree of penetration by the photons. Higher energy, shorter wavelength, high frequency photons will have a higher degree of penetrability than their lower energy counterparts.

10. B   The amount of energy converted into x-rays during the x-ray production process is about 500 times less than the amount of energy formed as heat or infrared radiation. It has been estimated that only about .2% of the incident energy in the electrons hitting the anode are converted into x-rays in the diagnostic energy region.

11. C   The use of a higher tube potential (kVp) will increase the speed (energy) of the electron stream across the x-ray tube. Since more energy is transferred to the atoms of the tungsten target, the photons produced will have higher frequency and a shorter wavelength than at lower energies.

12. C During the x-ray production process, the incident energy of the electron beam is converted into broad spectrums of low, medium and high energy photons. Because of the great variance in the nature of the beam, it is described as being heterogeneous (polychromatic) or consisting of many different photon wavelengths. White light that contains all the colors of the visible spectrum x-radiation is sometimes referred to as white radiation because it also contains photons with all different energies.

13. A Characteristic x rays are directly associated with the loss of energy resulting from the movement of electrons from a higher to lower orbital shell or energy level. In a given material, such as tungsten, these energies are highly consistent (59 and 69 keV) and depend primarily on the atomic number of the target.

14. D Since the energy of a photon is dependent on the frequency and wavelength of the wave, any change in these factors will effect the quality, energy and penetrability and ionization potential of the beam.

15. D There are three principal factors required in the production of x-rays. Ther first is a source of loosely bound electron, these are provided by the pr-heated filament of the x-ray tube. The second is the need to accelerate these loosely bound electrons at high speeds toward the anode dis. This is accomplished by a potential difference between the cathode and the anode. The last requirement is a target that is able to provide a series of tightly bound electrons that will release sufficient energy to produce x-ray photons when ionized. This requires that the target have a high atomic number.

16. A The formula $E = K \times Z \times kVp$ in the formula K is a constant, Z is the = atomic number of the target material, and the kVp is the potential difference applied across the tube, can be used to estimate the efficiency of the x-ray production process.

17. B A low quality beam that has a low penetrability and is sometimes described as being soft.

18. A Bremsstrahlung or Brems radiation results from the slowing down of the incident electron as it travels past the nucleus of the tungsten atom in the anode. This accounts for about 70-85% of all the diagnostic quality photons in a beam.

19. B Though a reduction of off target radiations and a constant tube potential would help to improve the beam quality, the basic heterogeneous nature of the bremsstrahlung process cannot be altered until the photons have exited from the tube.

20. C Bremsstrahlung or Brems radiation results from the slowing down of the incident electron as it travels past the nucleus of the tungsten atom in the anode. This accounts for about 70-85% of all the diagnostic quality photons in a beam.

21. C Because there are fewer photons produced at a lower milliampere setting, curve 3 is associated with lowest intensity.

22. C The higher intensity of Curve 1 is associated with the use of the highest milliampere setting.

23. D The electrons emitted from the filament of an x-ray tube are accelerated to speeds of about 1.5×10^{10} centimeters per second or half the speed of light in the narrow gap separating the cathode and anode.

24. A The main factor that controls the quality and penetrability of the beam is the kVp setting.

25. A In the diagnostic energy range the percentage of the electron steam which is converted into x-rays is normally about .2 - .9%. The remaining 99% of the energy of is converted into infrared radiation or heat.

26. C A graph of the x-ray distribution curves in diagram A indicates two curves that were taken at the same kVp(80) on a tungsten target. The difference in the two curves is related to the different milliampere setting used during the exposures.

27. B A graph of the x-ray distribution curves in diagram B indicates two curves that were taken at the same kVp (80) and mA on two different targets. The higher of the two curves shows the characteristic photons associated with tungsten: the lower curve shows the different characteristic photons of molybdenum.

28. B A graph of the x-ray distribution curves in diagram C indicates two curves that were taken on the same anode at different kVp values. This can be seen by the differences in the maximum energies reached by the two curves. The lower curve was taken at about 70 kVp while the other was taken at about 80 kVp.

29. C Bremsstrahlung or Brems radiation results from the slowing down of the incident electron as it travels past the nucleus of the tungsten atom in the anode. The energy of these photons is dependent upon three factors, the original energy or velocity of the incident (tube) electrons, the atomic number of the target and the distance between projectile electrons and the nucleus of the target.

30. C The collision of a high energy electron stream with the target material of the x-ray tube will result in the production of x-rays by two different processes. Bremsstrahlung or Brems radiation results from the slowing down of the incident electron as it travels past the nucleus of the tungsten atom in the anode. Characteristic x-ray production occurs when an incident electron hits and ejects one of the inner (K) shell electrons of the tungsten target.

1. Which of the following is most frequently used to increase the brightness levels during a fluoroscopic imaging procedure?

   A. Charged coupled device
   B. Photo-optics enhancer
   C. Image intensifier
   D. Steered array device

2. The electronic image in an image intensifier tube is first formed on the surface of the:

   A. Input phosphor
   B. Electronic lens
   C. Photocathode screen
   D. Output phosphor

3. All of the following are classified as types of vacuum tubes EXCEPT:

   A. Image intensifier
   B. Ionization chamber
   C. X-ray tube
   D. Vidicon television camera tube

4. The amount of brightness gain occurring in the image intensifier due to the acceleration of the electron stream between the input and output screens is termed the:

   A. Flux gain
   B. Edge spread gain
   C. Focal spot gain
   D. Reflection gain

5. The principal advantage of image intensified fluoroscopic technique compared to routine radiographic imaging is that it allows for:

   1. An increase in the degree of spatial resolution in the images
   2. A reduction in the amount of exposure the patient will receive
   3. The ability to visualize dynamic motion of the tissues

   A. 1 only
   B. 2 only
   C. 3 only
   D. 1, 2 & 3

6. The phosphor and photocathode layers, which convert the x-rays into electronic images in an image-intensified (II) tube, are located on the:

   A. Input side of the II tube
   B. Peripheral potions of the output side of the II tube
   C. Entire surface of the focusing lens
   D. Entire surface of the output screen

7. The electronic image formed on the surface of the photocathode will be converted into _____ by the output phosphor of the image intensified.

   A. Digital matrix
   B. Visible light image
   C. Video signal
   D. Laser image

8. The most common material used in the formation of the input phosphor of a modern image intensifier is:

   A. Calcium tungstate
   B. Cesium iodide
   C. Barium lead sulfate
   D. Zinc cadmium sulfide

9. The operating potential required to move the electronic image between the input and output sides of the image intensifier tube is approximately:

   A. 220 volts
   B. 25,000 volts
   C. 140,000 volts
   D. 280,000 volts

10. The squared ratio of the diameter of the input phosphor to the square of the diameter of the output phosphor in an image intensifier tube is termed:

    A. Magnification ratio
    B. Conversion ratio
    C. Modulation transfer ratio
    D. Minification ratio

11. To aid the radiologist during fluoroscopy, most modern units will automatically change radiation levels for thicker or thinner body parts. This control is termed:

A. Automatic brightness gain
B. Automatic adaptation time
C. Automatic tomography
D. Automatic subtraction

12. Which of the following layers are found on the input side of the image intensifier tube?

1. Incandescent screen   *2. Fluorescent screen*   *3. Photocathode*

A. 1 & 2 only
B. 1 & 3 only
C. 2 & 3 only
D. 1, 2 & 3

13. Which operating mode of a triple field image intensifier is associated with the highest minification gain?

A. 6 inch
B. 9 inch
C. 14 inch
D. All possess the same gain factor

14. The usual range of milliamperage required for image intensified fluoroscopy is about:

A. 1-5 mA
B. 6-10 mA
C. 25-100 mA
D. 100-600 mA

15. The International Commission on Radiologic Units has recommended the use of a standardized method of evaluating the light emission of an image intensifier called the:

A. Photoemission factor
B. Resonance factor
C. Bandwidth factor
D. Conversion factor

16. The ability of an input phosphor to capture and convert x-ray photons and convert them in to visible light in an image intensifier (II) is called the:

A. Fiberoptics ratio
B. Conversion efficiency
C. Modulation transfer function
D. Relaxation ratio

17. All of the following are normally associated with an increase in the brightness gain of an image intensifier (II) EXCEPT:

A. Increase of the minification ratio
B. Increase in the flux gain
C. Increase in the exposure rate to the input phosphor
D. Reduction of the milliampere setting

18. The improved ability to visualize small objects with image intensified fluoroscopy results from the human factor of vision associated with:

1. Peripheral vision   *2. Photopic (cone) vision*   *3. Scotopic (rod) vision*

A. 1 only
B. 2 only
C. 3 only
D. 1, 2 & 3

19. During the manual mode of operation image intensified fluoroscope, an increase in kilovoltage peak is normally associated with a/an:

A. Reduced penetration of the beam
B. Increase in the brightness of the image
C. Increase in the contrast of the image
D. Increase in the resolution of the image

20. The automatic brightness (gain) control in fluoroscopic system was developed to help maintain consistent image brightness as the:

1. Thickness of the patient varies
2. Kilovoltage peak and milliampere values change
3. Source-to-image receptor distance changes

A. 1 only
B. 2 only
C. 3 only
D. 1, 2 & 3

21. The principal advantage of the smaller field of a dual field image intensifier (II) is a/an:

    1. Improvement in the quality of the image
    2. Reduction in the amount of magnification appearing in the image
    3. Reduction in the amount of exposure the patient will receive

    A. 1 only   C. 3 only
    B. 2 only   D. 1, 2 & 3

22. When a fluoroscopic unit is used in the manual-operating mode which of the following technical changes would be associated with a reduction in the brightness of the image a/an?

    A. Reduction in the source-to-skin distance   C. Reduction in the timer setting
    B. Reduction in the milliampere setting   D. Increase in the kilovoltage peak setting

23. The total brightness gain for a modern image intensified fluoroscopic image is about _____ greater than a conventional fluorographic screen image.

    A. 5-20 times   C. 50-200 times
    B. 20-40 times   D. 5,000-20,000 times

24. If the size of the input phosphor in the image intensifier increased, which of the following is likely to occur?

    A. The amount of tissue appearing in the image (field of view) will decrease
    B. The potential needed to accelerate the electronic image will increase
    C. Less radiation will be needed to provide the desired image brightness
    D. The magnification within the image will increase

25. The reduction of brightness at the periphery of image display of an image intensified fluoroscope is termed:

    A. Vignetting   C. Lateral dispersion
    B. Diminution   D. Modulation transfer

26. The reduction of field of view (FOV) size in a dual field image intensifier (II) can be accomplished by:

    A. Increasing the size of the input phosphor
    B. Increasing the voltage on the electrostatic lenses
    C. Decreasing the size of the output phosphor
    D. Decreasing the potential difference across the tube

27. The point in the image intensifier (II) where the electrons crossover before reaching the output phosphor is termed the:

    A. Coupling point   C. Electronic focal spot
    B. Effective focal spot   D. Flux point

28. The major factors effecting the amount of brightness gain in an image intensifier are:

    A. Minification ratio and flux gains   C. Focal and scotopic vision gains
    B. Adaptation and vignetting gains   D. Dispersion and retinal gains

29. What is the minification ratio for an image intensifier that has an input phosphor with a diameter of 20 cm and an output phospho with a diameter of 2 cm?

    A. 2 times   C. 50 times
    B. 10 times   D. 100 times

30. When viewing an image-intensified image on a large field of view (FOV), the image brightness and resolution are found to be greatest:

   1. At the periphery of the displayed image
   2. Near the center of the displayed image
   3. When the intensity of the beam is increased

   A. 1 & 2 only   C. 2 & 3 only
   B. 1 & 3 only   D. 1, 2 & 3

31. The cesium iodide layer of an image intensifier which converts x-rays into visible light is one of the principal coatings found on the:

   A. Input side of the image intensifier tube
   B. Target assembly of the television camera tube
   C. Emission surface of the electron gun of the charged-coupled device
   D. Output phosphor of the image intensifier

32. All of the following will occur when dual field image intensifier tube with automatic brightness (gain) control is being used in the smaller image mode EXCEPT:

   A. A greater magnification of the tissues on the display
   B. An improved contrast in the displayed image
   C. A Reduction in the amount of noise appearing in the image
   D. A reduction in the exposure to the patient

33. Zinc cadmium sulfide an output phosphor of a modern image intensifiers serves to convert:

   A. X-rays into visible light   C. Electrons into visible light
   B. Electrons into x-rays   D. Visible light into electrons

34. A thin layer of silver is coated on the inside of the image intensifier tube to focus the electron stream on the output. This is referred to as a/an:

   A. Accelerating anode   C. Photo-converter
   B. Hooded anode   D. Electrostatic lens

35. During image intensified fluoroscopy in the automatic brightness mode, as the magnification of an image increases, the:

   A. Brightness of the image will increase   C. Field of view will increase
   B. Resolution of the image will decrease   D. Brightness of the image will remain constant

36. The ratio of the luminescence in candelas of the output phosphor to the input exposure rate in milliroentgens per second defines:

   A. Brightness gain   C. Intensification factor
   B. Lumination point   D. Conversion factor

Referring to the diagram of the image intensifier tube, answer questions 37-40.

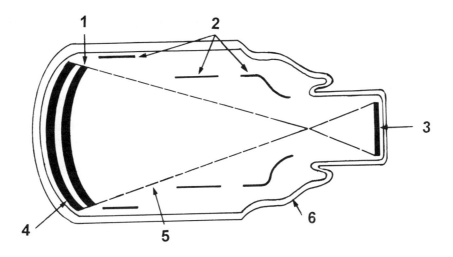

In the diagram of the image intensifier, place the number that corresponds to the following structures.

37. _____ Electrostatic lens
38. _____ Photocathode
39. _____ Output phosphor
40. _____ Input phosphor

41. During image intensified fluoroscopy, if the entrance skin exposure rate is estimated at 2 rem/mA-minute, what is the approximate entrance skin exposure for a 3-minute fluoroscopic study performed at an average of 2.4 mA?

    A. 14.4 rem   C. 3.3 rem
    B. 7.2 rem   D. 2.5 rem

42. During modern fluoroscopic examination of the gall bladder, using an iodinated contrast agent, the desired contrast can be maintained by use of a/an:

    A. Extremely high kVp value   C. Low mA setting
    B. Low to moderate kVp value   D. Long time value

43. The ability of the cone cells to perceive fine structural detail is most closely related to the:

    A. Speed at which the image is acquired   C. Adaptation time of the eye
    B. Intensity of light exiting the output phosphor   D. Time given for the integration of the image

44. The major source of noise in the image intensifier tube results from fluctuations of the number of absorbed x-ray photons by the:

    A. Input phosphor of the II tube
    B. Electrical interference by the electron gun
    C. Conversion inefficiency of the photocathode layer
    D. Ineffective focusing by the electrostatic lenses

45. During a fluoroscopic study if the entrance skin exposure rate is 2 rem/minute at a source-to-skin distance of 38 cm using 83 kVp and 3.2 mA. What new mA would be required to maintain the same exposure rate at a distance 76 cm using the same kVp?

    A. .8 mA   C. 6.4 mA
    B. 1.6 mA   D. 12.8 mA

46. During televised image intensified fluoroscopy, permanent static images or spot films can be produced and recorded using:

    1. Photo-spot imaging   ***2. Cassette-loaded spot imaging***   ***3. Video tape recordings***

A. 1 only
B. 2 only

C. 3 only
D. 1, 2 & 3

47. Because of the light emission from the output phosphor of most image intensifiers (Zn CdS:Ag), photo-spot film needs to have a maximum sensitivity to the visible light in the:

A. Red spectrum
B. Blue-violet spectrum

C. Yellow-orange spectrum
D. Yellow-green spectrum

48. The magnetic recording of a televised image is often performed with the use of a:

A. Cine camera
B. Video tape recorder

C. Digital analog recorder
D. Kymographic recorder

49. Which of the following spot film imaging systems will produce static images possessing the greatest recorded detail?

A. Photo-spot film camera
B. Video tape spot device

C. Cassette-loaded spot film
D. Digital fluoroscopic spot images

50. Which system is capable of the highest resolution?

A. Human eye
B. Radiographic film

C. Intensifying screen
D. Fluoroscopic screen

Explanations

1. C The main function of an image intensifier (II) tube is to increase the brightness level of the images produced during a fluoroscopic procedure. Most modern II's will increase the brightness by about 10.000 times compared to a conventional fluoroscopic system. Conventional radiography is limited to the production of still or static images. The main advantage of fluoroscopy is its ability to produce images in motion called dynamic or real time imaging.

2. C The x-rays that enter the input side of the image intensifier are first converted into a visible light image by the input phosphor. This light passes through to the photocathode screen that converts the visible light into an electrical pattern called the electronic image.

3. B The heart of the modern fluoroscopic system is a gas evacuated (vacuum) tube called the image intensifier tube. It shares this characteristic with the x-ray tube, all types of television camera tubes and all cathode ray tubes used to display the images.

   The main advance, which made modern fluoroscopy possible, was the development of the image intensifier tube and its associated electronics. This device dramatically increases the brightness level of the fluoroscopic image. An ionization chamber is a type of gas-filled tube.

4. A The energy increase as the electronic image passes from the input to output side of the image intensifier is responsible for about a 100-150 time increase in the image brightness. The 25 kV potential difference required to produce this effect is often referred to as the flux gain.

5. C The main advantage of fluoroscopic compared to radiographic is in the ability to visualize structures in motion or in real time. This type of imaging is often referred to as dynamic imaging while radiographic procedures provide only static of still images.

6. A After the imaging carrying (exit) radiation beam enters the input side of the image intensifier tube, a fluorescent screen (cesium iodide) is encountered which converts the x-rays into a visible light image. A photocathode layer immediately behind this screen transforms this visible image into a corresponding electronic image. Both of these screens are on the cathode input side of the II tube.

7. B The output screen of an image intensifier consists of a 1" diameter fluorescent screen made of zinc cadmium sulfide which converts accelerated electronic images into a bright yellow-green colored visible image.

8. B The concave input phosphor layer of the image intensifier is most often composed of a thin .1-.2 mm layer of sodium activated cesium iodide which absorbs a percentage of the exit beam and converts it into a visible light image.

9. B In order to move the electrons comprising the electronic image on the input side of the image intensifier across to the output side of the tube requires a potential difference of about 25,000 volts or 25 kVp.

10. D The increased brightness provided by an image intensifier is the result of two primary factors, the flux gain and the minification ratio. The flux gain relates to the 25 kVp potential difference placed between the input and output sides of the tube. The minification ratio is the amount of concentration of the electrons that occurs from focusing the electronic image from the photocathode (9") onto the smaller (1") output screen. Quantitatively, this minification ratio is proportional to the squared ratio of the input screen to the squared ratio of the output screen diameter or in this example 92 to 12 or 81 times. The total gain for the system is the product of the flux gain and minification ratio or about 10,000 in most modern systems.

11. A In order to compensate for the various differences in patient thickness and densities encountered during fluoroscopy, a system not unlike the automatic exposure control system in conventional radiography is employed by these units. These are most often called automatic brightness gain or control systems and will modify the kVp of mA to ensure a consistence brightness level during a fluoroscopic examination.

12. C   After the imaging carrying (exit) radiation the modified beam enters the input side of the image intensifier tube, a fluorescent screen (cesium iodide) is encountered which converts the x-rays into a visible light image. A photocathode layer immediately behind this screen transforms this visible image into a corresponding electronic image. Both of these screens are on the cathode of input side of the II tube.

13. C   Most modern image intensifiers can operate in two dual, or three triple modes. These modes or fields enable the operator to select an image that is close to the actual size of the region of interest or a magnified view of the region. The selection of the smallest mode (6") is associated with the highest amount of magnification, the highest minification gain occurs in the largest field (14").

14. A   Most texts give the operating range of an image intensifier as between 1-5 mA. Recent improvements have further reduced this value to the point where .5-3 mA range may be more appropriate for the average-sized patient.

15. D   In an attempt to standardize the amount of brightness gain, a unit called a conversion factor has been developed. The conversion factor is defined as the ratio of luminescence in candela for the output phosphor to input intensity as measured in milliroentgen/seconds.

16. B   The ability of an image intensifier tube to capture and convert x-rays into to visible light is termed the brightness gain or conversion efficiency. Over time this value may change. The deterioration for a modern II tube may be as high as 10% per year.

17. D   The increased brightness provided by an image intensifier is the result of two primary factors, the flux gain and the minification ratio. The flux gain relates to the 25 kVp potential difference placed between the input and output sides of the tube. The minification ratio is the amount of concentration of the electrons that occurs from focusing the electronic image from the photocathode (9") onto the smaller (1") output screen. In addition to a higher flux gain and a larger minification ratio a higher beam quality (kVp) or beam intensity (mA) can also increase the brightness gain of the system.

18. B   The low light levels associated with conventional fluorography allows only the use of scotopic or rod vision. These cells, which are active at low light levels, lack the fine discrimination of the cone cells that are more active when light levels are increased. The large increase in image brightness associated with the image intensifier enables the use of photopic (cone) vision and a substantial increase in visual acuity.

19. B   In the manual mode of a fluoroscopic unit (without automatic brightness control engaged), the fluoroscopic unit will respond in a similar fashion to an x-ray unit in the manual mode. Therefore an increase in the kVp or mA setting will increase the intensity and brightness of the image. The increased penetration will decrease the amount of differential attenuation within the tissue producing images with a lower or longer scale contrast.

20. D   The automatic control of a fluoroscopic system is designed to maintain a consistent brightness for changes to compensate for variations in the thickness of tissue density, and changes in the source-to-image receptor distance.

21. A   The use of smaller input field size with an image intensifier will provide images with a noticeable degree of magnification. Unlike radiographic magnification techniques, which degrades the image quality, a higher resolution and image quality are provided during fluoroscopic enlargement techniques. The main disadvantage of this technique is the substantial increase in the dose rate that is required to obtain the desired brightness level.

22. B   In the manual mode of a fluoroscopic unit (without automatic brightness control engaged), fluoroscopic unit will respond in a similar fashion to an x-ray unit in the manual mode. Therefore, a decrease in the kVp or mA setting will decrease the intensity and brightness of the image.

23. D   The total brightness gain of an image intensifier is dependent upon the product of minification ratio and the flux gain. In a modern unit this will increase to brightness over-conventional fluoroscope by about 5000 times for the small (6") input magnification mode and as much as 20,000 for larger format (14 inch) tubes.

24. C As the size of the input screen in an image intensifier increases the field of view (FOV) will also increase, this will reduce radiation that is needed to provide the desired image brightness due to a higher minification ratio.

25. A Because of the curved input screen and the flat output screen in the image intensifier, a higher density of electrons occurs near the center of the image. This effect, called vignetting, provides more detail and brightness in the center of the image and the monitor than its periphery.

26. B The adjustment from a larger to smaller input size field of view (FOV) in an image intensifier tube is accomplished electronically by a modification of the voltages on the electrostatic lens.

27. C The electrons from the top of the input screen are focused to the bottom of the output screen while the electrons on the bottom of the input screen are focused to the top of the output screen. The point at which they cross is called the electronic focal spot.

28. A The total gain for the system is the product of the flux gain and minification ratio or about 10,000 in most modern systems.

29. D The minification ration is found by the application of minification factor formula where minification ratio factor (100) = input screen diameter ^{2} (20^2) x output screen diameter^2 (2^2)

30. C Due to vignetting (explanation 25), the image quality is best at the center of the image. A higher intensity or dose rate improves the image brightness, reduces image noise and improves quality of the image.

31. A The cesium iodide phosphor layer is directly applied to the glass on the input side of the II tube.

32. D When a smaller input (magnification) mode is employed with an image intensifier, the reduction in the minification gain requires an increase in the intensity or dose rate to maintain the desired image brightness. This high dose rate decreases the noise of the image and provides a higher image contrast.

33. C The output screen of an image intensifier consists of a 1" diameter fluorescent screen made of zinc cadmium sulfide which converts accelerated electronic images into a bright yellow-green colored visible image.

34. D A thin uniform layer of silver on the inside of the image intensifier called the electronic lens is used for the focusing of the electron stream.

35. D When magnification mode is employed with an image intensifier, the reduction in the minification gain requires an increase in the intensity or dose rate to maintain the desired image brightness. This high dose rate decreases the noise of the image and provides a higher image contrast.

36. D In an attempt to standardize the amount of brightness gain, a unit called a conversion factor has been developed. The conversion factor is defined as the ratio of luminescence in candela for the output phosphor to input intensity as measured in milliroentgen/seconds.

37. 2 See corresponding diagram.

38. 1 See corresponding diagram.

39. 3 See corresponding diagram.

40. 4 See corresponding diagram.

41. A The exposure from a fluoroscopic examination can be estimated from the product of the average fluoro mA (2.4) x time of the exposure (3 minutes) x the constant 2 rem/mA minute = total exposure in rem (14.4).

42. B In order to prevent the overpenetration of an iodinated contrast media, the kilovoltage peak should be kept below the 80 kVp range when possible.

43. B The low light levels associated with conventional fluorography allows only the use of scotopic or rod vision. These cells, which are active at low light levels, lack the fine discrimination of the cone cells that are more active when light levels are increased. The large increase in amount of light exiting the output phosphor will enable the use of photopic (cone) vision and a substantial increase in perception of fine structural detail.

44. A The majority fo the noise in a televised fluoroscopic system appears caused by random absorbed x-rays by the input phosphor.

45. C The exposure from a fluoroscopic examination can be estimated from the product of the average fluoro mA (2.4) x time of the exposure (3 minutes) x the constant 2 rem/mA minute = total exposure in rem (14.4).

46. D The permanent static images can be obtained during a fluoroscopic examination in a number of ways. The oldest of these methods is the conventional or cassette-loaded spot film that is similar to a normal radiographic image. The main advantage of this type of spot film is its high spatial resolution. A photo-spot camera can take photographic images directly off the output phosphor of the II tube though the image quality is lower than that of a conventional cassette-loaded spot image. The image can be acquired 1/5th of the dose conventional spot film. Photo-spot can also be acquired at rates up to 12 images per second. Though it is also possible to record images on video disks or tapes the quality is far below that of the other types of spot films.

47. D Since the light from the output screen of the II tube is in the yellow-green spectrum, the films used for a photo-spot devices should have a maximum sensitivity to this color of light.

48. B A video tape recorder or videodisc recorder stores images as a pattern of magnetic fluctuations.

49. C The main advantage of the cassette-loaded spot films is their exceptional recorded detail of about 10-12 lp/mm. Photo-spots have a maximum resolution of about 2-4 lp/mm. Digital fluorographic spot images have a similar resolution to a phot0-spot.

50. B Radiographic film is capable of the highest spatial resolution (100 lp/mm). The human eye and the intensifying screens are capable of imaging 10-14 lp/mm. Fluoroscopic images can now be produced with 6-8 lp/mm of resolution.

1. Advances in semiconductor technology have made the development of solid state TV cameras possible. The usual name for this device is a:

   A. Signal plate reader
   B. Pulse width scanner

   C. Interlaced deflector
   D. Charge-coupled device

2. The main component of the television monitor is a large gas evacuated glass envelope that has been coated with a fluorescent screen called a:

   A. Fiber optics window
   B. Photo-ionization tube

   C. Cathode-ray tube
   D. Liquid crystal display device

3. The scanning of the electronic image at the signal plate of a television camera tube is most often accomplished with the grid controlled:

   A. Video signal
   B. Fiber optic bundle

   C. Electron gun
   D. Interlace framer

4. In order to accommodate human eye integration time, each frame of a television image must be presented in less than:

   A. 1 second
   B. ½ second

   C. 1/10 second
   D. 1/30 second

5. The horizontal resolution possible in a television imaging system is most closely related to the:

   A. Number of scan lines in the active trace
   B. Intensity of the beam that is used to modulate the video signal
   C. Number of times the electron gun fires each second
   D. Atomic number of the phosphor in the cathode ray tube

6. In a modern televised image chain, the device normally associated with the least amount of spatial resolution is/are:

   A. An image intensifier
   B. II-TV coupling devices

   C. A Video signal
   D. Television camera and monitor

7. All electronic devices possess a great deal of extraneous noise that serves to obscure the quality of the image. Which of the following signal-to-noise ratios would be preferable for a high-resolution television imaging chain?

   A. 1:1
   B. 200:1

   C. 400:1
   D. 1000:1

8. The scanned electronic image on the signal plate of a vidicon television camera tube is converted into a/an:

   A. Visible image
   B. Video signal

   C. Gamma ray beam
   D. Audible signal

9. One frame of a standard broadcast or closed circuit television system in this country is normally composed of:

   A. 262 lines of active trace per frame
   B. 525 lines of active trace per frame

   C. 685 lines of active trace per frame
   D. 1000 lines of active trace per frame

10. The visible light image that impinges upon the target of a television camera tube is converted into a/an:

    A. Latent image
    B. X-ray image

    C. Audible signal
    D. Electronic image

11. The target or photoconductive layer of a television camera is most often composed of a material called:

A. Cesium iodide
B. Lithium oxybromide
C. Antimony trisulfide
D. Calcium fluoride

12. In order to help eliminate image flicker during the viewing of a closed circuit television image, a technique called _____ is employed.

A. Depolarized light scanning
B. Image pre-amplification
C. Digital subtraction
D. Interlaced horizontal scanning

13. Modulation is the process by which the:

A. Blanking and retrace pulses are added to the video signal
B. Visible light of the II tube is converted into an electronic image
C. Light intensity of the TV camera is made proportional to the video signal
D. Two fields of a TV image are combined to form a single interlaced image

14. The size and position of the electron beam in a television camera tube is accomplished through the use of a number of sets of electromagnetic coils called:

A. Focus or steering coils
B. Optical and acoustic lenses
C. Fluorescent and luminescent screens
D. Ion and video coils

15. All of the following are the major components of a closed circuit television system EXCEPT:

A. Television camera tube
B. Video signal
C. Digital computer
D. CRT (monitor) tube

16. In a cathode ray tube (CRT), gas molecules that are ionized during image production are removed by a special component called a/an:

A. Electron gun
B. Ion trap
C. Control grid
D. Interlaced scanner

17. In order to coordinate the video signal between the television camera and the cathode ray tube (monitor) specialized electronic markers called:

A. Synchronization and blanking pulses are added to the video signal
B. Bandwidth and band pass signals are added to the line signal
C. Noise and quantum markers are added to the flicker pattern
D. Magnetic markers are added to the frame rate controls

18. The horizontal resolution of a television system can be greatly improved by increasing the:

1. Number of times the electron beam is modified each second in the TV camera tube
2. Thickness of the phosphor layers in the TV camera tube
3. Potential difference impressed across the TV camera tube

A. 1 only
B. 2 only
C. 3 only
D. 1, 2 & 3

19. The electronic signal that carries the information generated in the television camera to the cathode ray tube (monitor) is termed the:

A. Analog signal
B. Synchronization pulse
C. Video signal
D. Precession signal

20. When both the direct display of the intensified image and the recording cine spot films are desired:

A. Tandem mirror device is placed between the II tube and the TV camera
B. Retrace scanner is needed in between the TV camera and the CRT
C. Video recorder is needed between the II and the CRT
D. A digital processor is used in the imaging process

Referring to the diagram of the vidicon tube, answer questions 21-25.

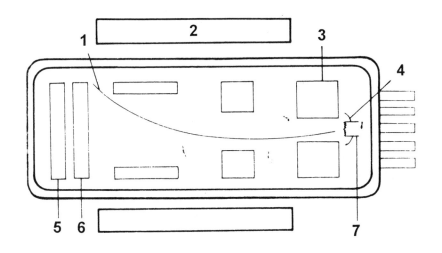

Place the number that corresponds to the following structures.

21._____ Electron gun

22._____ Signal plate

23._____ Electron beam

24._____ Accelerator grid

25._____ Steering (deflection) coils

26. A use of a smaller cathode ray tube (monitor) will have which of the following advantages compared to a fluoroscopic image that is viewed on a larger sized monitor?

A. The smaller monitor provides an improved vertical resolution
B. The larger monitor provides a higher amount of image persistence
C. The smaller monitor has a lower flicker effect
D. The larger monitor provides a smaller field of view (FOV)

27. The principal factor which effects the horizontal resolution is:

A. The number of vertical scan lines
B. The voltage used by the TV monitor
C. The bandwidth (band pass)
D. The number of synchronization pulses during the active trace

28. The band pass or bandwidth of the TV system used in a conventional fluoroscopic system is about.

A. 1.5 MHz   C. 4.5 MHz
B. 2.9 MHz   D. 18 MHz

29. Calculate the maximum resolution in line pairs per millimeter for a 6" (152 mm) image intensifier if the CRT monitor resolution is 185 line pairs.

A. .8 lp/mm   C. 28 lp/mm
B. 1.2 lp/mm   D. 31 lp/mm

30. If a bandwidth of about 3.5 MHz is required to provide an equal horizontal and vertical resolution in a 500 line system, what bandwidth would be required to provide an equal resolution in a 1000 line system?

A. 1 MHz   C. 8 MHz
B. 4 MHz   D. 14 MHz

Explanations

1. D   The rapid improvements that have been made in a charge-coupled device (CCD) now make this the most common type TV camera used in a modern radiography department. The CCD is considered more durable, sensitive and is now capable of providing a higher resolution than most types of TV camera tubes.

2. C   The cathode-ray tube (CRT) or television monitor is a large glass envelope that contains a series of electronic coils which serve to direct the electrons from the electron gun onto a fluorescent screen that is coated on the inside of the tube. The CRT serve for the display of nearly all type of analog and digital type video images.

3. C   In both vidicon and Plumicon TV camera tubes, the image on the signal plate is "read" by the use of a stream of electrons originating in the special designed cathode called the electron gun. The precise focusing of this beam is accomplished by a series of focusing deflection coils as well as an internal control grid.

4. D   When viewing a movie, each frame of the film must be presented in less than 1/30th of a second in order to see the images clearly without an annoying image flicker. In a similar fashion, each frame of a TV image must likewise be presented at speeds exceeding the eyes integration time of 1/20th second.

5. A   The vertical resolution of a video system is determined by the number of scan lines in the active trace. Conventional television system consisting of 525 active trace lines is considered to have a poor vertical resolution. Higher resolution system required in the medical imaging fields will normally have more than 1000 lines of active trace.

6. D   Though the image intensifier and television systems have improved fluoroscopic imaging in a number of important ways, the absolute spatial resolution of the images is degraded with the addition of each new electronic component. The various components required in the TV chain will normally posses the poorest resolution of all of the listed devices.

7. D   All electronic audio or imaging devices are associated with a large degree of background electrical noise which tends to degrade the image or sound quality of the original signal. This important imaging factor is normally expressed by the ratio of the maximum signal strength to the total electronic noise of the system, called the signal-to-noise ratio (SNR). In a conventional or home TV system, a SNR of 200:1 is considered acceptable. Most computer assisted or digital radiographic systems require a video system with a SNR of 500:1 or 1000:1 to maintain the image quality of these advanced techniques.

8. B   With either a vacuum type video TV camera tube or charge-coupled device TV camera, the output is in the form of an electronic video signal which carries the information presented on the signal plate of the camera to the cathode ray tube monitor.

9. B   Most TV camera tubes record images consisting of 525 lines of active trace which are presented on the cathode ray tube monitor. As with the signal-to-noise ratio, the TV systems required in most medical applications will require an increase to 875 or 1024 lines to provide the increased spatial resolution desired.

10. D   The principal purpose of the target-signal plate assembly of the vidicon camera tube is the conversion of the visible image on the input screen into an electric image which can be transformed into the video signal. This photoconductive target, which is usually made of antimony trisulfide, is suspended in a mica matrix. This is often considered the most important element of the TV camera tube.

11. C   This photoconductive target is usually made of antimony trisulfide suspended in a mica matrix.

12. D Image flicker, which may interfere with visualization of images from a television monitor can be eliminated by extremely rapid frame rates or by an electronic scanning pattern in which only every other line is scanned during each field called interlaced scanning. Improvements in the electronics now allows exceptional resolution in images that are acquired by the progressive scanning methods as well.

13. C The television camera tube creates a video signal that is directly proportional to the intensity of the light that strikes the target assembly. The term modulation refers to the process by which intensity of the electron beam in the CRT is adjusted to be proportional to the changes that occur in the magnitude of the video signal.

14. A The electron beam (gun) in a video camera is steered to its desired location on the target assembly by two sets of electrostatic devices known as the focusing and deflection (steering) coils.

15. C A closed-circuit television system is made up of three main components; the TV camera tube, the control unit or video signal and the cathode ray tube (monitor). No computer is needed for the operation of an analog TV system.

16. B Even with the best man-made vacuum some of remaining gasses will produce ions during the imaging process. A charged ion trap located at the cathode end of the CRT (monitor) helps to collect these gasses, increasing both the quality of the image and the longevity of the tube.

17. A To ensure the proper synchronization between the television camera and CRT (monitor), special electronic (synchronization) pulses and blanking pulses are added to the video signal at the end of each scanning line and the bottom of the fields.

18. A The two main factors that effect the image resolution of a television system are the bandwidth (bandpass) and the number of scan lines per frame of the image. The horizontal resolution is determined by a property called the bandwidth expressed in frequency (Hz) and describes the number of times the electron beam is changed (modulated). A value of 4.5 MHz is normal for the televisions used in most fluoroscopic systems. Some of the newer digital imaging systems will use high-resolution television systems having a bandwidth as high as 20 MHz. The vertical resolution is determined by the number of scan lines in the system. Conventional television system consisting of 525 active trace lines is considered to have a poor vertical resolution. Higher resolution systems required in the medical imaging fields will normally have more than 1000 lines of active trace.

19. C The electronic signal produced by a television camera or CCD camera is termed the video signal. When amplified, it is this signal that carries the video information to the CRT (monitor).

20. A Image coupling between an image intensifier and the television camera can be accomplished with a fiber optic coupler or tandem (beam splitting) mirror. The beam splitter or tandem mirror device is used when the monitoring of both the II image and the production of photo spot images is required.

21. 4   22. 5   23. 1   24. 3   25. 2

26. A Since the distance between the scan lines is directly related to the vertical resolution, a larger CRT monitor will display a greater scan line distance and a lower vertical resolution than with a smaller CRT monitor.

27. C The horizontal resolution of a video system is determined by the range of frequencies available in the video signal. The bandpass (bandwidth) is the range of frequencies that the video system can transmit. A higher bandpass will enable an improved horizontal resolution for a given imaging system.

28. B The vertical resolution of a video system is dependent on the number of scan lines and the size of the monitor. A standard 10" monitor using a 525-line system will provide a vertical resolution of about 370 lines or 185 line pairs. The horizontal resolution is determined by a property called the bandwidth expressed in frequency (Hz) and describes the number of times the electron beam is changed (modulated). A value of 4.5 MHz is normal for the televisions used in most fluoroscopic systems. Some of the newer digital imaging systems will use high-resolution television systems having a bandwidth as high as 20 MHz.

29. B   In order to calculate the resolution in line pairs per millimeter (lp/mm) for an image system requires the number of line pairs imaged on CRT system to be divided by the size of the image display in millimeters. In this example: $\frac{185 \text{ line pairs}}{152 \text{ mm}} = 1.21 \text{ lp/mm}$

30. D   In order to provide the same vertical and horizontal resolution with a 1000 line system will require a bandwidth which is about 4 times larger than the 3.5 MHz required with a 500 line system or 14 MHz.

1. When taking a radiograph during an ongoing surgical procedure i.e. c-arm mobile unit should be located on the:

   A. Same side as the surgical field
   B. Opposite side of the surgical field

   C. Either of the above

2. The termination of an exposure by stopping the discharge of a capacitor at some pre-selected point is termed:

   A. Field emission switching
   B. Wave-tail cutoff

   C. 3-phase switching
   D. Auto-rectification

3. In general, the type of unit with the lowest tube rating is a:

   1. Fixed installation high frequency unit
   2. Fixed installation 3-phase 6-pulse unit
   3. Battery-operated mobile

   A. 1 only
   B. 2 only

   C. 3 only
   D. 1, 2 & 3

4. After completion of a radiographic exposure, using a capacitor discharge portable unit, the generation of x-rays is possible without exposure switch activation. This can be avoided by:

   A. Discharging the capacitor after the exposure
   B. Unplugging the unit before the unit is moved

   C. Moving the tube head into the park position
   D. Closing the collimator after the exposure

5. A modern portable x-ray unit should have an exposure cord which permits the radiographer to be at least _____ from the tube assembly during the exposure.

   A. .5 meters
   B. 1 meter

   C. 2 meters
   D. 20 meters

6. Because of the somewhat limited output of many types of mobile radiographic units, the use of _____ is generally recommended to overcome this difficulty.

   A. High ratio grids
   B. Longer exposure times

   C. 400 speed screen-film combinations
   D. No tube filtration

7. Which type of portable unit is best suited for use in a hospital with a limited number of standard voltage outlets?

   A. High output mobile unit
   B. Capacitor discharge unit

   C. Battery-powered mobile unit
   D. Field emission mobile unit

8. A mobile fluoroscope is being utilized for an examination of the lumbar spine. The minimum source-to-image distance that can be used is:

   A. 7.5cm (3")
   B. 23cm(9")

   C. 30cm (12")
   D. 180cm (72")

9. The principal advantage of capacitor discharge and battery-powered mobile units is that they permit the use of:

   A. Standard voltage sources
   B. Multiple rapid exposures

   C. Unlimited technical capacities
   D. Tomographic techniques

10. The use of moderately high kVp is normally recommended when performing mobile or bedside exams for all of the following reasons EXCEPT:

A. To maintain a rapid exposure timer setting
B. To help insure an adequate penetration of the structures
C. To maintain a wide exposure latitude
D. To increase the contrast of the images

11. When radiographing patients with suspected spinal injuries, movement should be limited as much as possible but, when necessary, turning is accomplished using the:

A. Log-roll method
B. Pole lift method
C. Draw-sheet pull method
D. Sit up and bend

12. Which type of mobile x-ray unit requires "charging" before each radiographic exposure is attempted?

A. Battery-powered mobile unit
B. Fluoroscopic C-arm unit
C. Field emission mobile unit
D. Capacitor-discharge unit

13. Because of problems in obtaining proper distances and centering, the grid that is most acceptable for mobile or bedside techniques should have:

A. Low ratio and a focus grid pattern
B. Low ratio and a crossed grid pattern
C. High ratio and a focus grid pattern
D. High ratio and a crossed grid pattern

14. The termination of an x-ray exposure with a capacitor discharge mobile unit is often accomplished by a/an:

A. Grid-controlled x-ray tube
B. Synchronous timer
C. Impulse timer
D. Recycle timer

15. Manuel line voltage compensation adjustments are often necessary for proper exposures with many types of:

A. Fluorographic units
B. C-arm fluoroscopic units
C. Mobile radiographic units
D. Tomographic units

16. Recently, a new class of portable fluoroscopic units have become widely available. Tube and image receptors are attached to a movable arm and are often referred to as:

A. Fixed arm units
B. C-arm units
C. B-scan units
D. T-arm units

17. A radiographer often encounters difficulties when performing portable radiographs. Therefore, it is acceptable to avoid using:

A. Gonadal shielding
B. Protective aprons
C. Proper collimation
D. Optimal exposure techniques

18. The current waveform associated with a capacitor-discharge portable unit is most similar to that of a:

A. Half-wave rectified unit
B. Self-rectified unit
C. Full-wave rectified unit
D. High frequency unit

19. During a portable radiographic examination, a radiographer will receive the lowest amount of patient scatter by standing two meters:

A. Behind the patient
B. To the side of the patient
C. In front of the patient

20. The most commonly employed power sources for a battery-powered mobile unit are:

A. Field emission batteries
B. Alkaline batteries
C. Parallel plate capacitors
D. Nickel-cadmium cells

1. B During any surgical procedure the c-arm or mobile radiographic unit, used for the procedure, whenever possible, is placed on the opposite side of the surgical field to avoid compromising the field's sterility.

2. B As a capacitor (condenser) discharges its voltage drops in a predictable manner. The termination of the exposure at some pre-selected spot on the curve is called wave tail cutting or cutoff.

3. C Because of their smaller size and smaller available power supplies all mobile radiographic units have power ratings well below those of a 3-phase or high frequency fixed installation.

4. A A residual charge may remain in a capacitor following a radiographic exposure. This must be eliminated prior to moving the unit to avoid current leakage and the inadvertent production of x-rays without the activation of the exposure switch.

5. C In order to reduce the potential exposure to the operator during a mobile radiographic study the radiographer should wear a lead apron and stand as far from the patient and tube as possible. In order to accomplish this the exposure cord must be long enough to permit the operator to be at least 2 meters (6') from the patient or the tube during the exposure.

6. C The use of high speed (400-800) imaging system is advised for all mobile radiographic procedures to maintain low patient exposures and to provide acceptable images on large patients.

7. C In many older hospitals the availability of standard 110 or 220 volt outlet receptacles is limited. Capacitor discharge and low output mobile units which will obtain their power supplies from these receptacles often encounter difficulties. The use of self-contained battery-powered mobile units developed in the 1980's have all but eliminated this problem.

8. C The minimum source-to-image receptor distance allowable with any stationary or mobile radiographic or fluoroscopic unit shall not be less than 30 cm (12") and should not be less than 38 cm (15").

9. A In many older hospitals the availability of standard 110 or 220 volt outlet receptacles is limited. Capacitor discharge and low output mobile units which will obtain their power supplies from these receptacles often encounter difficulties. The use of self contained battery powered mobile units developed in the 1980's have all but eliminated this problem.

10. D The technical settings for most radiographic studies are similar to those used with fixed installations. With either unit the use of a high kVp value will increase the exposure latitude, ensure adequate penetration, and permit the use of shorter exposure times.

11. A With suspected injuries to the spinal cord, movement should be limited as much as possible. The log rolling method is one of the few means for safely accomplishing this task.

12. D A major drawback to the use of a capacitor discharge unit is the need to charge up the unit prior to making an exposure.

13. A In order to reduce the amount of grid cutoff, a low ratio linear focus grid with a wide distance and centering latitude is preferred.

14. A In addition to its use in cine fluorography a grid controlled (triode) x-ray tube may also be used to terminate the exposure in a capacitor discharge type mobile unit.

15. C Some older mobile (capacitor discharge) units require a line voltage compensation control to maintain a consistent voltage to the radiographic unit.

16. B Virtually all fluoroscopic units are designed with the tube and image receptor mounted on a C-shaped arm to ensure their alignment with each other.

17. D Regardless of any difficulties a radiographer may encounter with a mobile examination of a patient, it is never acceptable to ignore the basic radiation safety practices.

18. D Even though the voltage of a capacitor will drop off significantly after the first few milliseconds of an exposure, during the early part of the exposure, the voltage waveform is most similar to that of a high frequency unit.

19. B The intensity of radiation produced in a scatting object is greatest behind the object, least to the sides of the object.

20. D The rechargeable batteries most often used to power mobile radiographic units are made of nickel and cadmium compounds.

Computer Fundamentals

1. Which of the following class or categories of computers is normally associated with greatest storage capacity?

   A. Microprocessor
   B. Microcomputer

   C. Minicomputer
   D. Mainframe computer

2. The physical components of a computer system are generally described as:

   A. Software
   B. Hardware

   C. Analog devices
   D. Digital devices

3. The control center of the computer, the central processing unit (CPU), consists of:

   1. Modem   *2. Control unit*   *3. Memory unit*

   A. 1 & 2 only
   B. 1 & 3 only

   C. 2 & 3 only
   D. 1, 2 & 3

4. Most computers operate on a simple number system consisting of only the digits 0 and 1 called:

   A. Exponential system
   B. Duodecimal system

   C. Decimal system
   D. Binary system

5. The functional component of a computer involved with feeding information into the central processing unit is termed an:

   A. Input device
   B. Operating system

   C. Output device
   D. Optical printer

6. The mathematical computations or comparisons that can be made by a computer are performed by the:

   A. Cathode ray tube
   B. Read only memory

   C. Arithmetic/logic unit
   D. Random access memory

7. Thousands of individual circuit elements which are necessary for the operation of a computer, are incorporated onto a small piece of silicon called a/an:

   A. Integrated circuit chip (IC)
   B. Floppy disk

   C. CD-rom
   D. Flip-flop

8. In order for a computer to understand a set of instructions, they must first be translated into a form that is understood by the computer. These are termed:

   A. Registers
   B. Languages

   C. Addresses
   D. Compilers

9. Information that is processed by a computer is most often displayed on a device called a monitor or:

   A. Integrated circuit chip (IC)
   B. Cathode ray tube (CRT)

   C. Analog to digital converter (ADC)
   D. Central processing unit (CPU)

10. The display information as a series of letters, numbers or symbols is referred to as a/an:

    A. Graphic display
    B. Binary display

    C. Alphanumeric display
    D. Cryptic display

11. Permanent programs, dates and temporary information are normally stored in the_____ portion of the central processing unit.

    A. Input
    B. Output

    C. Memory
    D. Arithmetic/logic

Match the following group of computer hardware to the appropriate classification for questions 12-17.

A. Input device   **B. Output device**   **C. Input/output devices**

12. _____ Punch card
13. _____ Printer
14. _____ CD-rom
15. _____ Optical scanner
16. _____ Keyboard
17. _____ Cathode ray tube monitor with light pen

18. In computer terminology, a single binary digit is referred to as a:

A. Flip
B. Glitch

C. Bit
D. Chip

19. BASIC, FORTRAN, and COBOL are three of the most commonly employed:

A. Input devices
B. Computer languages

C. Hard disk drives
D. Retrieval systems

20. Each character (letter or number) of the keyboard is normally encoded by 8 bits of binary information. This will then constitute the basic unit by which the capacity of a computer is measured called a/an:

A. Mega-bit
B. Octet

C. Byte
D. Word

21. Permanent instructions (programs) that cannot be modified are stored by the:

A. Read only memory (ROM)
B. Random access memory (RAM)

C. Both of the above
D. Neither of the above

22. The interpretation of the user's instructions and the transfer and retrieval to the appropriate location is the primary function of the:

1. Control unit   2. Memory unit   3. Arithmetic unit

A. 1 only
B. 2 only

C. 3 only
D. 1, 2 & 3

23. In order to identify and direct data to the proper location in the computer memory, a unique identifying label, called a/an _____, is employed.

A. Cursor
B. Terminal

C. Bug
D. Address

24. The invisible portions of a computer consist of languages and programs that are normally referred to as:

A. Software
B. Resistors

C. Graphics
D. Buffers

25. In computer terminology, a kilobyte is the equivalent of 1024 characters that consist of _____ bits of information.

A. 128
B. 1024

C. 8192
D. 102,400

26. When information is to be added, deleted or changed by the computer, the information is stored by the:

A. Read only memory
B. Random access memory

C. Arithmetic/logic memory
D. Multiformat memory

27. When large amounts of information or data are to be stored, a secondary memory device, such as a magnetic _____, may be employed.

1. CD-rom   2. Floppy disk   3. Tape drive

A. 1 & 2 only
B. 1 & 3 only
C. 2 & 3 only
D. 1, 2 & 3

28. Electrical data transmission over conventional phone lines from computer to computer may be accomplished with the use of a/an:

A. Assembler
B. Dot printer
C. Modem
D. Daisy wheel

29. The most frequently used device for the production of a hard copy image in radiology is a:

A. Laser printer
B. Photo-spot camera
C. Floppy disk
D. Display terminal

30. Which of the following imaging modalities are largely dependent upon the computer for its acquisition and display of data?

1. Ultrasonography   2. Computed radiography   3. Magnetic resonance imaging

A. 1 only
B. 2 only
C. 3 only
D. 1, 2 & 3

Explanations

1. **D** Computers are made in three main sizes, the microcomputer (PC), the minicomputer used in many digital imaging applications and the main frame when extremely large data banks are required.

2. **B** The physical components including housing, circuits and keyboard and image display devices are termed the hardware. Software refers to the programs, languages and data processing methods that make the computer useful.

3. **C** The central processing unit (CPU) contains the primary control center (microprocessor) which manipulates the data, performs mathematical operations, and stores the information for retrieval.

4. **D** Most present day computers are only able to deal with a series of signals which translate information into on or off (flip-flop). A numerical system with only two digits (0 and 1) called the binary system is well suited for this type of operation.

5. **A** Information can be introduced into the computer by way of an input device; i.e., keyboard, scanner. CD-ROMs, floppy disks and hard drives can store or return information to the computer and are classified as input/output (I/O) devices. A printer is an example of a true output device.

6. **C** The arithmetical unit of the central processing unit (CPU) contains the mathematical programs and logic circuits required for math operations.

7. **A** Most modern computers have hundreds or thousands of tiny circuits which have been fused onto a small silicon block. These integrated circuits (IC) are normally called computer chips.

8. **D** The use of high level computer languages will require their translation into a more usable form. This is accomplished by a compiler (interpreter).

9. **B** The images related to the output are usually displayed on a cathode ray tube (CRT) or video monitor.

10. **C** The alpha numeric display consists of the letters and numbers that appear on the cathode ray tube (monitor).

11. **C** All of the new data that is entered into the computer is either stored as temporary or main memory portions of the central processing unit (CPU).

12. **A** Information can be entered into a computer by way of an input device. The main types of input devices at the this time are scanners and keyboards. Video monitors with touch screen capabilities, floppy and CD's and tape backup systems are all classified as input and output devices.

13. **B** Printers and standard CRT monitors are types of output devices.

14. **C** CD-ROM is a type of input and output.

15. **A** Optical scanners are capable of introducing printed information into the computer and are classified as input devices

16. **A** The keyboard is still the primary input device for most computers

17. **C** The cathode ray tube (monitor) that is used for the display of images is an out put devices the addition of the light pen or touch screen technology has added the ability to use this as an input device.

18. **C** The smallest quantity of computed information is carried by a single binary digit (bit). Since each alphanumeric character usually requires at least 8 bits or 1 byte of information, computer capabilities are usually based on the unit of the byte, kilobyte or megabyte.

19. B   A computer language provides the means for translating the more familiar alphanumeric system into the binary system which can be understood by the computer.

20. C   Since each alphanumeric character usually requires at least 8 bits or 1 byte of information, computer capabilities are usually based on the larger unit of the byte or kilobyte or megabyte.

21. A   The terms, random access memory (RAM) and read only memory (ROM) are applied to the data which is temporarily stored in the computer or the more or less permanent memory.

22. A   The principal control for the movement of data into other parts of a computer is the function of the control unit.

23. D   The unique label which identifies data and enables its storage or movement without effecting all other data is called an address.

24. A   Software refers to the programs, languages and data processing methods that make the computer useful.

25. C   If one Kilobyte is equivalent to 1024 x 8 bits and 1 character requires 1 byte then 1 kilobyte equals 8192 bits of information or 1024 characters.

26. B   Random access memory is used to store data that must be periodically modified.

27. B   In most computers a CD-ROM, hard disk, hard drives or tape device can be used to store large amounts of data.

28. C   The use of a modem enables the transmission of data from one computer to another over the phone lines. This device will soon enable transmission of virtually all types of diagnostic images to distinct viewing devices called teleradiology.

29. A   Most of the digital images in the radiology department are transferred into a hard copy form by a multiformat camera or laser printer.

30. D   Magnetic resonance imaging (MRI), computed tomography (CT), digital subtraction angiography (DSA), ultrasonography (US), computed radiography (CR), position emission tomography (PET), single photon emission computed tomography (SPECT), and diaphanography are among a rapidly growing list of imaging techniques that have emerged along with the modern computer.

1. The process by which a photostimulable luminescent plate is exposed to radiation and converted into a high quality two-dimensional digital image is called.

   A. Computed radiography
   B. Temporal imaging
   C. Digital subtraction
   D. Magnetic resonance imaging

2. After the initial acquisition of the image in computed radiography, which of the following is NOT one of the steps that is required for its evaluation?

   A. The identification of the image
   B. The reading or scanning of the image plate
   C. The digital processing of the image
   D. The acceptance testing of the image

3. The principal advantage of an area beam used in computed radiography as compared to a fan beam used in computed tomography is that:

   A. The area beam is associated with less scattered production
   B. The area beam provides a higher image contrast
   C. The area beam permits shorter exposure timer settings
   D. The area beam provides an improved low contrast detectability

4. Photostimulated luminescence refers to the process by which:

   A. The laser beam is converted into a usable electronic signal
   B. Radiation is absorbed into the lattice of a fluorescent crystal
   C. The photomultiplier tube converts the electronic signal into a laser image
   D. The trapped energy of the electronic latent image is released as a visible light

5. The energy of the x-ray beam is captured in a photostimulable phosphor as a series of electrons that are trapped in empty spaces in the crystal lattice called:

   A. Excited luminescence bands
   B. Meta-stable F centers
   C. Activated conduction bands
   D. Photomultiplier receptor sites

6. The release of trapped energy in a meta-stable F center of a BaFBr crystal is triggered by the exposure to visible light in the:

   A. Red spectrum
   B. Violet spectrum
   C. Green spectrum
   D. Ultraviolet spectrum

7. After a photostimulated luminescent phosphor has been scanned by the photomultiplier (PM) tube reader, the plate is exposed to a high intensity light to:

   A. Clean the transport system of any leftover positive ions
   B. Clear any of the remaining latent image from the image plate
   C. Realign the low-intensity laser mirror to the PM tube
   D. Amplify the electronic signal from the PM tube before it is passed into the CRT

8. The principal color of light emitted when a barium-fluorohalide crystal is stimulated by a helium neon laser is a/an:

   A. Yellow-green light
   B. Blue-violet light
   C. Ultraviolet light
   D. Orange light

9. The image on a photostimulable phosphor plate is not permanent and will fade at a rate of about:

   A. 10% per hour after the initial exposure
   B. 25% per hour after the initial exposure
   C. 10% within 8 hours of the exposure
   D. 25% within 8 hours of the exposure

10. The images in a computed radiographic system can be recorded or displayed in which of the following ways?

> *1. As soft-copies on CRT display*
> *2. As hard copies on a laser media*
> *3. As photographic copies of the photostimulated image*

A. 1 & 2 only   C. 2 & 3 only
B. 1 & 3 only   D. 1, 2 & 3

11. An increase in the amount of voltage that is applied to the photomultiplier (PM) tube in the CR reader will be associated with a/an:

A. Increase in the spatial resolution of the laser camera
B. Decrease in the amount of quantum noise in the CRT display
C. Improve the sensitivity of the signal
D. Decrease in the amount of scattered radiation in the BaFBr plate

12. The "reading" of the latent image on a photostimulable phosphor imaging plate is normally accomplished by a/an:

A. Electronically focused photomultiplier (PM) tube   C. Mirror deflected laser beam
B. Light guided spectrometer   D. Grid controlled magnetic beam

13. In a 12-bit analog to digital converter each pixel will have how many possible values?

A. 256   C. 1024
B. 512   D. 4096

14. In a laser or CR reader the galvanometer or rotating polygon is used to:

A. Modulate the radio frequency of the laser beam
B. Deflect the laser beam back and forth across the photostimulable phosphor plate
C. Amplify the signal detected by the light-channeling guide
D. Convert the analog signal from the photomultiplier tube into a digital signal

15. The axis of the movement of the laser across the image plate during the acquisition process is known as the:

A. Beam splitting direction   C. Integrated scan direction
B. Plate translation direction   D. Fast scan direction

16. What is the diameter of the laser beam at the surface of the photostimulable phosphor plate in a computed radiographic reader or processor?

A. 100 micrometers   C. 800 micrometers
B. 400 micrometers   D. 100 millimeters

17. The intensity of the blue-purple fluorescence emitted by the imaging plate when stimulated by a HeNe laser is directly proportional to:

A. The amount of quantum mottle present in the image plate
B. The heat that has been generated in the exposed tissue
C. The amount of the x-ray energy absorbed in the image plate
D. The speed of the image processor

18. During the reading or acquisition cycle, the direction in which the photostimulable phosphor plate moves during the scanning phase is called the:

A. Slow scan or sub scan direction   C. Interlaced or alternating scan axis
B. Spin-rotate-spin direction   D. Coordinated detection axis

19. After the activation of the F center by the laser beam, the emitted light is filtered. This enables photons with energies of:

A.100-200 nm to reach the beam splitter
B. 390-400 nm to reach the PM tube
C. 700-830 nm to reach the beam splitter
D. 800-900 nm to reach the PM tube

20. A photomultiplier (PM) tube is an important component of a computer radiographic (CR) reader that serves to:

A. Convert the red light of the laser into the blue-purple fluorescence
B. Redirecting the laser beam across the face of the imaging plate
C. Convert and amplify the photostimulated light into a measurable electronic signal
D. Filter the raw data from the image to eliminate any unwanted electronic noise

21. A picture archiving and communication system (PACS) can be used for the storage of digital images such as those in:

1. Computed radiography   *2. Magnetic resonance imaging*   *3. Computed tomography*

A. 1 & 2 only
B. 1 & 3 only
C. 2 & 3 only
D. 1, 2 & 3

22. The principal effect of the use of an insufficient amount of radiation when performing a computed radiographic examination is a/an:

A. Reduced number of gray scales in the image
B. Increase in the gray scale level of the image
C. Increase in the amount of quantum mottle on the image
D. Reduced speed of the photostimulable phosphor plate

23. The technique by which digital images may be transmitted over the phone lines for remote viewing is termed:

A. Computed radiology
B. Subtraction radiology
C. Teleradiology
D. Digital radiology

24. In order to employ the PACS system for the storage of conventional radiographic images will require their:

A. Conversion into an analog format
B. Conversion into a dynamic format
C. Conversion into a digital image format
D. Conversion into a subtraction mask form

25. With a digital examination employing energy subtraction, the principal difference in appearance of the two images is the result of acquiring images with:

A. Different milliampere-second settings
B. Different kilovoltage peak settings
C. Different source-to image receptor distances
D. Different radiographic grids

26. The adjustment of the image contrast with a digital or computed radiographic unit is accomplished through the use of the:

A. Frame integration controls
B. Window width controls
C. Mask controls
D. Time interval control

27. In order to retrofit a conventional fluoroscopic unit for digital fluoroscopy will normally require a/an:

1. Digital image processing unit
2. Upgraded image intensifier
3. Upgraded television system

A. 1 only
B. 2 only
C. 3 only
D. 1, 2 & 3

28. The major source of noise in any electronic device is the:

A. Signal-to-noise generator
B. Heated filament
C. Background radiation
D. Internal scattering

29. The organized arrangement of a digital image into columns and rows for its display on a cathode ray tube (monitor) is referred to as the:

A. Matrix
B. Signal plate
C. Noise pattern
D. Hybrid mode

30. The type of scan pattern required for the television monitor used in digital imaging systems is best described as a/an:

A. Interlaced scan
B. Skip scan
C. Progressive mode scan
D. Vertical mode scan

31. Which of the following is a type of digital image enhancement?

1. Filtering (smoothing)   *2. Edge enhancement*   *3. Mask re-registration*

A. 1 & 2 only
B. 1 & 3 only
C. 2, & 3 only
D. 1, 2 & 3

32. Most of the energy subtraction techniques are intended to demonstrate difference in the k-absorption edge between:

A. Iodine and soft tissue
B. Water and spinal fluid
C. Oxygen and intestinal gas
D. Barium and air

33. A change in the window level of a digital display is most closely related to a change in the:

A. Size distortion in the image
B. Noise level of the image
C. Shape distortion of the image
D. Digital density or brightness of the image

34. A failure to properly align the images during the digital subtraction process will result in a/an:

A. Interrogation artifact
B. Misregistration artifact
C. Archiving artifact
D. Windowing artifact

35. Which of the following type of laser is associated with the production of a coherent light with a wavelength of 633 nm?

A. Helium-neon
B. Chlorine-argon
C. Xenon-sodium
D. Hydrogen-krypton

36. A picture archiving and communication (PACS) system is best defined as a:

A. Computer assisted file and storage system for digital images
B. Digital method of recording radiographic images
C. Light stimulated method of image production
D. Satellite transmission network for analog images

37. Which of the following matrix size will be associated with the highest spatial resolution?

A. 128 x 128
B. 256 x 256
C. 512 x 512
D. 1024 x 1024

38. The individual picture elements, which contribute to the smallest segment of an image matrix, are termed:

A. Pixels
B. Bits
C. ROMs
D. Bytes

39. A digital technique which eliminates the majority of non-vascular structures by enhancement of the vascular structures is best termed:

A. Photostimulable radiography
B. Scanned projection radiography

C. Digital subtraction angiography
D. Teleradiography

40. Compared to conventional fluoroscopy, the mA employed for digital fluoroscopy needs to be somewhat higher. This is required to help:

A. Reduce patient exposures
B. Increase the size of the field of view

C. Reduce the effects of quantum mottle
D. Eliminate internal scattering in the laser

41. The sensitivity of most modern computed radiographic systems is about the same as a/an:

A. 1000 speed film-screen combination
B. 800 speed film-screen combination

C. 400 speed film-screen combination
D. 200 speed film-screen combination

42. The main advantage of a computed radiographic system is in its ability to image structures which:

A. Have a high atomic number
B. Have a relative contrast of only 1%

C. Are opaque to low energy x-rays
D. Are not penetrated by high energy x-rays

43. The gray scale value in the pixels is closely related to the:

A. Attenuation properties of the amount and density of tissue in the voxel
B. Absorption properties of the surrounding pixels
C. Atomic number of the phosphor plate
D. Type of printer used to record the image

44. During an iodinated contrast study using an energy subtraction technique the increased absorption at different energy levels is principally due to the:

A. K absorption edge of the contrast media
B. Thickness of the attenuating tissues

C. Acquisition rate of the system
D. Mass density of the tissue

45. In a computed radiographic system having a 1024 x 1024 pixel matrix what is the pixel size for a 23 cm x 23 cm field used for imaging of the hand?

A. 22 mm
B. 2.2 mm

C. .22 mm
D. .44 mm

1. A   Computed radiography is the process by which static radiographic like images can be acquired through the use of a photostimulable (luminescent) phosphor plate. In order to display this image in a digital form the plate is scanned by a laser to produce a visible light image that is collected and digitalized.

2. D   Unlike the conventional film imaging system process that is completed after the film has emerged from the processor, in a computed radiographic system, the reading of the image represents only the first of four steps that are needed to complete the imaging process that will eventually provide a completed image suitable for the interpretation by the physician. Though the various manufacturers of the CR equipment will use different names for their components the four steps are the same namely:

    1. Image identification (ID tablet)
    2. Reading of the image plate (digitizer)
    3. Digital processing station
    4. Hard and/or soft copy display (laser printer)

3. C   The main advantage of an area beam that is used for the production of normal diagnostic images is in its ability to cover a large area with an image forming radiation in a short period of time. The a fan beam, like those used with most computed tomographic units, is able to reduce nearly all scattered radiation. Therefore, often beams will provide a higher contrast image. The main disadvantage of the fan beam is its long acquisition time. The main disadvantage of the beam is its long acquisition time.

4. D   It has been found that a small number of barium-fluorohalide compounds; i.e., BaFBr, BaFCl possess a unique property called stimulated luminescence. Unlike an intensifying screen that converts the energy of the x-ray beam into visible light (fluorescence), a photostimulable phosphor stores the energy of the x-ray beam as an electronic latent image. This stores images released as visible light following its activation by a coherent light source laser beam.

5. B   When a pure crystalline compound such as barium-fluorobromide is "doped" or activated with small amounts of europium, the crystal develops a series of tiny defects called meta-stable sites or F centers in the crystal lattice. These F centers act like small "electronic holes" in the crystal that can capture or trap electrons that are formed when the exit radiation strikes the photostimulable phosphor imaging plate. In a process that is somewhat similar to the formation of latent image in a sensitized silver bromide crystal, the imaging plate stores the energy of the exit x-ray beam in the form of a latent electrical image within the F centers in the crystal lattice.

6. A   The latent electronic image of the photostimulable phosphor is excited by a laser to release a visible light that can be detected by a photomultiplier (PM) tube.

    When stimulated by a red light laser the image plate emits a blue-violet visible light phosphorescence. When the image carrying electrons of the meta-stable F centers are stimulated by a helium-neon laser the excited crystal releases its "trapped" energy in the form of a visible light. Because the intensity of the blue-violet light is directly proportional to amount of radiation reaching the imaging plate, emission spectra of varying intensities is produced.

7. B   Though the photostimulable plate is reusable it should not be reused until the pervious image has been completely removed from the plate. The exposure of the plate to a high intensity light is used to "erase" the image from the plate.

    Since the imaging plates are extremely sensitive to even small amount of background radiation a plate that has not been used for 24-72 hours may have collected a sufficient dose of radiation to cause the appearance of noise on the image. For this reason all manufacturers recommended that imaging plates that have not been used for more than 24 hours should to be erased before they are used to acquire additional images.

8. B When the image carrying electrons of the meta-stable F centers are stimulated by a red light (HeNe) laser the stimulated crystal releases its "trapped" energy in the form of blue-violet (400 nm) phosphorescence.

9. D It has be estimated that standard BaFBr:Eu imaging plate will retain up to 75% of the original electronic latent image as long as 8 hours after the exposure.

10. A Once a computed radiographic image has been acquired and has been converted into a to digital form it is normally displayed. Soft copy images are visible images that appear on the cathode ray tube or digital monitor. Soft copy images are most often stored as computer data in a PACS. Hard copy images are those that are printed as laser transparencies or on films in a multiformatted camera.

11. C An increased voltage setting of the photomultiplier tube will increase the sensitivity of the signal coming off of the phosphor plate.

12. C The reading photostimulable phosphor plate begins by removing the plate from the cassette. The plate is then moved by a series of rollers into the section where it is scanned by a mirror-deflected laser beam. The laser is rapidly moved across the plate causing the release of the stored latent image. This process in which the laser crosses the image plate laterally is termed the fast scan direction. The stimulated luminescent light given off when the laser moves across the plate is directed through a light-channeling guide on to a photomultiplier (PM) tube that records the relative intensity of the light emitted by the imaging plate.

13. D The number of gray scales that are assigned to a particular pixel is a function of the capabilities of the computer to process digital information. In most modern systems the ability to assign gray scale values can be approximated by the code values of the analog-to-digital converter (ADC). For example, in the early digital imaging system, the use of an 8 bit ADC would permit the 256 individual shades of grays for each of the pixels within the matrix. A typical CR system operating near the millennium will have a 10 bit ADC allowing for up to 1024 shades or a 12 bit ADC permitting as many as 4096 shades for each individual pixel.

14. B During the reading of the image plate, the laser is moved rapidly across the plate through the deflection by a rotating polygon or galvanometer causing the release of the stored latent image.

15. D This process, in which the laser crosses the image plate laterally, is termed the fast scan direction. The direction in which the imaging plate travels is the slow scan of sub-scan direction.

16. A The width of the laser beam in the reader is designed to be about the same dimension as the pixel or about 100 micrometers.

17. C The intensity of the light emitted by the imaging plate when stimulated by the laser is directly proportional to the amount of x-radiation absorbed in the plate.

18. A The direction in which the photostimulable phosphor plate travels through the reader is called the slow scan or sub-scan direction.

19. B Optical filters are used in the reader to filter the light reaching the photomultiplier tube this permits only photons having a wavelength of 390-400 nm (blue-violet region) to be recorded.

20. C A photomultiplier is an electronic tube that converts visible light into measurable electronic signal. Many devices such as scintillation counter radiation detectors and the detectors in CT unit will use this type of tube.

21. D Picture archiving and communication systems (PACS) are ideally suited for the storage of images which are generated in digital form such as CT, CR, MRI, DSA, SPECT, PET and US. This type of storage system is sometimes referred to as an electronic file room.

22. C An insufficient amount of radiation with a computed radiographic system will result in a high amount of quantum mottle (noise) that is seen as grainy image. The image quality or image scales should be maintained by the unit even at low dose levels.

23. C One of the most practical applications of digital images is in their ability to be transmitted electronically over the phone lines in a process known as teleradiography.

24. C   The standard hard copy analog radiographic image is not compatible with a digital storage system (PACS). In order to utilize the PAC system the radiograph must first be converted by a scanner or digitalizer into a digital form that can be understood by the computer.

25. B   Energy subtraction can be applied to two images that were taken at different kilovoltage peak levels to see the differences in the attenuation of the tissues

26. B   The gray scale of a digital image can be modified by the adjustment of the window level and window width controls. A narrow window width tends to increase the relative difference in the gray scales and create images that appear to have a higher contrast. The window level controls are used to bring different tissues into the visible range based on their attenuation properties.

27. D   Though in principle any conventional televised fluoroscopic system could be converted over into a digital system, in practice, the need for lower noise levels and improve a display quality will require the replacement of both the image intensifier and the entire television chain as well as the addition of the digital image processing unit.

28. B   Most electronic imaging devices such as television cameras, cathode ray tubes and image intensifiers, are associated with high levels of electronic noise. The use of high potential differences and heated filaments in these devices will result in small but detectable amount of electrical interference that is seen as noise (snow, grain) on the image display.

29. A   A digital image is displayed as a series of tiny picture elements (pixels) which are arranged in vertical and horizontal columns called the image matrix when displayed on a cathode ray tube (monitor).

30. C   The high resolution required in the television chain of a computed radiographic or digital system is in part accomplished by the use of a cathode ray tube that uses a progressive scanning mode rather than an interlaced mode.

31. D   A large number of techniques are used with digital imaging techniques to improve the quality of the image. These are sometimes referred to as post processing techniques because they involve manipulation of the image after it has been obtained. In most digital and CR systems these systems will include smoothing controls that help to eliminate image noise, edge enhancement controls for improving the appearance of small details and mask re-registration for reducing the motion or non-alignment between subtraction and post-injection images.

32. A   Temporal or energy subtraction is best suited for distinguishing between structures filled with iodinated contrast agents and soft tissue structures.

33. D   The gray scale of a digital image can be modified by the adjustment of the window level and window width controls. A narrow window width tends to increase the relative difference in the gray scales and create images that appear to have a higher contrast. The window level controls are used to bring different tissues into the visible range based on their attenuation properties. This is similar to the effects that are seen when the optical density or image brightness are changed.

34. B   Mask re-registration is related to the technique by which two images are aligned during a subtraction technique. The failure to maintain this alignment (non-alignment) between subtraction and post-injection images is called a misregistration artifact.

35. C   The red light (600 nm) required to scan the image plate is closely approximated by the 633 nm output of the helium neon laser.

36. A   Picture archiving and communication systems (PACS) are ideally suited for the storage of images which are generated in digital form such as CT, CR, MRI, DSA, SPECT, PET and US. This type of storage system is sometimes referred to as an electronic file room because of its ability store images and reports.

37. D   The size of the individual picture elements or pixels in a digital image is the most important factor in the ability to produce images that appear sharp or have what is called a high spatial resolution. Images composed of a large number of pixels with small dimensions will more accurately represent the tissues of interest than an image that is composed of a smaller number of larger pixels.

38. A Not unlike the dots that can be shown to make up the picture in a newspaper, the digital image is made up of thousands of tiny picture elements called pixels.

39. C The enhancement of vascular structure by a digital imaging system is normally accomplished through the use subtraction angiography.

40. C In both digital fluoroscopy and computed radiography there may be a considerable amount of quantum mottle present in images produced at low mA or mAs values. Therefore it is likely that the dose to the patient may need to be doubled during these procedures.

41. D At the present time the relative speed of a computed radiographic system is approximately the same as a 200 speed film-screen imaging system. The failure to increase the techniques during CR exposures will increase the grain.

42. B Though digital image cannot at the present time provide the high spatial resolution of a film-screen imaging system it can display much smaller differences in subject contrast of the tissues. This factor called low contrast delectability (contrast resolution) is digital imaging system's greatest advantage. A film-screen imaging system requires a contrast difference of at least 10% between adjacent tissues to be able to demonstrate these as separate structures. A digital imaging system to assign different gray scale values to tissues with a similar attenuation coefficient values. This enables the perception of tissue contrast differences as small as 1%. The minimum difference in tissue attenuation coefficient values that can be seen on an analog system is 10%.

43. A The gray scale value of the pixel is directly related to the attenuation properties of the thickness and physical density of the tissue within the voxel.

44. A The differences in absorption that takes place in tissues that are or are not filled with contrast agents is best demonstrated through the differences in the k absorption edge.

45. C The relationship of the pixel size to the size of the matrix and field of view is expressed by the relationship

$$\text{Pixel size} = \frac{\text{Field of view (FOV)}}{\text{Size of the matrix}}$$

This formula shows that the size of the pixel is directly related to the field of view and is inversely related to the size of the matrix.

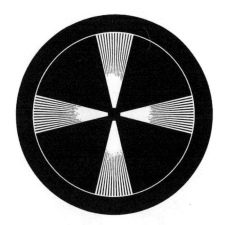

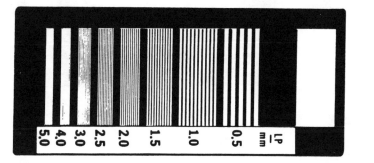

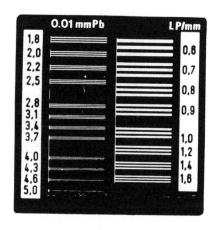

Radiographic Graffiti II

Can you identify these common patient or technical related artifacts? (Answers appear at the bottom of the page.)

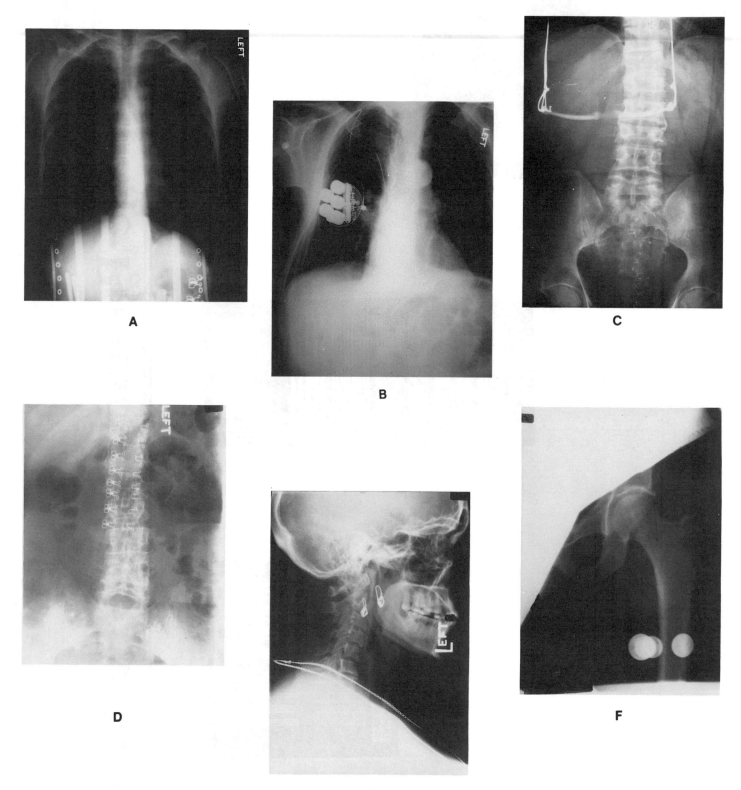

Radiographic Quality

1. The amount of darkening from the black metallic silver deposited in the film is termed:

    A. Radiographic contrast
    B. Optical density
    C. Attenuation coefficient
    D. Base plus fog level

2. The primary function of the milliampere-seconds during a radiographic exposure is:

    A. The regulation of the spatial resolution in the image
    B. The control of the scattered radiation produced in the patient
    C. The control of the penetrability of the beam
    D. The regulation the optical density in the image

Referring to the diagram, answer questions 3 & 4.

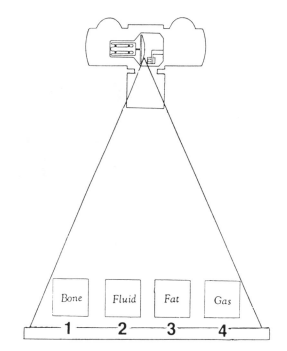

3. Which of the corresponding areas of the radiographic image is likely to show the highest optical density?

    A. Number 1
    C. Number 3
    B. Number 2
    D. Number 4

4. The greatest amount of attenuation has occurred in the region corresponding to tissue:

    A. Number 1
    C. Number 3
    B. Number 2
    D. Number 4

5. The primary exposure factor that is used for regulating radiographic contrast is:

    A. Source-to-image receptor distance
    B. Size of the focal spot
    C. Kilovoltage peak
    D. Milliampere-seconds

6. When a barium-type contrast agent is employed, the desired degree of penetration can be accomplished by using a kilovoltage setting in the range of:

    A. 50-60 kVp
    B. 60-80 kVp
    C. 75-85 kVp
    D. 90-110 kVp

7. A radiographic image is said to possess excessive optical density. This image would appear to be:

    A. Lighter than normal
    B. Darker than normal
    C. To have excessive contrast
    D. To have poor recorded detail

8. A reduction of scattered radiation on a radiographic image is likely to occur with a/an:

A. Increased kilovoltage peak
B. Increased collimation

C. Increased part thickness
D. Decreased focal spot size

9. In general, an increase of 15% in kilovoltage will necessitate a:

A. 15% decrease in the mAs to maintain the same optical density
B. 50% increase in the mAs to maintain the same optical density
C. 50% decrease in the mAs to maintain the same optical density
D. 15% increase in the mAs to maintain the same optical density

10. Which of the following is an example of a pathological change in a patient that is likely to require an increase in penetration to maintain the desired optical density:

1. Pulmonary effusion   **2. Acromegaly**   **3. Aseptic necrosis**

A. 1 & 2 only
B. 1 & 3 only

C. 2 & 3 only
D. 1, 2 & 3

Referring to the diagram of the step wedge film, answer questions 11 & 12.

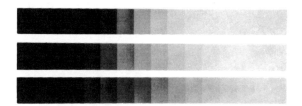

11. In the diagram of the step wedge film which number corresponds to the film taken at the highest kVp?

A. 1
B. 2

C. 3

12. Which of the step wedge films best demonstrates a short scale contrast?

A. 1
B. 2

C. 3

13. An abdominal radiograph is taken of a patient measuring 21 cm. If all the same conditions are maintained and a second image is taken on a different patient measuring 29 cm, the image will have a:

1. Reduced optical density   2. Reduced radiographic contrast   3. Reduced spatial resolution

A. 1 only
B. 2 only

C. 3 only
D. 1, 2 & 3

14. The dose a patient receives during a diagnostic exposure is directly proportional to the:

1. mAs employed   2. kVp employed   3. Focal spot size selected

A. 1 only
B. 2 only

C. 3 only
D. 1, 2 & 3

15. A decrease in the amount of scatter produced in a tissue will occur as:

A. Object-to-image receptor distance decreases
B. Kilovoltage peak setting is decreased

C. Source-to-image receptor distance decreases
D. Field of view increases

16. Which of the primary exposure factors has the greatest effect on beam quality, attenuation, and exposure latitude?

A. Milliampere setting
B. Timer setting
C. Milliampere-seconds value
D. Kilovoltage peak setting

17. A radiographic image is produced with insufficient contrast and excessive optical density. Which of the following changes in the exposure factors will best correct for this error?

A. Increase the mAs, decrease the kVp
B. Increase the mAs, increase the kVp
C. Decrease the mAs and increase the kVp
D. Decrease the mAs and kVp

18. An increase in the exposure factors to maintain the desired optical density is often required for patients with:

A. Emphysema
B. Hydrothorax
C. Osteoporosis
D. Leprosy

19. An increase in the number of x-ray photons reaching the image receptor is normally associated with a/an:

A. Higher image contrast
B. Increased optical density
C. Improved spectral matching
D. Reduced exposure latitude

20. Which of the following normal human tissues will have the lowest absorption coefficient to x-ray photons in the 50-120 kVp range?

A. Compact bone tissue
B. Connective tissue
C. Free air or gas
D. Muscle tissue

21. As the wavelengths of the photons in an x-ray beam are increased the:

A. Scatter produced by the beam increases
B. Energy of the beam increases
C. Beam penetration increases
D. Beam penetration decreases

22. Radiographic contrast can be defined as the:

A. Difference in the optical densities between adjacent areas of the radiograph
B. Overall blackening appearing on the radiographic image
C. Base plus fog blackening of the radiographic image
D. Ratio of film speed to screen speed of the imaging system

23. The kilovoltage peak setting is most closely related to the:

A. Quantity of the radiation and spatial resolution appearing on the radiographic image
B. Radiographic contrast of the image and penetration of the beam
C. Motion and focal spot blur on the radiographic image
D. Field of view and the dynamic range of the radiographic image

24. Many skeletal disorders will require an increase in technique due to an overgrowth of bone tissue. These are referred to as:

A. Osteoblastic conditions
B. Osteolytic conditions
C. Osteoporotic conditions
D. Osteonecrotic conditions

25. An example of a pathologic change in a patient that will require an increase in penetration to maintain the desired image quality is:

1. Pneumoconiosis   *2. Empyema*   *3. Ascites*

A. 1 & 2 only
B. 1 & 3 only
C. 2 & 3 only
D. 1, 2 & 3

26. Which of the following will give the same mAs as a 400 mA, .2-second exposure?

A. 1000 mA, 800 milliseconds
B. 500 mA, 50 milliseconds

C. 200 mA, 200 milliseconds
D. 100 mA, 80 milliseconds

27. A radiographic image is produced using 20 mAs and 74 kVp. Which of the following are likely to occur on a second image taken with 15% increase in the kVp value?

A. The optical density of the first image will exceed that of the second
B. The radiographic contrast of the first image will exceed that of the second
C. The optical density will be similar for both images
D. The focal spot blur of the image will increase

Referring to the pair of radiographic images, answer questions 28 & 29.

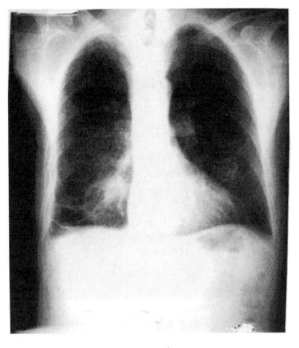

A

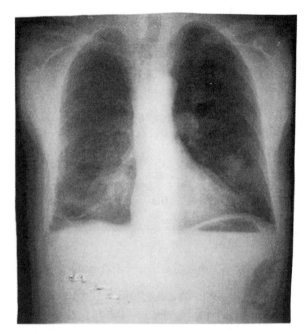

B

28. Which radiographic image appears to have the highest radiographic contrast?

29. Which radiographic image appears to have the least optical density?

30. The milliampere-seconds value employed during an exposure is:

A. Directly proportional to the quantity of x-rays produced in the exposure
B. Inversely proportional to the quantity of x-rays produced in the exposure
C. Directly proportional to the radiographic contrast of the image
D. Inversely proportional to the speed of the electron stream in the tube

31. A radiographic image is produced of a high tissue density structure. The area of the radiographic image corresponding to this structure will show a:

A. High optical density
B. Low optical density

C. Low radiographic contrast
D. Low spatial resolution

32. A radiograph that shows a short scale contrast often results from the use of:

A. Breathing techniques
B. Tilted anode techniques

C. Low kilovoltage techniques
D. Falling load techniques

33. Which of the following sets of technical factors will result in a radiographic image with the least amount of involuntary motion?

	mA	Time	kVp
A.	200	100 ms	80
B.	400	.05 sec.	80
C.	800	.03 sec.	80
D.	800	75 ms	70

34. The utilization of high kVp (above 90 kVp) for radiographs of the abdomen will result in:

A. High exposure latitude and short scale contrast C. High exposure latitude and long scale contrast
B. Low exposure latitude and short scale contrast   D. Low exposure latitude and long scale contrast

35. When a fixed kVp technique chart is employed for the selection of exposure factors what is the expected result from underestimating the physical density of tissue in the patient?

A. A radiographic image with an insufficient optical density
B. A radiographic image with an excessive optical density
C. A radiographic image with an excessive radiographic contrast
D. A radiographic image with an increased amount of motion

36. The appropriate degree of penetration for a small body part (finger-hand) is normally accomplished by using a kilovoltage setting in the range of:

A. 55-65 kVp
B. 85-95 kVp

C. 95-105 kVp
D. 115-125 kVp

37. Which of the following is likely to occur if kilovoltage is increased for a radiographic exposure?

1. Increased optical density   *2. Lower penetration*   *3. Longer scale contrast*

A. 1 & 2 only
B. 1 & 3 only

C. 2 & 3 only
D. 1, 2 & 3

38. A beam with a low penetrability is often referred to as a soft beam or:

A. High intensity beam
B. High quality beam

C. Low quality beam
D. High-resolution beam

39. Which of the following substances possesses the highest degree of radiopacity?

A. Calcium
B. Aluminum

C. Lead
D. Oxygen

40. A radiographic image produced showing a small degree differential attenuation, is best described as having a:

A. Low radiographic contrast
B. High optical density

C. High spatial resolution
D. High focal spot blur

41. In order to see a noticeable change in the optical density, a change of _____ in the mAs is required.

A. 5%
B. 10%

C. 15%
D. 30%

42. A radiographic image of a child has been obtained using 100 mA 20 ms at 74 kVp, but due to movement, the radiographic image is unacceptable. In order to reduce motion, a 60% reduction in time is desired. What new timer setting should be employed?

A. 8 ms
B. 12 ms

C. 60 ms
D. 120 ms

43. Which of the following sets of technical factors will result in a radiographic image with the greatest amount of involuntary motion?

	mA	Time	kVp
A.	200	.02 sec.	86
B.	300	10 ms	84
C.	400	.13 sec.	79
D.	800	50 ms	76

44. An increase in technique is usually required to maintain normal optical density for patients with:

A. Giant cell sarcoma
B. Leprosy

C. Osteoporosis
D. Acute kyphosis

45. When a substance is exposed to diagnostic x-rays, the amount of attenuation is primarily dependent upon the _____ of the attenuator.

A. Atomic number and thickness
B. Electron density and valence

C. Viscosity and clarity
D. Proton-neutron ratio

46. The optical density corresponds to the amount of light from the view box that is _____ the radiographic film.

A. Transmitted through
B. Scattered by

C. Reflected by
D. Absorbed by

Referring to the image of the wrist, answer questions 47-49.

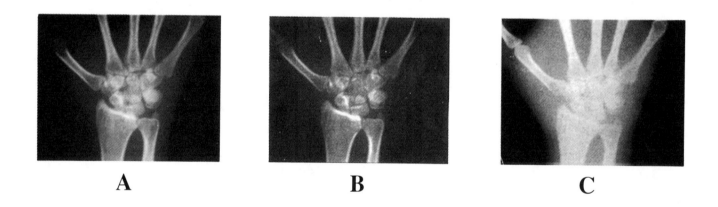

A   B   C

47. Which radiographic image was taken using the highest mAs value?

48. Which radiographic image was produced from the lowest exit mAs value?

49. Which radiographic has the most desirable optical density?

50. A technique of 200 mA, 100 milliseconds and 86 kVp is required to produce a radiographic image with the desired optical density at a 180 cm source-to-image receptor distance. What new mAs could be used to maintain the same optical density on a second exposure if a 800 mA station was desired?

A. 25 milliseconds
B. 50 milliseconds

C. 75 milliseconds
D. 150 milliseconds

51. Which set of technical factors should be used on a small child's abdomen in order to limit the amount of involuntary motion?

	mA	Time	kVp
A.	100	.2 sec.	70
B.	200	100 ms	70
C.	400	.05 sec.	70
D.	800	20 ms	70

52. When developing a variable-kilovoltage technique chart, an increase of _____ should be used for every centimeter of thickness difference.

A. 2 kVp
B. 15% in kVp

C. 39% of mAs
D. 50% of mAs

53. Any material that greatly reduces the number of incident x-ray photons passing through the material is classified as:

A. Isotropic
B. Radiopaque

C. Radiotransparent
D. Radiolucent

54. The principal method of controlling voluntary motion is by the use of:

A. Sandbags and adhesive tape
B. Proper patient instructions

C. Restraining devices
D. Longer exposure times

55. An acceptable anteroposterior abdominal radiograph is taken using 50 mAs at 80 kVp. If the patient is placed in the oblique position, with no change in technique, the radiograph will possess:

A. Excessive optical density
B. Insufficient optical density

C. Excessive radiographic contrast
D. Insufficient radiographic contrast

56. A satisfactory radiographic image of the thoracic spine was obtained using an exposure of 400 mA, 100 ms, 83 kVp and a source-to-image receptor distance of 100 cm. Assuming no change to the other factors, which set of factors will produce a second image showing a similar optical density?

A. 200 mA, 20 ms, 96 kVp
B. 400 mA, 20 ms, 83 kVp

C. 800 mA, 50 ms, 83 kVp
D. 800 mA, 50 ms, 96 kVp

57. A radiographic image was taken on a patient with ascites (fluid in the abdominal cavity). What is the most likely cause of the an insufficient optical density on the image?

A. An excessive penetration by the beam due to the fluid
B. The excessive scattering of the beam by the fluid
C. The excessive attenuation of the beam by the fluid
D. An excessive movement of the fluid

58. The principal controlling factor for the amount of photons produced during a radiographic exposure is the:

A. Milliampere-seconds value
B. Amount of filtration used for the beam

C. Kilovoltage peak setting
D. Area covered by the beam

59. Which of the normal body tissues is usually associated with highest amount of beam attenuation for an 80 kVp exposure?

A. Adult bone tissue
B. Adipose tissue

C. Cardiac muscle tissue
D. Air-filled lung tissue

60. The amount of penetration of an x-ray beam is dependent upon all the following EXCEPT:

A. The thickness of the tissue being imaged
B. The energy of the photons in the beam

C. The number of x-ray photons in the beam
D. The physical density of the tissue being imaged

61. An increase in the optical density on a radiographic image will result from a:

    A. 20 cm increase in the source-to-image receptor distance
    B. 15% decrease in the kilovoltage peak setting
    C. 50% increase in the milliampere setting
    D. 50% increase for filtration

62. What is the most likely cause of excessive optical density of the radiographic image taken on a patient with an abdominal obstruction?

    A. Excessive scatter due to increased amount of fecal material
    B. Excessive absorption of the x-rays by the intestines
    C. Excessive attenuation by the obstructing tissues
    D. Excessive penetration of the intestinal gas

63. Which of the following exposure settings will provide a value of 40 mAs?

    A. 200 mA and 100 ms   C. 400 mA and 100 ms
    B. 100 mA and 40 ms   D. 400 mA and 10 ms

64. The principal controlling factors for the amount of photons produced during a radiographic exposure are the:

    A. kVp and source-to-image receptor distance
    B. mA and the exposure time
    C. mAs and kVp
    D. Exposure time and source-to image receptor distance

65. In order to maintain a similar optical density, what change would be required to compensate for a 50% reduction in the exposure timer setting?

    A. A 50% reduction in the source-to-image receptor distance
    B. A 15% reduction in the kilovoltage peak
    C. A doubling of the milliampere setting
    D. A doubling of the kilovoltage peak setting

66. The region of a radiographic image that corresponds to a tissue with a low tissue density will normally have:

    A. High optical density or appear dark   C. Low optical density or appear light
    B. High optical density or appear light   D. Low optical density or appear dark

67. All of the following may effect the x-ray absorption characteristic of a tissue EXCEPT:

    A. The age of the patient
    B. The degree of muscular development of the patient
    C. The type of pathologic condition the patient may have
    D. The type of imaging system to record the image

68. Which of the following changes is likely to increase the percentage of scatter produced by a structure?

    A. A decrease in the kilovoltage setting
    B. An increase in the source-to-image receptor distance
    C. A decrease in the physical density of the tissue
    D. An increase in the thickness of the tissue

69. The appropriate degree of penetration for a large body part (lateral lumbar spine) is normally accomplished by using a kilovoltage setting in the range of:

    A. 45-55 kVp   C. 85-95 kVp
    B. 65-75 kVp   D. 105-115 kVp

70. The various shades or tones on a radiographic image that represent the tissues are called:

A. Absorption coefficients
B. Image contrasts
C. Optical densities
D. Penetration paths

71. All of the following are likely to require a decrease in the exposure to maintain the desired optical density on the radiographic image EXCEPT:

A. Osteoporosis
B. Emphysema
C. Paget's disease
D. Pneumothorax

72. Which of the following change in the exposure factors is normally associated with modification in the scale contrast?

A. Kilovoltage peak setting
B. Source-to-image receptor distance
C. Milliampere setting
D. Size of the focal spot

73. A satisfactory radiographic image of the abdomen was obtained using an exposure of 300 mA, 100 ms, 76 kVp and a source-to-image receptor distance of 100 cm. Which set of exposure factors could be used to produce a second image showing a similar optical density but with a reduced exposure?

A. 300 mA, 200 ms, 65 kVp
B. 600 mA, 100 ms, 87 kVp
C. 300 mA, 50 ms, 87 kVp
D. 600 mA, 50 ms, 65 kVp

74. A radiographic image that possesses a small number of widely varying optical density shades is described as having a:

A. Low subject (tissue) contrast
B. High window level
C. Short scale contrast
D. Long scale contrast

75. The percentage of scatter produced during an imaging process is dependent upon all of the following EXCEPT:

A. Kilovoltage peak value
B. The physical density of the tissue
C. Thickness of the body part
D. The timer setting employed

76. A PA chest radiograph is taken of a patient measuring 26 cm. If all the same conditions are maintained and a second image is taken on a different patient measuring 20 cm, the image will appear to have:

A. An increased optical density
B. A reduced amount of voluntary motion
C. A reduced radiographic contrast
D. Longer scale contrast

77. A technique of 600 mA, .2 seconds and 79 kVp has been used to provide an image with the desired optical density. What new timer setting will provide a similar optical density if a 300 mA station is selected?

A. 100 milliseconds
B. 400 milliseconds
C. 800 milliseconds
D. 1600 milliseconds

78. The timer setting is the principal controlling factor for:

A. Radiographic contrast and spatial resolution
B. Image brightness and penetration
C. Optical density and motion
D. Focal spot blur and magnification

79. The use of 15% kilovoltage rule in the maintenance of optical density as mAs is halved offers the advantage of:

1. Smaller patient dose   *2. Greater exposure latitude*   *3. Shorter exposure times*

A. 1 only
B. 2 only
C. 3 only
D. 1, 2 & 3

80. A radiographic image is produced of a homogeneous object. The resultant image will show a:

A. Long scale contrast
B. Short scale contrast
C. It will show no contrast

81. Most gasses such as nitrogen, oxygen and hydrogen, have a low attenuation for x-rays and are classified as:

A. Radiogeneous   C. Radiopaque
B. Radiolucent   D. Radioactive

Referring to the following statement, answer questions 82-87.

A technique of 300 mA, 100 ms and 75 kVp is employed to produce an image with the desired optical density. Using the principles stated in the 15% kVp rule, fill in the values below that will maintain a similar optical density.

	mA	Time	kVp
82.	300	_____	86
83.	600	_____	64
84.	300	200 ms	_____
85.	100	150 ms	_____
86.	_____	25 ms	86
87.	_____	600 ms	64

Using the information provided, answer questions 88-90.

According to the principles of the 15% kVp rule, all three exposures will produce a similar optical density.

	mA	Time	kVp
A.	300	50 ms	92
B.	100	300 ms	80
C.	600	100 ms	68

88. Which set of techniques would be associated with an image possessing the lowest radiographic contrast?

89. Which set of techniques would be associated with a patient receiving the highest patient exposure?

90. Which set of techniques should be employed to reduce the effects caused by involuntary motion?

1. B The optical density or the amount of blackness in the image is directly proportional to the amount of black metallic silver that formed when the silver bromide crystals in the film are exposed to the light emitted by the intensifying screens and the direct action of the x-rays. The action of the reducing agent in the developer solution will speed the conversion of the exposed silver bromide crystals into the black metallic silver that make up darker regions of the radiographic image.

2. D The milliampere-seconds (mAs) value determines the amount of electrons that move across the x-ray tube. The mAs and is directly related to the quantity of photons produced in the x-ray beam. The mAs value is strongly related to the exposure received by the patient and the optical density of the radiographic image.

3. D The optical density on any portion of the radiographic image is related to the degree to which the bean is attenuated by a tissue. An equal thickness of the four normal human tissues will attenuate the beam based on their relative physical density and the effective atomic number. The air that fills the normal lung or the gas found in the intestines have a low physical density and atomic number. Air and gas attenuates only a small amount of radiation. The low attenuation of gas corresponds to a region of the image that will have a high optical density.

4. A Normal adult bone has a relatively high density and atomic number and will absorb or attenuate more of the beam than fluids fats or gasses. A tissue that attenuates a large portion of the beam will appear to have a low optical density on the image.

5. C The kilovoltage peak (kVp) value controls the speed at which the electrons pass across the x-ray tube and the energy and penetrability of the resulting photons in the beam. A higher kVp setting is associated with an increased optical density, greater degree of penetration, and the reduction in the relative attenuation that occurs between the adjacent tissues. This last factor results in a reduction of the radiographic contrast in the image.

6. D In order to see through a barium type contrast agent will require a high amount of penetration. This is normally accomplished by the use of kilovoltage peak setting in the 90 - 110 kVp range.

7. B A radiographic image or region of an image that appears dark is said to have a high or excessive optical density. These regions are associated with areas of the film that have been struck by large amounts of radiation.

8. B Scattered radiation is an unavoidable consequence of the interaction of x-rays and/or other types of radiation with matter. The percentage of radiation that is scattered is related to three principal factors. The first factor that effects the percentage of scattered radiation produced in the tissue is the kVp setting. It has been found that a higher kVp setting is associated with an increase in the amount scattered radiation production. The two others factors that are related to the amount of scattered radiation production are the physical density and the thickness of the tissues. Tissues that have a high physical density and greater thickness will create a much greater percentage of scattered radiation than low density or thinner tissues.

9. C The 15% kVp rule is used to equate the change required in the amount of kVp that will result in the same change in optical density as a doubling or halving (50%) of the mAs value. An increase of 15% in the kVp will require about a 50% reduction in the mAs to maintain the same optical density.

10. A Many pathological conditions have a considerable effect on the physical density of a tissue and the resulting optical density. Conditions associated with excessive bone growth such as acromegaly or those associated with the production and buildup of fluids like pulmonary effusion will often require an increase of about 30-50% in the mAs value to maintain the desired optical density. Aseptic necrosis is a condition in which the bone begins to die since this reduces the physical density of the tissue. Therefore a reduction in the exposure setting is required to maintain the desired optical density.

11. C   The increased number of density steps in the image 3 indicates this film has a long scale or low radiographic contrast. This appearance is associated with the greater penetration that occurs from a higher kVp setting.

12. A   The decreased number of density steps in image 1 is associated with less penetration, a lower kVp and a short scale contrast.

13. D   An increase in the thickness of the tissue will have a number of negative imaging consequences, including an increase in the amount of scattered radiation, a reduction in the radiographic contrast of the image, a higher primary beam attenuation and the resulting loss of optical density. The increased object-to image receptor distance in the larger patient will also increase the amount of image unsharpness and magnification reducing the spatial resolution of the image.

14. A   The exposure to a patient (patient dose) is directly proportional to the number of x-ray photons to which the patient is exposed. Since the milliampere-seconds value is the primary controlling factor for the number of x-rays produced the mAs is also directly proportional to the exposure.

15. B   Though scattered radiation is directly effected by the patient's thickness and physical density, the main exposure factor that effects the amount of scattered radiation is the kVp kilovoltage peak (kVp).

16. D   The kilovoltage peak is perhaps the single most important exposure factor since its modification effects beam quality, penetrability, attenuation, image contrast, optical density, and the exposure latitude.

17. D   A lower radiographic contrast combined with an excessive optical density will require a reduction in penetration (kVp), provide a higher contrast and a decrease in mAs value to reduce the optical density of the image.

18. B   Any pathological condition associated with excessive fluid production such as a hydrothorax or buildup of fluid in the tissues will require a 30 - 50% increase in the mAs value to compensate for the increased attenuation of the beam. Emphysema that results from the destruction of the lung tissue, osteoporosis that is associated with a loss of bone mass and leprosy will result in the reduction of the body mass. Therefore, all of these conditions would require a decrease in the exposure to maintain the usual optical density.

19. B   As the number of x-ray photons passing through the patient and hitting the imaging system increases, more energy is transferred into the film emulsion. This will increase the amount of darkness or optical density of the image.

20. C   Tissues with relatively low atomic and physical density such as air and gas are more easily penetrated by an x-ray beam. These tissues will therefore have the lowest absorption or attenuation coefficients to photons in the diagnostic energy region.

21. D   The distance between the crests of a photon, the wavelength, is inversely proportional to the energy of the photon. Therefore, as the wavelength increases, the energy decreases causing less scatter production, less penetration, higher beam attenuation and an image having a higher radiographic contrast.

22. A   The relative difference in optical densities between adjacent areas of a radiograph is termed radiographic contrast.

23. B   Both the radiographic contrast and beam penetrability are related to the energy of the photon. the primary controlling factor for the energy of the beam is the kVp.

24. A   Because most osteoblastic disorders are related to a substantial increase of bone growth they will tend to increase the attenuation of the tissues. Therefore, an increase in the mAs value is required to maintain the appropriate optical density of the image. Osteolytic, osteoporotic, and necrotic conditions reduce the physical density orf the bone and are more easily penetrated by the beam.

25. D   The excess fluids seen in pneumoconiosis, empyema, ascites are all conditions that will benefit from an increase in the usual exposure to compensate for a greater radiation attenuation. An increase in the mAs or kVp will provide the appropriate optical density.

26. D The mAs (80) is the product of the mA (400) and the time (.2).

   A= 800 mAs B= 25 mAs C= 40 mAs

27. B An increase in the kVp is associated with increase in the penetration of the beam, an increase in the optical density and a reduction in the radiographic contrast. Therefore the original image taken at a lower kVp will have the higher contrast.

28. A The image labeled A has a short scale or higher image contrast.

29. B Image B has a great deal of scattered radiation that has caused a loss in the radiographic contrast of the image. This image also has a lower optical density than image A.

30. A The milliampere-seconds (mAs) value is the primary controlling factor for the number or quantity of x-ray photons produced during the exposure. The mAs value is directly proportional to the quantity of the x-ray photons that are produced.

31. B The optical density on any portion of the radiographic image is related to the degree to which the bean is attenuated by a tissue. Normal adult bone has a relatively high density and atomic number and will absorb or attenuate a large percentage of the beam. A tissue that attenuates a large portion of the beam will appear to have a low optical density on the image.

32. C A high contrast image consisting of a small number of density shades (short scale) with a wide density difference between them is most commonly seen when low kVp exposures are employed.

33. C The shortest exposure time setting will be associated with the least involuntary motion appearing on the radiograph. In order to answer this question all the timer settings must be converted into milliseconds A= 100 ms B= 50 ms C= 30 ms D= 75 ms

34. C A higher kilovoltage setting is associated with increased exposure latitude and a longer scale contrast.

35. A The exposure settings selected from a technique chart will provide the desired optical density for the patient based on the assumption that the patient has no pre-existing pathological condition. If the radiographer fails to adjust for a condition that increases the density of the tissues, the resulting image will have an insufficient optical density.

36. A The use of lower kilovoltage settings for smaller body parts will help prevent over-penetration and maintain the high contrast desired in these images. The usual kVp setting for an adult finger or hand is in the range of 55-65 kVp.

37. B A higher kilovoltage (kVp) setting will result in a higher penetration, wider exposure latitude, lower contrast, long scale contrast, increased scattered radiation production and a higher optical density.

38. C An x-ray beam with a low quality (low kVp) has a low penetrability and is sometimes referred to as a soft x-ray beam.

39. C Because the substance with the highest atomic number will attenuate the greatest amount of radiation, lead, with an atomic number of 86, will have the higher degree of radiopacity than calcium (20), Aluminum (13), or oxygen (8).

40. A A small difference in the attenuation properties of a tissue will provide only small differences in the optical densities and produce a low or long scale contrast.

41. D A significant change in optical density will require a change of 25-30% in the mAs value or about 6% change in the kVp.

42. A In order to accomplish a 60% reduction in time, the initial timer setting is multiplied by .6. This value is then subtracted from the original timer value. 20 ms - 12 ms = 8 ms

43. C The longest timer setting would be associated with the greatest amount of involuntary motion. In order to answer this question all the timer settings must be converted into milliseconds A= 20 ms B= 10 ms C= 130 ms D= 50 ms

44. D The marked expansion in the diameter of the chest cavity in patients with acute kyphosis (humpback) will often necessitate an increase in technique to maintain the appropriate optical density. Leprosy, giant cell tumors, and osteoporosis are all associated with a decrease in overall physical density of the tissues.

45. A The principal factors effecting the amount of attenuation of a tissue are its effective atomic number and thickness of the substance.

46. A The log value of the light transmitted through a film is called the optical density.

47. B The image labeled letter B has the highest optical density.

48. C The image labeled letter C posses the lowest exit mAs or lowest optical density.

49. A Image A posses the desired optical density.

50. A The original techniques require 20 mAs and 86 kVp. In order to provide a similar image at 800 mA will require a timer setting of 25 milliseconds. 20 mAs/800 ms = .025 seconds or 25 milliseconds.

51. D The shortest exposure time setting will be associated with the least involuntary motion appearing on the radiograph. In order to answer this question all the timer settings must be converted into milliseconds A= 200 ms B= 100 ms C= 50 ms D= 20 ms

52. A When a variable kVp technique chart is used, a 2 kVp change is recommended for every 1 cm change in tissue thickness to maintain a consistent optical density on the image.

53. B Any material that has a high physical density and/or a high atomic number will attenuate a large number of x-rays. These are often classified as radiopaque substances.

54. B Voluntary motion in most adult patients is controllable through the use of proper patient instructions. Sand bags, restraining devices and short exposure times are some of the ways in which involuntary motion can be reduced.

55. B The increased thickness of the tissue in the oblique position would result in less radiation reaching the image receptor, and a lower optical density. The higher degree of scattered radiation produced by this thicker tissue will also reduce the radiographic contrast of the radiographic image.

56. C An original image of the T-spine was taken at 40 mAs and 83 kVp. Only answer C (40 mAs, 83 kVp) will provide a similar optical density.

57. C The added physical density of the fluid from ascites will tend to attenuate more radiation reducing the amount of radiation that reaches the image receptor. This will reduce the optical density of the image.

58. A The combination of the milliampere station and the timer value milliampere-seconds (mAs) is used to give an estimation of the amount of radiation used during the imaging process.

59. A Tissues with relatively high atomic and physical density such as bone is more likely to absorb the x-ray in the beam. These tissues will therefore have higher absorption or attenuation coefficients to x-ray photons in the diagnostic energy region.

60. C The amount of penetration of the x-ray beam is dependent upon the wavelength and frequency (energy) of the photons in the x-ray beam and the atomic number, physical density and the thickness of the tissues.

61. C An increase in the optical density can be accomplished by an increase in kVp setting, mA setting, timer setting or a decrease in source-to-image receptor distance.

62. D Large amounts of intestinal gas that are often present in obstructed patients will result in a decrease in the average physical density of the tissue. Therefore, the resulting image will show excessive penetration and an increase in optical density.

63. C The problem is solved by using the formula mAs = mA x time.

A= 20 mAs B= 4 mAs C= 40 mAs D= 4 mAs

64. B The number or amount of photons produced during a radiographic exposure is directly related to the mA and time (milliampere-seconds value).

65. C Because of the mAs relationship, a 50% reduction in the time can be compensated for by a doubling of the milliampere setting.

66. A The optical density on any portion of the radiographic image is related to the degree to which the beam is attenuated by a tissue. Air-filled lungs and intestinal gas have a relatively low physical density and atomic number and will absorb or attenuate a small percentage of the beam. A tissue that attenuates a small portion of the beam will appear to have a high optical density and will appear dark on the image.

67. D The subject contrast of an individual is related to one's general state of health, age, degree of development, and number of pathological conditions. All of these can modify the physical density of the tissue.

68. D An increase in the amount of scattered radiation is likely to occur in thicker tissues, tissues with a higher physical density and at higher kilovoltage levels.

69. C In order to demonstrate the vertebra of the lumber spine will usually require a moderate to high amount of penetration. This is accomplished by the use of kilovoltage peak setting in the 85 - 95 kVp range.

70. C The various shades appearing in the radiographic image are termed optical densities.

71. C The inflammation caused by Paget's disease will often cause an excessive over-growth of bone. Because of this increase in the amount of bone tissue an increase in the exposure factors is required to maintain the desired optical density. Emphysema, osteoporosis, and pneumothorax are all associated with a decrease in the physical density of the tissues and therefore would require a decrease in the exposure factors to maintain the desired optical density of the image.

72. A A high contrast image consisting of a small number of density shades (short scale) with a wide density difference between them is most commonly seen when low kVp exposures are employed. Contrast and scale contrast are primarily controlled by changes in the kilovoltage peak (kVp).

73. C One of the best uses of the 15% kVp rule is to help maintain the optical density of the image when the mAs is reduced to reduce the exposure to the patient. The exposure in answer C reduces the mAs by 50% through an increase of 15% (11 kVp) of the kVp.

74. C By definition, short scale contrast is described as having an image consisting of a small number of widely varying optical density shades.

75. D An increase in the amount of scattered radiation is likely to occur in thicker tissues, tissues with a higher physical density and at higher kilovoltage levels. The timer setting has no effect on the percentage of scattered radiation produced.

76. A A decrease in the thickness of the tissue will have a number of positive imaging consequences, including a decrease in the amount of scattered radiation, an increase in the radiographic contrast of the image, a lower primary beam attenuation and an increase in the optical density.

77. B The mAs (120) is the product of the mA (600) and the time (.2).

A= 30 mAs B= 120 mAs C= 240 mAs

78. C The effects related to a change in the timer setting are most closely relate to the amount of motion and the optical density of the image.

79. C   A reduction of the mAs value by ½ combined with a15% increase in the kVp should result in an image that has nearly the same optical density as the original factors. This modification is often used to reduce the patient dose and the exposure time while increasing the exposure latitude and penetration of the beam.

80. D   Since a homogenous object shows no differences in the ability to attenuate radiation, it will show only a single optical density and no radiographic or subject contrast.

81. B   Substances with a low atomic number (nitrogen, oxygen, hydrogen) are only poorly attenuated by x-rays and are classified as radiolucent or radiotransparent.

82. 50 ms   15 mAs is needed at 86 kVp to maintain the optical density therefore: 15 mAs/300 mA = .05 seconds or 50 ms

83. 100 ms   60 mAs is needed at 64 kVp to maintain the optical density therefore: 60 mAs/600 mA = .1 seconds or 100 ms

84. 64 kVp   At 60 mAs the kVp must be set at 64 kVp to maintain the optical density.

85. 86 kVp   At 15 mAs the kVp must be set at 86 kVp to maintain the optical density.

86. 600 mA   15 mAs is needed at 86 kVp to maintain the optical density therefore: 15 mAs/.025 seconds = 600 mA

87. 100 mA   60 mAs is needed at 64 kVp to maintain the optical density therefore: 60 mAs/.1 seconds = 100 mA

88. A   The lowest kVp provides the highest contrast.

89. C   The highest mAs results in the greatest exposure.

90. A   The shortest time provides for the least involuntary motion.

Radiographic Quality Advanced

1. Besides the variables of kilovoltage peak time and milliampere setting when preparing a technique chart, what other factors must be standardized?

   1. Screen speed   *2. Grid usage*   *3. Degree of collimation*

   A. 1 & 2 only
   B. 1 & 3 only
   C. 2 & 3 only
   D. 1, 2 & 3

2. A satisfactory radiograph is made at a 160 cm (64") source-to-image receptor distance using 40 mAs at 87 kVp. What new factors will produce a similar optical density if an 80 cm (32") source-to-image receptor distance is employed?

   A. 5 mAs at 87 kVp
   B. 10 mAs at 87 kVp
   C. 20 mAs at 87 kVp
   D. 40 mAs at 87 kVp

3. Which pair of tissue interfaces will possess the highest differential attenuation to diagnostic x-rays?

   A. Muscle to bone
   B. Water to muscle
   C. Soft tissue to bone
   D. Soft tissue to muscle

4. The percentage of scatter reaching the image receptor during a non-grid exposure of an abdomen comprises about _____ of the image.

   A. 10%
   B. 20%
   C. 50%
   D. 90%

5. A technique of 10 mAs at 78 kVp is used at a 140 cm (56") source-to-image receptor distance. What new source-to-image receptor distance should be used to decrease this exposure to 5 mAs at 78 kVp?

   A. 70 cm (28")
   B. 99 cm (40")
   C. 198 cm (80")
   D. 280 cm (112")

6. According to the law of reciprocity, which set of technical factors should give the same optical density as one using 200 mA, 40 ms at 80 kVp?

	mA	Time	kVp
A.	200	20 ms	80
B.	400	20 ms	80
C.	800	80 ms	80
D.	1000	80 ms	80

7. A satisfactory radiograph is made using 15 mAs at 75 kVp at a 140 cm source-to-image receptor distance . Assuming no other changes in the exposure factors are made, what new mAs value will produce the same optical density at a 180 cm source-to image receptor distance?

   A. 5 mAs
   B. 9 mAs
   C. 19 mAs
   D. 25 mAs

8. An acceptable radiograph is made at 10 mAs and 75 kVp on full wave radiographic unit. In order to maintain the same optical density on a 3-phase unit would require a technique of about:

   A. 5 mAs, 68 kVp
   B. 5 mAs, 75 kVp
   C. 20 mAs, 75 kVp
   D. 20 mAs, 85 kVp

9. The production of scattered radiation when a diagnostic x-ray photon strikes an object is primarily due to an event called:

   A. Compton's scattering
   B. Bremsstrahlung scattering
   C. Photoelectric scattering
   D. Characteristic scattering

10. The amount of energy absorbed by an irradiated object is primarily related to all the following EXCEPT:

A. The atomic number of the tissue
B. The atomic mass of the tissue

C. The energy of the photon in the beam
D. The direction of the beam

11. The incident photon is best defined as x-ray that:

A. Has not yet hit any electron in the scattering object
B. Has been directed through a primary barrier
C. Has been remitted from a scattering object
D. Has been totally absorbed by an object

12. The exposure a patient receives during a radiographic examination is dependent upon:

1. Compton effect   *2. Photoelectric effect*   *3. Pair production event*

A. 1 & 2 only
B. 1 & 3 only

C. 2 & 3 only
D. 1, 2 & 3

Referring to the diagram, answer questions 13-14.

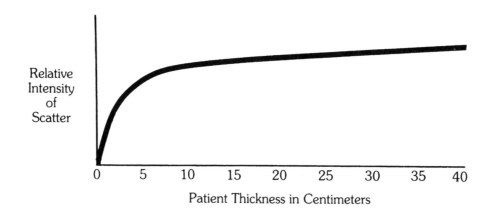

13. According to the graph the intensity of scattered radiation is:

A. Directly proportional to the thickness of the patient
B. Inversely proportional to the thickness of the patient
C. Directly proportional to the quality of the beam
D. Inversely proportional to the optical density of the image

14. The intensity of the scattered radiation will be the greatest for a patient that measures:

A. 7.5 cm
B. 17 cm

C. 24 cm
D. 38 cm

15. Which of the following devices can be used to reduce the amount of scattered radiation produced in the abdomen of a patient that measures 45 cm?

A. A smaller focal spot
B. A higher speed imaging system

C. A compression band
D. A gradient screen

16. An x-ray generator has a maximum capacity of 600 mA. What is the shortest possible exposure timer setting that can be used to obtain 30 mAs at a kVp of 80?

A. 20 milliseconds
B. 50 milliseconds

C. 200 milliseconds
D. 500 milliseconds

17. A step wedge (penetrometer) film is taken showing a small number of widely varying shades of optical densities. Choose the set of factors that most likely produced this image.

	mA	Time	kVp	FSS
A.	100	50 ms	98	1 mm
B.	200	30 ms	94	1 mm
C.	300	150 ms	68	2 mm
D.	300	100 ms	81	2 mm

18. When an air gap technique is used, the amount of scatter reaching the image receptor is decreased because:

A. Of the filtering effects of the air
B. More scatter misses the image receptor
C. Less scatter is produced at longer source-to-image receptor distances
D. The tissues absorb more radiation

19. To help insure that the desired optical density is maintained on images taken after the introduction of a positive contrast media (barium) it is usually necessary to:

A. Decrease the kilovoltage peak setting
B. Decrease the timer setting
C. Increase the amount of collimation of the object
D. Increase the milliampere setting

20. It is normally possible to perform radiographic images of the chest without the use of a radiographic grid. This is possible because of the:

A. High physical density of the heart and lungs
B. Natural air gap that is formed between the heart and the imaging system
C. High subject contrast between the structures in the chest cavity
D. Low tissue density of the structures in the thoracic cavity

21. The majority of the x-rays that are attenuated during the exposure to a 50 keV beam result from an event called:

A. Pair production
B. Photoelectric effect
C. Compton's scattering
D. Nuclear fission

22. Which of the following exposure fields is likely to produce the greatest amount of scattered radiation for a given patient and exposure setting?

A. 35 cm x 43 cm
B. 28 cm x 35 cm
C. 25 cm x 30 cm
D. 18 cm x 43 cm

23. A scoliosis filter is to be used to help image the entire spine. It should be inserted into the x-ray beam so the thickest of the filter corresponds to the:

A. Cervical spine
B. Lower thoracic spine
C. Lumbar spine
D. Sacrum-coccyx

24. A film -screen type imaging system is used to make the following exposures:

	mA	Time	kVp	SID
1.	600	100 ms	68	100 cm
2.	300	200 ms	68	100 cm
3.	100	600 ms	68	100 cm

Though the same optical density should be expected, #3 possesses a higher optical density. This is an example of:

A. The failure of the inverse square law
B. The failure of the reciprocity law
C. The failure of the spectral efficiency law
D. The failure of the 15% kVp law

25. The majority of scattered radiation produced in the tissues of the body when exposed to diagnostic x-rays occurs at energies that are:

A. Too low to exit from the scattering object
B. Similar to energy of the photons in the primary beam
C. Well above the energy of the photons in the primary beam
D. At an energy that is about 50% of that in the primary beam

Explanations

1. D   In order to provide the desired optical density when using a technique chart, it is important for the radiographer to make sure that the appropriate imaging system, grid ratio and degree of collimation are employed.

2. B   The problem can be solved by using the formula:
New mAs (10 mAs) = Original mAs (40 mAs) x New SID^2 (6,400) /Original SID^2 (25,600).

3. C   An equal thickness of the four normal human tissues will attenuate the beam based on their relative physical density and the effective atomic number. The pair of tissues that have the highest differences will provide the highest differential attenuation. In this question the greatest difference is between the soft tissue and the bone.

4. D   According to Selman the amount of scattered radiation produced during an abdominal exposure is estimated at 90% of the total. For a chest, this value drops to about 50%.

5. B   The problem can be solved by using the formula
New distance (99 cm) $= \sqrt{Original\ dis\tan ce^2\ (19,600) \times New\ mAs\ (5)}$ / Original mAs (10)

6. B   According to the law of reciprocity, using the same mAs value assuming all other variables are maintained, should provide an image with a similar optical density. Since an exposure of 200 mA and 40 milliseconds provides an exposure of 8 mAs, a similar optical density would be obtained by the set of factors that also provides 8mAs (400 mA at 20 ms).

7. D   The problem can be solved by using the formula:
New mAs (25 mAs) = Original mAs (15 mAs) x New SID^2 (32,400) /Original SID^2 (19,600).

8. B   Since a unit operating with 3-phase current will produce about twice the amount of effective mAs as a single-phase unit, only about half the mAs setting is required to produce a similar optical density.

9. A   Most of scattered radiation occurring during the exposure of tissues to diagnostic x-rays is the result of what is called the Compton's scatter. When a high energy x-ray photon strikes an outer orbital electron and the electron is ionized along with the production of a scattered photon, this is called the Compton event.

10. A   The amount of energy absorbed by an irradiated object is related to the atomic number of the object, its atomic mass and its thickness and the energy of the beam.

11. A   The photons in beam before they strike any object are referred to as the incident beam. After the photons emerge from an object they are referred to as the exit or remnant beam.

12. A   The exposure received by the patient is related to two interactions Compton's and photoelectric effect that cause ionization in the tissues. Pair production does not occur at diagnostic energy levels.

13. A   The chart indicates that the intensity of scatter is directly proportional to the thickness of the patient; i.e., a larger patient creates more scattered radiation.

14. D   The amount of scattered radiation is much greater in tissue that has a greater thickness.

15. C   Because the amount of scattered radiation is directly effected by the thickness of the object, a compression band can be used to reduce the amount of scattered radiation produced in larger patients.

16. B   In order to obtain 30 mAs at a 600 mA station will require a timer setting of .05 second or 50 milliseconds.

17. C   A penetrometer or step wedge showing a small number of widely varying optical densities is most likely produced using the lowest kVp setting.

18. B An increase in the object-to-image receptor distance creates a space called an air gap between the patient and the imaging system. This allows the divergent scatter radiation produced in the patient to miss the imaging system. This reduces the amount of scattered radiation appearing in the image and improves the contrast of the image.

19. D Since the number of x-rays reaching the image receptor is reduced by a positive contrast agent, an increase in the mAs is often desirable to maintain the desired optical density.

20. C The relatively low atomic and physical density of the air filled lungs produce relatively small amounts of scattered radiation as well as providing a high subject contrast. This enables the chest images to be obtained with out the use of a grid.

21. D The photoelectric effect is responsible for virtually all of the attenuation that occurs in the diagnostic energy region.

22. A The amount of scattered radiation is greatest in a larger field and least in a smaller field. Therefore 35 cm by 43 cm field is associated with greatest amount of scatter production and the lowest image contrast.

23. A All compensating filters are designed to prevent the full intensity of the beam from reaching tissues that have a low physical density and to enable higher amount of radiation to reach areas of a high tissue density. During scoliosis the thickest part of the filter needs to be placed over the thinner areas associated with the cervical spine.

24. B According to the law of reciprocity, the same mAs value, assuming all other variables are maintained, should provide a similar optical density. At long timer setting, this relationship can fail resulting in an increase in the optical density of images taken at the same mAs. This is known as the failure of the reciprocity law. This effect is only seen with film scree combinations.

25. B Nearly all the scattered photons that are produced in diagnostic radiology are created at energies that are nearly identical to those photons in the incident beam.

1. A radiograph is taken using 300 mA, 20 ms at 90 kVp at a 200 cm source-to-image receptor distance (SID). What new timer setting can be used to maintain the same optical density at a 100 cm source-to-image receptor distance (SID), assuming all other factors remain unchanged?

2.-9. State the new timer setting that can be used to maintain the same optical density, given the original timer setting, original (SID), and new SID, assuming all other factors remain unchanged.

	Original Timer Setting	Original SID	New SID	New Timer Setting
2.	40 ms	100 cm	50 cm	_____
3.	160 ms	200 cm	100 cm	_____
4.	160 ms	180 cm	90 cm	_____
5.	250 ms	150 cm	100 cm	_____
6.	300 ms	100 cm	200 cm	_____
7.	60 ms	60 inches	30 inches	_____
8.	1 sec	200 cm	150 cm	_____
9.	1 sec	300 cm	150 cm	_____

10. A radiograph is taken using 400 mA, 50 ms at 90 kVp at a 200 cm source-to-image receptor distance (SID). What new mA setting can be used to maintain the same optical density at a 100 cm source-to-image receptor distance (SID), assuming all other factors remain unchanged?

11.-16. State the new mA setting that can be used to maintain a same optical density given the original mA, original SID, and new SID, assuming all other factors remain unchanged.

	Original mA Setting	Original SID	New SID	New mA Setting
11.	400 mA	40 inches	20 inches	_____
12.	100 mA	100 cm	300 cm	_____
13.	800 mA	200 cm	100 cm	_____
14.	200 mA	36 inches	72 inches	_____
15.	600 mA	180 cm	90 cm	_____
16.	900 mA	180 cm	60 cm	_____

17. A radiograph is taken using 600 mA, 150 ms at 60 kVp at a 180 cm source-to-image receptor distance (SID). What new mAs can be used to maintain the same optical density at a 90 cm source-to-image receptor distance (SID), assuming all other factors remain the same?

18.-25. State the new mAs that can be used to maintain the same optical density given the original mAs, and original SID and the new SID, assuming all other factors remain unchanged.

	Original mAs	Original SID	New SID	New mAs
18.	8 mAs	100 cm	50 cm	_____
19.	8 mAs	200 cm	100 cm	_____
20.	18 mAs	180 cm	60 cm	_____
21.	20 mAs	30 inches	60 inches	_____
22.	28 mAs	72 inches	36 inches	_____
23.	16 mAs	40 cm	60 cm	_____
24.	30 mAs	100 cm	300 cm	_____
25.	100 mAs	90 cm	180 cm	_____

The following formulas have been used in solving these problems.

$$\text{New Timer Setting} = \frac{\text{Original Timer Setting} \times \text{New SID}^2}{\text{Original SID}^2}$$

$$\text{New mA Setting} = \frac{\text{Original mA Setting} \times \text{New SID}^2}{\text{Original SID}^2} \qquad \text{New mAs} = \frac{\text{Original mAs} \times \text{New SID}^2}{\text{Original SID}^2}$$

Time - Distance

1. A radiograph is taken using 300 mA, 240 ms at 90 kVp at a 200 cm source-to-image receptor distance (SID). What new timer setting can be used to maintain the same optical density at a 100 cm source-to-image receptor distance (SID), assuming all other factors remain unchanged?

2.-9. State the new timer setting that can be used to maintain the same optical density given the original timer setting, original SID and new SID, assuming all other factors remain unchanged.

	Original Timer Setting	Original SID	New SID	New Timer Setting
2.	120 ms	120 cm	60 cm	_____
3.	320 ms	400 cm	200 cm	_____
4.	260 ms	90 cm	50 cm	_____
5.	200 ms	150 cm	300 cm	_____
6.	100 ms	100 cm	200 cm	_____
7.	600 ms	30 inches	60 inches	_____
8.	8 sec	300 cm	150 cm	_____
9.	1 sec	300 cm	60 cm	_____

10. A radiograph is taken using 400 mA, 50 ms at 90 kVp at a 200 cm source-to-image receptor distance (SID). What new mA setting can be used to maintain the same optical density at a 100 cm source-to-image receptor distance (SID), assuming all other factors remain unchanged?

11.-16. State the new mA setting that can be used to maintain the same optical density given the original mA, original SID and new SID, assuming all other factors remain unchanged.

	Original mA Setting	Original SID	New SID	New mA Setting
11.	200 mA	20 inches	40 inches	_____
12.	900 mA	300 cm	100 cm	_____
13.	800 mA	400 cm	200 cm	_____
14.	100 mA	36 inches	72 inches	_____
15.	600 mA	60 cm	120 cm	_____
16.	900 mA	150 cm	50 cm	_____

17. A radiograph is taken using 800 mA, 250 ms at 60 kVp at a 180 cm source-to-image receptor distance (SID). What new mAs can be used to maintain the same optical density at a 90 cm source-to-image receptor distance (SID), assuming all other factors remain unchanged?

18.-25. State the new mAs that can be used to maintain the same optical density given the original mAs, and original SID, and the new SID, assuming all other factors remain unchanged.

	Original mAs	Original SID	New SID	New mAs
18.	15 mAs	100 cm	200 cm	_____
19.	80 mAs	400 cm	100 cm	_____
20.	36 mAs	180 cm	60 cm	_____
21.	120 mAs	40 inches	60 inches	_____
22.	24 mAs	72 inches	36 inches	_____
23.	8 mAs	30 cm	60 cm	_____
24.	80 mAs	150 cm	300 cm	_____
25.	10 mAs	9 cm	180 cm	_____

The following formulas have been used in solving these problems.

$$\text{New Timer Setting} = \frac{\text{Original Timer Setting} \times \text{New SID}^2}{\text{Original SID}^2}$$

$$\text{New mAs} = \frac{\text{Original mAs} \times \text{New SID}^2}{\text{Original SID}^2}$$

$$\text{New mA Setting} = \frac{\text{Original mA Setting} \times \text{New SID}^2}{\text{Original SID}^2}$$

1. A radiographic image is taken using 400 mA, 10 ms at 90 kVp at a 200 cm source-to-image receptor distance (SID). If a new timer setting of .04 sec, is desired, what new source-to-image receptor distance (SID) can be used to maintain the same optical density, assuming all other factors remain unchanged?

2.-9. State the new source-to-image receptor distance (SID) that can be used to maintain the same optical density given the original timer setting, new timer setting, and the original source-to-image receptor distance(SID), assuming all other factors remain unchanged.

	Original Timer Setting	New Timer Setting	Original SID	New SID
2.	20 ms	80 ms	100 cm	_____
3.	50 ms	200 ms	200 cm	_____
4.	160 ms	40 ms	180 cm	_____
5.	300 ms	600 ms	40 inches	_____
6.	900 ms	100 ms	300 cm	_____
7.	2 sec	500 ms	40 inches	_____
8.	40 ms	160 ms	100 cm	_____
9.	60 ms	240 ms	200 cm	_____

10. A radiograph is taken using 100 mA, 350 ms at 80 kVp at a 60cm source-to-image receptor distance (SID). If a new mA setting of 400 mA is desired, what new source-to-image receptor distance (SID) can be used to maintain the same optical density, assuming all other factors remain unchanged?

11.-16. State the new source-to-image receptor distance (SID) that can be used to maintain the same optical density given the original mA setting, new mA setting, and original source-to-image receptor distance (SID), assuming all other factors remain unchanged.

	Original mA Setting	New mA Setting	Original SID	New SID
11.	100 mA	25 mA	100 cm	_____
12.	400 mA	100 mA	200 cm	_____
13.	600 mA	1200 mA	80 cm	_____
14.	900 mA	100 mA	150 inches	_____
15.	300 mA	600 mA	40 inches	_____
16.	800 mA	200 mA	100 cm	_____

17. A radiograph is taken using 200 mA, 400 ms, 80 kVp at 100 cm source-to-image receptor distance (SID). If a new mAs of 20 mAs is desired, what new source-to-image receptor distance (SID) can be used to maintain the same optical density, assuming all other factors remain unchanged?

18.-25. State the new source-to-image receptor distance (SID) that can be used to maintain the same optical density given the original mAs, the new mAs and original source-to-image receptor distance (SID), assuming all other factors remain unchanged.

	Original mAs	New mAs	Original SID	New SID
18.	160 mAs	40 mAs	100 cm	_____
19.	10 mAs	40 mAs	200 cm	_____
20.	30 mAs	270 mAs	100 cm	_____
21.	50 mAs	100 mAs	40 inches	_____
22.	80 mAs	50 mAs	180 cm	_____
23.	15 mAs	200 mAs	100 cm	_____
24.	20 mAs	30 mAs	150 cm	_____
25.	15 mAs	45 mAs	200 cm	_____

Advanced Time - Distance

1. A radiographic image is taken using 800 mA, 20 ms at 80 kVp at a 200 cm source-to-image receptor distance (SID). If a new timer setting of .08 sec, is desired, what new source-to-image receptor distance (SID) can be used to maintain the same optical density, assuming all other factors remain unchanged?

2.-9. State the new source-to-image receptor distance (SID) that can be used to maintain the same optical density given the original timer setting, new timer setting, and original source-to-image receptor distance (SID), assuming all other factors remain unchanged.

	Original Timer Setting	New Timer Setting	Original SID	New SID
2.	10 ms	40 ms	100 cm	_____
3.	5 ms	20 ms	300 cm	_____
4.	240 ms	60 ms	150 cm	_____
5.	800 ms	200 ms	60 inches	_____
6.	900 ms	100 ms	300 cm	_____
7.	1.0 sec	250 ms	40 inches	_____
8.	330 ms	660 ms	100 cm	_____
9.	500 ms	800 ms	200 cm	_____

10. A radiograph is taken using 200 mA, 150 ms at 80 kVp at a 60 cm source-to-image receptor distance (SID). If a new mA setting of 800 mA is desired, what new source-to-image receptor distance (SID) can be used to maintain the same optical density, assuming all other factors remain unchanged?

11.-16. State the new source-to-image receptor distance (SID) that can be used to maintain the same optical density given the original mA setting, the new mA setting, and source-to-image receptor distance (SID), assuming all other factors remain unchanged.

	Original mA Setting	New mA Setting	Original SID	New SID
11.	25 mA	100 mA	100 cm	_____
12.	100 mA	400 mA	180 cm	_____
13.	300 mA	1200 mA	100 cm	_____
14.	900 mA	100 mA	40 inches	_____
15.	150 mA	300 mA	36 inches	_____
16.	300 mA	100 mA	100 cm	_____

17. A radiograph is taken using 400 mA, 200 ms, 90 kVp at 200 cm source-to-image receptor distance (SID). If a new mAs of 20 mAs is desired, what new source-to-image receptor distance (SID) can be used to maintain the same optical density, assuming all other factors remain unchanged?

18.-25. State the new source-to-image receptor distance (SID) that can be used to maintain the same optical density given the original mAs, new mAs, and source-to-image receptor distance (SID), assuming all other factors remain unchanged.

	Original mAs	New mAs	Original SID	New SID
18.	180 mAs	45 mAs	200 cm	_____
19.	25 mAs	100 mAs	200 cm	_____
20.	15 mAs	135 mAs	36 inches	_____
21.	60 mAs	100 mAs	40 inches	_____
22.	90 mAs	50 mAs	100 cm	_____
23.	15 mAs	105 mAs	50 cm	_____
24.	120 mAs	30 mAs	150 cm	_____
25.	25 mAs	65 mAs	200 cm	_____

Answers

Time - Distance
Exercise 1

1. 5 ms
2. 10 ms
3. 40 ms
4. 40 ms
5. 110 ms
6. 120 ms
7. 15 ms
8. 560 ms
9. 250 ms
10. 100 mA
11. 100 mA
12. 900 mA
13. 200 mA
14. 800 mA
15. 150 mA
16. 100 mA
17. 22.5 mAs
18. 2 mAs
19. 2 mAs
20. 3 mAs
21. 80 mAs
22. 7 mAs
23. 36 mAs
24. 270 mAs
25. 400 mAs

Time - Distance
Exercise 2

1. 60 ms
2. 30 ms
3. 80 ms
4. 80 ms
5. 800 ms
6. 400 ms
7. 2.4 sec.
8. 20 ms
9. 40 ms
10. 100 mA
11. 800 mA
12. 100 mA
13. 200 mA
14. 400 mA
15. 2400 mA
16. 100 mA
17. 50 mAs
18. 60 mAs
19. 5 mAs
20. 4 mAs
21. 270 mAs
22. 6 mAs
23. 24 mAs
24. 240 mAs
25. 40 mAs

Advanced Time - Distance
Exercise 1

1. 400 cm
3. 400 cm
2. 200 cm
4. 90 cm
5. 56.6 in.
6. 100 cm
7. 20 in.
8. 200 in.
9. 400 in.
10. 120 cm
11. 50 cm
12. 100 cm
13. 113.1 cm
14. 50 in.
16. 50 cm
15. 56.6 in.
17. 50 cm
18. 50 cm
19. 400 cm
20. 300 cm
21. 56.6 in.
22. 142.3 cm
23. 365.1 cm
24. 183.7 cm
25. 346.4 cm

Advanced Time - Distance
Exercise 2

1. 400 cm
3. 600 cm
2. 200 cm
4. 75 cm
5. 30 in.
6. 100 cm
7. 20 in.
8. 141 cm
9. 252.9 cm
10. 120 cm
11. 200 cm
12. 360 cm
13. 200 cm
14. 13.3 in.
15. 50.9 in.
16. 57.7 cm
17. 100 cm
18. 100 cm
19. 400 cm
20. 108 in.
21. 51.6 in.
22. 74.5 cm
23. 132.3 cm
24. 75 cm
25. 322.5 cm

1. Which of the following techniques can be used during an imaging procedure to reduce the superimposition tissues in a given anatomic region?

    1. Tomography   *2. Oblique projections*   *3. Tube angulations*

    A. 1 only   C. 3 only
    B. 2 only   D. 1, 2 & 3

2. The region of unsharpness found at the edges of the images appearing at the imaging plane is called the:

    A. Focal spot blur   C. Quantum mottle
    B. Pseudo point   D. Heel effect

Referring to the diagram, answer question 3.

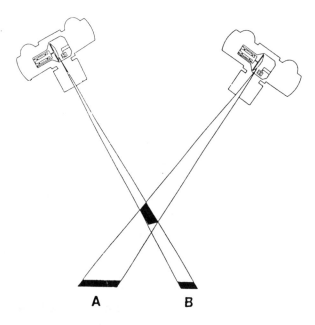

A   **B**

3. In the diagram the image seen at the area labeled A indicates a type of:

    A. Distortion known as foreshortening   C. Distortion known as elongation
    B. Distortion known as quantum blur   D. Distortion known as vignetting

4. Which of the following sets of factors would be associated with a radiographic image possessing greatest amount of focal spot blur?

	mAs	kVp	SID	OID	FSS
A.	20	83	100 cm	20 cm	.8 mm
B.	40	78	180 cm	10 cm	.3 mm
C.	15	87	100 cm	20 cm	.3 mm
D.	35	74	180 cm	20 cm	.6 mm

5. The amount of size distortion (magnification) appearing in a radiographic image can be determined by the formula:

A. Magnification = $\dfrac{SID}{SOD}$

B. Magnification = $\dfrac{OID}{SOD}$

C. Magnification = $\dfrac{SOD}{OID}$

D. Magnification = $\dfrac{OID^2}{SOD^2}$

6. Elongation and/or foreshortening of a radiographic image will most likely occur from a/an:

A. Increased source-to-image receptor distance
B. Angulation of the central ray

C. Increase in the anode-heel effect
D. Decreased object-to-image receptor distance

7. Which of the following changes will be associated with an increase in the size distortion (magnification) that will occur in a radiographic image?

1. A decreased source-to-image receptor distance
2. An increase in the size of the effective focal spot
3. An increased object-to-image receptor distance

A. 1 & 2 only
B. 1 & 3 only

C. 2 & 3 only
D. 1, 2 & 3

8. A radiograph is performed using a 50 mAs, 80 kVp exposure with a 100 cm (40") source-to-image receptor distance, a 40 cm (16") object-to-image receptor distance, and a 2 mm focal spot. The magnification factor for this image is approximately:

A. .09
B. 1.6

C. 2.5
D. 3.3

9. Improper tube, object and image receptor alignment with no change in source-to-image receptor distance will result in:

A. Size distortion
B. Shape distortion

C. Motion distortion
D. Screen distortion

10. Which of the following set of factors will produce a radiographic image that possess the least amount of focal spot blur?

	mA	Time	kVp	SID	OID	FSS
A.	300	10 ms	82	300 cm	20 cm	.8 mm
B.	200	30 ms	84	100 cm	20 cm	1 mm
C.	600	10 ms	86	300 cm	20 cm	.3 mm
D.	400	20 ms	85	150 cm	20 cm	2 mm

11. The image of a spherical shaped object that is exposed by the central portion of an x-ray beam will appear to have a:

A. Cylindrical shape
B. Circular shape

C. Elliptical shape
D. Irregular shape

12. Grids are NOT normally employed during a macro-radiographic (magnification) technique. Grids are NOT generally employed because of the reduced amount of scatter that reaches the image receptor when a:

A. Smaller effective focal spot is used
B. Longer object-to-image receptor distance is used

C. Larger tube angulation is used
D. Greater amount of diagnostic filtration is used

13. Tube angulations in many radiographic examinations are useful for the visualization of:

A. Tissues with a low physical density
B. Tissues which are superimposed

C. Tissues with a high atomic numbers
D. Tissues with a high degree of motion

14. Which of the following sets of technical factors will produce a radiograph showing the greatest magnification?

	mAs	kVp	SID	OID	FSS
A.	25	86	140 cm	10 cm	2 mm
B.	35	86	180 cm	20 cm	1 mm
C.	15	93	180 cm	30 cm	1 mm
D.	25	93	100 cm	30 cm	2 mm

15. As the distance from a radiation source increases, the object receives less radiation. This occurs principally because x-ray photons:

   1. Are absorbed by the air
   2. Lose energy at long distances
   3. Are distributed over a larger area

   A. 1 only
   B. 2 only

   C. 3 only
   D. 1, 2 & 3

16. An increase in source-to-image receptor distance with no other changes made, will result in a radiograph showing:

   A. Shorter scale of contrast
   B. Lower optical density

   C. Lower recorded detail
   D. More magnification distortion

17. Magnification radiography is being performed using a 140 cm source-to-image receptor distance (SID), 40 cm object-to-image receptor distance (OID), and .3 mm focal spot. The magnification factor for this set of factors is:

   A. 4.2
   B. 3.5

   C. 1.8
   D. 1.4

18. The formation of the focal spot blur around a radiographic image is effected by changes in:

   1. Object-to-image receptor distance
   2. Source-to image receptor distance
   3. Degree of collimation

   A. 1 & 2 only
   B. 1 & 3 only

   C. 2 & 3 only
   D. 1, 2 & 3

19. If a heart measures 12.5 cm from side to side at its widest point, and its image on a chest radiograph measures 15.0 cm, the magnification factor for this image is:

   A. 1.2
   B. 1.5

   C. 1.8
   D. 2.4

20. Which of the following can be used to maintain the same relative size of a radiographic image, if an increase in the object-to-image receptor distance is required?

   A. Increase focal spot size
   B. Decrease focal spot size

   C. Increase source-to-image receptor distance
   D. Decrease source-to-image receptor distance

21. Which of the following set of factors will result in the radiographic image possessing the least amount of focal spot blur?

	mAs	kVp	Grid Ratio	Screen Speed	Focal Spot Size
A.	20	87	18:1	(400)	2 mm
B.	50	83	16:1	(50)	1 mm
C.	40	74	16:1	(400)	1 mm
D.	70	73	12:1	(200)	2 mm

Chapter 23   Focal Spot Blur

22. The image of a disk-shaped object that is exposed on the central axis of the beam will appear as a/an _____ shaped image.

A. Cylindrical
B. Circular

C. Elliptical
D. Irregular

23. Which of the following changes will increase the amount of shape distortion appearing on a radiographic image?

1. Decreasing the SID   **2. Employing tube angulations**   **3. Increasing OID**

A. 1 only
B. 2 only

C. 3 only
D. 1, 2 & 3

24. An object measuring 18 cm in length is radiographed at a 100 cm source-to-image receptor distance and a 10 cm object-to-image receptor distance. How long will the resulting radiographic image be?

A. 16 cm
B. 20 cm

C. 24 cm
D. 34 cm

25. To insure an acceptable amount of focal spot blur during macro-radiographic or mammographic studies, a/an _____ should be employed.

A. Fractional focal spot
B. Large focal spot

C. Air gap technique
D. Subtraction technique

26. Which of the following set of factors will produce a radiographic image with the least amount of focal spot blur?

	mA	Time	kVp	SID	OID	FSS
A.	100	50 ms	83	100 cm	12 cm	1 mm
B.	300	20 ms	76	300 cm	40 cm	.3 mm
C.	100	100 ms	64	150 cm	20 cm	1 mm
D.	500	50 ms	79	100 cm	6 cm	.3 mm

27. Which of the following radiographic images of the skull is likely to have the greatest amount of shape distortion?

A. AP with no tube angulation
B. AP with a 15 degree caudal angulation

C. PA with a 15 degree cephalic angulation
D. PA with a 30 degree caudal angulation

28. The amount of size distortion in an object is effected by changes in all of the following EXCEPT:

A. The size of focal spot
B. The source-to-image receptor distance

C. The object-to image receptor distance
D. The source-to-object distance

29. Which of the following set of factors will produce a radiograph with the greatest recorded detail (definition)?

	mA	Time	kVp	SID	OID	FSS
A.	200	100 ms	85	300 cm	20 cm	1 mm
B.	400	50 ms	85	200 cm	10 cm	2 mm
C.	200	100 ms	85	200 cm	10 cm	1 mm
D.	400	50 ms	85	100 cm	20 cm	.5 mm

30. If an x-ray beam was produced from a single point source, the resultant image would consist of:

A. Umbra only
B. Focal spot blur only

C. Distortion only
D. Magnification only

31. A 10 cm object is radiographed using a 90 cm source-to-image receptor distance and 60 cm object-to-image receptor distance. The image of the object at the image plane should measure:

A. 7.5 cm
B. 12.5 cm

C. 23 cm
D. 30 cm

32. The recorded detail of a radiographic image can be improved by employing a:

    1. Longer source-to-image receptor distance
    2. Smaller effective focal spot size
    3. Shorter object-to-image receptor distance

    A. 1 & 2 only   C. 2 & 3 only
    B. 1 & 3 only   D. 1, 2 & 3

33. The degree of target (anode) angulation will principally effect the amount of:

    A. Film coverage   C. Secondary scatter production
    B. Thermionic emission   D. Image magnification

34. Three identical thin objects are located on the same plane and at the same distance from the imaging system. The object showing the greatest magnification is located:

    A. Toward the cathode side of the tube   C. On the central axis of the tube
    B. Toward the anode side of the tube   D. There is no uneven magnification of the object

35. The true image shadow of a radiographic image is termed the:

    A. Blur edge   C. Umbra
    B. Edge gradient   D. Point source

Explanations

1. D The reduction of the superimposition of tissues can be accomplished by three methods. The first, called tomography, allows for the visualization of a single tissue plane through the blurring of images above and below the desired plane. The second is oblique views and the third is tube angulations which are able to separate superimposed tissues through the introduction of shape distortion.

2. A The true image border or umbra in all radiographic images is surrounded by an area of unsharpness called the focal spot blur.

3. C Depending on the shape and orientation of the original object, a tube angulation may cause a shape distortion called elongation or foreshortening.

    In this diagram the image labeled A shows elongation while image B shows foreshortening.

4. A The degree of focal spot blur or image unsharpness can be obtained from the unsharpness formula: Focal spot blur = Focal spot size (FSS) x OID)/SID - OID. Using this formula the following values are obtained.   A=.2, B=.017, C=.075, D=.075

5. A Size distortion or magnification caused by the relationship of the distance factors is determined by the formula: Magnification factor (MF) = SID/SOD or SID/(SID - OID).

6. B Shape distortion or the visualization of an image which does not conform in shape to the actual object is commonly seen in radiography due to the prevalence of 3-dimensional objects being presented on a 2-dimensional image display and through the use of angulated techniques. Depending on the shape and orientation of the original object, a tube angulation may cause a shape distortion called elongation or foreshortening.

7. B Size distortion (magnification) is increased by employing a long object-to-image receptor distance , shorter source-to-object distance, and a shorter source-to-image receptor distance. The size of the focal spot does not alter the size of the image.

8. B Using the magnification formula:
    Magnification factor (1.66)= SID(100 cm)/SOD (60 cm) where SOD = SID-OID

9. B Improper alignment of the tube, object and image receptor may expose the object to the divergent x- rays in the beam. Since these are off of the central axis, some degree of shape distortion will be imported in the image.

10. C The degree of focal spot blur or image unsharpness can be obtained from the unsharpness formula: Focal spot blur = focal spot size (FSS) x OID)/SID - OID. The exposure factors kVp and mAs do not effect image blur. Using this formula the following values are obtained A=.057, B=.25, C=.021, D=.307

11. B All disks or spheres on the central axis will form images that will have a circular shape. When the disk-shaped object is lateral to the central axis and is imaged by the peripheral portion of the beam, the image will appear elliptical.

12. B The long object-to-image receptor distance that needs to be employed in all magnification techniques provides an air gap. Because the distance between the scattering object and the imaging system is increased, much of the divergent scattered radiation will miss the image receptor.

13. B Even though tube angulations will introduce some degree of shape distortion to the image, they are extremely beneficial for the separation of superimposed tissues.

14. D Size distortion or magnification caused by the relationship of the distance factors is determined by the formula: magnification factor (MF) = SID/SOD or SID/(SID - OID).Using this formula the following values are obtained.   A=1.07, B=1.125, C=1.2, D 1.42

15. C The reduction of the radiation received by a given object as the distance varies is directly related to spread or divergence of the beam that distributes the radiation over a larger area.

16. B An increased source-to-image receptor distance is associated with less magnification, less unsharpness, less focal spot blur, less optical density and greater amount of recorded detail.

17. D Using the magnification formula:
Magnification factor (1.4)= SID(140 cm)/SOD (100 cm) where SOD = SID-OID

18. D The unsharp borders or focal spot blur surrounding the true image shadow or umbra are effected by changes in a number of factors. These include the distance factor, (SID, OID, SOD),the effective focal spot size, the image receptor and the degree of beam limitation.

19. A This problem can be solved by the use of the second magnification formula: magnification factor (1.2 = Image width (15 cm)/Object width (12.5 cm).

20. C Magnification (size distortion) can be reduced by increasing the source-to-image receptor distance (SID) and/or source-to-object distance (SOD).

21. B A combination of the smallest focal spot and the slowest imaging system will normally provide for the least unsharpness or least amount of focal spot blur.

22. B All disks or spheres on the central axis will form images with a circular shape, when the disk (object) is lateral to the central axis, its image would appear elliptical.

23. B Shape and size distortion are the two common forms of image distortion. Shape distortion is effected by tube angulations and patient rotations, where size distortion (magnification) is controlled by the distance factors. The size of the focal spot that has a great effect on geometrically recorded detail has little or no effect on the degree of shape or size distortion.

24. B This problem is solved using the combined magnification formula:
Image width (20 cm) = (Object width (18 cm) x SID (100 cm))/SOD (90 cm)

25. A During any technique requiring a small degree of image unsharpness or focal spot blur can be maintained through the use of an extremely small (fractional) focal spot of less than .3 mm.

26. D The degree of focal spot blur or image unsharpness can be obtained from the unsharpness formula: Focal spot blur = Focal spot size (FSS) x OID)/SID - OID. Using the formula the following values can be obtained.   A= .136, B=.033, C=.153, D=.019

27. D The amount of shape distortion or the visualization of an image which does not conform in shape to the actual object is related to the amount of tube angulation that is used. Therefore the larger angle the greater the distortion.

28. A Magnification (size distortion) is principally controlled by the distance factors, the source-to-image receptor (SID), object-to-image receptor (OID), and source-to-object (SOD) distances. The size of the focal spot has a great effect on geometrically recorded detail, but little or no effect on the degree of shape and size distortion.

29. C The degree of focal spot blur or image unsharpness can be obtained from the unsharpness formula: Focal spot blur = Focal spot size (FSS) x SID)/SID - OID. Using this formula the following values are obtained A=.375, B=1.06, C=2.5, D=.338

30. A If an image were produced by theoretical point source, the image would consist entirely of umbra. No image unsharpness or focal spot blur would be produced regardless of the degree of magnification.

31. D This problem is solved using the combined magnification formula:
Image width (30 cm) = (Object width (10 cm) x SID (90 cm))/SOD (30 cm)

32. D The amount of image unsharpness reducing the amount of focal spot blur in an image can be improved by the use of a longer SID, a longer SOD, a shorter OID, a slower imaging system, a smaller focal spot size and a smaller collimated field.

33. A   The anode angulation and the resulting anode heel effect has a direct influence on the degree of coverage by beam.

34. D   Since there is no change in the distance, no uneven magnification of the object and its corresponding images is seen.

35. C   The image appearing on a radiograph consists of a clearly defined central region called the umbra surrounded by an area of unsharpness called the geometric or focal-spot blur.

Magnification

1. An object measuring 20 cm in width is radiographed and appears as a 50 cm in width radiographic image. What is the magnification factor for this image?

2.-8. State the magnification factors for the following objects when they are radiographed and result in the given image widths.

	Object Width	Image Width	Magnification Factor
2.	10 cm	40 cm	_____
3.	15 inches	30 inches	_____
4.	20 cm	30 cm	_____
5.	30 inches	40 inches	_____
6.	10 cm	15 cm	_____
7.	25 cm	50 cm	_____
8.	35 cm	105 cm	_____

9. An object is radiographed at a 100 cm source-to-image receptor distance (SID) and 20 cm source-to-object distance (SOD). What is the magnification factor for this object?

10.-17. State the magnification factors which will result from the use of the following source-to-image receptor (SID) and source-to-object distances (SOD).

	SID	SOD	Magnification Factor
10.	40 cm	10 cm	_____
11.	40 inches	20 inches	_____
12.	40 inches	30 inches	_____
13.	100 cm	40 cm	_____
14.	100 cm	50 cm	_____
15.	120 cm	60 cm	_____
16.	180 cm	60 cm	_____
17.	72 inches	36 inches	_____

18. An object is radiographed at a 200 cm source-to-image receptor distance (SID) and 40 cm object-to-image receptor distance (OID). What is the magnification factor for this object?

19.-25. State the magnification factors which will result for the use of the following source-to-image receptor (SID) and object-to-image receptor distances (OID).

	SID	OID	Magnification Factor
19.	40 cm	10 cm	_____
20.	40 inches	20 inches	_____
21.	40 inches	30 inches	_____
22.	100 cm	60 cm	_____
23.	100 cm	40 cm	_____
24.	180 cm	18 cm	_____
25.	300 cm	60 cm	_____

The following formulas have been used in solving these problems.

$$\text{Magnification Factor} = \frac{Image\ width}{Object\ width} \qquad SOD = SID - OID$$

$$\text{Magnification factor} = \frac{SID}{SOD}$$

Magnification

1. An object measuring 30 cm in width is radiographed and appears as a 60 cm in width radiographic image. What is the magnification factor for this image?

2.-8. State the magnification factors for the following objects when they are radiographed and result in the given image widths.

	Object Width	Image Width	Magnification Factor
2.	20 cm	40 cm	_____
3.	10 inches	30 inches	_____
4.	30 cm	80 cm	_____
5.	30 inches	50 inches	_____
6.	10 cm	25 cm	_____
7.	25 cm	75 cm	_____
8.	30 cm	90 cm	_____

9. An object is radiographed at a 200 cm source-to-image receptor distance (SID) and 20 cm source-to-object distance (SOD). What is the magnification factor for this object?

10.-17. State the magnification factors which will result from the use of the following source-to-image receptor (SID) and source-to-object distances (SOD).

	SID	SOD	Magnification Factor
10.	30 cm	10 cm	_____
11.	80 inches	20 inches	_____
12.	60 inches	30 inches	_____
13.	100 cm	60 cm	_____
14.	150 cm	50 cm	_____
15.	160 cm	80 cm	_____
16.	180 cm	90 cm	_____
17.	72 inches	18 inches	_____

18. An object is radiographed at a 200 cm source-to-image receptor distance (SID) and 40 cm object-to-image receptor distance OID). What is the magnification factor for this object?

19.-25. State the magnification factors which will result for the use of the following source-to-image receptor (SID) and object-to-image receptor distances (OID).

	SID	OID	Magnification Factor
19.	40 cm	30 cm	_____
20.	60 inches	20 inches	_____
21.	90 inches	30 inches	_____
22.	180 cm	60 cm	_____
23.	100 cm	60 cm	_____
24.	100 cm	20 cm	_____
25.	300 cm	50 cm	_____

The following formulas have been used in solving these problems.

$$\text{Magnification Factor} = \frac{Image\ width}{Object\ width} \qquad SOD = SID - OID$$

$$\text{Magnification factor} = \frac{SID}{SOD}$$

Advanced Magnification

1. An object measuring 35 cm in width is radiographed at 100 cm source-to-image (SID) receptor distance and a 50 cm object-to-image receptor distance (OID). The radiographic image of the object will have a width of:

2.-6. State the length of the radiographic image that will result from using the following source-to-image (SID) receptor and object-to-image receptor distances (OID) and object lengths.

	Object Width	SID	OID	Image Size
2.	15 cm	100 cm	50 cm	_____
3.	20 inches	150 cm	50 cm	_____
4.	30 inches	200 cm	50 cm	_____
5.	30 cm	100 cm	40 cm	_____
6.	40 cm	100 cm	60 cm	_____

7. An object measuring 10 cm in length is radiographed at 200 cm source-to-image receptor distance (SID) and at 100 cm object-to-image receptor distance (OID). The radiographic image of the object will have a length of:

8.-12. State the length of the radiographic image which will result from the use of the following source-to-image (SID) receptor and object-to-image receptor distances (OID) and object lengths.

	Object Width	SID	OID	Image Size
8.	25 cm	100 cm	40 cm	_____
9.	30 cm	150 cm	50 cm	_____
10.	15 inches	200 cm	150 cm	_____
11.	20 inches	100 cm	60 cm	_____
12.	30 cm	180 cm	18 cm	_____

13. A radiograph is taken of an object that measures 10 cm in width, if the resulting radiographic image measures 20 cm in width. If a 100 cm source-to-image receptor distance (SID) was used, what was the source-to-object distance (SOD) employed?

14.-18. State the source-to-object distance that will result for the following source-to-image receptor distance (SID) and object and images sizes.

	Object Width	Image Width	SID	SOD
14.	25 cm	50 cm	100 cm	_____
15.	15 cm	20 cm	180 cm	_____
16.	50 cm	75 cm	100 cm	_____
17.	5 cm	5 cm	300 cm	_____
18.	12 cm	20 cm	100 cm	_____

19. An object is radiographed at a 200 cm source-to-image receptor distance (SID) and 100 cm OID. What is the percentage of magnification for the resulting radiographic image?

20. An object is radiographed at a 200 cm source-to-image receptor distance (SID) and 150 cm SOD. What is the percentage of magnification?

21. An object undergoes a two time linear magnification. The area of the radiographic image covered is _____ greater than the original object.

22. A 4 cm x 8 cm object undergoes a 1.5 magnification factor. What are the dimensions of the radiographic image?

23. A 6 cm diameter circular object undergoes a 2.5 magnification factor. What is the diameter of the radiographic image?

24. An 8 cm x 10 cm object undergoes a 2.0 magnification factor. What is the total area covered by the radiographic image?

25. A circular object covering 30 cm2 undergoes a 1.5 magnification factor. What is the total area covered by the radiographic image?

Advanced Magnification

1. An object measuring 25 cm in width is radiographed at 100 cm source-to-image receptor distance (SID) and a 50 cm object-to image receptor distance (OID). What is the width of the radiographic image of the object?

2.-6. State the length of the radiographic image that will result from using the following source-to-image receptor (SID) and object-to-image receptor distances (OID) and object lengths.

	Object Width	SID	OID	Image Size
2.	25 cm	100 cm	40 cm	_____
3.	30 inches	150 cm	50 cm	_____
4.	10 inches	100 cm	50 cm	_____
5.	80 cm	200 cm	40 cm	_____
6.	10 cm	180 cm	60 cm	_____

7. An object measuring 20 cm in length is radiographed at 200 cm source-to-image receptor distance and at 150 cm object-to-image receptor distance. What is the length of the radiographic image of the object?

8.-12. State the length of the radiographic image which will result from the use of the following source-to-image receptor and object-to-image receptor distances and object lengths.

	Object Width	SID	OID	Image Size
8.	5 cm	100 cm	60 cm	_____
9.	20 cm	250 cm	50 cm	_____
10	5 inches	300 cm	150 cm	_____
11.	10 inches	200 cm	40 cm	_____
12.	40 cm	180 cm	60 cm	_____

13. A radiograph is taken of an object measuring 10 cm in width if the resulting radiographic image measures 20 cm in width. If a 200 cm source-to-image receptor distance was used, what was the source-to-object distance employed?

14.-18. State the source-to-object distance that will result for the following source-to-image receptor distance and object and images sizes.

	Object Width	Image Width	SID	SOD
14.	15 cm	45 cm	100 cm	_____
15.	25 cm	30 cm	180 cm	_____
16.	60 cm	120 cm	40 cm	_____
17.	5 cm	25 cm	300 cm	_____
18.	12 cm	36 cm	100 cm	_____

19. An object is radiographed at a 200 cm source-to-image receptor distance (SID) and 150 cm object-to-image receptor distance (OID). What is the percentage of magnification for the resulting radiographic image?

20. An object is radiographed at a 300 cm source-to-image receptor distance (SID) and 150 cm source-to-object distance (SOD). What is the percentage of magnification?

21. An object undergoes a four time linear magnification. The area of the radiographic image covered is _____ greater than the original object.

22. A 6 cm x 9 cm object undergoes a 2.5 magnification factor. What are the dimensions of the radiographic image?

23. A 12 cm diameter circular object undergoes a 1.5 magnification factor. What is the diameter of the radiographic image?

24. An 7 cm x 15 cm object undergoes a 3.0 magnification factor. What is the total area covered by the radiographic image?

25. A circular object covering 24 cm2 undergoes a 1.5 magnification factor. What is the total area covered by the radiographic image?

Answers

Magnification
Exercise 1

1. 2.5	14. 2
2. 4	15. 2
3. 2	16. 3
4. 1.5	17. 2
5. 1.33	18. 1.66
6. 1.5	19. 1.33
7. 2	20. 2
8. 3	21. 4
9. 5	22. 2.5
10. 4	23. 1.66
11. 2	24. 1.11
12. 1.33	25. 1.25
13. 2.5	

Magnification
Exercise 2

1. 2	14. 3
2. 2	15. 2
3. 3	16. 2
4. 2.66	17. 4
5. 1.66	18. 1.25
6. 2.5	19. 4
7. 3	20. 1.5
8. 3	21. 1.5
9. 10	22. 1.5
10. 3	23. 2.5
11. 4	24. 1.25
12. 2	25. 1.2
13. 1.66	

Advanced Magnification
Exercise 1

1. 70 cm	14. 50 cm
2. 30 cm	16. 135 cm
3. 30 in.	15. 66.7 cm
4. 39.9 in.	17. 100 cm
5. 49.9 cm	18. 60 cm
6. 100 cm	19. 100%
7. 20 cm	20. 33%
8. 41.5 cm	21. Four
9. 45 cm	22. 6 cm by 12 cm
10. 60 in.	23. 15 cm
11. 50 in.	24. 16 cm by 20 cm
12. 33.3 cm	25. 45 cm^2
13. 50 cm	

Advanced Magnification
Exercise 2

1. 50 cm	14. 33.3 cm
2. 41.5 in.	15. 150 cm
3. 45 in.	16. 20 cm
4. 20 in.	17. 60 cm
5. 100 cm	18. 33.3 cm
6. 15 cm	19. 300%
7. 80 cm	20. 100%
8. 12.5 cm	21. 16 times
9. 25 cm	22. 15 cm by 22.5 cm
10. 10 in.	23. 18 cm
11. 12.5 in.	24. 21 cm by 45 cm
12. 60 cm	25. 36 cm^2
13. 100 cm	

Focal Spot Blur (Image Unsharpness)

1.-7. Calculate the amount of focal-spot blur (image unsharpness) given the following focal spot sizes, object-to-image receptor distance (OID) and source-to-image receptor distances (SID).

	Focal Spot Size	OID	SID	Focal Spot Blur
1.	1 mm	50 cm	50 cm	_____
2.	2 mm	30 cm	100 cm	_____
3.	1 mm	30 cm	60 cm	_____
4.	2 mm	10 cm	90 cm	_____
5.	.3 mm	30 cm	70 cm	_____
6.	2 mm	30 cm	70 cm	_____
7.	.3 mm	30 cm	90 cm	_____

8.-14. Calculate the amount of focal-spot blur (image unsharpness) given the focal spot size, object-to-image receptor distance (OID) and source-to-image receptor distance (SID).

	Focal Spot Size	OID	SID	Focal Spot Blur
8.	1 mm	50 cm	100 cm	_____
9.	2 mm	15 cm	60 cm	_____
10.	.3 mm	25 cm	100 cm	_____
11.	2 mm	5 cm	60 cm	_____
12.	1 mm	60 cm	100 cm	_____
13.	.3 mm	10 cm	180 cm	_____
14.	5 mm	15 cm	150 cm	_____

15.-20. Given the following pairs of factors, select the one that will show the least amount of focal spot blur (image unsharpness).

		SID	OID	FSS
15.	A.	100 cm	20 cm	1 mm
	B.	100 cm	20 cm	2 mm
	C.	100 cm	20 cm	.3 mm
16.	A.	200 cm	10 cm	1 mm
	B.	100 cm	10 cm	1 mm
	C.	180 cm	10 cm	2 mm
17.	A.	180 cm	5 cm	1 mm
	B.	180 cm	10 cm	1 mm
	C.	200 cm	5 cm	2 mm
18.	A.	180 cm	20 cm	.3 mm
	B.	100 cm	10 cm	1 mm
	C.	200 cm	10 cm	.3 mm
19.	A.	180 cm	40 cm	1 mm
	B.	100 cm	20 cm	2 mm
	C.	200 cm	80 cm	1 mm
20.	A.	200 cm	30 cm	2 mm
	B.	180 cm	60 cm	.3 mm
	C.	100 cm	30 cm	2 mm

The following formula has been used in calculating these problems.

$$\text{Focal Spot Blur} = \text{FSS} \times \frac{\text{OID}}{\text{SOD}} \qquad \text{SOD} = \text{SOD} - \text{OID}$$

Focal Spot Blur (Image Unsharpness)

1.-7. Calculate the amount of focal spot blur (image unsharpness) given the following focal spot sizes, object-to-image receptor (OID) and source-to-image receptor distances (SID).

	Focal Spot Size	OID	SID	Focal Spot Blur
1.	2 mm	100 cm	200 cm	_____
2.	1 mm	60 cm	160 cm	_____
3.	1 mm	20 cm	80 cm	_____
4.	2 mm	20 cm	200 cm	_____
5.	.3 mm	30 cm	100 cm	_____
6.	2 mm	40 cm	110 cm	_____
7.	.3 mm	35 cm	105 cm	_____

8.-14. Calculate the amount of focal spot blur (image unsharpness) given the focal spot size, object-to-image receptor distance (OID) and source-to-image receptor distance (SID).

	Focal Spot Size	OID	SID	Focal Spot Blur
8.	2 mm	50 cm	100 cm	_____
9.	1 mm	20 cm	80 cm	_____
10.	.3 mm	35 cm	135 cm	_____
11.	1 mm	15 cm	60 cm	_____
12.	2 mm	40 cm	100 cm	_____
13.	.3 mm	100 cm	180 cm	_____
14.	.5 mm	30 cm	300 cm	_____

15.-20. Given the following sets of factors, select the one that will show the least amount of focal spot blur (least image unsharpness).

		SID	OID	FSS
15.	A.	200 cm	20 cm	1 mm
	B.	200 cm	20 cm	2 mm
	C.	200 cm	20 cm	.3 mm
16.	A.	100 cm	20 cm	1 mm
	B.	200 cm	20 cm	1 mm
	C.	100 cm	10 cm	2 mm
17.	A.	180 cm	5 cm	1 mm
	B.	180 cm	10 cm	1 mm
	C.	90 cm	10 cm	1 mm
18.	A.	180 cm	20 cm	.3 mm
	B.	100 cm	10 cm	2 mm
	C.	200 cm	10 cm	.3 mm
19.	A.	180 cm	40 cm	1 mm
	B.	100 cm	20 cm	2 mm
	C.	100 cm	40 cm	1 mm
20.	A.	200 cm	30 cm	2 mm
	B.	180 cm	60 cm	.3 mm
	C.	100 cm	40 cm	2 mm

The following formula has been used in calculating these problems.

Focal Spot Blur = FSS x $\frac{OID}{SOD}$ SOD = SOD - OID

Answers

Focal Spot Blur
Exercise 1

1. 1	11. .18
2. .86	12. 1.5
3. 1	13. .02
4. .25	14. .05
5. .23	15. C
6. .02	16. A
7. .003	17. A
8. 1	18. C
9. .66	19. A
10. .1	20. B

Focal Spot Blur
Exercise 2

1. 2	11. .33
2. .60	12. 1.3
3. .33	13. .38
4. .22	14. .06
5. .13	15. C
6. 1.14	16. B
7. .15	17. A
8. 2	18. C
9. .33	19. A
10. .11	20. B

1. The main purpose of a beam limiting device such as the collimator is:

   A. The reduction of patient exposures and the improvement of image quality
   B. An increase in the beam intensity and beam quality
   C. The reduction in the number of low energy photons reaching the patient
   D. To improve the alignment of the film and the cassette

2. A modern positive beam limiting collimation system will need to contain all of the following EXCEPT:

   A. A Bucky tray that contains a series of electronic sensors
   B. A series of motor driven high density (lead) shutters
   C. Radiolucent mirror to redirect the light from the lamp
   D. A series of ionization type radiation sensors

Referring to the diagram, answer questions 3 & 4.

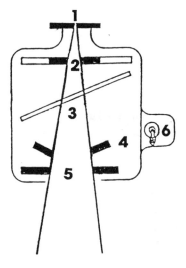

3. The mirror that reflects the light from the lamp toward the patient is represented by number:

   A. 1   C. 3
   B. 2   D. 5

4. The structures labeled number 4 represent the:

   A. Primary shutters   C. Entrance shutters
   B. Added filtration   D. Internal cones

5. The anode heel effect, which results from absorption of x-rays by the anode itself, results in the:

   A. Increase of the beam quality on the cathode side of the tube
   B. Elimination of off-focus radiation from the collimator
   C. Reduction in the beam intensity on the anode side of the tube
   D. Reduction of secondary scatter reaching the image receptor

6. In order to improve the visualization of a chest or abdominal image, the utilization of the anode heel effect the thickest portion of the body is placed under the:

   A. Anode side of the tube   C. Bucky side of the table
   B. Cathode side of the tube   D. Back side of the table

7. The upper or entrance shutters in a collimator serve principally to:

   A. Reduce the primary radiation reaching the image receptor
   B. Decrease the amount of off-focus radiation reaching the image receptor
   C. Increase the amount of scatter absorbed by the image receptor
   D. Reduce the appearance of the anode heel effect on the image receptor

8. If the focal spot to mirror distance in a collimator is 5 cm, the light to mirror distance should be:

   A. 5 cm   C. 10 cm
   B. 7.5 cm   D. 15 cm

9. The use of beam limiting devices will result in a smaller field coverage that, in turn, will tend to produce a:

1. Higher radiographic contrast   **2. Higher optical density**   **3. Higher beam intensity**

A. 1 only
B. 2 only
C. 3 only
D. 1, 2 & 3

10. Which of the following exposure fields should be employed in order to expose the least amount of tissue?

A. 12 cm diameter circular field
B. 8 cm x 8 cm square field
C. 10 cm x 7 cm rectangular field
D. 9 cm x 8 cm rectangular field

11. In order to best utilize the anode heel effect during upright chest radiography, the anode should be placed:

A. Laterally to the x-ray window
B. Medially to the x-ray window
C. Inferiorly to the cathode
D. Superiorly to the cathode

12. All types of currently employed beam limiting devices are located between the:

A. Patient and the image receptor
B. X-ray source and the patient
C. Patient and Bucky tray
D. Buck tray and AEC sensor

13. The Bureau of Radiologic Health has mandated that all new radiographic units must have a type of collimation known as:

A. Fixed collimation
B. Incident collimation
C. Automatic collimation
D. Negative collimation

14. An acceptable full abdomen radiograph of the gallbladder is obtained on a 35 cm x 43 cm image receptor. If a cone-down gallbladder view is taken using the same exposure and technical factors the likely result will be an image with a/an:

A. Higher optical density and higher radiographic contrast
B. Decreased optical density and higher radiographic contrast
C. Increased optical density and lower radiographic contrast
D. Higher optical density and lower radiographic contrast

15. The use of positive (automatic) collimation systems that are required on all new radiographic units will serve to:

A. Decrease the amount of scattered radiation produced in the patient
B. Increase the radiation quality of the x-ray beam
C. Increase the optical density of the radiographic image
D. Limit the entrance skin exposure to the tissues within the useful beam

16. The anode heel effect is more pronounced when images are acquired:

A. At long source-to-image receptor distances
B. When smaller sized image receptors are used
C. At short source-to-image receptor distance
D. When low grid ratio techniques are attempted

17. A 20 cm x 8 cm aperture diaphragm is located 10 cm below the x-ray source at a 40 cm source-to-image receptor distance. What is the size of the exposed field?

A. 5 cm x 2 cm
B. 10 cm x 4 cm
C. 40 cm x 16 cm
D. 80 cm x 32 cm

18. If the diameter of the circular x-ray field is 24 cm at 100 cm source-to-image receptor distance, what diameter would the beam cover at 200 cm source-to-image receptor distance?

A. 12 cm
B. 36 cm
C. 48 cm
D. 96 cm

Referring to the diagram, answer questions 19 & 20.

19. Pertaining to the diagram, which of the following numbers corresponds to the region with the least radiation intensity?

A. Number 1
B. Number 3
C. Number 5
D. All possess the same intensity

20. Pertaining to the diagram, number 3 corresponds to a relative exposure intensity rate of approximately:

A. 70%
B. 90%
C. 100%
D. 120%

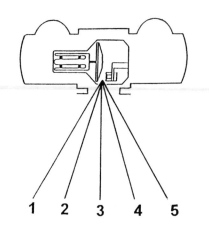

1 2 3 4 5

21. The width of the image unsharpness focal spot blur can be reduced by placing a beam limiting device as:

A. Close to x-ray source as possible
B. Close to the patient as possible
C. Close to the focal spot as possible
D. Close to the sensor as possible

22. When a variable aperture light illuminated collimator is employed, which of the following components must remain in the path of the x-ray beam?

A. Collimator light bulb
B. Positioning mirror
C. Motion trimmers
D. Induction motor

23. The collimator shutter leaves must have thickness that is equal to the lead equivalency of a/the:

A. Protective (lead) apron
B. Secondary barrier
C. Protective housing of the tube
D. Diagnostic filtration in the beam

24. Which of the following devices is NOT classified as a type of beam limiting device?

A. Collimator
B. Aperture diaphragm
C. Hooded anode
D. Extension cylinder

25. In order to provide the appropriate alignment with the x-ray beam, the positioning mirror should be mounted at an angle of:

A. 20 degrees to the central ray
B. 30 degrees to the image plane
C. 45 degrees to the central ray
D. 90 degrees to the image plane

Calculate the area covered for the following circular fields.

	Field Shape	Radius	Diameter	Area
26.	Circular	10 cm	20 cm	_____
27.	Circular	23 cm	46 cm	_____

Calculate the area covered for the following rectangular fields.

	Field Shape	Width	Length	Area
28.	Rectangular	18 cm	25 cm	_____
29.	Rectangular	12 cm	30 cm	_____
30.	Rectangular	40 cm	15 cm	_____

Chapter 24

Beam Limiting Devices

Explanations

1. A   Beam limiting or restricting devices; i.e., cones and collimators, tend to both reduce the scattered radiation produced in the patient and reduce the secondary scattered radiation reaching the image receptor. This will improve both the image quality and reduce the exposure to the patient. The reduction in the amount of scattered radiation reaching the image receptor will decrease the optical density of the resulting image unless an increase in incident beam intensity (mAs) is made prior to the exposure.

2. D   A positive beam limiting (automatic) collimator consists of two sets of motor driven shutters, a light localizing mirror, light source, and series of electronic sensors located in or under the Bucky tray. The sensors help to automatically collimate the beam to the borders of the image receptor.

3. C   A modern collimator consists of two sets of motor driven shutters (4), a light localizing mirror (3), light source (6), and series of electronic sensors located in or under the Bucky tray. The top of the collimator contains a fixed aperture that limits the maximum size of the radiation field to 17" by 17" at an SID of 100 cm.

4. A   The primary shutters are labeled number 4.

5. C   The tilting of the anode in the x-ray tube results in a beam intensity that varies from the head to the foot of the x-ray table. This effect, known as anode heel effect, provides a beam with a higher intensity towards the cathode side of the tube and a lower intensity on the anode side of the tube. In most modern rooms the cathode is on the right of the tube closest to the foot of the table.

6. B   Since the greatest radiation intensity is found at the cathode side of the tube, the heaviest body parts should be placed under the cathode or toward the foot end of the table.

7. B   The entrance shutters, which extend from the collimator into an opening in the bottom of the tube housing, serve to reduce the widely divergent, off-focus radiations produced from the anode disk or stem.

8. D   In order to produce light and radiation fields with the same physical size, the distance from the x-ray source and the mirror must be the same as the distance from light source and the mirror. This requires that the mirror be oriented to an angle of 45 degrees to the central ray.

9. A   All beam limiting devices serve to reduce the size and amount of the tissue exposed to the beam, thus reducing the scattered radiation produced in the patient. This results in an image consisting of improved resolution, decreased optical density, a higher contrast and reduced exposure to the patient.

10. B   Using the formulas: area = length x width and area = πr^2 ,the following values can be obtained A=113 cm2, B=64 cm2, C=70 cm2, D=72 cm2.

11. D   In order to provide a higher intensity to the lower thoracic cavity in an erect patient, the anode is placed toward the head or superior to the cathode in the upright projection.

12. B   In order to restrict the radiation beam before it reaches the patient, all beam limiting devices must be located between the source of the radiation and the patient.

13. C   Positive beam limitation (automatic collimation) refers to a device that can collimate the beam to the size of the cassette. In order to accomplish this process will require a series of sensors in the Bucky tray and incorporation of motor driven shutters in the collimator.

14. B   The use of tight collimation or use of an accessory cylinder will reduce both the scattered radiation produced in the patient and the percentage of radiation reaching the image receptor. This will produce a higher radiographic contrast and a reduced optical density in the image. A 30-50% increase in mAs or a 6-10 increase kVp setting is recommended to compensate for this expected drop in optical density.

15. A Beam limiting or restricting devices; i.e., cones and collimators, tend to both reduce the scattered radiation produced in the patient and reduce the secondary scattered radiation reaching the image receptor. This will improve both the image quality and reduce the exposure to the patient. The reduction in the amount of scattered radiation reaching the image receptor will decrease the optical density of the resulting image unless an increase in incident beam intensity (mAs) is made prior to the exposure.

16. B The anode heel effect is maximized when the entire length of the anode (large exposure field) is employed. This normally occurs when shorter (40") source-to-image receptor distances, largest image receptors and large collimator openings are used.

17. D This problem is solved by using the field coverage formula: Field Size(80 cm x 32 cm = source-to-image receptor distance (40 cm) x Size of the aperture (20 cm x 8 cm)/source-to-aperture distance (10).

18. C This problem is solved by using the field dimension formula:
New field dimension (48 cm) = (original field dimension (24 cm) x New SID (200 cm)
                                                    original SID (100 cm).

19. A The variation in radiation intensity across the x-ray beam, the anode heel effect, reduces the relative intensity of the beam on the anode side (1) and increases the relative intensity on the cathode side (5).

20. C The intensity at central ray or axis is designated as 100%.

21. B The maximum reduction in off-focus radiation occurs from a beam limiting device which is placed as close to the patient or as far from the source as possible.

22. B In order to provide light and radiation field coincidence, the mirror must be positioned within the beam equidistant from the x-ray and light sources at an angle of 45 degrees.

23. C In order to meet current regulations, the collimator shutters, like the tube housing, are considered to be primary barriers and are therefore required to have a lead equivalent thickness of 1.5-2 mm.

24. C The four main types of beam limiting devices include the aperture diaphragm, radiographic cone, radiographic cylinder and the variable aperture diaphragm or collimator.

25. C In order to provide light and radiation field coincidence, the mirror must be in the beam equidistant from the x-ray and light sources at an angle of 45 degrees.

26. 314 cm² This problem is solved using the area = πr^2 formula: 3.14 x 10 x 10 = 314 cm²

27. 1661 cm² This problem is solved using the area = πr^2 formula: 3.14 x 23 x 23 = 1661 cm²

28. 450 cm² This problem is solved using the area = length x width formula: 18 x 25 = 450 cm²

29. 360 cm² This problem is solved using the area = length x width formula: 12 x 30 = 360 cm²

30. 600 cm² This problem is solved using the area = length x width formula: 40 x 15 = 600 cm²

Radiographic Grids

1. The principal reason for using a radiographic grid in a diagnostic examination is:

   A. Reduction in the amount of secondary scatter reaching the image receptor
   B. Increase in intensity of the radiation reaching the image receptor
   C. The reduction of the amount of backscatter radiation reaching the image receptor
   D. Reduction of patient exposure

2. The reduction of secondary radiation reaching the image receptor can be accomplished by using:

   1. An increased OID   2. Radiographic grids   3. Beam Limiting devices

   A. 1 & 2 only   C. 2 & 3 only
   B. 1 & 3 only   D. 1, 2 & 3

3. Grid ratio is represented by the formula:

   A. $r = h \times D$ C. $r = \dfrac{h}{D}$

   B. $r = \dfrac{25.4}{D + d}$ D. $r = \dfrac{D}{h}$

Referring to the diagram, answer questions 4 & 5.

4. Which grid cross-section demonstrates the grid with the highest grid ratio?

   A. Grid 1   C. Grid 3
   B. Grid 2   D. All have the same ratio

5. Which of these grids if used for a non-Bucky application would show the most objectionable grid pattern on the radiographic image?

   A. Grid 1   C. Grid 3
   B. Grid 2   D. All would have the same radiographic
   appearance

6. Which type of grid pattern will result in a radiograph with the least amount of primary beam cutoff?

   A. Focus grid   C. Crossed grid
   B. Parallel grid   D. Rhombic grid

7. The principal function of the Potter-Bucky diaphragm (Bucky tray) is to provide a simple means for moving the imaging device without moving the patient. The Bucky tray also servers to:

   A. Reduce the field of view
   B. Reduce the exposure to patients
   C. Move the grid during the exposure
   D. Reduce the amount of backscattered radiation produced

8. Which of the following combination of factors is most likely to be associated with a radiograph image possessing the lowest radiographic contrast?

	mA	Time	kVp	Grid Ratio
A.	200	25 ms	86	16:1
B.	200	20 ms	94	12:1
C.	400	10 ms	102	8:1
D.	300	10 ms	64	8:1

9. Which of the following focus grids will allow for the greatest amount of centering and distance latitude:

A. 6:1 ratio

B. 8:1 ratio

C. 12:1 ratio

D. 16:1 ratio

10. In order to select a grid with the proper ratio, which of the following factors should be considered?

1. Positioning latitude   *2. Degree of cleanup*   *3. Kilovoltage range*

A. 1 only

B. 2 only

C. 3 only

D. 1, 2 & 3

Referring to the radiographic images, answer questions 11 & 12.

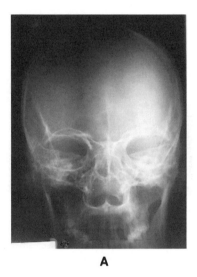

A

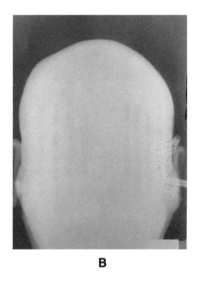

B

11. In the radiographic image (A), the decreased optical density seen in the lateral portion of the image mostly resulted from a/an:

A. Excessive collimation

B. Improperly aligned grid

C. Patient motion

D. Excessive focal film distance

12. In the radiographic image (B), the decreased optical density seen throughout the image was caused by an:

A. Improper centering

B. Excessive Bucky movement

C. Upside down gird

D. Improper technical compensation

13. In order to avoid the appearance of grid lines using a moving grid, the Bucky tray motion must commence:

A. Before the exposure begins

B. After the exposure begins

C. Grid movement is not necessary

14. A grid that is constructed with all the grid strips running parallel to each other is an example of a:

A. Linear focus grid   C. Rhombic grid
B. Linear parallel grid   D. Cross-hatched grid

15. When grids are utilized, the technical factors must be adjusted to maintain the optical density of the image because:

1. The amount of scatter produced in the patient decreases
2. The amount of scatter reaching the image receptor decreases
3. The source-to-image receptor distance must be increased

A. 1 only   C. 3 only
B. 2 only   D. 1, 2 & 3

Using the following diagram, answer questions 16 & 17.

16. This grid will have a ratio of:

A. 16:1
B. 12:1
C. 8:1
D. 6:1

17. The frequency of the grid strips is:

A. 100 lines/in.
B. 66 lines/in.
C. 33 lines/in.
D. 11 lines/in.

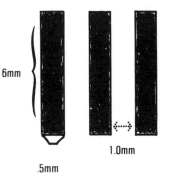

6mm

1.0mm

.5mm

18. Which of the following combination of factors is most likely to be associated with a radiographic image possessing the highest radiographic contrast?

	mA	Time	kVp	Grid Ratio
A.	300	30 ms	68	12:1
B.	400	25 ms	87	8:1
C.	200	50 ms	84	10:1
D.	600	15 ms	98	12:1

19. The distance at which a focus grid can be used is most closely related to the:

A. Grid ratio   C. Convergence line
B. Grid frequency   D. Interspacing material

20. If grid ratio was increased, and no technical changes made, the resulting radiograph would show an increase in:

A. Optical density   C. Contrast
B. Magnification   D. Shape distortion

21. The use of high ratio grids are most often recommended when:

A. A high kilovoltage technique must be employed
B. A number of different source-to-image distances need to be employed
C. The dose reduction to the patient is of great concern
D. A considerable degree of off-centering may be required

22. In a modern 8:1 focus grid, the majority of scatter reduction results from the:

    A. Lead grid strips
    B. Aluminum grid strips
    C. Lead interspacing material
    D. Aluminum interspacing material

23. What is the ratio of a grid that has a grid strip thickness of 4 mm, a grid interspace thickness of 20 mm and height of 160 mm.?

    A. 5:1
    B. 6:1
    C. 8:1
    D. 16:1

24. The amount of primary beam cutoff by grids can be greatly reduced by employing a:

    A. Crossed grid
    B. Focused grid
    C. Linear grid
    D. Moving grid

25. A type of grid in which the grid lines are situated parallel to the divergent rays of the x-ray beam is termed a:

    A. Stationary grid
    B. Parallel grid
    C. Linear grid
    D. Focused grid

26. The ability of a grid to absorb scattered radiation after it exits from the patient, is termed the:

    A. Contrast improvement factor
    B. Grid efficiency
    C. Grid uniformity
    D. Grid sensitivity

Referring to the diagram, answer questions 27 & 28.

27. Which of the following grid cross-sections is associated with the highest grid ratio?

    A. Grid 1
    B. Grid 2
    C. Grid 3
    D. All have the same ratio

28. Which of the following grids if used in a Bucky application would be associated with the lowest contrast improvement factor?

    A. Grid 1
    B. Grid 2
    C. Grid 3
    D. All will provide a similar contrast improvement factor

29. An increase in the amount of scatter pickup or contrast improvement factor is associated with a/an:

    1. Increase in the grid ratio   2. Increase in the grid frequency   3. Decrease in the selectivity

    A. 1 only
    B. 2 only
    C. 3 only
    D. 1, 2 & 3

30. When a focus type grid is used the appearance of grid cutoff is associated with all of the following EXCEPT:

A. Tube angulations directed along the grid strips
B. A central ray that is directed to the side of the grid
C. The use of the wrong source-to-image receptor distance
D. The excessive tilting of the grid to the tube

31. The main advantage involved with the use of a radiographic grid is:

A. Improved radiographic contrast of the image
B. Reduced exposure to the patient
C. Reduced amount of scatter production in the patient
D. An increased beam intensity

32. What is the ratio of a grid that is constructed with .125 mm thick lead strips, 8 mm in height and separated by .5 mm aluminum spacers?

A. 4:1 ratio   C. 12:1 ratio
B. 8:1 ratio   D. 16:1 ratio

33. A radiograph is produced in a table Bucky. The tube was properly centered to the image receptor in the Bucky tray yet grid lines were seen. A probable cause is:

1. Excessive grid speed   2. Insufficient grid speed   3. Insufficient collimation

A. 1 only   C. 3 only
B. 2 only   D. 1, 2 & 3

34. The man most often credited with the construction of the first practical radiographic grid is:

A. W.K. Roentgen   C. G. Bucky
B. H. Potter   D. J.P. Focus

35. The main advantage of a moving grid compared to a stationary grid of the same ratio and frequency is:

1. The ability to reduce the appearance of the grid lines
2. The ability to reduce the amount of grid cutoff
3. The ability to employ larger tube angulations

A. 1 only   C. 3 only
B. 2 only   D. 1, 2 & 3

36. A crossed grid is formed by 8:1 and 6:1 linear grids. This combined grid will have distance and centering latitudes proportional to a/an:

A. 6:1 linear grid   C. 13:1 linear grid
B. 8:1 linear grid   D. 16:1 linear grid

37. The amount of scatter reduction when a 6:1 crossed grid is used is nearly equivalent to that of a/an:

A. 6:1 fine line grid   C. 8:1 linear grid
B. 12:1 linear grid   D. 5:1 focus grid

Referring to the radiographic images, answer questions 38 & 39.

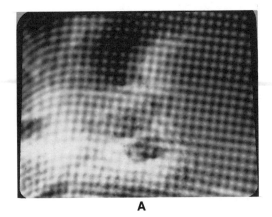

A

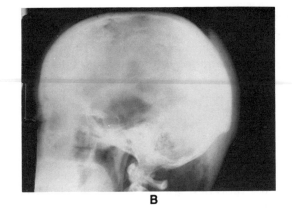

B

38. The radiographic image of film A demonstrates a grid artifact called the:

    A. Moire effect
    B. Quantum mottle effect

    C. Rhomboid effect
    D. Cross hatch effect

39. In the radiograph image of film B, a decrease of optical density in the lateral areas of the image were caused by a/an:

    A. Off focus decentering
    B. Upside down focus grid

    C. Improperly aligned grid
    D. Tube angulation across the grid

40. When an 8:1 grid is used with a film screen image receptor (cassette), in order to maintain the desired optical density, the technique should be:

    A. Increased by 1 ½ times the mAs value
    B. Increased by 3 times the mAs value

    C. Decreased by 1 ½ times the mAs value
    D. Decreased by 3 times the mAs value

41. A radiograph is taken that shows peripheral grid cutoff on both sides of a radiographic image. The possible cause includes a/an:

    1. Upside down focus grid
    2. Excessive source-to-image distance
    3. Deficient source-to-image distance

    A. 1 only
    B. 2 only

    C. 3 only
    D. 1, 2 & 3

42. In general, when tube potential exceeds 90 kVp the minimum grid ratio that should be employed is:

    A. 5:1
    B. 6:1

    C. 12:1
    D. 16:1

43. Which of the following combination of factors is most likely to be associated with a radiograph image possessing the highest radiographic contrast?

	mA	Time	kVp	Grid Ratio
A.	300	15 ms	110	8:1
B.	200	35 ms	90	5:1
C.	200	25 ms	100	6:1
D.	400	5 ms	84	8:1

44. The first moving grid was designed by:

    A. H. Potter
    B. T. Edison

    C. G. Bucky
    D. M. Pupin

45. Which of the following focus grids would be associated with the least amount of centering and distance latitude?

   A. 5:1 ratio grid
   B. 6:1 ratio grid

   C. 8:1 ratio grid
   D. 12:1 ratio grid

46. The appearance of grid lines, when a focused grid is used in a Bucky tray, is likely in all of the following situations EXCEPT?

   A. When an excessive source-to-image receptor distance is used
   B. When the grid is stopped after the exposure commences
   C. When the grid begins to move before the exposure commences
   D. When the grid is not centered to the tube

47. The ratio of the height of the lead strips to the distance between the lead strips defines:

   A. Grid ratio
   B. Grid radius

   C. Grid latitude
   D. Lines per inch

48. Which of the grid factors listed below will result in the greatest amount of scattered radiation attenuation?

   A. 6:1 ratio/60 lp/in
   B. 8:1 ratio/60 lp/in

   C. 6:1 ratio/100 lp/in
   D. 8:1 ratio/100 lp/in

49. The placement of two identical parallel grids on top of each other with the grid strips running in the same direction may cause:

   A. Parallel effect
   B. Strobe effect

   C. Moire effect
   D. Heel effect

50. The contrast improvement that can be obtained with a particular grid is dependent upon the:

   1. Kilovoltage employed   *2. Size of the field*   *3. Patient thickness*

   A. 1 only
   B. 2 only

   C. 3 only
   D. 1, 2 & 3

Referring to the radiographic images, answer questions 51 & 52.

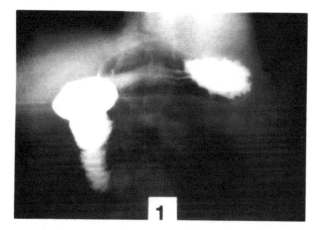
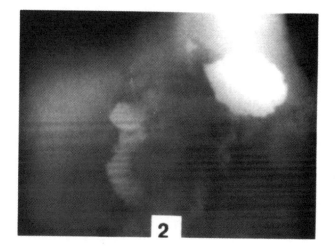

51. These radiographic images were taken on the same patient within a minute of each other. Comparing two images, image 1 is to have a/an:

    A. Lower radiographic contrast than image 2
    B. Higher recorded detail than image 2
    C. Lower optical density than image 2
    D. Improved spatial resolution than image 2

52. The use of which of the following factors was responsible for the improved image quality of image 2?

    A. The application of automatic collimation
    B. The use of a stationary grid
    C. The use of a higher kVp
    D. The use of a higher speed imaging system

53. If a 12:1 ratio grid is substituted for an 8:1 ratio grid, and the same technical factors are employed, the image will possess a higher:

    1. Optical density   2. Radiographic contrast   3. Percentage of scatter radiation

    A. 1 only
    B. 2 only
    C. 3 only
    D. 1, 2 & 3

54. Because tube angulations are required during many normal radiographic procedures the least commonly employed grid for a general purpose room is a:

    A. Parallel grid
    B. Crossed grid
    C. Focused grid
    D. Linear grid

55. When the greatest reduction of scattered radiation is required, a/an _____ ratio grid should be employed.

    A. 6:1
    B. 8:1
    C. 12:1
    D. 16:1

1.   A   In addition to holding the image receptor in the center of the table for an exposure, the Bucky contains a stationary grid and a spring or motor driven device which moves the grid to blur the appearance of the grid lines. The primary purpose of the grid is to reduce the amount of secondary scattered radiation that reaches the image receptor.

2.   D   A reduction in scattered radiation can be accomplished by the restriction of the radiation source prior to its reaching the object through the use of beam limiting devices or after it leaves the object by air gap techniques and radiographic grids.

3.   C   The grid ratio (r) which is the single best factor for predicting grid efficiency and latitude is determined by the ratio of the height of the grid strips (h) to the distance between the grid strips (D): $r = h/D$.

4.   A   Since all of the grids have the same height, the grid with the highest ratio will have the smallest distance between the grid strips of grid 1.

5.   C   During a non-Bucky exposure, using a stationary grid, the grid having the thickest grid strips (grid 3) will produce the most noticeable grid lines on the radiographic image.

6.   A   The type of grid pattern which most parallels the divergence of the primary beam, the focus grid, will also absorb the least amount of primary radiation to which it is exposed.

7.   C   Hollis Potter developed a device for the movement of a grid during the exposure. This device was later incorporated in the Potter-Bucky diaphragm (Bucky tray) to eliminate the appearance of the grid strips on the radiographic image.

8.   C   Radiographic contrast is primarily controlled by the kilovoltage peak of the primary beam and the ratio of the radiographic grid. Low contrast or long scale contrast will result from a combination of a high kVp technique and low ratio grid.

9.   A   The amount of off-centering permitted before noticeable grid cutoff is seen, is termed centering latitude. Distance latitude defines the range of source-to-image receptor distances that can be used before causing a noticeable amount of cutoff. Both the distance and centering latitude ranges are increased as the grid ratio is reduced. High ratio grids are associated with more scatter cleanup and narrow latitude or range of latitudes.

10.   D   The appropriate grid for a given situation is based on the degree of cleanup required, the centering and distance latitudes that may be necessary and the range of kilovoltages that will need to be used to provide the desired penetration.

11.   B   The reduced optical density seen on the right side of this image is caused by an uneven attenuation or grid cutoff due to failure to direct the central ray to the center of the grid.

12.   D   When grids are employed during a radiographic examination, they serve to improve contrast by the absorption or attenuation of the scattered radiation produced in the patient. Since this will also reduce the total number of photons reaching the image receptor, an increase in the mAs value selected is necessary to maintain the appropriate image quality. This image was made without the necessary technical compensation and shows a marked reduction in the radiographic density.

13.   A   The movement of the grid within the Bucky tray must begin prior to the exposure and continue until the exposure is terminated to avoid the appearance of grid lines on the image. The movement of the grid in most modern units is started by the activation of the prep or boost stage of the exposure switch.

14.   B   Any grid which is designed with grid strips which run parallel with each other are called linear parallel grids.

15. B The use of a grid with increased attenuation or absorption of the primary beam is necessary to increase the technique factors to avoid a reduction in the optical density of the image.

16. D The problem is solved using the grid ratio formula: *grid ratio r = h/D*. The grid ratio (r) which is the single best factor for predicting grid efficiency and latitude is determined by the ratio of the height of the grid strips (h) to the distance between the grid strips (D): *r=h/D*.

17. B This problem is solved using the grid frequency formula: *frequency = 100/(D+d)*, when D is the distance between the strips and d is the thickness of the lead strip.

18. A Radiographic contrast is primarily controlled by the kilovoltage peak of the primary beam and the ratio of the radiographic grid. Low contrast or long scale contrast will result from a combination of a high kVp technique and low ratio grid.

19. C A linear grid is designed to have a series of grid strips running lengthwise parallel to the x-ray table. The term focus applies to the titling of this strip to match the divergence of the primary beam. The inability of a linear focus grid to maintain its alignment with the primary beam results in an uneven absorption of the primary radiation called grid cutoff. This occurs when the central ray is improperly centered, the wrong distance is used or the tube is angled across the grid strips.

20. C The efficiency or scatter cleanup of a grid is directly proportional to grid ratio. Therefore, grids with a higher ratio eliminate more scattered radiation and have a higher contrast improvement factor than grids with lower ratios.

21. A The design of many high ratio grids (16:1) requires the use of a relatively high 90-120 kilovolt peak range in order to provide acceptable attenuation factors.

22. A In all grids, it is the high atomic number (82) of the lead strips that makes the attenuation of scattered radiation possible.

23. C The grid ratio (r) which is the single best factor for predicting grid efficiency and latitude is determined by the ratio of the height of the grid strips (h) to the distance between the grid strips (D): *r=h/D*.

24. B Because the design of a focus grid will more readily follow the divergence of the primary beam, it has the lower amount of primary beam cut than a parallel grid.

25. D A focus grid is designed with grid strips that parallel the divergence of the x-ray beam. The angled grid strips converge at a point over the center of the grid; a line drawn perpendicular from this point to the grid defines the optimum distance for the use of the grid.

26. B The grid efficiency is defined as the ability of the grid to absorb scattered radiation. Cleanup from a grid is primarily effected by two factors: grid ratio and the grid frequency. The highest cleanup or grid efficiency will result from a grid design having high ratio and the lowest grid frequency.

27. B The use of a higher grid ratio is associated with a lower radiographic density, higher radiographic contrast, improved scatter cleanup and higher radiation exposure to the patient when the appropriate technical compensations are made to maintain the optical density.

28. B Crossed grids which are comprised of two individual grids with their grid strips directed at right angles to each other have a major disadvantage of not permitting any tube angulated techniques. These grids otherwise have an excellent scatter clean-up, wide distance, and centering latitudes.

29. A The grid efficiency is defined as the ability of the grid to absorb scattered radiation. cleanup from a grid is primarily effected by two factors: grid ratio and the grid frequency. The highest cleanup or grid efficiency will result from a grid design having high ratio and the lowest grid frequency.

30. A Grid cutoff with a focus grid may occur from the use of the wrong source-to-image receptor distance, improper centering, off-leveling of the grid and angulating across the grid strips.

31. A Though grids are associated with higher patient exposures, the reduction in the amount of scattered radiation reaching the image receptor dramatically improves the radiographic contrast of the image. The use of the grid will not decrease the amount of scattered radiation produced in the patient or object, only that which reaches the image receptor.

32. D The grid ratio (r) which is the single best factor for predicting grid efficiency and latitude is determined by the ratio of the height of the grid strips (h) to the distance between the grid strips (D): $r=h/D$.

33. B Grid lines associated with moving the (Bucky) grid, though rarely seen, may result from failing to place the grid in motion prior to the exposure, having the grid moving too slowly, or having the grid stop during the exposure.

34. C The first practical radiographic grid was developed by Dr. Gustave Bucky in the early 1910's and though many design changes have been made, the basic form of the grid is still close to the original design.

35. A The movement of the grid during the exposure will blur the grid strips, thus eliminating their appearance on the image.

36. B The combined or crossed grid latitude produced by placing two grids on top of each other, has a centering latitude of the grid with the highest ratio (8:1 grid).

37. B A cross grid consisting of two individual 6:1 grid at right angels with each other has the latitude ranges of a 6:1 grid but grid efficiency of a 12:1 grid.

38. A The unusual (moire effect) appearance of this image was due to the placement of two separate grids on top of each other with their grid strips parallel to each other.

39. B A reduction of optical density on the lateral edges of the image and a near normal density in the center of the image may occur from the use of an excessive SID, insufficient SID or, as it was in this case, an upside-down grid.

40. B Though many different references exist, a general consensus finds that a 3 or 4 times increase in mAs is required to maintain appropriate optical density when an 8:1 grid is employed.

41. D A reduction of optical density on the lateral edges of the image and a near normal density in the center of the image may occur from the use of an excessive SID, insufficient SID or, as it was in this case, an upside-down grid.

42. C High ratio grids (16:1) are recommended for use with kilovoltage above 90 kVp. Low ratio grids (5:1 -6:1) are usually limited to kilovoltages below 80 kVp.

43. D Radiographic contrast is primarily controlled by the kilovoltage peak of the primary beam and the ratio of the radiographic grid. Low contrast or long scale contrast will result from a combination of a high kVp technique and low ratio grid.

44. A A major advance in the application of grid was made in the 1920's by Dr. Hollis Potter and his design of a diaphragm capable of moving the grid during the exposure. Now commonly called a Bucky tray, its proper name is the Potter-Bucky diaphragm.

45. C The grid with the highest ratio will have the least amount of distance or centering latitude.

46. C The movement of the grid within the Bucky tray must begin prior to the exposure and continue until the exposure is terminated to avoid the appearance of grid lines on the image. The movement of the grid in most modern units is started by the activation of the prep or boost stage of the exposure switch.

47. A The grid ratio (r) which is the single best factor for predicting grid efficiency and latitude is determined by the ratio of the height of the grid strips (h) to the distance between the grid strips (D): $r=h/D$.

48. B The amount of scattered radiation a grid eliminates is controlled by the grid ratio, total lead content and the number of lines per inch (frequency). The highest efficiency or amount of scatter reduction occurs with a lower frequency, high lead content and high grid ratio.

49. C The unusual (moire effect) appearance of this image was due to the placement of two separate grids on top of each other with their grid strips parallel to each other.

50. D The radiographic contrast of an image is dependent upon the patient factors of tissue density and thickness, degree of beam limitation, the kilovoltage peak selected and the use of air gap techniques.

51. A Image 1 taken without a radiographic grid has a lower radiographic contrast than image 2.

52. B The use of the radiographic grid was the main reason for the greater contrast in image 2.

53. B Since a higher ratio grid will absorb more scattered radiation the optical density will be reduced and the radiographic contrast will increase.

54. B Because crossed grids do not permit the use of tube angulations they are rarely used in a general purpose room.

55. D The grid with the highest ratio will reduce the greatest amount of radiation.

Grid Conversion

1. A radiograph is taken with 200 speed imaging system, non-Bucky, using 10 mAs at 80 kVp. If a 6:1 grid is employed, what new mAs should be used to maintain the same optical density?

2.-10. State the new mAs value that should be employed to maintain the same optical density when the following grid ratio changes are made.

	Original mAs	Original Grid Ratio	New Grid Ratio	New mAs
2.	21 mAs	No grid	8:1	_____
3.	20 mAs	8:1	12:1	_____
4.	15 mAs	6:1	12:1	_____
5.	60 mAs	No grid	12:1	_____
6.	100 mAs	12:1	16:1	_____
7.	50 mAs	6:1	8:1	_____
8.	70 mAs	No grid	6:1	_____
9.	5 mAs	No grid	16:1	_____
10.	12 mAs	5:1	8:1	_____

11. A radiograph is taken with a 400 speed imaging system with 12:1 Bucky using 40 mAs at 80 kVp. If a 6:1 grid is substituted, what new mAs should be used to maintain the same optical density?

12.-20. State the mAs value that should be employed to maintain the same optical density when the following grid ratio changes are made.

	Original mAs	Original Grid Ratio	New Grid Ratio	New mAs
12.	100 mAs	12:1	8:1	_____
13.	20 mAs	6:1	No grid	_____
14.	50 mAs	12:1	6:1	_____
15.	60 mAs	16:1	No grid	_____
16.	42 mAs	8:1	No grid	_____
17.	12 mAs	16:1	12:1	_____
18.	24 mAs	8:1	5:1	_____
19.	16 mAs	12:1	No grid	_____
20.	60 mAs	6:1	5:1	_____

All answers on this form were calculated using the following grid conversion factors.

No grid = 1 x mAs

5:1 = 1.5 x mAs

10:1-12:1 = 4 x mAs

6:1 = 2 x mAs

16:1 = 5 x mAs

8:1 = 3 x mAs

The following formula has been used in calculating these problems.

New mAs = $\dfrac{\text{Original mAs x New Grid Conversion Factor}}{\text{Original Grid Conversion Factor}}$

Grid Conversion

1. A radiograph is taken with 200 speed imaging system, non-Bucky, using 5 mAs at 80 kVp. If an 8:1 grid is employed, what new mAs should be used to maintain the same optical density?

2.-10. State the new mAs value that should be employed to maintain the same optical density when the following grid ratio changes are made.

	Original mAs	Original Grid Ratio	New Grid Ratio	New mAs
2.	30 mAs	No grid	12:1	_____
3.	10 mAs	No grid	16:1	_____
4.	25 mAs	6:1	16:1	_____
5.	30 mAs	12:1	16:1	_____
6.	7 mAs	No grid	8:1	_____
7.	15 mAs	8:1	12:1	_____
8.	12 mAs	6:1	16:1	_____
9.	5 mAs	5:1	8:1	_____
10.	100 mAs	No grid	12:1	_____

11. A radiograph is taken with a 400 speed imaging system with 12:1 Bucky using 60 mAs at 85 kVp. If a 6:1 grid is substituted, what new mAs should be used to maintain the optical density?

12.-20. State the mAs value that should be employed to maintain the same optical density when the following grid ratio changes are made.

	Original mAs	Original Grid Ratio	New Grid Ratio	New mAs
12.	10 mAs	6:1	No grid	_____
13.	5 mAs	16:1	No grid	_____
14.	7 mAs	12:1	8:1	_____
15.	16 mAs	8:1	No grid	_____
16.	100 mAs	6:1	5:1	_____
17.	24 mAs	12:1	6:1	_____
18.	36 mAs	8:1	5:1	_____
19.	80 mAs	12:1	No grid	_____
20.	30 mAs	16:1	12:1	_____

All answers on this form were calculated using the following grid conversion factors.

No grid = 1 x mAs

5:1 = 1.5 x mAs

10:1 or 12:1 = 4 x mAs

6:1 = 2 x mAs

16:1 = 5 x mAs

8:1 = 3 x mAs

The following formula has been used in calculating these problems.

New mAs = $\dfrac{\text{Original mAs x New Grid Conversion Factor}}{\text{Original Grid Conversion Factor}}$

Density Analysis Grid

In each of the following sets of exposure and technical factors, select the set of factors that would provide the greatest optical density.

		mA	Time	kVp	Grid Ratio
1.	A.	300	30 ms	94	5:1
	B.	300	10 ms	89	12:1
	C.	300	20 ms	84	8:1
	D.	300	1 ms	86	16:1
2.	A.	200	15 ms	76	8:1
	B.	100	25 ms	74	12:1
	C.	300	20 ms	78	8:1
	D.	200	6 ms	78	12:1
3.	A.	600	50 ms	90	12:1
	B.	400	50 ms	92	8:1
	C.	800	50 ms	88	6:1
	D.	600	50 ms	90	No grid
4.	A.	600	35 ms	105	12:1
	B.	300	12 ms	103	6:1
	C.	300	33 ms	98	12:1
	D.	600	15 ms	95	8:1
5.	A.	200	250 ms	88	6:1
	B.	300	330 ms	98	12:1
	C.	100	500 ms	88	8:1
	D.	200	15 ms	86	6:1
6.	A.	300	500 ms	96	12:1
	B.	200	750 ms	92	16:1
	C.	100	1.5 sec	90	5:1
	D.	400	600 ms	88	8:1
7.	A.	400	1 ms	86	12:1
	B.	800	50 ms	89	No grid
	C.	400	70 ms	94	6:1
	D.	800	100 ms	94	No grid
8.	A.	300	50 ms	72	No grid
	B.	600	.15 sec	80	16:1
	C.	100	.15 sec	82	12:1
	D.	300	1 ms	75	8:1
9.	A.	200	.3 sec	92	6:1
	B.	100	.4 sec	92	12:1
	C.	200	.75 sec	88	8:1
	D.	100	.25 sec	82	6:1
10.	A.	800	.025	86	5:1
	B.	600	.033	84	8:1
	C.	400	.05	92	12:1
	D.	600	.008	86	8:1

Density Analysis Grid

In each of the following sets of exposure and technical factors, select the set of factors that will result in the greatest optical density.

		mA	Time	kVp	Grid Ratio
1.	A.	100	30 ms	84	6:1
	B.	100	10 ms	76	16:1
	C.	100	20 ms	82	8:1
	D.	100	10 ms	76	No grid
2.	A.	200	150 ms	86	12:1
	B.	400	330 ms	89	12:1
	C.	300	250 ms	82	8:1
	D.	100	150 ms	84	12:1
3.	A.	600	150 ms	92	6:1
	B.	300	70 ms	82	No grid
	C.	800	50 ms	84	12:1
	D.	1000	10 ms	74	6:1
4.	A.	300	15 ms	96	16:1
	B.	100	25 ms	92	8:1
	C.	100	33 ms	87	16:1
	D.	400	15 ms	76	12:1
5.	A.	200	10 ms	86	5:1
	B.	300	10 ms	89	8:1
	C.	300	10 ms	92	16:1
	D.	200	10 ms	84	12:1
6.	A.	800	50 ms	86	6:1
	B.	800	70 ms	86	8:1
	C.	600	150 ms	92	12:1
	D.	600	100 ms	88	12:1
7.	A.	300	25 ms	87	8:1
	B.	100	33 ms	76	No grid
	C.	600	33 ms	89	6:1
	D.	30	15 ms	86	8:1
8.	A.	400	7 ms	83	8:1
	B.	600	4 ms	92	No grid
	C.	400	100 ms	91	12:1
	D.	600	70 ms	70	16:1
9.	A.	800	.15 sec.	76	12:1
	B.	400	.75 sec.	82	16:1
	C.	300	.66 sec.	94	16:1
	D.	800	.33 sec.	94	8:1
10.	A.	100	.016 sec.	82	6:1
	B.	300	.016 sec.	84	8:1
	C.	200	.016 sec.	82	8:1
	D.	100	.008 sec.	82	6:1

Answers

Grid Conversion
Exercise 1

1. 20 mAs	11. 20 mAs
2. 63 mAs	12. 75 mAs
3. 27 mAs	13. 10 mAs
4. 30 mAs	14. 25 mAs
5. 240 mAs	15. 12 mAs
6. 125 mAs	16. 14 mAs
7. 75 mAs	17. 10 mAs
8. 140 mAs	18. 12 mAs
9. 25 mAs	19. 4 mAs
10. 24 mAs	20. 45 mAs

Grid Conversion
Exercise 2

1. 15 mAs	11. 30 mAs
2. 120 mAs	12. 5 mAs
3. 50 mAs	13. 1 mAs
4. 63 mAs	14. 5 mAs
5. 38 mAs	15. 5 mAs
6. 21 mAs	16. 75 mAs
7. 20 mAs	17. 12 mAs
8. 30 mAs	18. 18 mAs
9. 10 mAs	19. 20 mAs
10. 400 mAs	20. 24 mAs

Density Analysis Grid
Exercise 1

1. A
2. C
3. D
4. A
5. B
6. C
7. D
8. B
9. C
10. A

Density Analysis Grid
Exercise 2

1. A
2. B
3. A
4. A
5. B
6. C
7. C
8. C
9. D
10. B

1. The principal action of the crystals found in intensifying screens is the conversion of:

   A. Visible light into electrons
   B. X-rays into visible light

   C. Visible light into x-rays
   D. Electrons into visible light

2. The process by which a crystalline material gives off visible light in response to x-radiation is termed:

   A. Luminescence
   B. Phosphorescence

   C. Incandescence
   D. Photo-electrolysis

3. Gadolinium and lanthanum are two common materials used in a new class of high speed screens called:

   A. Rare earth phosphors
   B. Parallax phosphors

   C. Gamma phosphors
   D. Quantum phosphors

Referring to the diagram, answer questions 4 & 5.

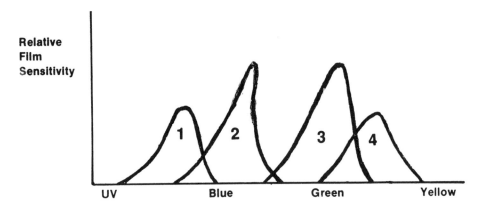

4. The diagram shows the relative light emission of four different phosphors to the same amount x-radiation. Which phosphor should be used with a green sensitive x-ray film?

   A. Phosphor number 1
   B. Phosphor number 2

   C. Phosphor number 3
   D. Phosphor number 4

5. Which of these most likely represents the emission from rare earth type phosphors?

   A. 1 & 2 only
   B. 1 & 4 only

   C. 2 & 3 only
   D. 1, 2, 3, & 4

6. The effect of grainy appearance on a radiographic image caused by quantum mottle is most commonly seen when:

   A. Cardboard type holder image receptors are used
   B. Extremely high speed imaging systems are used

   C. Relative slow speed imaging systems are used
   D. Single emulsion type imaging systems are used

7. The main reason that rare earth phosphors have all but replaced the older calcium tungstate phosphors in modern intensifying screens is their:

   A. Higher conversion efficiency
   B. More precise spectral matching

   C. Reduced amount of phosphorescence
   D. Reduced appearance of image noise

8. When all other factors remain unchanged, an increase in the size of the crystals in the phosphor layer will be associated with all of the following EXCEPT:

A. Increase in the speed of the phosphor
B. Decrease in the recorded detail of the radiographic image
C. Increase in the light emitted by the phosphor layer
D. Increase in the resolution of the imaging system

9. The speed of a film-screen image receptor can be increased by making the following modification to the phosphor layer of the intensifying screens.

   1. Increase the thickness of the phosphor layer
   2. Add a reflective backing to the phosphor layer
   3. Add light absorbing dyes to the phosphor layer

A. 1 & 2 only
B. 1 & 3 only
C. 2 & 3 only
D. 1, 2 & 3

10. The primary advantage of the use of rare earth phosphors in modern film-screen type image receptors is:

A. The improved resolution the system provides
B. The reduced cost of the image receptor
C. The reduction in the exposure received by the patient
D. The reduced amount of focal spot blur the system provides

11. Many of the phosphors employed in intensifying and fluoroscopic screens were first discovered by:

A. Michel Pupin
B. Thomas Edison
C. Wilhelm Roentgen
D. George Eastman

12. The intensification factor for film-screen imaging system can be determined by the formula:

A. Intensification factor = screen speed x phosphor thickness
B. Intensification factor = crystal size x phosphor thickness
C. Intensification factor = $\dfrac{\text{exposure without screens}}{\text{exposure with screens}}$
D. Intensification factor = $\dfrac{\text{screen speed}}{\text{phosphor thickness}}$

13. Which film-holder image receptor is normally associated with the highest intensification factor?

A. Non-screen film in a cardboard holder
B. A 50 speed film-screen image receptor
C. A 200 speed computed radiographic image receptor
D. A 400 film-screen image receptor

14. Which of the following photoemissive chemicals (phosphors) is associated with the highest conversion efficiency?

A. Gadolinium oxysulfide
B. Calcium tungstate
C. Silver bromide
D. Copper oxybromide

15. The resolution (recorded detail) of an intensifying screen can be improved by:

A. Using a thicker phosphor layer
B. Using light-absorbing dyes in the phosphor layer
C. Employing larger crystals in the phosphor layer
D. Adding a reflective screen backing to the phosphor layer

16. In order to obtain the maximum efficiency from a film-screen image receptor, the light emitted by the screens should correspond to the maximum light sensitivity of the film. This process is termed:

A. Spectral matching
B. Homogeneity
C. Opacity matching
D. Synergism

17. The emission of light from a phosphor after the incident radiation exposure is terminated is called:

A. Fluorescence
B. Solarization
C. Phosphorescence
D. Incandescence

18. The speed of a film-screen image receptor can be increased by:

A. Employing a system that uses a single intensifying screen
B. A reflective backing behind each screen
C. Increasing the distance between the film and the screen
D. Placing a piece of lead foil in the back lid of the cassette

19. The emission of light equally in all directions when phosphors are exposed to a stimulus is termed:

A. Quantum mottle
B. Selectivity
C. Spectral scattering
D. Isotropic propagation

20. As the speed of a film-screen imaging system increases:

1. Image resolution decreases   *2. Quantum mottle increases*   *3. Noise increases*

A. 1 & 2 only
B. 1 & 3 only
C. 2 & 3 only
D. 1, 2 & 3

21. The amount of x-ray absorption by an intensifying screen phosphor is primarily dependent upon the:

A. Pair production effect
B. Photoelectric effect
C. Compton effect
D. Thomson effect

22. A thin layer of lead foil is incorporated into the back lid of most cassettes to:

A. Reduce the amount of backscatter that reaches the image receptor
B. Increase the distance between the film and the screen
C. Focus the x-ray beam to cover the entire image receptor
D. Increase the spatial resolution of the image receptor

23. The manufacturer of slow speed film-screen imaging system may add a pink or yellow dye to the crystal layer to:

A. Increase the amount of visible light emitted by the screen
B. Increase the recorded detail of the imaging system
C. Reduce the amount of scattered radiation produced by the cassette
D. Increase the amount of differential attenuation in the screens

Referring to the diagram answer questions 24 & 25.

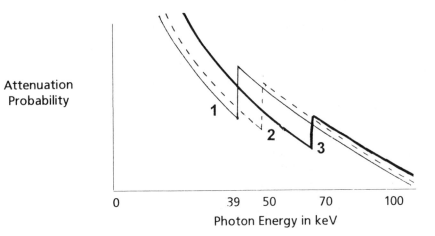

Chapter 26
Intensifying Screens

24. Which of the following phosphors is associated with the least sensitivity to x-rays with energies between 39-70 keV?

    A. Phosphor number 1
    B. Phosphor number 2

    C. Phosphor number 3
    D. Unable to determine

25. The vertical lines seen at 39 keV, 49 keV and 63 keV are associated with the:

    A. Conversion bands of the tungsten atoms in the x-ray tube
    B. K-shell absorption edges of the phosphors
    C. Spectral matching lines of the film and screen materials
    D. Color separation lines of the phosphor materials

26. The grainy look or areas of uneven densities that may appear from the random distribution of x-rays when a small amount of radiation reaches the image receptor is called?

    A. Quantum mottle
    B. Isotropic propagation

    C. Vignetting
    D. Convergence ratio

27. Reflection from successive underlying layers of luminescent crystals in slow intensifying screens can be lessened by:

    A. Using intensifying screens that are pure white
    B. Using a larger number crystals in phosphor layer
    C. Adding a light-absorbing dye to the phosphor layer
    D. Placing a piece of dark paper into the film-screen image receptor

28. All of the following changes will effect the speed of a film-screen image receptor EXCEPT:

    A. Size of the crystals in the phosphor layer
    B. Thickness of the crystal layer

    C. Temperature of the phosphor layer
    D. The barometric pressure

29. When a film-screen image receptor is employed, about what percentage of the x-ray film's exposure is due to the light emitted by the phosphors in the screens?

    A. 2-5%
    B. 10-13%

    C. 47-50%
    D. 95-98%

30. The use of asymmetric (different speed) front and back intensifying screens can result in images that have a:

    A. Wider latitude and higher contrast
    B. Reduced amount of patient motion

    C. Reduced optical density
    D. Reduced penetration and focal spot blur

31. The resolution (resolving power of an intensifying screen) is generally measured in units of:

    A. Light units per millimeter
    B. Lines pairs per millimeter

    C. Lead strips per millimeter
    D. Half value layers per millimeter

32. The ratio of the x-ray energy absorbed by a phosphor to the visible light energy emitted is termed:

    A. K-shell absorption edge
    B. Spectral matching

    C. Conversion efficiency
    D. Intensification factor

33. The most common material used in older 100 speed film-screen image receptor was the blue light-emitting phosphor called:

    A. Calcium tungstate
    B. Lanthanum oxybromide

    C. Silver bromide
    D. Sodium thiosulfate

34. The principal advantage that rare earth image receptor have over calcium tungstate screens is a greater speed without an appreciable loss of:

1. Optical density   **2. Radiographic contrast**   **3. Spatial resolution**

A. 1 only
B. 2 only

C. 3 only
D. 1, 2 & 3

35. Poor screen contact in a film-screen imaging system or computed radiographic imaging system can be determined by using a:

A. Pinhole camera
B. Spin top test

C. Wire mesh tester
D. Test for lag

36. The front of any type of film-screen cassette or computed radiographic cassette holder must be covered by a light-tight substance that is both:

A. Homogeneous and radiopaque
B. Heterogeneous and radiolucent

C. Homogeneous and radiolucent
D. Heterogeneous and radiopaque

37. In a film-screen cassette, the thin layer of foam rubber (felt) behind each intensifying screen serves to:

1. Increase speed of the image receptor
2. Improve the contact between the film-screen
3. Decrease the amount of scattered radiation that reaches the film

A. 1 only
B. 2 only

C. 3 only
D. 1, 2 & 3

38. A rare earth phosphor intensifying screen is to be manufactured with a thicker crystal layer. This will:

1. Increase the speed of the image receptor
2. Increase the resolving power of the screen
3. Decease the inherent noise of the image receptor

A. 1 only
B. 2 only

C. 3 only
D. 1, 2 & 3

39. The light emitted by a phosphor during its exposure to x-radiation is termed:

A. Photoelectrolysis
B. Thermionic emission

C. Incandescence
D. Fluorescence

40. A non-screen (cardboard) image receptor requires a technique of 200 mA, 500 ms, 75 kVp. If a 400 speed screen requires an exposure of 200 mA, 10 ms, 75 kVp to give a similar optical density, what is the intensification factor for this exposure?

A. 5 times
B. 10 times

C. 25 times
D. 50 times

Explanations

1. B   Visible light is given off by a luminescent crystal or phosphor whenever it is struck by x-rays, gamma radiation or high speed electrons. This characteristic light is emitted by excited electrons in the outer shell as they return to their ground state.

2. A   The process by which a phosphor or crystalline material gives off light by a process is called luminescence. All of the light emitted by the intensifying screens is known as luminescence. Two forms of luminescence are fluorescence, the portion of light emitted during the actual exposure to the radiation source and phosphorescence (lag), the light emitted after the exposure has ended. In modern phosphor, the preferred materials for a film-screen imaging systems will have a minimal amount of phosphorescence.

3. A   Lanthanum, gadolinium, and yttrium are the three main elements employed in the manufacture of rare earth intensifying screens.

4. C   The relative emission spectrum of four different phosphors are shown. Number 1 shows the emission of blue light emitting calcium tungstate. Number 2 is the blue light emission of the rare earth phosphor lanthanum oxybromide. Number 3 is the green emitting rare earth phosphor Gadolinium oxysulfide. Number 4 is a hypothetical yellow-red emitting phosphor. A green sensitive film should be used with a green emitting phosphor.

5. C   The rare earth phosphor numbers 2 and 3 are associated with a higher emission of light than do other two materials.

6. B   During x-ray production, the number of photons emitted by the tube will vary from one area to the next. This unavoidable random distribution of photons is called quantum mottle. Quantum mottle can lead to the variation of optical densities on an imaging system called grain. This effect becomes more noticeable and objectionable at lower mAs values. Therefore, the use of the higher speed rare earth intensifying screens and computed radiographic imaging systems are more often associated with quantum mottle.

7. A   For a given crystal size and layer thickness a rare earth phosphor emits about four times more light than a calcium tungstate phosphor for a given exposure. The higher atomic number of the rare earth elements not only stop more radiation but also converts more of the absorbed radiation into light. The ratio of the incoming radiation to the amount of light produced is termed the conversion efficiency.

8. B   A larger crystal produces more light and has a larger corona than smaller crystals. The recorded detail of the image is in part related to this factor since no structure smaller than the corona of the crystals light can be imaged. Therefore, larger crystals are associated with a loss of recorded detail and reduced spatial resolution.

9. B   Three principal intrinsic factors can be used to increase the speed of the intensifying screen; namely, an increase in the crystal size, an increase in the thickness of the crystal layer and the use of a reflective backing for the screen. Light absorbing yellow or pink dyes reduce the amount of light emitted and therefore reduce the speed of the screen.

10. C   The higher speed rare earth phosphors that are now universally employed in modern intensifying screens will substantially reduce the patient exposure while maintaining the same recorded detail of the calcium tungstate (CaW) screens they replaced.

11. B   Nearly 2000 phosphors were discovered or categorized by Thomas Edison including the first practical intensifying screen phosphor calcium tungstate (CaW) which was used extensively in radiography from the 1930's through the mid-1970's.

12. C   The formula for calculating the intensification factor is:

$$\text{intensification factor} = \frac{\text{exposure without screens}}{\text{exposure with screens}}$$

13. D The intensification factor or the ratio of exposures required to produce a similar optical density with or without the use of intensifying screens is most closely related to the relative screen speed numbers. An intensifying screen with the highest screen speed number will also have the highest intensification factor.

14. A The principal factor that effects the conversion of x-rays into visible light in an intensifying screen (conversion efficiency) is the chemical nature of the material. The rare earth phosphors such as gadolinium oxysulfide (Lanex), and lanthanum oxybromide (Quanta), have about a 15% conversion efficiency or about four times that of calcium tungstate screens.

15. B A combination of a small crystal, a thinner crystal layer and the use of a light absorbing dye will reduce the relative speed of the intensifying screen while improving the recorded detail and resolution of the resulting image.

16. A In order to maximize the speed of a film-screen imaging system, the light emitted by the screen should be matched to the highest sensitivity of the film (spectral matching). A film-screen combination with a poor spectral match may be as much as four times slower than those with appropriate components.

17. C The process by which a phosphor or crystalline material gives off light, is called luminescence. Two forms of luminescence are fluorescence, the portion of light emitted during the actual exposure to the radiation source and phosphorescence (lag), the light emitted after the exposure has ended. In modern phosphor, the preferred materials for a film-screen imaging systems will have a minimal amount of phosphorescence.

18. B The speed of a film-screen imaging system can be increased by a number of factors. These include the use of larger crystals in the phosphor, a thicker phosphor layer, the use of a reflective backing and the use of a double screen.

19. D Like all other sources of electromagnetic radiations, x-rays are emitted with equal intensity in all directions. The term which best describes this property is isotropic propagation.

20. D The use of a high speed imaging system is associated with lower radiation exposure to achieve the desired radiographic density or fluoroscopic brightness level. As the exposures are decreased, the effect of quantum mottle and electronic noise levels rise proportionally.

21. B The principal factor which effects x-ray absorption by a phosphor is the effective atomic number of the compound. Elements with high atomic numbers are associated with more photoelectric interactions and higher absorption of the incident photons.

22. A A high atomic number material such as lead is normally placed in the back lid of a cassette to eliminate the majority of the secondary scatter (back scatter) radiation that is directed toward the back of the image receptor from nearby walls or floors.

23. B The size of the corona of light around the crystal can be reduced by the use of light-absorbing dyes. Light absorbing yellow or pink dyes reduce the amount of light emitted by the screen and improve the recorded detail of the imaging system.

24. C In the usual range of energies employed for most diagnostic studies (39-70 keV) the rare earth phosphors (numbers 1 & 2) have the highest attenuation. Curve number 3 that represents calcium tungstate has the lowest attenuation in this range of energies.

25. B The vertical lines on the curves relate to the increased attenuation that occurs from the K-shell absorption edges of the phosphors.

26. A During x-ray production, the number of photons emitted by the tube will vary from one area to the next. This unavoidable random distribution of photons is called quantum mottle. Quantum mottle can lead to variation of optical densities on an imaging system called grain. This effect becomes more noticeable and objectionable at the lower mAs values used. Therefore, the use of the higher speed rare earth intensifying screens and computed radiographic imaging system are more often associated with quantum mottle.

27. C Light absorbing dyes are used to absorb the light photons emitted by the screens that are directed towards the image receptor medium at steep angles. These dyes will therefore improve the recorded detail of the image.

28. D In addition to the intrinsic factors of crystal size and phosphor layer thickness which are directly related to the screen speed, a colder phosphor will emit substantially more light than a warmer phosphor. This effect is rarely significant in actual practice because of the opposite effect that the temperature will have on the speed of the film.

29. D In a film-screen imaging system, about 95-98% of the resulting optical density is due to the blue or green color light emitted by the intensifying screens. The remaining 2-5% is caused by the direct action of the x-rays with the film.

30. A The improved latitude and higher contrast of the Kodak InSight imaging system is largely due to the use of asymmetric screens. This system uses a slower front screen and a faster back screen to accomplish this effect.

31. B The simplest means for determining the resolution of an imaging system is by taking an exposure on a specially designed test pattern formed by alternating lead and radiolucent strips called a line pair or resolution phantom. Most of these devices will determine the resolution in the units of line pairs per millimeter.

32. C The higher atomic number of the rare earth elements not only stops more radiation but also converts more of the absorbed radiation into light. The ratio of the incoming radiation to the amount of light produced is termed the conversion efficiency. In calcium tungstate, this rate of conversion is about 4%, in rare earth phosphors a 15-20% conversion efficiency is expected.

33. A Before the switch to the rare earth phosphors in the mid-1970's, the blue light emitting calcium tungstate was the main screen material employed between the 1930's and 1970's.

34. D The main advantage of the rare earth phosphors is their ability to maintain the desired optical density at a much lower exposure. In most other respects the images in terms of their contrast and detail(resolution) are nearly identical.

35. C Though blurring of the radiographic image due to poor screen contact is rare, in a modern cassette, an improperly mounted screen or damaged cassette can be evaluated by placing a wire screen or mesh over the cassette and examining the radiograph for aberrations.

36. C In order to enable x-rays to reach the x-ray film with only minimal attenuation, the front of the cassette should be made of a low atomic number material (carbon fiber) that is homogeneous and radiolucent so that no differences in the optical density are recorded by the image receptor.

37. B Good film-screen contact is critical in maintaining an acceptable recorded detail. Among the factors that are responsible for this are overall cassette design, status of the latching devices, and the compressibility of the foam rubber backing that helps to apply pressure to the film when it is placed between the screens.

38. A The speed of a film-screen imaging system can be increased by a number of factors. These include the use of larger crystals in the phosphor, a thicker phosphor layer, the use of a reflective backing and the use of a double screen.

39 D All of the light emitted by the intensifying screens is known as luminescence. Two forms of luminescence are fluorescence, the portion of light that is emitted during the actual exposure to the radiation source and phosphorescence (lag), the light emitted after the exposure has ended.

40. D The intensification factor is the ratio of exposures that will produce a similar optical density with and without the use of intensifying screens. In this case 100 mAs /2 mAs or a factor of 50 times.

Screen Conversion

1. A radiograph is taken with a 50 speed image receptor and non-Bucky, using 40 mAs at 80 kVp. If a 400 speed image receptor is substituted, what new mAs should be used to maintain a similar optical density?

2-10. State the mAs value that should be employed to maintain a similar optical density when the following changes are made to the speed of the image receptor.

	Original mAs	Original image receptor speed	New image receptor speed	New mAs
2.	10 mAs	50	100	_____
3.	20 mAs	100	50	_____
4.	16 mAs	100	200	_____
5.	60 mAs	50	200	_____
6.	100 mAs	200	100	_____
7.	50 mAs	160	320	_____
8.	80 mAs	50	400	_____
9.	6 mAs	50	150	_____
10.	600 mAs	No screen	300	_____

11. A radiograph is taken with a 400 speed imaging system, non-Bucky, using 3 mAs at 70 kVp. If a 100 speed imaging system is substituted, what new mAs should be used to maintain a similar optical density?

12-20. State the mAs values that can be employed to maintain a similar optical density when the following imaging speed system changes are made.

	Original mAs	Original image receptor speed	New image receptor speed	New mAs
12.	24 mAs	200	100	_____
13.	12 mAs	100	50	_____
14.	30 mAs	320	160	_____
15.	20 mAs	200	100	_____
16.	70 mAs	200	50	_____
17.	50 mAs	320	80	_____
18.	90 mAs	480	160	_____
19.	15 mAs	180	60	_____
20.	10 mAs	400	100	_____

The following formula has been used to calculate these problems.

New mAs = Original mAs x $\dfrac{\text{Original speed of the image receptor}}{\text{New speed of the image receptor}}$

Screen Conversion

1. A radiograph is taken with a 100 speed image receptor and 12:1 Bucky, using 40 mAs at 90 kVp. If a 400 speed image receptor is substituted, what new mAs should be used to maintain a similar optical density?

2-10. State the mAs value that should be employed to maintain a similar optical density when the following changes are made to the speed of the image receptor.

	Original mAs	Original image receptor speed	New image receptor speed	New mAs
2.	40 mAs	300	100	_____
3.	30 mAs	100	50	_____
4.	24 mAs	50	200	_____
5.	46 mAs	200	400	_____
6.	100 mAs	50	100	_____
7.	80 mAs	160	320	_____
8.	8 mAs	100	400	_____
9.	10 mAs	200	50	_____
10.	15 mAs	50	150	_____

11. A radiograph is taken with a 320 speed imaging system, non-Bucky, using 5 mAs at 60 kVp. If a 160 speed image receptor is substituted, what new mAs should be used to maintain a similar optical density?

12-20. State the mAs values that can be employed to maintain a similar optical density when the following changes are made to the speed of the image receptor.

	Original mAs	Original image receptor speed	New image receptor speed	New mAs
12.	30 mAs	200	100	_____
13.	24 mAs	100	50	_____
14.	100 mAs	100	200	_____
15.	50 mAs	320	160	_____
16.	36 mAs	200	400	_____
17.	20 mAs	100	200	_____
18.	270 mAs	50	300	_____
19.	100 mAs	240	480	_____
20.	40 mAs	50	100	_____

The following formula has been used to calculate these problems:

New mAs = $\dfrac{\text{Original mAs x Original image speed of the receptor}}{\text{New speed of the image receptor}}$

Density Analysis Screens

In each set of exposure and technical factors select the set of factors that will provide the greatest optical density.

	mA	Time	kVp	Screen Speed
1. A.	200	10 ms	92	200
B.	200	5 ms	74	400
C.	200	25 ms	84	50
D.	200	25 ms	74	200
2. A.	100	30 ms	84	200
B.	100	300 ms	87	100
C.	200	10 ms	74	400
D.	300	1 ms	74	100
3. A.	300	100 ms	96	400
B.	200	150 ms	88	100
C.	100	200 ms	90	200
D.	300	200 ms	96	400
4. A.	400	40 ms	90	100
B.	800	40 ms	96	50
C.	300	1 second	105	No screen
D.	300	40 ms	94	No screen
5. A.	100	150 ms	83	400
B.	400	250 ms	85	200
C.	600	50 ms	86	100
D.	300	50 ms	88	100
6. A.	300	1.5 seconds	72	No screen
B.	200	1 second	64	50
C.	400	750 ms	66	50
D.	300	600 ms	72	No screen
7. A.	200	50 ms	62	200
B.	100	35 ms	74	200
C.	200	40 ms	86	400
D.	200	35 ms	74	400
8. A.	200	.5 second	76	200
B.	300	500 ms	88	200
C.	100	.4 second	88	400
D.	600	250 ms	86	400
9. A.	400	500 ms	88	400
B.	800	125 ms	90	200
C.	200	50 ms	89	800
D.	400	18 ms	88	200
10.A.	300	.1 second	82	400
B.	200	.05 second	91	200
C.	300	.03 second	86	100
D.	200	.05 second	74	100

In each set of exposure and technical factors select the set of factors that will provide the greatest optical density.

		mA	Time	kVp	Screen Speed
1.	A.	150	100 ms	64	200
	B.	300	100 ms	68	100
	C.	600	100 ms	66	100
	D.	400	100 ms	66	50
2.	A.	600	150 ms	82	160
	B.	800	50 ms	87	100
	C.	1000	40 ms	93	320
	D.	800	15 ms	86	160
3.	A.	300	60 ms	91	50
	B.	200	150 ms	84	100
	C.	200	250 ms	88	100
	D.	300	10 ms	86	50
4.	A.	500	35 ms	87	320
	B.	300	60 ms	83	50
	C.	100	150 ms	83	100
	D.	300	5 ms	89	100
5.	A.	400	250 ms	92	50
	B.	800	150 ms	88	No screen
	C.	200	250 ms	82	100
	D.	600	150 ms	84	No screen
6.	A.	50	.3 second	74	160
	B.	100	.1 second	68	400
	C.	300	50 ms	70	100
	D.	500	10 ms	65	100
7.	A.	600	50 ms	94	100
	B.	600	10 ms	80	320
	C.	600	70 ms	92	100
	D.	600	20 ms	84	50
	A.	300	1.5 seconds	80	200
	B.	150	1.5 seconds	66	400
	C.	50	5.0 seconds	82	200
	D.	50	3.0 seconds	74	400
9.	A.	800	80 ms	79	100
	B.	400	160 ms	76	50
	C.	300	.1 second	72	200
	D.	400	.25 second	70	400
10.	A.	300	150 ms	77	320
	B.	200	.08 second	87	200
	C.	200	600 ms	90	100
	D.	300	.12 second	84	50

Grid and Screen Conversion

1-7. Use the following information to answer questions 1-7.

A radiograph is taken at 30 mAs, 12:1 Bucky, using a 100 speed image receptor. What new mAs should be employed to maintain the same optical density for the following screen and grid ratio changes?

	New image receptor speed	New Grid Ratio	New mAs
1.	50	6:1	_____
2.	200	8:1	_____
3.	300	5:1	_____
4.	50	16:1	_____
5.	200	No grid	_____
6.	300	6:1	_____
7.	400	12:1	_____

Use the following information to answer questions 8-14.

8-14. A radiograph is taken at 40 mAs using 8:1 Bucky and a 50 speed image receptor. What new mAs should be employed to maintain the same optical density for the following screen and grid ratio changes?

	New image receptor speed	New Grid Ratio	New mAs
8.	50	No grid	_____
9.	400	5:1	_____
10.	100	8:1	_____
11.	30	No grid	_____
12.	100	12:1	_____
13.	300	8:1	_____
14.	200	No grid	_____

15-20. A radiograph is taken at 50 mAs using 6:1 grid using a 100 speed image receptor. What new mAs should be employed to maintain the same optical density for the following screen and grid ratio changes?

	New image receptor speed	New Grid Ratio	New mAs
15.	400	No grid	_____
16.	100	12:1	_____
17.	50	6:1	_____
18.	400	12:1	_____
19.	100	16:1	_____
20.	50	8:1	_____

The following formula has been used in calculating these problems.

New mAs = $\dfrac{\text{Original mAs x Original speed of the image receptor x New grid conversion factor}}{\text{New imaging speed of the image receptor x Original grid conversion factor}}$

Grid and Screen Conversion

1-7. Use the following information to answer questions 1-7.

A radiograph is taken at 10 mAs, non-Bucky, using a 50 speed image receptor. What new mAs should be employed to maintain the same optical density for the following screen and grid ratio changes?

	New image receptor speed	New Grid Ratio	New mAs
1.	100	6:1	_____
2.	200	8:1	_____
3.	100	12:1	_____
4.	400	12:1	_____
5.	200	8:1	_____
6.	100	5:1	_____
7.	No screen	No grid	_____

Use the following information to answer questions 8-14.

8-14. A radiograph is taken at 30 mAs using 12:1 Bucky and a 400 speed image receptor. What new mAs should be employed to maintain the same optical density for the following screen and grid ratio changes?

	New image receptor speed	New Grid Ratio	New mAs
8.	100	16:1	_____
9.	50	6:1	_____
10.	200	No grid	_____
11.	200	6:1	_____
12.	50	No grid	_____
13.	200	8:1	_____
14.	100	12:1	_____

15-20. A radiograph is taken at 50 mAs using 6:1 grid using a 100 speed image receptor. What new mAs should be employed to maintain the same optical density for the following screen and grid ratio changes?

	New image receptor speed	New Grid Ratio	New mAs
15.	50	5:1	_____
16.	200	12:1	_____
17.	300	12:1	_____
18.	50	8:1	_____
19.	400	16:1	_____
20.	200	No grid	_____

The following formula has been used in calculating these problems.

New mAs = $\dfrac{\text{Original mAs x Original speed of the image receptor x New grid conversion factor}}{\text{New speed of the image receptor x Original grid conversion factor}}$

Answers

Screen Conversion
Exercise 1

1. 5 mAs	11. 12 mAs
2. 5 mAs	12. 48 mAs
3. 40 mAs	13. 24 mAs
4. 8 mAs	14. 60 mAs
5. 15 mAs	15. 40 mAs
6. 200 mAs	16. 280 mAs
7. 25 mAs	17. 200 mAs
8. 10 mAs	18. 270 mAs
9. 2 mAs	19. 45 mAs
10. 2 mAs	20. 40 mAs

Screen Conversion
Exercise 2

1. 10 mAs	11. 10 mAs
2. 120 mAs	12. 60 mAs
3. 60 mAs	13. 48 mAs
4. 6 mAs	14. 50 mAs
5. 23 mAs	15. 100 mAs
6. 50 mAs	16. 18 mAs
7. 40 mAs	17. 10 mAs
8. 2 mAs	18. 45 mAs
9. 40 mAs	19. 50 mAs
10. 3 mAs	20. 20 mAs

Density Analysis Screens
Exercise 1

1. A
2. B
3. D
4. B
5. B
6. C
7. C
8. D
9. A
10. A

Density Analysis Screens
Exercise 2

1. C
2. D
3. C
4. A
5. A
6. A
7. C
8. A
9. D
10. C

Grid and Screen Conversion
Exercise 1

1. 30 mAs	11. 2 mAs
2. 11 mAs	12. 27 mAs
3. 4 mAs	13. 7 mAs
4. 75 mAs	14. 3 mAs
5. 4 mAs	15. 15 mAs
6. 5 mAs	16. 240 mAs
7. 8 mAs	17. 240 mAs
8. 13 mAs	18. 60 mAs
9. 3 mAs	19. 300 mAs
10. 20 mAs	20. 360 mAs

Grid and Screen Conversion
Exercise 2

1. 10 mAs	11. 30 mAs
2. 8 mAs	12. 60 mAs
3. 20 mAs	13. 45 mAs
4. 5 mAs	14. 45 mAs
5. 8 mAs	15. 75 mAs
6. 8 mAs	16. 50 mAs
7. 500 mAs	17. 33 mAs
8. 150 mAs	18. 150 mAs
9. 120 mAs	19. 31 mAs
10. 15 mAs	20. 13 mAs

Radiographic Films

1. The majority of screen-type radiographic films consist of a/an:

   A. Opaque film base and a single film emulsion   C. Transparent film base and a dual film emulsion
   B. Opaque film base and multiple film emulsions   D. Transparent film emulsion and opaque base

Referring to the radiographic images, answer questions 2 & 3.

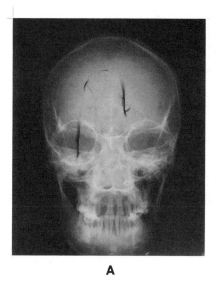

A

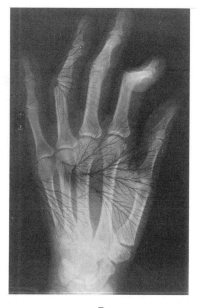

B

2. In the radiographic image in Film A, the increased optical density marks seen on the skull most likely resulted from:

   A. The improper handling of the x-ray film in the darkroom
   B. A damaged or warped cassette with a light leak
   C. An improperly cleaned intensifying screen
   D. Fog resulting from an improperly filtered safelight

3. In the radiographic image of Film B, the increased optical density artifact was caused by:

   A. A water stain on the film emulsion
   B. Solution spills that have effected the emulsion
   C. A static electrical discharge to the film emulsion
   D. The improper bending of the x-ray film before processing

4. The useful range of optical densities on a characteristic (H&D) curve is approximately:

   A. 0.1-.25   C. 0.2-2.5
   B. 0.1-1.5   D. 0.5-5.0

5. Many of the characteristic properties of a film were first described by:

   A. Hurter and Driffield   C. Hittorf and Day
   B. Hinze and Downs   D. Edison and Potter

6. The major factor that will cause an increase in film speed is a/an:

   A. Increase in the amount of silver in the film emulsion
   B. Increase in the thickness of the film base
   C. Decrease in the temperature of the film
   D. Decrease in the size of the silver grain in the emulsion

Referring to the diagram of the characteristic curve, answer questions 7 & 8.

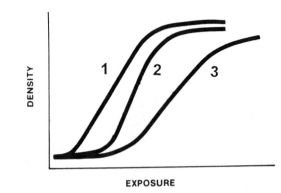

7. Which of the three films shows the greatest radiographic contrast?

   A. Film 1   C. Film 3
   B. Film 2

8. Which of the three films would be associated with the highest optical density for a given exposure?

   A. Film 1   C. Film 3
   B. Film 2

9. Any unwanted marks that appear on the film after it is processed are called film:

   A. Glitches   C. Artifacts
   B. Pi lines   D. Scatter

10. The graphic relationship between the intensity of exposure to a film and the resulting optical density is demonstrated by a/an:

    A. Characteristic curve   C. Spectral curve
    B. Threshold curve   D. Analog to density curve

11. The latent image is the invisible image formed by the action of x-rays and the:

    A. Polyester in the film base   C. Silver crystals in the film emulsion
    B. Gelatin of the film emulsion   D. Adhesive layer of the film

12. Exceeding the film manufacturer's temperature recommendation levels during storage will result in all of the following EXCEPT:

    A. A reduced amount of radiographic contrast
    B. An increase in the optical density appearing on the film
    C. An increase in the amount of fog appearing on the film
    D. An increase in the static appearing on the film

13. A film that transmits 1/100th of the light from a view box will have optical density:

    A. .01   C. 1.0
    B. .25   D. 2.0

14. A normal screen-type radiographic film will normally have a base plus fog optical density of about:

    A. 0.1   C. 0.8
    B. 0.5   D. 1.6

15. The optical properties of a film base can be improved by the incorporation of blue dye into the:

A. Film emulsion
B. Film base
C. Gelatin layer
D. Protective coating

16. In general, radiographic film possesses the greatest sensitivity to light in the:

1. Red-orange spectrum **2. Yellow-orange spectrum** **3. Blue-green spectrum**

A. 1 only
B. 2 only
C. 3 only
D. 1, 2 & 3

17. A characteristic or Hurter and Driffield curve that expresses relationship between optical density and relative exposure is used to determine the:

A. Opacity of the film base
B. Relative speed of sensitivity of the film emulsion
C. Resolution of the film
D. Relative energy of the x-ray beam

18. Black metallic silver is formed in a chemical reaction in which:

A. Photons are removed from the film base
B. Neutrons are liberated from the silver crystal lattice
C. Protons are neutralized by defects in the crystal lattice
D. Electrons are added to the silver bromide crystals

19. A characteristic or Hurter and Driffield curve is made by plotting the resulting:

A. Optical density and with the log relative exposure to the film
B. Radiographic contrast with the kilovoltage peak setting
C. Base plus fog density with the log relative latitude
D. Log relative exposure with the mAs setting

20. The inherent ability of film emulsion to react to radiation and record a range of optical densities is termed the:

A. Radiographic contrast
B. Subject contrast
C. Exposure latitude
D. Inherent contrast

21. The clear, porous substance of the film that permits the uniform dispersion of the silver halogen crystals throughout the film emulsion is called:

A. Polyester
B. Collagen
C. Gelatin
D. Cycon

Referring to the diagram of the characteristic curve, answer questions 22-24.

22. Which number corresponds to the log relative exposure axis of the curves?

A. Number 1
B. Number 2
C. Number 4
D. Number 7

23. Which number corresponds to the latitude of diagnostically useful density range?

A. Number 1
B. Number 2
C. Number 4
D. Number 7

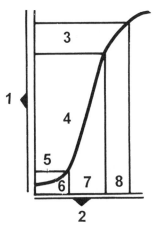

24. Number 5 corresponds to the area of the film possessing the least amount of:

   A. Focal spot blur
   B. Radiographic contrast
   C. Optical density
   D. Image sensitivity

25. A radiograph comes out of a processor with numerous dark, treelike markings. These most likely are caused by:

   A. Bending the film before processing
   B. A static electrical discharge to the film emulsion before processing
   C. An exposure to white light fog during processing of the image
   D. The movement of the film during the exposure

26. All of the following methods can be employed to reduce post-exposure fog in the darkroom EXCEPT:

   A. An increase in the distance between safelight and the film
   B. A decrease of the wattage of the bulbs used in the safe lights
   C. A reduction in the number of safelights used in the darkroom
   D. An increase in the time the film is left on the counter before processing

27. Small amounts of impurities such as sulfur are added to the silver halogen crystals in the film to aid in formation of:

   A. Beneficial artifacts
   B. Sensitivity specks
   C. Reticulation specks
   D. Surfactant centers

28. The active light-sensitive components of a radiographic film are normally found in the:

   A. Adhesive layer of the film
   B. Emulsion layer of the film
   C. Super-coating of the film
   D. Tinted base layer of the film

29. The slope or steepness of the characteristic curve of a radiographic film is used in the determination of its:

   A. Radiographic contrast
   B. Subject contrast
   C. Film resolution
   D. Recorded detail

30. A typical radiographic film emulsion is composed of:

   1. Gelatin　　　　**2. Polyester**　　　　**3. Silver halogen crystals**

   A. 1 & 2 only
   B. 1 & 3 only
   C. 2 & 3 only
   D. 1, 2 & 3

31. An optical density of 3.0 corresponds to portion of the film that has transmitted _____ of the light from a view box.

   A. 1/10th of the light from standard view box light
   B. 1/30th of the radiation from a standard phantom
   C. 1/300th of the light from a standard collimator
   D. 1/1000th of the light from a standard view box

   Referring to the diagram of the characteristic curve, answer questions 32-34.

32. Which of the numbers corresponds to the useful density exposure region of the film?

   A. Number 1
   B. Number 4
   C. Number 5
   D. Number 6

33. Which of the numbers above corresponds to the latitude of the overexposure region?

   A. Number 2
   B. Number 3
   C. Number 6
   D. Number 8

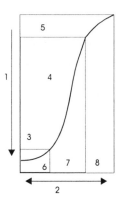

34. Number 2 corresponds to the:

    A. Exposure axis of the films   C. Contrast axis of the films
    B. Optical density axis of the films   D. Latitude axis of the films

35. The principal silver halogen compound used in the formation of the crystal in the film emulsion is made of:

    A. Silver bromide   C. Silver sulfate
    B. Silver iodide   D. Silver sulfite

Referring to the diagram of the characteristic curve, answer questions 36 & 37.

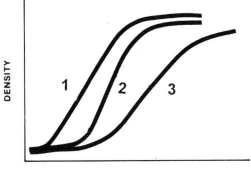

EXPOSURE

36. Which of the numbered films above will have the greatest exposure latitude?

    A. Film 1   C. Film 3
    B. Film 2

37. Which of the numbered films above will demonstrate the greatest sensitivity?

    A. Film 1   C. Film 3
    B. Film 2

38. The majority of processed radiographic images that are recorded on film are described as:

    A. Positive images   C. Subtracted images
    B. Negative images   D. Duplicated images

39. Since the 1960's, the universally accepted film base material for virtually all types of films is a plastic called:

    A. Baser   C. Estrogen
    B. Cellulose acetate   D. Polyester

40. The optical density of a film is expressed by the formula;

    A. $D = \dfrac{I_o}{I_t}$ C. $D = \dfrac{I_o^2}{I_t^2}$

    B. $D = I_o \times I_t$ D. $D = I_t^2 \times I_o^2$

41. At the region above the shoulder of an H & D curve, continued exposures may cause the slope of the line begin to curve downward. This area is known as the:

    A. Solarization region   C. Polarization region
    B. Low optical density region   D. Post exposure region

42. The property of the film base to resist stretching or shrinking, thus resulting in more consistent radiographic images, is termed:

A. Neutrality
B. Permeability
C. Dimensional stability
D. Retentivity

43. Which of the manufacturing processes of a radiographic film must occur in total darkness to avoid the development of light fog?

A. The formation of the silver bromide crystal from the metallic silver
B. The coating of the adhesive layer to film base
C. The formation of the film base from the liquid polyester
D. The removal of the impurities from the adhesive layer

44. Which is NOT one of the useful properties of gelatin?

A. Its ability to swell in warm solution
B. Its ability to chemically unite with the silver crystals
C. Its ability to harden when dried
D. Its ability to suspend the silver bromide crystals

45. The ideal storage temperature for radiographic film is:

A. 5-10° C
B. 15-20° C
C. 35-40° C
D. 65-70° C

46. The amount of image blur that occurs due to the parallax effect when a double emulsion film is employed during a tube angulated technique can be minimized:

A. By adding a blue dye to the film base
B. By decreasing the thickness of the film base
C. By increasing the thickness of the emulsion layer
D. By decreasing the size of the silver bromide crystals

47. The principal controlling factor for the speed of the film is the:

A. Size of the lattice network in the silver bromide crystals
B. Location of the sensitivity specks in the silver bromides crystals
C. Concentration of the gelatin in the film emulsion
D. Concentration of the silver halide crystals in the film emulsion

48. A protective (super) coating placed on most modern film helps to:

A. Increase sensitivity of the silver bromide crystals in the emulsion
B. Increase the lucency of the film base
C. Prevents scratches and finger prints on the film emulsion
D. Prevent the transfer of light through the film base

49. During latent image formation, the concentration of metallic silver in the exposed crystal will normally occur in the region of the:

A. Bromide pool
B. Sensitivity speck
C. Adhesive speck
D. Super coating

50. A static discharge is most likely to occur to a film most on a:

A. Hot, humid day
B. Cold, humid day
C. Hot, rainy day
D. Cold, dry day

Explanations

1. C The most commonly employed radiographic film has a transparent film base that is coated with a light sensitive emulsion. When placed in a cassette each intensifying screen is in contact with an emulsion surface. This design helps to reduce the exposure to the patient by maximizing the speed of the imaging system.

2. A The high optical density artifact appearing in this image was caused by the pressure associated with the bending of the film before the film was processed. These artifacts often appear as small half moon shaped areas and are referred to as crescent marks.

3. C This branch-like increased optical density artifact is called tree static. These are normally caused by a static discharge to the film emulsion. These marks can also have the appearance of a smudge mark (smudge static) or a king's crown (crown static). The grounding of a loading bench and the maintenance of higher humidity levels in the darkroom will help in the prevention of these artifacts.

4. C The full range of densities recorded by an x-ray film is from about .1 - 3. The human eye being less sensitive can only visualize a smaller range, between .25 -2.5, constituting the useful range of optical densities.

5. A In the late 1800's, two students, Hurter and Driffeld, developed a method for expressing the properties of a photographic image by the plotting of the optical density of the film as a function of its log relative exposure to light. This characteristic or sensitometric (H+D) curve can used to determine the sensitivity (speed), contrast and exposure latitude of a film.

6. A Similar to intensifying screens, the size of the silver halide crystals and the thickness of the crystal layer in the films emulsion are the main determining factors for the speed of radiographic film. In recent years the development of the T-mat process has improved the sensitivity and resolution of modern films.

7. B The two most important relationships of the characteristic (H+D) curve are the distance of the curve from the density axis which indicates the film's sensitivity (speed), and the slope of the useful density region of the film which is used in the determination of the relative contrast and exposure latitude. Film II is the most vertical and has the highest contrast and film III has the greatest latitude (lowest contrast).

8. A Comparing the curves, one finds that film I closest to the density axis has the highest sensitivity or speed.

9. C The term artifact is used in radiography to describe any unwanted mark on a radiographic image. Artifacts can result from improper film handling, processing problems, equipment errors or patient related factors.

10. A The plotting of the optical density of the film to the relative exposure of light provides a relationship known as the characteristic or sensitometric curve. This curve can be used to determine the sensitivity (speed), contrast and exposure latitude of a film.

11. C After a film is exposed to light or radiation, a chemical change in the silver halogen crystals of the film occurs. Because this change is imperceptible to the human eye, it is called a latent image. This image is made visible by the film's immersion into the processing (developer) solutions that completes the conversion of the silver halide crystals into the black metallic silver appearing in the dark regions of the film.

12. D All films are unavoidably exposed to heat, moisture and natural background radiation that over time will cause an unwanted fog on the film. For this reason, each box of film is given an expiration date to eliminate the loss of speed, contrast and the unwanted fog this may cause.

13. D The optical density of a film is found to be a logarithmic function of the incident light from the view box to the light transmitted through the film: Optical density = Log IO/IT where IO is the original intensity of light from the view box and it is the amount of light transmitted through the film. The corresponding optical density values for a number of common fractions of original to transmitted intensities are provided. 1/10 = 1; 1/100 = 2; 1/1000 = 3; 1/10,000 = 4.

14. A Even under ideal conditions, an unexposed film after it has been through the processor is not perfectly clear but shows a small but measurable density of about .1 - .15. This so-called base plus fog density is caused by the absorption of light by the blue tint used in most x-ray film bases and the unavoidable chemical fog produced during film processing.

15. B The bright light from the view boxes has been known to cause eye strain in persons that were required to look at these images for prolonged periods of time. This problem is helped by the incorporation of a blue tint into the base of most modern x-ray films.

16. C In order to provide the maximum speed, the intensifying screens must produce light in the blue-green spectrum to which film is most sensitive.

17. B The plotting of the optical density of the film as a function of its log relative exposure to light provides a relationship known as the characteristic or sensitometric (H+D) curve. This curve can used to determine the sensitivity (speed), contrast and exposure latitude of a film.

18. D The main chemical reaction by which silver bromide crystals are converted into the black metallic silver image involves the neutralization of the silver ion in the crystal lattice by an electron. The developer solution is made of chemicals that add additional electrons to the silver bromide (AgBr) crystals to complete the conversion of the latent image into the visible image.

19. A Hurter and Driffeld developed a method for expressing the properties of a photograph through the plotting of the optical density of the film as a function of its log relative exposure to light called the characteristic curve.

20. A The ability of film to record and display a series of varying optical densities is termed film contrast. This inherent factor of the film is different for each type of film and the chemical make-up of processing solutions in the development of the film.

21. C The active component of all films are the light sensitive silver halide (bromide) crystals. To be effective, these must be evenly coated and disbursed over the entire film. A clean porous substance called gelatin serves as the mechanical support medium for the silver bromide (AgBr) crystals.

22. B Pertaining to the corresponding diagram, 1=density axis; 2=log exposure axis; 3=over-exposure region; 4=useful density region; 5=under-exposure region; 6=latitude of the under-exposed region; 7=Latitude of the useful density region; 8=latitude of the over-exposure region.

23. D See explanation number 22 of this section.

24. C See explanation number 22 of this section.

25. B The most likely cause of tree branch-like increased optical density artifacts is a static discharge to the film emulsion.

26. D A fog which a film obtains after it has been exposed to radiation is called post-exposure fog. The most common cause of this exposure is improper shielding of the safe light fixtures or improper seals around the processor opening allowing white light leakage into the darkroom. Other causes can be the use of bulbs with too much wattage, excessive number of safe light fixtures or a insufficient distance between the safe light and the film.

27. B Because a perfect silver bromide crystal will not permit a chemical change to occur, a contaminant such as sulfur (mustard oil) is added during the manufacturing process to create small faults on the crystals called sensitivity specks or developmental centers.

28. B After the light sensitive silver bromide crystals are mixed with gelatin and coated onto the film base, this layer is referred to as the film emulsion.

29. A   The slope of the straight line portion of the characteristic curve is used in the determination of the film contrast and latitude. The more vertical the slope, the higher the contrast and the more narrow the film latitude.

30. B   The light sensitive emulsion of a radiographic film is composed of two separate elements, silver bromide crystals and gelatin. The silver bromide crystals are mixed with the gelatin and coated onto the film base.

31. D   The corresponding optical density values for a number of common fractions of original to transmitted intensities are provided. $1/10 = 1$; $1/100 = 2$; $1/1000 = 3$; $1/10,000 = 4$.

32. B   Pertaining to the corresponding diagram, 1=density axis; 2=log exposure axis; 3=over-exposure region; 4=useful density region; 5=under-exposure region; 6=latitude of the under-exposed region; 7=Latitude of the useful density region; 8=latitude of the overexposure region.

33. D   See explanation number 32 of this section.

34. A   See explanation number 32 of this section.

35. A   Any of the compounds formed by silver and the halogen family of elements will form light sensitive compounds suitable for films. At the present time, the vast majority of films' crystals are composed of silver bromide ($AgBr$).

36. C   Film I possesses the greatest sensitivity (speed) because of its location nearest the density axis. Film II being the most vertical has the highest contrast and lowest latitude. Film III has the lowest contrast and speed but possesses the widest exposure latitude.

37. A   Film I possesses the greatest sensitivity (speed) because of its location nearest the density axis.

38. B   A radiograph is most similar to a photographic negative. Unlike a photographic image which is normally viewed only after the image is printed to make it into positive form, a radiographic is viewed in this transparent negative form.

39. D   Films have employed a number of transparent base materials. Until the late 1800's, film bases were made of glass plates that were fragile and difficult to work with. In the early 1900's, cellulose nitrate and cellulose triacetate, though improved, failed to meet all of the desirable properties possessed by the present day polyester base.

40. B   The optical density of a film is found to be a logarithmic function of the incident light from the view box to the light transmitted through the film: Optical density = $\mathrm{Log}\ I_O/I_T$ where I_O is the original intensity of light from the view box and I_T is the amount of light transmitted through the film.

41. A   The characteristic (H&D) curve has been described as consisting of four basic regions: the low optical density region called the toe, the straight line portion and the high optical density region called the shoulder. The fourth and least commonly depicted is the near mirror image of the first three portions called the reversal or solarization region.

42. C   The foundation of the film, the base, must resist shrinkage or expansion and be able to maintain shape (dimensional stability) under various conditions.

43. A   The process by which silver is formed into silver bromide crystal must take place in total darkness to prevent activation of the crystals prior to their radiographic applications.

44. B   The gelatin which suspends the silver halide crystals will swell in a warm solution enabling the processing solution access to all of the silver crystals. When dried the gelatin shrinks and hardens forming the film's familiar texture. A chemical union with the $AgBr$ crystals and the gelatin would deactivate the compound.

45. B   Films must be stored in a cool dry area ideally at temperatures in the 50-70°F (10-20°C) range. Exposure to excessive heat may cause the film to be fogged resulting in its loss of contrast.

46. B   Radiographic parallax is an artifact that occurs from the use of angulated techniques on a double emulsion film. The use of a tube angle produces a distinct image on each of the two emulsions. When viewed in the normal fashion these images are not perfectly aligned forming the so-called parallax effect. Films with a thin base are used to reduce this effect.

47. D   Until the advent of the T-grain process for making silver grains, the main controlling factor for film speed was the concentration of the crystals. A new technique for producing T-grain crystals by Kodak enables a higher speed, better contrast with a smaller amount of silver.

48. C   A protective super-coating covering the emulsion of the film is employed to help prevent scratches, pressure and fingerprint artifacts.

49. B   The secondary electrons produced when x-ray photons are absorbed into the crystal lattice of a silver bromide crystal migrate toward a silver sulfide fault called a sensitivity speck. These free electrons begin the conversion of the AgBr into neutral black metallic silver.

50. D   In the northern states, the low humidity conditions of the winter provide the ideal dryness for static buildup to occur.

1. What is the proper order of steps that are required for the automatic processing of a radiographic film?

   A. Fixation, development, washing, drying
   B. Development, fixation, washing, drying
   C. Development, washing, fixation, drying
   D. Development, drying, washing, fixation

2. The reducing agents in the developer solution are responsible for the:

   A. Conversion of exposed silver bromide into black metallic silver
   B. Neutralization of the clearing agents in the fixer solution
   C. Elimination of the oxidized silver grains in the film
   D. Removal of exposed silver grain from the film emulsion

3. Radiographic films are normally placed into the automatic processor widthwise or crosswise to help insure that the:

   A. Emulsion side of the film enters the processor first
   B. Correct amount of solution replenishment occurs
   C. Both sides of the film are processed to the same degree
   D. Solution reaches the edges of the emulsion layer

4. Which of the following systems in an automatic processor is most closely related to the time the film remains in the processing solutions?

   A. Replenisher system
   B. Dryer system
   C. Recirculation system
   D. Transport system

5. In most fast access (90 second) automatic processors, it is important to maintain the developer temperature at about:

   A. 80 degrees F (28 degrees C)
   B. 95 degrees F (35 degrees C)
   C. 110 degrees F (68 degrees C)
   D. 125 degrees F (95 degrees C)

6. The addition of restrainers in the developer solutions will tend to prevent the development of unexposed silver bromide crystals and help to:

   A. Prevent the rapid breakdown of the processing solutions
   B. Prevent the softening of the film emulsion
   C. Prevent the production of unwanted chemical fog on the image
   D. Reduce the oxidation of the reducing agents

Referring to the diagram of the characteristic curves of the three identical films, answer question 7.

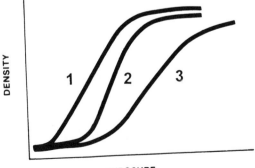

7. The three identical films were processed using the same processor and solutions. Which film shows the effects of excessive developer temperature?

A. Film 1
B. Film 2

C. Film 3
D. Unable to determine

Referring to the radiographic images, answer questions 8 & 9.

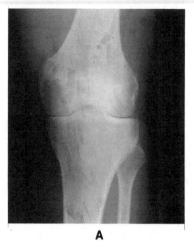

A

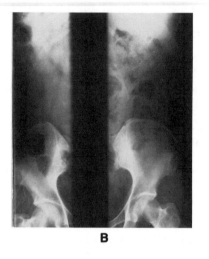

B

8. In the radiographic image in Film A, the increased density marks seen throughout the image were caused by:

A. Static electrical discharge
B. Solutions on the feed tray

C. Bending of the film before processing
D. Insufficient fixation

9. In the radiographic image in Film B, the large area of increased optical density in the center of the radiograph was most likely caused by:

A. A safelight light leak
B. The inadvertent exposure of the film to light during development
C. The processing of the film in a developer solution that was too hot
D. The placement of the film into processor with dirty wash water

10. The most commonly employed reducing agents used in a modern automatic processing solution are:

A. Water and glutaraldehyde
B. Hydroquinone and phenindione

C. Sodium sulfate and acidic acid
D. Thiosulfate and glutaraldehyde

11. Which of the following is NOT one of the components that would be found in the transport system?

A. Drive motor
B. Transport rollers

C. Drying tubes
D. Cross-over rack

12. Which stage of the automatic processor is most closely related to the conversion of the latent image into the manifest image?

A. Development stage
B. Solarization stage

C. Fixation stage
D. Digestion stage

13. The most common result of excessive developer temperature would be:

A. An increase in the radiographic contrast of the radiographic image
B. An increase in the optical density of the radiographic image
C. A decrease in the amount of chemical fog appearing in the image
D. An increase in the speed at which the film is transported

14. A deficient developer temperature and an under-replenished developer solution are two of the most common causes for a radiographic image that has an:

A. Insufficient optical density
B. Excessive radiographic contrast
C. Insufficient recorded detail
D. Excessive amount of image magnification

15. The most common cause of a film that comes out of the automatic processor that is wet and tachy is:

A. A deficient temperature of the developer solution
B. An excessive temperature of the dryer section of the processor
C. A reduction in the rate at which the film is transported
D. A developer or fixer solution that contains a depleted hardener agent

16. During automatic processing, solution carryover between the developer and fixer tanks is prevented by the:

A. Squeegee rollers
B. Rinse bath
C. Overflow valves
D. Microswitches

17. Any solution that is capable of giving up negative ions (electrons) is classified as a/an:

A. Reducing agent
B. Oxidizing agent
C. Bleaching agent
D. Washing agent

18. To insure consistency of the quality of images passing through an automatic processor, the unit should be evaluated:

A. Once each day
B. Once each week
C. Once each month
D. Twice each year

19. The rate of solution replenishment for both the fixer and developer solutions is effected by the:

1. Thickness of the film emulsion 2. Size of the films processed 3. Number of films processed

A. 1 & 2 only
B. 1 & 3 only
C. 2 & 3 only
D. 1, 2 & 3

20. During automatic processing, the activation of the replenisher pump microswitch is accomplished as the:

A. Film enters the developer/fixer crossover assembly
B. Film enters the entrance roller assembly
C. Film enters the fixer squeegee assembly
D. Film enters the pre-wash section

Referring to the radiographic images, answer questions 21 & 22.

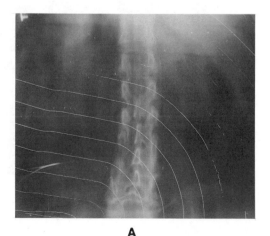

A

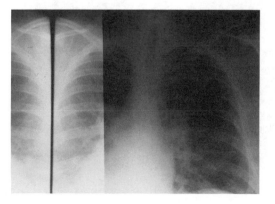

B

21. The decreased optical density artifacts seen on the radiographic image of Film A were most likely caused by:

    A. Scratches on the surface of the x-ray table
    B. Scratches caused by processor guide shoes
    C. Static electrical discharges associated with a faulty processor
    D. Film defects caused by a radiographic parallax

22. The unusual appearance of the radiographic image of Film B most likely resulted from:

    A. A film that improperly fed into the automatic processor
    B. A film that was processed in a cold developer solution
    C. A film that was exposed more than one time (double exposure)
    D. A film that was improperly loaded into the cassette

23. One of the most common causes of the failure of the film to be transported in an automatic processor is:

    A. An excessive amount of dirt in the wash water
    B. An under-replenished developer solution
    C. An over-replenished fixer solution
    D. An insufficient temperature of the dryer section

24. The developing agents, when placed together, have a chemical action stronger than the sum of the two working independently. This chemical action is termed:

    A. Syntax
    B. Synchronism
    C. Synarthrosis
    D. Synergism

25. All of the following are common causes of an under-replenished developer solution in an automatic processor EXCEPT:

    A. The failure of the entrance roller microswitch
    B. The failure of the replenisher pump
    C. The presence of air in replenisher lines
    D. The blockage of the circulation pump

26. The most common clearing agent used in fixing solution of an automatic processor is:

    A. Sodium sulfite
    B. Ammonium thiosulfate
    C. Aluminum chloride
    D. Hydroquinone

27. The principal solvent used in all wet automatic processing solutions is:

    A. Hydroquinone
    B. Thiosulfate
    C. Water
    D. Sodium sulfate

Referring to the radiographic images answer questions 28 & 29.

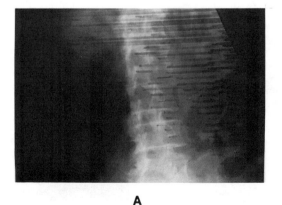

A

B

28. The increased density streaks appearing on the radiographic image of Film A were most likely caused by:

    A. Light exposure to a film while it is being fed into the processor
    B. The misalignment of the processor guide shoes of the crossover rack
    C. The improper seating of the developer/fixer squeegee rollers
    D. Developer solutions on the entrance rollers

29. The decreased optical density marks seen on the radiographic image of Film B resulted from:

    A. The radiation fog due to the improper storage of the film
    B. The exposure to the back side of strap type cassette
    C. A cassette with a warped cassette front
    D. The use of the an upside down focus grid

30. One of the common causes of processor jam up is the depletion of the hardener in the developer solution called:

    A. Glutaraldehyde
    B. Aluminum chloride
    C. Silver iodide
    D. Hydroquinone

31. The solution that is maintained at the temperature of approximately 80 degrees F (27 degrees C) is the:

    A. Wash
    B. Fixer
    C. Developer
    D. Stop bath

32. The most common chemical agent used to restrain the action of the reducing agent in an automatic processing solution is:

    A. Silver iodide
    B. Sodium sulfite
    C. Aluminum chloride
    D. Potassium bromide

33. The temperature stabilization in an automatic processor is most closely related to the operation of the:

    A. Circulation system
    B. Replenisher system
    C. Dryer system
    D. Transport system

34. All of the following conditions are likely to cause fog on a radiographic film EXCEPT:

    A. An increase in the temperature of the fixer solution
    B. An increase in the time the film is immersed in the developer solution
    C. An increase in the developer solution activity
    D. An increase in the temperature of the developer solution

35. The numerous lines that appear on the radiograph opposite the direction of film travel caused by the dirty processor rollers are known as:

    A. Guideshoe marks
    B. Wet processing artifacts
    C. Crescent marks
    D. Pi lines

36. Which of the following factors will effect the immersion time for a modern x-ray film?

    1. The thickness of the film base
    2. The activity of the developer solution
    3. The temperature of the developer solution

    A. 1 & 2 only
    B. 1 & 3 only
    C. 2 & 3 only
    D. 1, 2 & 3

37. The loss of radiographic contrast that is likely to occur on a film processed in a hotter than normal developer solution is related to:

A. The excessive amount of chemical fog produced during processing
B. The decreased speed of the film at higher temperatures
C. The increased amount of x-ray absorption in the emulsion of the film
D. The inactivation of the fixer solution by the hot developer solution

38. Which is a TRUE statement concerning the replenisher rates for the solutions in an automatic processor?

A. The replenisher rate for the developer is greater than that of the fixer
B. The fixer solution does not require solution replenishment
C. The replenishment is the same for both the fixer and developer
D. The replenisher rate for the fixer is greater than that for the developer

39. A developer temperature that exceeds recommended values will lead to all of the following EXCEPT:

A. An excessive amount of chemical fog in the film emulsion
B. A marked decrease in radiographic contrast of the image
C. A rapid deterioration of the of the developer solution
D. A reduced optical density of the image

40. The termination of development process can be accomplished by placing the film in a/an:

A. Brine solution   C. Acid solution
B. Alkaline solution   D. Acrid solution

41. The normal developer replenisher rate per 43 cm (14 inches) of film is approximately:

A. 10-30 ml   C. 60-70 ml
B. 30-40 ml   D. 90-100 ml

42. The specific gravity of the developer solution is often used to give an estimation of the:

A. Chemical activity or strength of the developer solution
B. Opacity of the developer solution
C. Electrical conductivity of the developer solution
D. Acidity of the developer solution

43. The principal cause of oxidation in the developer solution of an automatic processor is the exposure to:

A. X-rays   C. Air
B. Ultraviolet radiation   D. Visible light

44. A film comes out of an automatic processor with numerous scratches in the direction that the film travels. These are caused by improperly seated:

A. Entrance rollers   C. Drying tubes
B. Guideshoes   D. Replenisher pumps

Referring to the radiographic images, answer questions 45 & 46.

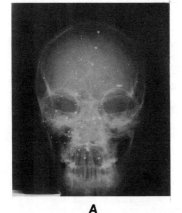

A

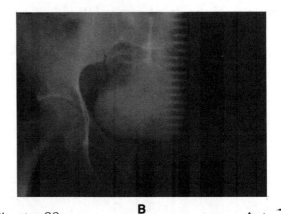

B

45. In the radiographic image of Film A, the decreased optical density marks appearing throughout the film were most likely caused by:

A. Dirty processing rollers
B. Dirty intensifying screens
C. Cracks in the safelight filter
D. Light leaks in the cassette

46. The serrated appearance near the center of Film B was caused by a white light exposure during:

A. Drying of the film
B. Fixation of the film
C. Washing of the film
D. Feeding of the film

47. The mixing and agitation of the processing solution during automatic film processing is accomplished by the:

1. Circulation system   2. Replenisher system   3. Dryer system

A. 1 only
B. 2 only
C. 3 only
D. 1, 2 & 3

48. The hardening agents (aldehydes) which are added to the automatic processing solutions also aid in the reduction of:

A. Rapid oxidation
B. Excessive acidity
C. Fungus growth
D. Viral growth

49. A silver reclaimer unit serves to collect the silver that is removed from the:

1. Developer solution   2. Fixer solution   3. Wash water

A. 1 only
B. 2 only
C. 3 only
D. 1, 2 & 3

50. The major function of the clearing agent in the fixer solution is the:

A. Removal of unexposed silver bromide from the film emulsion during processing
B. Removal of black metallic silver from the film emulsion during processing
C. Conversion of silver into silver bromide in the film emulsion during processing
D. Removal of the silver from the fixer solution

Explanations

1. B The normal sequence of the film passage in a fast access or 90 second automatic processing is development, fixation, washing and drying.

2. A The reducing agents, hydroquinone and phenidone, in the developer solution are responsible for conversion of the unexposed silver bromide crystal of the film emulsion into black metallic silver grains that make up the darker regions of the radiograph. The action of the reducing agent is greatly enhanced in the presence of the low PH, alkali agents such as sodium carbonate or sodium hydroxide.

3. B The replenisher rates in an automatic processor are adjusted for the crosswise placement of the film on the feed tray. The actual replenishment of the solutions is accomplished by two pumps that are triggered by a microswitch which is activated as the film enters the processor. The placement of the film lengthwise is wasteful and may lead to the over-replenishment of the developer and fixer solutions.

4. D The processing or immersion time in an automated unit is directly related to the speed of the transport system. This speed is critical to maintaining image quality and should be maintained to within a ± 2% limit of the optimal time.

5. B Strict temperature control in an automatic processor is especially important for the developer solution because of its direct relationship to the resulting optical density and contrast of the image. In most automated systems, the developer is maintained between 90-95°F (33-35°C). Precise regulation is required to prevent temperature fluctuations more than ± 2°F which can noticeably effect the optical density and radiographic contrast of the image.

6. C During rapid processing the high activity of the developer solution may cause the development of even the unexposed silver bromide crystals. The addition of potassium bromide to this solution helps restrain the action of the reducing agents so that only the unexposed silver bromide crystals are converted into black metallic silver. This will reduce the amount of unwanted chemical fog.

7. A An excessive rise in the temperature of the developer solution will increase the optical density of the image while producing film with a lower contrast. Film I shows these effects on the characteristic curve.

8. B Water or developer solutions that are spread over parts of the film before processing will often produce this type of positive optical density artifact.

9. C The artifact appearing in radiograph B was caused by a momentary opening of the processor lid while the film was in the developer section.

10. B The process by which exposed silver bromide is converted in the black metallic silver is termed reduction. The most common reducing agents are hydroquinone, phenidone and metol.

11. C The transport system which consists of the drive motor, the transport rollers, and transport racks, also contain a series of grooved metal plates called guide shoes. Guide shoes assist the films in changing its direction as it passes into the turn around or crossover racks during normal processing.

12. A The invisible latent is converted into a manifest or visible image during the development stage of film processing. This occurs as the exposed silver bromide crystals obtain free electrons from the reducing agents, hydroquinone and phenidone, and are converted into the black metallic silver grains.

13. B Excessive optical density combined with a lower radiographic contrast is most commonly associated with an excessive temperature of the developer solution.

14. A   The two principal reducing agents, hydroquinone and phenidone, work together in maintaining the desired optical density on the film. The faster acting phenidone begins the development process and is most responsible for lighter shades of optical density seen in the toe portion of the H + D curve. Hydroquinone acts more slowly but provides the darkest shades of optical density on the film. A reduced activity or a reduced temperature of these solutions will lead to a film with less contrast and a reduced optical density.

15. D   Improperly dried films may be caused by directly improperly seated drying tubes or problems with the heater or thermostat. Chemical causes of wet or tacky films include excessive processing solution temperatures and depleted developer or fixer solutions.

16. A   The neutralizing effect of the developer solution with the fixer can have detrimental effects on the fixation (clearing) process. For this reason, a pair of closely spaced squeegee rollers are placed at the end of the developer solution tank to remove excess developer prior to its immersion in the fixer section of the unit.

17. A   The reducing agents of the developer solution provide an excess of electrons to the film that aid in the reduction of exposed silver bromide crystals. These electrons neutralize the positively charged silver ions forming a compound termed black metallic silver.

18. A   In order to maintain consistent image quality, an automatic processor should be checked daily for proper solution temperature (thermometer), chemical activity (hydrometer) and have sensitometric images made and interpreted.

19. C   When a 35 cm-43 cm (14"-17") film is processed correctly, about 60 cc of developer is replenished and 100 cc of fixer is replenished. This will help to maintain the chemical activity of the solutions. The microswitchs in the entrance roller assembly will help control the appropriate replenishment rate. This rate is dependent on the size and varies based on the number of film processed.

20. B   The microswitch within the entrance roller assembly serves both to regulate the developer and fixer replenishment rates and provide a visual or audible signal that assists with proper film spacing when feeding films into the processor.

21. B   The decreased optical density of radiograph A is due to scratches in the film emulsion caused by the pulling of a jammed film over the guide shoes of the fixer-wash crossover rack.

22. D   The image of radiograph B resulted from a bent film that was improperly loaded into the cassette.

23. B   The most common chemical cause for processor jam-ups is an exhausted developer or fixer solution. This occurs because these solutions will contain insufficient hardener enabling the excessive swelling of the film emulsion. Mechanical jam-ups are nearly all related to problems of the transport system.

24. D   The action of the two reducing agents (hydroquinone and phenidone) independently would be substantially less than their action together. This chemical phenomenon is termed synergism or superaddidity.

25. D   Any malfunction of the replenisher system such as air in the replenisher lines, faulty micro-switches, and faulty replenisher pumps, results in a decrease in the chemical activity of the solution. This lowers the speed (optical density) and radiographic contrast on the film. This becomes progressively worse over a series of images as the developer and fixer solutions are slowly depleted.

26. B   The clearing or fixing agent most commonly employed in most automatic processors is ammonium thiosulfate, which serves to remove the unexposed silver bromide crystals from the film's emulsion during the second processing step.

27. C   The mixing of the processing solutions has changed substantially over the years. In the past, solutions were prepared by hand mixing dry chemicals in water (solvent). Most modern solutions are provided in 3 - 5 gallon concentrates that are pre-mixed with the water so as to achieve the desired activity level.

28. D This radiograph shows a series of evenly spaced streaks which were caused when an improperly fed film was pulled backwards out of the developer solution processor. Since portions of the film were in the developer, excess solution was dragged over the entrance rollers and redeposited on the film when processing was attempted the second time.

29. C The low optical density areas of this radiograph correspond to the straps and latching devices of an older type strap cassette. These are seen because the cassette was inadvertently exposed with its back surface facing the x-ray tube.

30. A A hardening agent called glutaraldehyde is added to the developer solution to prevent excessive swelling of the film emulsion when it is immersed into a warm solution. A depleted or exhausted developer solution is unable to prevent the over-expansion of the film emulsion causing it to adhere to the rollers and jamming the processor.

31. A In a 90-second automatic processor, the developer and fixer solution temperatures are in part stabilized by a series of heat exchange tubes from the wash tank. The wash water in most processors is maintained at 80-85 degrees F (27-33 degrees C) or about 10 degrees F (5 degrees C) below the temperature of the developer solution.

32. D During the rapid processing, the high activity of the developer solution may cause the development of even the unexposed silver bromide crystals. The addition of potassium bromide to this solution helps restrain the action of the reducing agents so that only the unexposed silver bromide crystals are converted into black metallic silver.

33. A In a 90-second automatic processor, the circulation system serves to agitate and maintain the solution temperature. In most processors, a heat exchanger is provided at the bottom of the developer, fixer and wash tanks to maintain the temperature within the 1 degree F accuracy desired.

34. A Excessive developer temperature, increased developer (immersion) time, over-replenished developer solution, and exhausted fixer can all lead to an increase in the production of chemical fog on the radiograph.

35. D Evenly spaced lines usually 3.14 inches apart are most often caused by dirt or chemical stains deposited by the revolution of the 1" diameter processor rollers. The term pi lines is derived from the formula for the circumference of a circle = π d. Pi line occurs in a direction that is opposite the direction of the film.

36. C The proper development time is dependent on the type of film emulsion employed, its compatibility with the solution, the developer activity and temperature.

37. A An increase in the temperature of the developer solution will increase that amount of chemical fog and the optical density of the radiographic image. The hotter solution will also reduce the radiographic contrast appearing in the image.

38. D When a 35 cm-43 cm (14"-17") film is processed correctly, about 60 cc of developer is replenished and 100 cc of fixer is replenished. This will help to maintain the chemical activity of the solutions. The microswitchs in the entrance roller assembly will help control the appropriate replenishment rate. This rate is dependent on the size and varies based on the number of film processed.

39. D Excessive developer temperature can lead to numerous processor related problems such as chemical fog formation, rapid chemical oxidation and increased optical density and lower contrast of the image. The main physical sign of an excessive developer temperature is the wet and tacky feel of the films coming out of the processor.

40. C Film development which requires a highly alkaline medium is rapidly terminated by its placement into the strongly acid fixer solution.

41. C When a 35 cm-43 cm (14"-17") film is processed correctly, about 60 cc of developer is replenished and 100 cc of fixer is replenished. This will help to maintain the chemical activity of the solutions. The microswitchs in the entrance roller assembly will help control the appropriate replenishment rate. This rate is dependent on the size and varies based on the number of film processed.

42. A   The specific gravity provides a measure of the concentration of a solution. This will provide an indirect means for the estimation of the strength or chemical activity of the developer and fixer solutions.

43. C   Both the developer and fixer solutions are highly active chemical solutions which are not limited to their action with the film emulsion but also readily combine with the oxygen atoms of the air. The addition of sodium sulfite to the processing solutions slows this process by more readily combining with the oxygen than the balance of the solution, thereby extending its life.

44. B   Improperly seated guide shoes in the crossover or turnaround racks of an automatic processor may result in evenly spaced scratches on the film in the direction of the film travel, called guide shoe marks.

45. B   The most common causes of low optical density artifacts are related to the inability of light emitted from the intensifying screen to reach the film. Dirty screens are the most common cause of this effect.

46. D   The alternating normal and high optical density on the center of this image correspond to the light penetrating through the spaces of the entrance roller when the darkroom light was turned on before the film completely entered the developer section of the processor.

47. A   In a 90-second automatic processor, the circulation system serves to agitate and maintain the solution temperature.

48. C   The glutaraldehyde hardening agent that is added to the developer also has antifungal properties that helps to restrain the growth of bacteria or fungi in the developer.

49. B   In order to prevent the loss of silver and avoid the hazards of its disposal into the environment, all automatic processors should have the unexposed silver bromide removed by placing a silver reclaimer unit to the fixer tank of the processor.

50. A   The clearing or fixing agent most commonly employed in most automatic processors is ammonium thiosulfate, which serves to remove the unexposed silver bromide crystals from the film's emulsion during the second processing step.

Tomography

1. In order to visualize a single plane of tissue, an image technique called tomography is employed that requires the movement of the:

   A. Tube and image receptor at right angles to each other
   B. Tube and image receptor in opposite directions of each other
   C. Tube and image receptor vertically to each other
   D. Tube and image receptor in the same direction

   Referring to the following diagram answer questions 2 & 3.

2. The tomographic exposure angle or arc is represented by:

   A. Number 1
   B. Number 2
   C. Number 4
   D. Number 5

3. The point where the beams cross represented by number 4 is most closely related to the:

   A. Thickness of the tomographic section
   B. Height of the tomographic plane
   C. Tube travel speed
   D. Angle of reflection

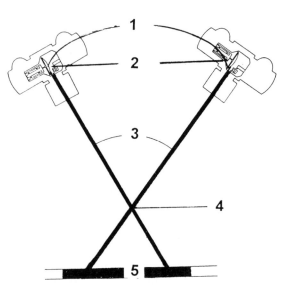

4. During tomography, all tissue planes:

   A. Above and below the focus plane will show a marked degree of blurring
   B. On the focus plane will show a marked degree of blurring
   C. Lateral to the focus plane will show a marked degree of blurring
   D. Medial to the focus plane will show a marked degree of blurring

5. The single thickness of tissue that remains in focus during tomography is called the:

   A. Objective plane   C. Tomographic angle
   C. Fulcrum plane   D. Grossman angle

6. The distance traveled by the tube travels during the time the tomographic exposure is being made is termed:

   A. Fulcrum arc distance   C. Total travel distance
   B. Amplitude distance   D. Source-to-image receptor distance

7. In order to obtain a tomographic section at a cut level further from the tabletop will require:

   A. An increase in the tomographic arc
   B. An increase in the height of the fulcrum
   C. An increase in the source-to-image receptor distance
   D. A decrease in the tube travel time

8. In order to maximize the blurring during rectilinear tomography, the tube movement should be at a:

   A. 30 degree angle to the transverse axis of the x-ray table
   B. 90 degree angle to the dominant axis of the part radiographed
   C. 60 degree angle to the transverse axis of the x-ray table
   D. 180 degree angle to the dominant axis of the part radiographed

9. Which type of tomographic movement is generally described as a cloverleaf pattern?

   A. Spiral
   B. Sinusoidal
   C. Rectilinear
   D. Hypocycloidal

Referring to the diagram, answer question 10.

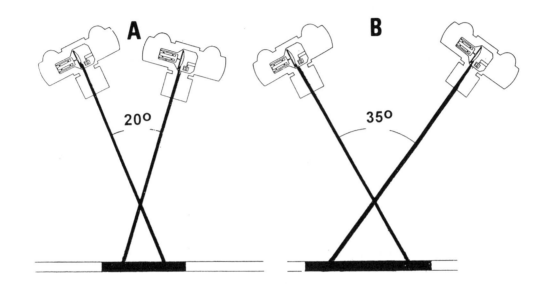

10. Which of the following statements is TRUE concerning the example above?

    A. The tomographic setup A will produce a thinner slice than setup B
    B. The tomographic setup B will produce a thinner slice than setup A
    C. The tomographic setup A will be at a higher fulcrum level than setup B

11. Which of the following is TRUE for a thinner tomographic section?

    A. Less blurring occurs in the tissues above and below the tissue plane
    B. Less tissue density is found in the thinner tomographic sections
    C. Image definition is reduced because of the longer source-to-object distance
    D. A thinner tomographic section possess a higher radiographic contrast

12. All of the following factors will have an effect on the thickness of the tomographic section EXCEPT:

    A. The distance the tube travels during the actual exposure
    B. The height of the fulcrum
    C. The source-to-image receptor distance
    D. The amount of the arc formed by the tube during the exposure

13. Which is NOT a common term that has been employed to describe the visualization of a single tissue plane?

    A. Body section radiography
    B. Tomography
    C. Zonography
    D. Scanography

14. Which of the following exposure angles will produce a tomographic section of less than one millimeter?

A. 15 degrees   C. 48 degrees
B. 36 degrees   D. 64 degrees

15. The exposure angle or arc during tomography can be increased by:

A. Decreasing the source-to-image receptor distance
B. Decreasing the distance the tube travels during the exposure
C. Decreasing the height of the fulcrum
D. Decreasing the speed at which the tube moves

16. Which of the following technical settings would create the thinnest tomographic cut?

	SID	Amplitude	Exposure arc
A.	100 cm	50 cm	30 degrees
B.	100 cm	75 cm	40 degrees
C.	125 cm	75 cm	35 degrees
D.	125 cm	50 cm	20 degrees

Using the following information, answer questions 17-20.

A tomographic examination is performed using a 90 cm amplitude (30-degree arc) and a tube travel time of 3 seconds of a tumor that is found 6 cm anterior from the posterior chest wall. The patient measures 30 cm anteroposterior diameter.

17. The tube travel rate for this setup must be at least:

A. 10 centimeter/second   C. 90 centimeters/second
B. 30 centimeters/second   D. 180 centimeters/second

18. In order to visualize the tumor with patient in the supine position, the fulcrum should be placed at a height of:

A. 6 cm   C. 23 cm
B. 17 cm   D. 30 cm

19. If the fulcrum was set at 6 cm, the point of maximum blurring would be found at the level of _____ in this supine patient.

A. 23 cm   C. 6 cm
B. 17 cm   D. 3 cm

20. In order to obtain a midline section of this patient, the fulcrum should be placed at height of:

A. 30 cm   C. 11.5 cm
B. 15 cm   D. 3 cm

Using the following information, answer questions 21 & 22.

The gallbladder is located 10 cm posterior to the anterior abdominal wall in a patient that has a 24 centimeter anteroposterior diameter.

21. At what height should the fulcrum be set to visualize the gallbladder if the patient is in the supine position?

A. 24 cm   C. 12 cm
B. 14 cm   D. 10 cm

22. At what height should the fulcrum be set to best visualize the cystic duct?

A. 18 cm   C. 14 cm
B. 16 cm   D. 12 cm

23. During a tomographic study, if the height of the fulcrum is changed from 5 cm to 9 cm with the exposure arc remaining the same for both exposures, the result will be a:

1. Different cut level   2. Thicker tomographic section   3. Higher tube travel speed

A. 1 only
B. 2 only

C. 3 only
D. 1, 2 & 3

24. The procedure for producing a thick tomographic section using 5-10 degree exposure arc is called:

A. Zonography
B. Stratigraphy

C. Thermography
D. Scanography

25. The amount of stereo-shift distance is based on the ratio of optimum viewing distance and:

A. Cornea-pupil distance
B. Interpupillary distance

C. Pupil diameter
D. Retinal diameter

26. A method of producing a subjective three-dimensional representation is termed:

A. Stereoradiography
B. Scanography

C. Tomography
D. Plesiotomography

27. A stereographic examination is performed at 100 cm source-to-image receptor distance. The total amount of shift required is:

A. 5 cm
B. 10 cm

C. 15 cm
D. 20 cm

28. Stereoradiography requires:

A. One film and one exposure
B. One film and two exposures

C. Two films and one exposure
D. Two films and two exposures

29. The total shift distance during stereoradiography is found by taking:

A. 5% of the object-to-image receptor distance
B. 10% of the source-to-image receptor distance

C. 15% of the object-to-image receptor distance
D. 30% of the source-to-image receptor distance

30. A stereographic exam of the chest in the erect position is requested at an 180 cm distance. The proper total tube shift and the direction is:

A. 18 cm vertical
B. 18 cm horizontal

C. 9 cm vertical
D. 9 cm horizontal

Explanations

1. B   In order to radiographically demonstrate a single plane of tissue within the body, requires the movement of the tube and image receptor in equal but opposite directions. This technique is called tomography.

2. A   The total distance the tube travels during the exposure is called the exposure angle or arc. It is the curved distance measured from the points at which the central ray intersects with the image receptor.

3. C   The point at which the beams cross number 4 is most closely related to height of the tissue that will remain in focus. The height of this level is called the fulcrum. The plane of tissue that remains in focus is called the tomographic or objective plane.

4. A   During tomography the tissue plane that remains in focus is called the objective or tomographic plane. All tissues above or below this plane will be blurred. The greater the distance from the tomographic plane, the greater the amount of image blur.

5. A   The objective plane which is found at the height of the fulcrum is the only region of the object to remain in sharp focus during tomography.

6. B   The linear distance the tube travels during the tomographic exposure is termed the amplitude. In most newer units, the term amplitude is replaced by the term exposure arc which more properly describes the arc-like movement of the tube above the table.

7. B   The height of the point at which the beams cross during tomography is known as the height of the fulcrum. An increase in the height of the fulcrum will move the plane of tissue that remains in focus to a level farther above the table.

8. B   During tomography, there is a direct relationship between the amount of blurring and the orientation of the object. When the tube movement is at a 90-degree or right angle to the major lines of interest (bony structures) a maximum degree of blurring is seen. Correspondingly, as tube moves parallel to the dominant lines of the anatomic region of interest, a reduced amount of blurring is noted.

9. D   The movement pattern called hypocycloidal which looks similar to a cloverleaf involves the formation of 3 overlapping ellipses providing a very long amplitude. This pattern is associated with a thin tomographic section and exceptional uniformity of blurring.

10. B   The thickness of the tomographic section is inversely proportional to the amplitude, and exposure arc. Since the setup in B is associated with a larger exposure arc it will produce a thinner tomographic section. Because the height of the fulcrum has not been changed the cut level will remain the same.

11. D   Wide-angle tomography which produces very thin tomographic sections is used to visualize structures normally obscured by overlaying tissues; i.e., cochlea, semicircular canals. The main advantage of these techniques is the higher short scale contrast that results between various tissues in the image plane.

12. B   The tomographic cut thickness is effected by changes in the amplitude, source-to-image receptor distance, the exposure arc. The height of the fulcrum should not effect cut thickness.

13. D   Over the last few years, a number of terms have been used to describe the visualization of a single tissue plane including body section radiography, tomography, zonography and plesiotomography. Scanography is a technique used for the measurement of the long bones of the legs.

14. D   The principal controlling factor for the cut thickness is the exposure or tomographic arc. In order to obtain images in the 1 mm thickness range requires an arc in excess of 60 degrees. This large arc is most easily obtained through the use of pluridirectional movement patterns such as those provided with a polytomographic unit. Polytomography is the name given to any technique that uses circular, elliptical, hypocycloidal or spiral movement patterns to provide extremely thin tomographic sections.

15. A   The exposure arc or tomographic arc is the angle through which the central ray moves during the exposure. This exposure arc can be increased by reduction of the source-to-image receptor distance, increasing the amplitude or increasing the exposure angle. The larger arc will produce the thinner tissue section.

16. B   The thinnest tomographic section is obtained through a combination possessing the longest amplitude and the shortest source-to-image receptor distance and greatest exposure arc.

17. B   In order to cover the 90 centimeter distance in the 3 seconds selected the tube must travel at a speed of 30 centimeters/second.

18. A   The tissue plane that remains in focus, the objective plane, during tomography is determined by the height of the fulcrum. Measured from the top of the table, fulcrum can be set at any level within the patient to be sectioned. In this supine patient measuring 30 cm the tumor is found using a 6 cm fulcrum level since this is its height above the table.

19. A   The point of maximum blurring is simply the point most distant from the fulcrum level or at 23 cm in this question.

20. B   A midline section is obtained by simply measuring the patient (30 cm) and dividing that number in half. A midline section for a patient measuring 30 cam would e 15 cm.

21. B   In order to place the gall bladder at the fulcrum level in this supine patient measuring 24 cm the fulcrum should be set at 14 centimeters above the tabletop.

22. D   Since the cystic duct projects posteriorly from the gall bladder it would be best demonstrated in this supine patient by reducing the fulcrum setting by about 2 cm to 12 cm.

23. A   The fulcrum is exclusively related to the height or level of the tomographic image and will have no effect on the cut thickness or speed at which the tube travels.

24. A   Narrow-angle tomography, also known as zonography, is used to produce sharply defined images of the objective plane with little interference from the overlaying structures. A rectilinear movement pattern is used for this technique.

25. B   Due to normal human physiology, the eyes are best able to focus on objects which are seen at a viewing distance of about 25 inches. The measured distance between eyes (interpupillary) is 2.5" providing a 10:1 viewing to distance ratio. This ratio is used in calculating the appropriate amount of stereographic shift that is required during this imaging procedure. This can be simplified to become 1/10 or 10% of the source-to-image receptor distance.

26. A   In order to add depth to the two dimensional radiographic image requires a stereoradiographic technique. During this technique, two radiographs are taken from different locations of a single object. The radiographs are optically merged through the use of a stereoscopic viewing device to form a single image with the added dimension of depth.

27. B   The total tube shift at a 100 cm distance is 10 cm. During a stereoradiographic technique the x-ray tube is shifted one half of this distance to each side object prior to exposing each of the two images.

28. D   A stereoradiographic technique requires two radiographs that are taken from different locations of a single object. The radiographs are optically merged through the use of a stereoscopic viewing device to form a single image with the added dimension of depth.

29. B   The 10:1 ratio is used in calculating the appropriate amount of stereographic shift. In practice this is accomplished by taking 10% of the source-to-image receptor distance. This will provide the desired total shift distance.

30. A Similar to a tomographic procedure, the major lines of interest (body structures) are used in the determination of the shift direction. In an upright chest exam, the tube should be shifted at right angles to the ribs or vertically. A vertical shift of 18 cm is employed for a 180 cm chest radiography.

Density Analysis (All factors)

In each set of technical factors listed, select the set of factors that will provide the greatest optical density.

	mA	Time	kVp	Screen Speed	Grid Ratio	SID in cm
1. A.	200	150 ms	86	100	6:1	150
B.	300	150 ms	94	400	6:1	150
C.	100	200 ms	82	200	6:1	200
D.	200	50 ms	84	100	6:1	100
2. A.	300	250 ms	82	400	8:1	100
B.	200	500 ms	78	400	16:1	100
C.	100	250 ms	72	400	12:1	100
D.	400	150 ms	78	400	12:1	200
3. A.	100	150 ms	86	200	8:1	100
B.	100	100 ms	88	100	6:1	200
C.	200	100 ms	84	200	5:1	100
D.	300	750 ms	76	400	12:1	100
4. A.	300	330 ms	74	100	12:1	200
B.	400	50 ms	78	100	No grid	100
C.	100	500 ms	76	300	8:1	200
D.	300	150 ms	76	50	12:1	200
5. A.	200	50 ms	96	400	12:1	100
B.	200	70 ms	84	100	8:1	200
C.	200	150 ms	86	200	8:1	200
D.	200	150 ms	96	400	6:1	100
6. A.	400	50 ms	90	200	5:1	100
B.	600	30 ms	88	200	8:1	100
C.	400	70 ms	94	50	6:1	100
D.	300	150 ms	96	200	No grid	100
7. A.	300	25 ms	78	400	8:1	100
B.	800	25 ms	78	100	8:1	100
C.	600	33 ms	72	400	12:1	200
D.	300	15 ms	84	200	16:1	100
8. A.	1000	1 ms	90	100	6:1	100
B.	800	1 ms	92	200	12:1	150
C.	400	7 ms	94	400	8:1	100
D.	800	2 ms	88	100	16:1	200
9. A.	300	5 ms	80	50	16:1	200
B.	200	1 second	90	No screen	No grid	100
C.	500	1 ms	82	50	12:1	200
D.	300	10 ms	87	No screen	6:1	100
10. A.	50	15 ms	67	50	6:1	200
B.	25	50 ms	60	100	8:1	100
C.	100	15 ms	67	200	No grid	100
D.	50	75 ms	62	320	6:1	100

Density Analysis (All factors)

In each set of exposure and technical factors select the set of factors that will provide the greatest optical density.

	mA	Time	kVp	Screen Speed	Grid Ratio	SID in cm
1. A.	600	50 ms	72	200	8:1	100
B.	400	50 ms	88	200	6:1	200
C.	300	50 ms	92	200	6:1	100
D.	200	50 ms	86	200	8:1	200
2. A.	200	150 ms	78	100	8:1	100
B.	300	250 ms	86	400	6:1	200
C.	200	150 ms	86	200	12:1	100
D.	300	250 ms	84	160	5:1	200
3. A.	500	50 ms	82	320	12:1	200
B.	500	150 ms	84	160	8:1	100
C.	100	600 ms	80	100	No grid	100
D.	300	50 ms	82	50	No grid	100
4. A.	50	35 ms	62	200	6:1	200
B.	100	15 ms	70	400	8:1	200
C.	50	25 ms	64	200	6:1	200
D.	200	15 ms	72	400	6:1	100
5. A.	100	300 ms	89	160	12:1	100
B.	300	300 ms	91	320	16:1	100
C.	400	200 ms	87	320	8:1	100
D.	400	100 ms	86	100	8:1	100
6. A.	300	70 ms	105	200	12:1	100
B.	800	30 ms	103	400	16:1	150
C.	500	40 ms	100	200	12:1	100
D.	300	150 ms	103	400	6:1	100
7. A.	600	70 ms	86	50	12:1	200
B.	100	150 ms	84	320	12:1	200
C.	800	1.0 sec.	87	No screen	6:1	200
D.	100	150 ms	92	100	8:1	100
8. A.	200	50 ms	80	100	No grid	100
B.	200	10 ms	82	200	12:1	200
C.	200	70 ms	80	400	8:1	200
D.	200	70 ms	82	400	No grid	100
9. A.	400	250 ms	78	100	6:1	100
B.	800	150 ms	76	160	8:1	100
C.	600	300 ms	74	320	8:1	100
D.	400	150 ms	82	100	6:1	100
10. A.	300	750 ms	92	300	12:1	200
B.	200	500 ms	88	400	8:1	150
C.	50	2.5 sec.	94	320	6:1	100
D.	600	300 ms	86	100	6:1	200

Answers

Density Analysis (All Factors)
Exercise 1

1. B
2. A
3. C
4. B
5. D
6. D
7. A
8. C
9. B
10. C

Density Analysis (All Factors)
Exercise 2

1. C
2. B
3. C
4. D
5. C
6. D
7. D
8. D
9. C
10. C

Section IV
Anatomy and Physiology

Radiographic Graffiti III

Can you identify these common patient related artifacts?   Answers appear at the bottom of the page.

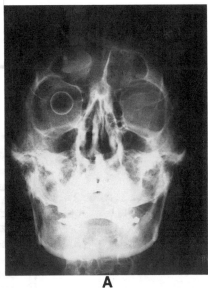

A

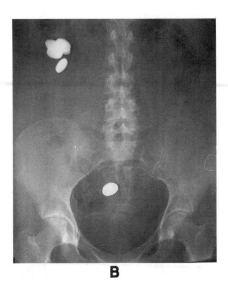

B

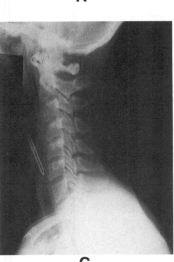

C

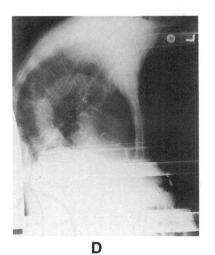

D

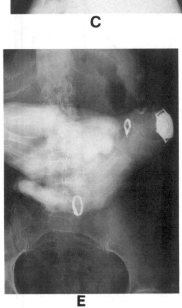

E

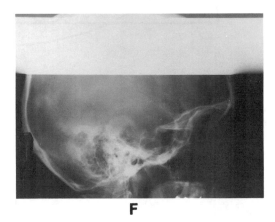

F

Film A - Patient with a false eye. Film B - Patient who has swallowed coins and other metallic objects. Film C - Patient with a razor blade in the upper esophasus. Film D - Side rails of the wheelchair. Film F - Patient with his hands over his abdomen. Film F - Support bar located under the top of the x-ray table.

Bone Development and Articulations

1. The primary ossification center on the shaft of a bone is the point at which bone replacement begins. This region is called the:

   A. Diaphysis
   B. Epiphysis

   C. Metaphysis
   D. Synapse

2. The tough, fibrous covering of the bones that aids in the repair of bones is called the:

   A. Periodontal membrane
   B. Articular cartilage

   C. Periosteum
   D. Epiphysis

3. Bone cells receive nourishment and eliminate wastes from a series of microscopic blood vessels contained within small bony canals called:

   A. Haversian canals
   B. Lacunae

   C. Canaliculi
   D. Lamella

4. The bone-forming cells that serve to replace the embryonic skeleton by building up the bony matrix are termed:

   A. Osteoclasts
   B. Osteoblasts

   C. Osteocytes
   D. Osteons

5. The roughened areas on a bone that serve for the attachment of tendons are called:

   A. Condyles
   B. Foramen

   C. Meatus
   D. Tubercles

6. The flat bones of the adult skull are usually joined together along irregular articulations called:

   A. Sutures
   B. Fontanels

   C. Sinuses
   D. Diaphyses

7. The entire normal human adult skeleton consists of _____ distinct bones.

   A. 238
   B. 206

   C. 184
   D. 126

8. The hollowed out central cavity that is found in the shaft of the long bones is called the:

   A. Nutrient cavity
   B. Diploë

   C. Medullary cavity
   D. Haversian canal

9. The flexibility of bone tissue is primarily due to the action of the chief organic component of bone called:

   A. Collagen
   B. Calcium

   C. Vitamin D
   D. Marrow

10. Bones that form in a joint capsule and serve to reduce the friction created at the joint are termed:

    A. Flagella
    B. Fontanels

    C. Sesamoid bones
    D. Carbonates

11. With the exception of the wrist and foot bones, all the bones of the upper and lower extremities are classified as:

    A. Short bones
    B. Irregular bones

    C. Flat bones
    D. Long bones

12. The gland that has the greatest single effect on bone development and growth is the:

A. Pituitary gland   C. Reproductive gland
B. Thyroid gland   D. Adrenal gland

13. The cranial sutures and other immovable joints are classified as:

A. Diarthrodial   C. Amphiarthrodial
B. Synarthrodial   D. Enarthrodial

14. The movement involving the turning of a part toward the midline of the body is termed:

A. Circumduction   C. Abduction
B. Flexion   D. Inversion

15. The stability of synovial joints is dependent, in large part, upon the action of:

1. Ligaments   2. Tendons   3 Muscles

A. 1 & 2 only   C. 2 & 3 only
B. 1 & 3 only   D. 1, 2 & 3

16. Most of the bones of the body are pre-formed by a special type of cartilage. This embryonic skeleton is made of:

A. Articular cartilage   C. Cancellous cartilage
B. Hyaline cartilage   D. Sesamoid cartilage

17. Which of the following is considered a function of the skeletal system?

1. Protection for internal organs   2. Framework for the body   3. Houses blood forming tissues

A. 1 only   C. 3 only
B. 2 only   D. 1, 2 & 3

18. The action in which the angle between two bones increases is termed:

A. Adduction   C. Extension
B. Luxation   D. Rotation

19. The outer layer (cortical layer) of bone is made up of tightly packed bone cells and is given the name:

A. Spongy bone   C. Cartilaginous bone
B. Compact bone   D. Epiphyseal bone

20. The most common type of bone marrow found within the bones of the adult skeletal system is:

A. Yellow bone marrow   C. White bone marrow
B. Red bone marrow   D. Reticular bone marrow

21. The connection of two bones at a joint is accomplished by a specialized dense connective tissue called:

A. Tendons   C. Plasma
B. Cartilage   D. Ligaments

Match the individual bone to its classification for questions 22-36.

22. _____ Tenth dorsal vertebra   A. Long bone
23. _____ Metacarpal bone   B. Short bone
24. _____ Sternum   C. Flat bone
25. _____ Patella   D. Irregular bone
26. _____ Navicular bone   E. Sesamoid bone

27. Bone is the most rigid of the connective tissues. Its hardness is principally due to the presence of mineral salts such as:

A. Collagen sulfate
B. Calcium phosphate
C. Lead phosphate
D. Radium carbonate

28. In a newborn infant, the soft spot (fontanel) located at the junction of the frontal and parietal bones is termed the:

A. Anterior fontanel
B. Posterior fontanel
C. Lateral fontanel
D. Medial fontanel

29. The true bone cell (osteocyte) can be distinguished from an osteoblast by the presence of the surrounding:

A. Bone marrow
B. Hyaline cartilage
C. Lacunae
D. Skeletal muscle

30. The ends of many long bones contain several branching bony plates which increase the strength of the bone. This type of bone is classified as:

A. Compact bone
B. Osteoporotic bone
C. Volkmann's bone
D. Cancellous bone

31. In freely movable joints, the ends of the bone are covered by _____ which prevents direct bone to bone contact during movement.

A. Ligaments
B. Articular cartilage
C. Yellow bone marrow
D. Periosteum

32. The secondary growth centers that appear in the ends of the long bones in the first few years after birth are called:

A. Metaphyses
B. Articular cartilage
C. Epiphyses
D. Sutures

33. In the developing infant, large amounts of red bone marrow are found in both the flat bones and at the ends of the long bones. This is important for the production of:

1. Growth hormones   *2. Blood cells*   *3. Calcium salts*

A. 1 only
B. 2 only
C. 3 only
D. 1, 2 & 3

Match the given joint to its proper classification for questions 34-38.

34. _____ Pivot joint
35. _____ Hinge joint
36. _____ Ball and socket joint
37. _____ Saddle joint
38. _____ Gliding joint

A. First carpal metacarpal joint
B. Intervertebral (apophyseal) joints
C. First and second cervical vertebrae
D. Metacarpal phalangeal joint
E. Hip joint

39. Many bones of the adult skull possess a layer of cancellous bone between the inner and outer layers of bone. This is called the:

A. Diploë
B. Canaliculi
C. Suture
D. Haversian canal

40. Most freely movable joints have a special lining membrane which secretes a lubricating fluid called:

A. Meniscus
B. Mucus
C. Synovial fluid
D. Cerumen

41. Broad, flat bones of the skill develop from a special connective tissue and are termed:

   A. Endochondral bones
   B. Intramembranous bones

   C. Trabecular bones
   D. Cancellous bones

42. The medullary cavity is in large part formed by specialized cells which reabsorb bone tissue called:

   A. Osteoblasts
   B. Osteomites

   C. Osteocytes
   D. Osteoclasts

43. Individual bone cells are able to communicate and pass nutrients between each other by a series of:

   A. Phorocytes
   B. Endosteum

   C. Canaliculi
   D. Hematomas

44. Muscle tissue is attached to bone by a specialized, connective tissue known as (a):

   A. Tendon
   B. Ligament

   C. Cartilage
   D. Epiconium

45. Aging, especially in women, may lead to an abnormally large amount of calcium loss in the bones causing a condition termed:

   A. Osteoporosis
   B. Osteopetrosis

   C. Osteomyelitis
   D. Osteitis

46. The improper intake of calcium and Vitamin D can lead to a softening of developing bones called:

   A. Rickets
   B. Osteodermia

   C. Acromegaly
   D. Gout

47. Chronic rheumatoid arthritis often damages the joint membranes and may lead to an abnormal joint stiffening called:

   A. Achondroplasia
   B. Ankylosis

   C. Hemarthrosis
   D. Osteonecrosis

48. The shoulder joint is the most common of the larger joints to be injured. The term employed for a dislocation is:

   A. Hyperextension
   B. Subluxation

   C. Luxation
   D. Sprain

49. The overproduction of growth hormones may result in a condition called:

   1. Gigantism   2. Dwarfism   3. Acromegaly

   A. 1 only
   B. 2 only

   C. 1 & 3 only
   D. 1, 2 & 3

50. Tennis elbow and housemaid's knee are two common disorders involving a specialized sac near the joint capsule called the:

   A. Meniscus
   B. Cruciate ligament

   C. Tendon
   D. Bursa

Explanations

1. A   Osteogenesis is the process by which bones develop. The involves the conversion of hyaline cartilage to bone in the primary ossification centers. These ossification centers are called the diaphysis.

2. C   The blood supply to bone which is required for normal growth and repair is augmented by a fibrous membrane covering called the periosteum.

3. A   In order to provide nourishment for the bone cell, a dense matrix of blood vessels forms within tiny vascular cavities of the bone. These Haversian and Volkmann canals extend horizontally and vertically through all of the dense bone tissue of the skeletal system.

4. B   The hyaline cartilage of the embryonic skeleton contains numerous immature bone cells called osteoblasts. The osteoblast develops into a true bone cell in small open spaces called lacunae by surrounding itself with numerous layers of calcium.

5. D   From the Latin for "little swelling" a tubercle provides a roughened surface for the attachment of ligaments and tendons.

6 A   The synarthrodial or immovable joints of the adult skull which provide the fixed connections between the skull bones are called sutures.

7. B   The normal adult skeleton, consisting of the axial skeleton (skull and spine), contains 80 distinct bones and the appendicular skeleton containing 126 bones. These numbers exclude the teeth, sesamoid bones and most of those bones which are not completely fused before puberty.

8. C   The hollowed out central (medullary) cavity which serves to reduce the weight of the bone houses the blood forming (hemopoietic) red bone marrow. In an adult most of the red bone marrow is replaced with the fat rich yellow bone marrow.

9. A   Bone consists of two primary substances: the inorganic calcium salts, which provide the brittle strength and the organic rubber-like collagen giving the bone its resiliency.

10. C   Where considerable pressure is exerted on an area such as the ball of the foot, small (sesamoid) bones may develop within the tendons which help to dissipate energy and reduce friction at the joint.

11. D   The long bones characterized by a shaft and two epiphyseal ends make up the majority of the bones of the arm, legs, hands and feet.

12. A   Though a number of glands and their hormonal secretions play a secondary role in bone growth none is as important as the human growth hormones produced in the pituitary gland.

13. B   The small irregular (wormian) bone which develops in the sutures of the skull aids in the fixation of these immovable or synarthrodial joints.

14. D   Inversion and its counterpart eversion describe the turning of the foot or hand toward or away from the midline of the body.

15. D   Freely movable joints contain an articular joint capsule and its associated synovial membrane which bathes the articulating bones with an oil-like fluid. This capsule is penetrated by the numerous tendons, ligaments, cartilages, and muscles which provide additional support for the joint.

16. B   The replacement of cartilage by bone (endochondral ossification) in the fetus primarily involves a special type of cartilage (hyaline). The flat bones of the skull and mandible, which are made of fibrous tissue, undergo a different process termed intramembranous ossification.

17. D   The bony framework known as the internal skeleton protects the critical organs of the body, provides the attachments for muscles, stores mineral salts and houses the hemopoietic or blood forming tissues.

18. C   Flexion, which decreases the angle between the bones and its counterpart, extension, are the two main motions occurring at a hinge joint. In the axial skeleton (neck) a forward bending is called flexion with a backward bending being referred to as extension.

19. B   The dense or compact bony layer covering the outer layer of most bone provides the support and strength to resist the mechanical stresses placed on them during normal activities.

20. A   Red bone marrow which is found in nearly all bones of a child is to a large extent replaced with the fatty yellow bone marrow by adulthood.

21. D   The principal support between two bones at an articulation is provided by dense connective tissues called ligaments. Ligaments which connect two or more bones will help to provide stability in all amphiarthrodial and diarthrodial joints.

22. D   Long bones: phalanges, metacarpals, tibia, fibula, femur, and humerus, ulna and radius.

23. A   Short bones: wrist and foot bones.

24. C   Irregular bones: all vertebra and most facial bones.

25. E   Flat bones: Ribs, sternum, scapula, ilium and bones of the cranial vault.

26. B   Sesamoid bones: patella, metatarsophalangeal.

27. B   The majority of the intercellular substance of a bone is composed of the mineral salts, calcium phosphate and calcium carbonate.

28. A   At birth, the skull bones are incompletely fused to enable overlapping of the sutures during the birth process. Until their final closure at about 1-2 years of age, four membranous filled spaces called fontanels can be felt on the neonatal head. The anterior fontanel found between the frontal and parietal bones is called the bregma in adulthood. The posterior fontanel found between the occipital and parietal bones is called the lambda in adulthood. The anterolateral and posterolateral fontanels are the two soft spots located on the side of the skull.

29. C   An osteoblast or bone building cell develops in the lake-like lacunae by depositing layers of calcium salts. Now unable to move, the remaining cell is now called an osteocyte or true bone cell.

30. D   The lattice type network seen at the end of the long bones which helps to both maintain its strength and reduce its weight, is called cancellous or spongy bone.

31. B   In order to prevent the damage of bone to bone contact, the articulating surfaces of the bone are covered by a tough fibrous membrane called articular cartilage.

32. C   In most long bones the secondary ossification centers are found at the proximal and distal ends of the bone. These epiphyses or growth plates will normally close at or shortly after puberty.

33. B   The precursor cell for all type of blood cells erythrocytes (RBC), leukocytes (WBC), and thrombocytes (platelets), is the hemocytoblast which resides in the red bone marrow.

34. C   Pivot joint (monoaxial); e.g., atlantoaxial, and the joint at the proximal end of radius and ulna.

35. D   Hinge joint (monoaxial); e.g., elbow, knee, ankle and interphalangeal joint.

36. E   Ball and socket joint (triaxial); e.g., shoulder and hip joints.

37. A   Saddle joint (biaxial); e.g., trapezium and first carpal metacarpal joint.

38. B   Gliding joint (biaxial); e.g., carpal and tarsal articulations, costotransverse, SC, AC, and apophyseal joints.

39. A   The sandwich-like layer of spongy bone between the two layers of dense cranial (diploë) bone provides considerable protection to the brain from a blunt trauma.

40. C   The lubrication and nourishment of the joint capsule is largely derived from the synovial membrane and its oil-like secretion, synovial fluid.

41. B   The most direct form of ossification involving osteoblasts formed within a fibrous membrane (intramembranous), principally occurs in the flat bones of the skull.

42. D   The formation of the hollow medullary cavity and the spaces within cancellous bone is a function of the bone-eating cells called osteoclasts.

43. C   As osteoblasts develop their surrounding calcium walls, small connecting extensions called canaliculi will enable adjacent osteocytes to exchange nutrients and chemical signals with each other.

44. A   The special dense fibrous connective tissue which attaches muscle tissue to the periosteum of the bone is called a tendon.

45. A   A decrease in osteoblastic activity with advancing age may lead to the reduced bone mass seen in osteoporosis. Recent studies indicate that even light daily exercise can delay or even reverse this common condition more often seen in females.

46. A   Rickets though rare in this country, is caused by a severe vitamin D deficiency. This condition is associated with the over-development of the cartilage, reduced ossification which will lead to characteristic bowing of the leg bones in patients with rickets.

47. B   Severe arthritic changes which destroy bone and joint tissues will often lead to the formation of scar tissue and eventual loss of movement. This condition involving the stiffening of the joints is called ankylosis.

48. C   Luxation is the term for a full dislocation, subluxation for partial dislocation and hyper-extension for the over-extension or spraining of a joint. These are the three common types of soft tissue injuries.

49. C   Human growth hormone (somatotropin) which is excreted from the pituitary gland or hypophysis is largely responsible for the growth of tissues; i.e., bone. Overproduction of this hormone leads to gigantism in children and acromegaly in adults. Dwarfism is caused by a reduction in this critical hormonal secretion.

50. D   Trauma to the fluid-filled bursal sacs of the knee, elbow, hip or first metatarsal bone may lead to a local inflammation called bursitis. This painful condition is normally treated by rest and anti-inflammatory drugs.

Chapter 31

1. The bony framework for the body of the hand is formed by the:

   1. Proximal phalanges   2. Wrist bones   3. Metacarpal bones

   A. 1 only   C. 3 only
   B. 2 only   D. 1, 2 & 3

2. The tip of the shoulder joint is formed by a large, flat, roughened area of the lateral most part of the scapular spine called the:

   A. Glenoid process   C. Coracoid process
   B. Acromion   D. Coronoid process

3. The second digit of the normal adult hand contains:

   A. One interphalangeal joint   C. Three interphalangeal joints
   B. Two interphalangeal joints   D. Four interphalangeal joints

4. The pointed extensions on the distal ends of the ulna and radius, which serve for ligament attachments with the wrist bones, are called the:

   A. Styloid processes   C. Coronoid processes
   B. Tuberosities   D. Epicondyles

5. Many of the muscles for lateral rotation of the humerus attach to a tubercle found on the lateral surface of the proximal humerus called the:

   A. Greater tubercle   C. Radial tubercle
   B. Lesser tubercle   D. Deltoid tubercle

6. The hook is a characteristic extension found on the wrist bone called the:

   A. Navicular   C. Trapezoid
   B. Capitate   D. Hamate

7. The epiphysis of the proximal humerus, after its closure near the time of puberty, forms a bony ridge called the:

   A. Bicipital groove   C. Anatomic neck
   B. Humeral head   D. Trochlea

8. The large, cup-shaped structure on the lateral border of the scapula that articulates with the humeral head is termed the:

   A. Acromion   C. Semilunar notch
   B. Glenoid fossa   D. Subscapular fossa

9. The head of the first proximal phalanx articulates with the:

   A. Head of the metacarpal bone   C. Base of the distal phalanx
   B. Base of the metacarpal bone   D. Greater multangular bone

10. The shoulder girdle is a structure formed by the:

    1. Clavicle   2. Scapula   3. Sternum

    A. 1 & 2 only   C. 2 & 3 only
    B. 1 & 3 only   D. 1, 2 & 3

Referring to the diagram of the hand, place the number that corresponds to the following structures for questions 11-18.

11. _____ Head of the proximal phalanx
12. _____ Distal phalanx
13. _____ Ulnar styloid process
14. _____ Navicular bone
15. _____ Radial styloid process
16. _____ Base of the first metacarpal
17. _____ Head of the fifth metacarpal
18. _____ Middle phalanx

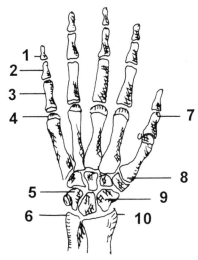

19. In anatomical position, the most lateral bone of the forearm is the:

A. Tibia
B. Ulna

C. Radius
D. Humerus

20. Projecting anteriorly from the upper scapula is a fingerlike process that serves for numerous muscle attachments called the:

A. Coracoid process
B. Acromial process

C. Coronoid process
D. Lesser tubercle

21. The wrist is formed by _____ distinct bones which are collectively known as the _____ bones.

A. 6/tarsal
B. 8/carpal

C. 6/carpal
D. 8/tarsal

22. The rotation of the radius around the ulna during the act of pronation occurs between the radial head and the:

A. Capitulum
B. Acromion

C. Trochlea
D. Semilunar notch

23. The depression on the proximal humerus separating the greater and lesser tubercles is called the:

A. Intercondylar groove
B. Nutrient fossa

C. Bicipital groove
D. Glenoid fossa

24. During acute flexion of the elbow, the coronoid process of the ulna articulates with the coronoid fossa which is located on the:

A. Anterior humerus
B. Posterior humerus

C. Posterior radius
D. Medial radius

25. The common abbreviation used for the articulation formed by the medial end of the clavicle and the sternum is:

A. PIP joint
B. AC joint

C. SC joint
D. TM joint

Referring to the diagram of the shoulder, place the letter that corresponds to the following structures.

26. _____ Coracoid process
27. _____ Humeral head
28. _____ Axillary border of the scapula
29. _____ Acromion
30. _____ Inferior angle of the scapula
31. _____ Vertebral border of the scapula
32. _____ Deltoid tubercle
33. _____ Greater tubercle
34. _____ Superior angle of the scapula

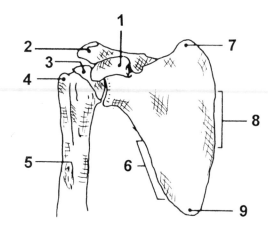

35. The glenohumeral joint is more commonly referred to as the:

A. Elbow
B. Interphalangeal joint
C. Shoulder
D. Acromioclavicular joint

36. The supraspinous fossa is a structure located on the:

A. Anterior aspect of the clavicle
B. Posterior aspect of the scapula
C. Lateral aspect of the humerus
D. Inferior aspect of the scapula

37. The largest and most lateral bone in the proximal row of wrist bones is the:

A. Os magnum
B. Hamate
C. Greater multangular
D. Navicular

38. The normal adult hand contains a total of:

A. Fourteen phalanges
B. Fifteen phalanges
C. Twenty-four phalanges
D. Twenty-six phalanges

39. During the act of pronation, the_____ moves from lateral to medial.

1. Humerus   2. Ulna   3. Radius

A. 1 & 2 only
B. 1 & 3 only
C. 2 & 3 only
D. 1, 2 & 3

40. The greater multangular bone of the wrist articulates with the:

A. Head of the first metacarpal
B. Head of the fifth metacarpal
C. Base of the first metacarpal
D. Base of the fifth metacarpal

41. The proximal end of the ulna articulates with a large fossa on the posterior humerus called the:

A. Olecranon fossa
B. Condyloid fossa
C. Glenoid fossa
D. Acromial fossa

42. The large, bony prominence of the distal humerus sitting medial to the trochlea is called the:

A. Anterior condyle
B. Posterior condyle
C. Medial epicondyle
D. Lateral epicondyle

43. The biceps muscles attach at the proximal end of the radius to a structure called the:

A. Bicipital groove
B. Coronoid process
C. Radial tuberosity
D. Styloid process

44. The semilunar notch of the proximal ulna articulates with a smooth, rounded condyle called the:

A. Trochlea
B. Styloid process

C. Capitulum
D. Olecranon process

45. Each normal human hand contains a total of interphalangeal joints:

A. Four interphalangeal joints
B. Five interphalangeal joints

C. Nine interphalangeal joints
D. Ten interphalangeal joints

46. On the anterolateral surface of the mid-humerus is a tubercle that serves for the insertion of the:

A. Deltoid tendon
B. Pectoralis tendon

C. Biceps tendon
D. Teres minor tendon

47. The medial border of the scapula derives its name from its proximity to the:

A. Vertebral column
B. Axilla

C. Clavicle
D. Diaphragm

48. The intertubercular groove of the humerus is formed by the action of the:

A. Deltoid tendon
B. Biceps tendon

C. Supraspinous tendon
D. Subscapular tendon

49. The largest wrist bone in the normal adult is the:

A. Capitate
B. Lunate

C. Pisiform
D. Hamate

50. The clavicle has articulations with which of the following bones?

1. Humerus   *2. Sternum*   *3. Scapula*

A. 1 & 2 only
B. 1 & 3 only

C. 2 & 3 only
D. 1, 2 & 3

Explanations

1. C The majority of the hand is made-up of the metacarpal bones and the soft tissue and skin that covers the bones. The bones of the fingers are called the phalanges.

2. B The roughened area on the lateral portion of the scapular spine called the acromion, serves as the point of attachment for the tendon of the deltoid muscle and coracoacromial ligaments. These structures help to provide stability for the shoulder joint.

3. B Each finger (numbers 2-5) consists of 3 phalanges articulating over two interphalangeal (IP) joints.

4. A The concave surfaces formed by the ulnar and radial styloids conform well to the convex border of the proximal row of wrist bones. This arrangement aids in the stability of this joint.

5. A The supra, infraspinous, and teres minor muscles which enable lateral rotation of the humerus, attach on the posterior surface of the greater tubercle of the humerus.

6. D The hook-like hamulus on the inferior surface of the hamate bone serves for the attachment and origin for 4 muscles and tendons of the wrist and fingers.

7. C The proximal epiphysis, between the head and shaft of the humerus, closes shortly after puberty at about the age of 20.

8. B The oval articular surface of the scapula called the glenoid fossa is directed forward and laterally toward its junction with the humeral head.

9. C Since the head of the phalanx is always distal and the base is proximal, the head of a proximal phalanx will articulate with the base of the more distal phalanx.

10. A The support structures for the upper extremity are the scapula and clavicle which collectively form the shoulder girdle. The shoulder joint consists of all three bones: the clavicle, scapula and humerus.

11. 7   12. 1   13. 6   14. 9   15. 10   16. 8

17. 4   18. 2

19. C When the hand is supinated and the forearm is in the anatomical position, the ulna is found to be more medial and the radius to be more lateral.

20. A The beak-like coracoid process projecting anterolaterally from the neck of the scapula provides ligamentous attachments for the humerus, acromion, clavicle and tendinous attachments for the pectoralis minor, coracobrachialis and biceps muscles.

21. B The normal adult human has 8 carpal bones arranged in two rows which are held together by dozens of strong ligaments.

22. A The elbow consists of two separated joints. The first, the humeroradial articulation is the connection between the head of the radius and the capitulum. This articulation enables the rotation of the forearm. The second is the radio-ulnar articulation that connects the radial head and radial notch of the ulna. This articulation enables the flexion of the elbow.

23. C The longer of the two heads of the biceps muscle creates a deep groove separating the tubercles of the humerus called the bicipital or intertubercular groove.

24. A The coronoid fossa located on the anterior humerus is a large depression that accepts the coronoid process of the ulna during acute flexion of the elbow.

25. C The sternoclavicular (SC) joint, the only bony connection between the trunk and the upper extremity, is contained within a joint capsule that is continuous with the cartilage of the first rib.

26. 1 27. 3 28. 6 29. 2 30. 9 31. 8

32. 5 33. 4 34. 7

35. C The articulation between the glenoid cavity of the scapula and the head of the humerus is commonly called the shoulder joint.

36. B The posterior surface of the scapula is separated into upper and lower sections by the large scapular spine. The smaller upper section, the supraspinous fossa, serves as the origin for the muscle of the same name. The infraspinous fossa forms the majority of the lower scapula and houses the infraspinous muscle.

37. D The scaphoid or navicular bone is the largest and lateral most bone in the proximal row of the wrist. It articulates with the radius, trapezium, trapezoid, capitate and lunate bones.

38. A The fingers of the hand consist of 14 phalanges: 5 distal, 5 proximal, and 4 middle. The first finger (thumb) has no middle phalanx.

39. C The act of turning the hand to display its posterior surface (pronation) involves the medial rotation of the radius across the ulna.

40. C The greater multangular or trapezium which is located on the lateral aspect of the distal row of wrist bones articulates with the base of the first metacarpal bone.

41. A The thick projection on the proximal ulna which forms the point of the elbow, the olecranon process, articulates with the olecranon fossa on the posterior humerus.

42. C The distal humerus contains two articular surfaces, the larger and more medial condyle, the trochlea and the more lateral rounded condyle called the capitulum. A prominent epicondyle extends from the medial surface of the trochlea and the lateral surface of the capitulum.

43. C The roughened surface on the proximal radius, the radial tuberosity, serves for the insertion of the distal end of the biceps muscle.

44. A The larger of the two articulating surfaces of the distal humerus, the trochlea, joins with the semilunar notch of the proximal ulna to form the hinge joint of the elbow.

45. C Each finger (numbers 2-5) contains 3 phalanges separated by 2 interphalangeal joints (IP). The first finger (thumb) contains only 2 phalanges and a single (IP) joint.

46. A The deltoid muscle which originates from the clavicle and scapula passes laterally over the shoulder to its insertion on the deltoid tuberosity of the lateral humerus.

47. A The more medial vertebral border of the scapula serves for the attachment of the rhomboid and levator scapulae muscles.

48. B One of the tendinous heads of the biceps muscle of the anterior arm extends between the greater and lesser tubercles forming the bicipital or intertubercular groove.

49. A The largest wrist bone, the capitate, due to its central location, articulates with four of the other wrist bones and three middle metacarpal bones.

50. C The clavicle articulates with the acromion process of the scapula laterally and the manubrium of the sternum medially.

1. The total number of phalanges in the normal adult foot is:

   A. Five
   B. Seven
   C. Fourteen
   D. Fifteen

2. The expanded portion of the proximal end of the fibula (fibular head) articulates with the tibia at the:

   A. Proximal medial surface
   B. Distal medial surface
   C. Proximal lateral surface
   D. Distal lateral surface

3. When compared to the male pelvis, the female pelvis is characterized by a:

   1. Wider transverse diameter   *2. Large pelvic outlet*   *3. Larger pubic arch*

   A. 1 only
   B. 2 only
   C. 3 only
   D. 1, 2 & 3

4. Located on the posterior surface of the femur is a prominent ridge of bone serving for the attachment of many of the muscles of the lower extremity. This process is called the:

   A. Intercondylar eminence
   B. Epicondylar ridge
   C. Linea aspera
   D. Intertrochanteric crest

5. The most superior portion of the tibia (tibial eminence) projects upwards from the plateau into the:

   A. Intercondylar space
   B. Acetabulum
   C. Mortal space
   D. Obturator foramen

6. The two large bones that help to form the majority of the adult pelvis are termed the:

   A. Sesamoid bones
   B. Innominate bones
   C. Styloid bones
   D. Cruciate bones

7. The two small, rounded bones that are found on the plantar surface of the first metatarsophalangeal joint to help reduce friction are termed:

   A. Sesamoid bones
   B. Pisiform bones
   C. Pre-metatarsal bones
   D. Pre-metacarpal bones

8. The largest bone in the normal adult foot is called the:

   A. Talus
   B. Navicular
   C. Calcaneus
   D. Cuboid

9. The base of the patella is found on the _____ surface of the bone.

   A. Inferior
   B. Superior
   C. Medial
   D. Lateral

10. The junction of the two innominate bones at the anteroinferior margin is termed the:

    A. Symphysis menti
    B. Acetabulum
    C. Symphysis pubis
    D. Sacroiliac joint

11. The smooth, upper surfaces of the proximal tibia (plateau) serves as the articulating surface for the:

    A. Talus bones
    B. Lateral malleoli
    C. Fibular condyles
    D. Femoral condyles

Referring to the diagram of the foot, place the number that corresponds to the following structures for questions 12-20.

12. _____ Cuboid bone
13. _____ Head of the metatarsal
14. _____ Distal phalanx
15. _____ Calcaneus
16. _____ Base of the metatarsal
17. _____ Navicular
18. _____ Medial cuneiform
19. _____ Proximal phalanx
20. _____ Talus

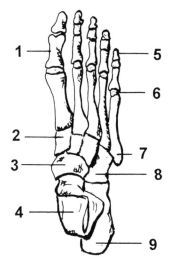

21. The larger and more medial bone of the lower leg is the:

A. Tibia
B. Fibula
C. Femur
D. Tibula

22. The rounded condyles of the femur are separated by a deep depression termed the:

A. Symphysis femoral
B. Intercondylar fossa
C. Bicipital groove
D. Ischial notch

23. The head of the femur articulates with a cup-shaped socket on the lateral surface of the pelvis called the:

A. Obturator foramen
B. Ischial spine
C. Olecranon fossa
D. Acetabulum

24. In the normal sitting position, most of the weight of the trunk is supported by the:

A. Iliac crest of the pelvis
B. Ischial spine of the pelvis
C. Ischial tuberosity of the pelvis
D. Pubic ramus of the pelvis

25. The tendon of the quadriceps femoris attaches on the anterior surface of a raised process of the proximal tibia called the:

A. Tibial spine
B. Tibial eminence
C. Tibial malleolus
D. Tibial tuberosity

26. The majority of the greater or false pelvis is formed by all or part of the:

A. Sacrum
B. Ischium
C. Ilium
D. Pubis

27. The largest and strongest bone of the appendicular skeletal system is the:

A. Tibia
B. Femur
C. Fibula
D. Humerus

Referring to the diagram of the ankle, place the number that corresponds to the following structures for questions 28-31.

28. _____ Lateral malleolus
29. _____ Medial lateral malleolus
30. _____ Medial malleolus
31. _____ Talus bone

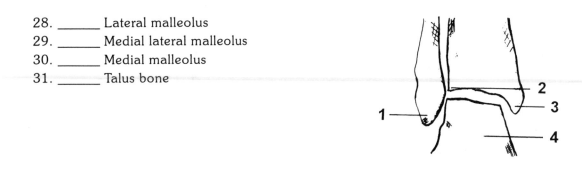

32. During development, the pelvis consists of six separate bones that fuse during childhood. These paired bones are the:

   1. Ilium   2. Ischium   3. Pubis

   A. 1 & 2 only
   B. 1 & 3 only

   C. 2 & 3 only
   D. 1, 2 & 3

33. The large, smooth articulating surfaces located on the distal femur are termed the:

   A. Femoral condyles
   B. Greater trochanters

   C. Linea asperae
   D. Epicondyles

34. The astragalus or talus bone of the foot is normally located:

   A. Anterior to the cuboid bone
   B. Posterior to the tibia

   C. Superior to the calcaneus or os calcis
   D. Inferior to the navicular bone

35. The protuberance that is felt on the lateral surface of the lower leg distal to the knee joint is the:

   A. Medial epicondyle
   B. Lateral epicondyle

   C. Fibular head
   D. Tibial tuberosity

36. The acetabulum is composed of the:

   1. Ilium   2. Ischium   3. Pubis

   A. 1 & 2 only
   B. 1 & 3 only

   C. 2 & 3 only
   D. 1, 2 & 3

37. The cuneiform bones articulate anteriorly with the:

   A. Calcaneus
   B. Base of the metatarsals

   C. Head of the metatarsals
   D. Navicular bone

38. Fallen arches or flat feet result from the weakening of the ligaments in the foot. Damage to these ligaments will result in the loss of the:

   A. Longitudinal arch
   B. Vertical arch

   C. Transverse arch
   D. Neural arch

39. The pointed, bony prominence located on the distal fibula is called the:

   A. Head of the fibula
   B. Medial malleolus

   C. Lateral malleolus
   D. Fibula eminence

Referring to the diagram of the hip joint, place the number that corresponds to the following structures for questions 40-49.

40. _____ Ischial tuberosity
41. _____ Anatomic neck
42. _____ Lesser trochanter
43. _____ Surgical neck
44. _____ Iliac crest
45. _____ Anterior superior iliac spine
46. _____ Obturator foramen
47. _____ Pubic bone
48. _____ Head of the femur
49. _____ Greater trochanter

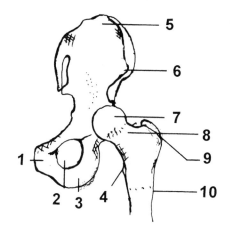

50. The navicular or scaphoid bone of the foot is located:

    1. Inferior to the cuboid   **2. Posterior to the cuneiform**   **3. Anterior to the talus**

    A. 1 & 2 only   C. 2 & 3 only
    B. 1 & 3 only   D. 1, 2 & 3

51. The large bony ridges of the proximal femur that serve to attach many of the muscles for moving of the hip are called the:

    A. Greater and lesser malleoli   C. Greater and lesser condyles
    B. Greater and lesser trochanters   D. Greater and lesser styloids

52. In the normal adult, the angle formed by the neck and shaft of the femur is about:

    A. 90 degrees   C. 120 degrees
    B. 105 degrees   D. 145 degrees

53. The knee joint is formed by the articulation of the:

    1. Femur   **2. Tibia**   **3. Fibula**

    A. 1 & 2 only   C. 2 & 3 only
    B. 1 & 3 only   D. 1, 2 & 3

54. The cuboid bone of the foot articulates posteriorly with the:

    A. Navicular   C. Calcaneus
    B. Talus   D. Cuneiform

Referring to the diagram of the knee, place the number that corresponds to the following structures for questions 55-62.

55. _____ Lateral epicondyle of the femur
56. _____ Shaft of the fibula
57. _____ Tibial tuberosity
58. _____ Lateral condyle of the femur
59. _____ Head of the fibula
60. _____ Intercondylar eminence
61. _____ Medial condyle of the femur
62. _____ Medial epicondyle of the tibia

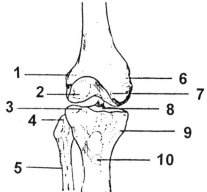

63. The obturator foramen of the pelvis is formed by the junction of the following bones:

1. Ilium   2. Ischium   3. Pubis

A. 1 & 2 only
B. 1 & 3 only

C. 2 & 3 only
D. 1, 2 & 3

64. The ankle joint is formed by all or part of the:

1. Fibula   2. Tibia   3. Talus bone

A. 1 & 2 only
B. 1 & 3 only

C. 2 & 3 only
D. 1, 2 & 3

65. The deep notch located on the posterior margin of the innominate bone between the posterior inferior iliac spine and the ischial spine is called the:

A. Greater sciatic notch
B. Pubic angle

C. Symphysis pubis
D. Ischial tuberosity

Explanations

1. C   Each foot contains five distal, five proximal and four middle phalanges. The first (big) toe has no middle phalanx.

2. C   The shorter and thinner of the two lower leg bones, the fibula, articulates with the tibia on the lateral surface at both its proximal and distal ends. Unlike the larger tibia, the fibula is not a weight-bearing bone.

3. D   The female pelvis has a larger inlet, larger outlet diameter and wider pubic arc than the male pelvis. These aid in the ability of the woman to carry and bear children.

4. C   The thickened ridge, the linea aspera, which extends down the middle two thirds of the posterior femur provides for the attachment for the tendons of the adductor muscles of the thigh.

5. A   Extending upwards from the superior articulating surfaces of the tibia are two eminences called the tibia plateaus. These provide the attachments for the anterior and posterior cruciate ligaments of the knee.

6. B   The innominate or coxa bones are the two large bones that form the adult pelvis. During early childhood the innominate bones developed from three individual bones that fuse in the mid teens called the ilium, pubis, and ischial bones.

7. A   The two most commonly found sesamoid bones of the foot are located on the inferior surface of the first metatarsal phalangeal joint. These bones are contained within the tendon that helps increase the leverage between the bones.

8. C   The posterior portion of the foot is formed by a large bone called the calcaneus or os calcis. This serves as the point of origin or insertion for a number of the tendons of the muscles that move the foot and the toes.

9. B   The patella located on the anterior surface of the femur is oriented with its apex inferior to the broad base on the superior aspect of the bone. This results in the apex being inferior to the base.

10. C   The two halves of the pelvis are united anteriorly across the superior rami of the pubic bones by a fibrocartilage interpubic disc. Posteriorly the innominate bones articulate laterally with the sacrum.

11. D   The upper articulating condyles of the tibia, the tibial plateaus and the femoral condyles at the knee are separated by the two C-shaped cartilages called the menisci.

12. 8   13. 6   14. 5   15. 9   16. 7   17. 3

18. 2   19. 1   20. 4

21. A   The larger of the two bones of the lower leg, the tibia, is the only weight-bearing bone of the lower leg.

22. B   The deep depression found between the inferoposterior portion of the femoral condyles is called the intercondylar fossa.

23. D   The deep hemispheric cavity on the lateral surface of the innominate bone which receives the head of the femur is called the acetabulum.

24. C   When seated, the pelvis is tilted so that the majority of weight is transferred onto the large roughened ischial tuberosities.

25. D   The roughened area on the upper anterior surface of the tibia called the tibial tuberosity serves for the insertion of the broad tendon of the quadriceps femoris muscle which encapsulates the patella.

26. C   The pelvis is subdivided into a true pelvis containing the inlet and outlet and the ala wing. The upper portion of the ilium called the wing is within the abdominal cavity as is referred to as the false pelvis.

27. B   The arch shape of the neck and large diameter bone of the femur enables it to support the weight of the entire torso and upper extremities.

28. 1   29.   2   30.   3   31.   4

32. D   Before its final fusion in the middle teen years, the pelvis consists of 3 pairs of bones: the ilium, ischium and pubis. In the adult, the pelvis is formed by the two innominate bones.

33. A   The knee joint is formed by the articulation of the tibial plateaus with the medial and lateral condyles of the femur. The tibia does not contribute to the knee joint.

34. C   The most superior bone of the foot, the talus (astragalus) sits above the calcaneus and just posterior to the navicular bone. It forms the lower portion of the ankle joint and provides for numerous tendon and ligaments attachments to the tibia, fibula and other bones of the foot.

35. C   The expanded upper end of the fibula, the fibula head, articulates with the lateral surface of the tibia just distal to the knee joint.

36. C   The more or less central location of the acetabulum on the lateral surface of the pelvis enables its formation by all three of the bones of the immature pelvis: the ilium, ischium and pubis.

37. B   The three wedge-shaped cuneiform bones of the foot articulate posteriorly with the navicular and anteriorly with the base of the first three metatarsal bones.

38. A   Pes planus or flat feet often results from excessive stress on the longitudinal arch of the foot and is common in occupations requiring long periods of standing or walking.

39. C   The longer of the two distal processes of the ankle joint extending about 1 cm below the superior aspect of the talus bone is the lateral malleolus of the fibula.

40. 3   41.   8   42.   4   43.   10   44.   5   45.   6

46. 2   47.   1   48.   7   49.   9

50. C   The flattened oval navicular bone is centrally located in the foot lying anterior to the talus, posterior to the cuneiforms, and medial to the cuboid bone.

51. B   The large bony prominence felt on the lateral aspect of the hip, the greater trochanter provides the attachments for the gluteus maximus and minimus muscles. The greater trochanter is also a useful landmark for finding the level of the symphysis pubis which lies on the same axial plane.

52. C   The normal angle between the neck and shaft of the femur is between 115-130 degrees. Deformities causing an excessive angle, coxa valga or insufficient angle, coxa vera will seriously effect the gait of an individual.

53. A   The knee joint consists of two true bones, the femur and tibia, and a sesamoid bone, the patella. The fibula, which articulates with the lateral surface of the tibia, does not contribute to the knee joint.

54. C   The cuboid bone articulates anteriorly with the fourth and fifth metatarsals, posteriorly with the calcaneus, and laterally with the navicular and lateral cuneiform.

55. 1   56.   5   57.   10   58.   2   59.   4

60. 8   61.   7   62.   9

63. C   The junction of the ischium and pubic bones of the lower pelvis forms the large obturator foramen found on the lower aspect of the innominate bone.

64. D   The medial malleolus of the tibia, the lateral malleolus of the fibula, and the superior portion of the talus all contribute in the formation of the ankle joint.

65. A   Just above the ischial spine is a large depression formed by the passage of the pelvic vessels and nerves as they pass in the lower limb called the greater sciatic notch.

1. The largest number of vertebrae are found in the _____ portion of the vertebral column.

   A. Cervical
   B. Thoracic

   C. Lumbar
   D. Sacrum

2. The segment of the vertebral column that is characterized by foramina in the transverse process is termed the:

   A. Sacral vertebrae
   B. Lumbar vertebrae

   C. Thoracic vertebrae
   D. Cervical vertebrae

3. The fibrous cartilage wedge that separates the vertebral bodies of the spinal column is termed the:

   A. Meningeal spacer
   B. Intervertebral disc

   C. Laminal wedge
   D. Transverse disc

4. In the normal adult, the ribs are subdivided into three types based upon their:

   A. Sternal attachments
   B. Type of costal cartilage

   C. Vertebral attachments
   D. Direction of articulation

5. The odontoid process is a tooth-like extension arising from the anterior aspect of the:

   A. First cervical vertebrae
   B. Second cervical vertebrae

   C. Seventh cervical vertebrae
   D. Fifth lumbar vertebrae

6. The intervertebral foramina, which are found between all the upper segments of the vertebral column, serve to allow passage of the:

   A. Vertebral arteries
   B. Spinal cord

   C. Spinal nerves
   D. Cauda equina

7. The first cervical vertebra that serves to support the weight of the skull is often referred to as the:

   A. Odontoid
   B. Hyoid

   C. Axis
   D. Atlas

8. The greatest amount of flexibility is permitted by the vertebrae in the:

   A. Cervical region
   B. Thoracic region

   C. Lumbar region
   D. Sacral region

9. The posterior rib articulation with the body of a thoracic vertebra is called a/an:

   A. Costotransverse joint
   B. Costovertebral joint

   C. Apophyseal joint
   D. Intervertebral joint

10. In the normal adult, the space for the spinal cord extends to the vertebral level of:

    A. First lumbar vertebra
    B. Fourth lumbar vertebra

    C. Tenth thoracic vertebra
    D. Fourth sacral vertebra

11. The terminal segments of the vertebral column are formed by the fusion of four rudimentary vertebrae and, in an adult, are known collectively as the:

    A. Ala
    B. Lamina

    C. Coccyx
    D. Facets

Referring to the diagram of the cervical vertebrae, place the number that corresponds to the following structures for questions 12-18.

12. _____ Articulating condyle
13. _____ Spinous process
14. _____ Vertebral foramen
15. _____ Vertebral body
16. _____ Transverse process
17. _____ Transverse foramen
18. _____ Lamina

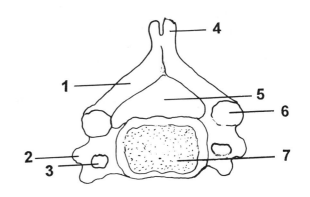

19. The exaggerated lip on the body of the first sacral vertebra is often termed the:

    A. Sacral promontory   C. Lateral mass
    B. Sacral spine   D. Ala

20. The sacrum articulates with the innominate bones on the sacrum's:

    A. Anterior medial surface   C. Inferior lateral surface
    B. Posterior medial surface   D. Superior lateral surfaces

21. The sternum has a small, pointed process (xiphoid) on its inferior surface. This is a useful landmark in the location of the level of the:

    A. Seventh cervical vertebra   C. Tenth thoracic vertebra
    B. Second thoracic vertebra   D. Third lumbar vertebra

22. The body of a vertebra serves to support the majority of weight impressed upon it and is normally located:

    A. Posterior to the spinous process   C. Posterior to the pedicle
    B. Anterior to the vertebral foramen   D. Anterior to the transverse process

23. The spinal cord is protected along its course by passing within a large foramen of the vertebrae called the:

    A. Vertebral foramen   C. Spinal foramen
    B. Intervertebral foramen   D. Transverse foramen

24. In the normal adult, the lumbar vertebrae form a convex forward or _____ curve.

    A. Lordotic   C. Scoliotic
    B. Intervertebral foramen   D. Spinal

25. The dorsal vertebrae are distinguished from the other vertebrae by the presence of the:

    A. Lamina   C. Articular processes
    B. Rib facets   D. Transverse foramina

26. The vertebral prominence that is found on the seventh cervical vertebra derives its name from its:

    A. Long transverse process   C. Long spinous process
    B. Large vertebral body   D. Large articular process

27. The spinal cord is protected along its course by a bony foramen of the vertebrae. The spinal cord proper usually ends at the vertebral level of the:

A. Tenth dorsal vertebra
B. Second lumbar vertebra

C. Fifth lumbar vertebra
D. Second sacral vertebra

28. The joint that allows the rotation or turning of the skull around a pivot point occurs between the:

A. Occipital bone & the first cervical vertebra
B. Occipital bone & the second cervical vertebra

C. First & second cervical vertebrae
D. Second & third cervical vertebrae

29. The last two pairs of ribs are classified as floating ribs because they have:

A. No vertebral attachment
B. No sternal attachment

C. Both of the above
D. Neither of the above

30. The first cervical vertebra is characterized by its lack of a/an:

A. Neural arch
B. Pedicle

C. Vertebral body
D. Articulating process

31. The sternal or manubrial notch is often employed for the localization of the:

A. Fifth cervical vertebra
B. Seventh cervical vertebra

C. Second thoracic vertebra
D. Tenth thoracic vertebra

32. The most anterior portion of a normal vertebra is the:

A. Arch
B. Pedicle

C. Spinous process
D. Body

33. The vertebral foramen is formed by all or part of the:

1. Vertebral body   2. Cornu   3. Pedicles

A. 1 & 2 only
B. 1 & 3 only

C. 2 & 3 only
D. 1, 2 & 3

34. The manubrium serves for the point of attachment for the costal cartilage of the first:

A. Two pairs of ribs
B. Three pairs of ribs

C. Five pairs of ribs
D. Seven pairs of ribs

35. The semi-fluid central portion of the intervertebral disc is called the:

A. Nucleus pulposus
B. Nucleus fibrous

C. Annulus fibrous
D. Annulus pulposus

36. The first seven pairs of ribs are classified as:

A. False ribs
B. Ribbon ribs

C. Cartilaginous ribs
D. True ribs

37. An example of an zygapophyseal joint is the articulation between the:

A. Bones of the skull
B. Ribs & the vertebrae

C. Superior & inferior articular processes
D. First vertebra & the occipital bone

Referring to the diagram of the sternum, place the number that corresponds to the following structures for questions 38-42.

38. _____ Xiphoid
39. _____ Sternal notch
40. _____ Gladiolus
41. _____ Costal cartilage
42. _____ Manubrium

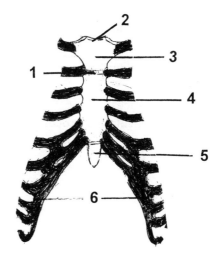

43. Which foramina of the vertebral column are formed by portions of more than one individual vertebra?

A. Transverse foramina
B. Vertebral foramina

C. Optic foramina
D. Intervertebral foramina

44. The atlanto-occipital joint of the upper cervical region allows for:

1. Rotation of the head   *2. Flexion of the head*   *3. Extension of the head*

A. 1 & 2 only
B. 1 & 3 only

C. 2 & 3 only
D. 1, 2 & 3

45. The two laminae project posteriorly from the transverse process to their junction in the midline. They help to form the:

A. Spinous process
B. Superior articulating process

C. Inferior articulating process
D. Vertebral body

46. The usual points of articulation between the ribs and a thoracic vertebra are the:

1. Body & spinous process   *2. Lamina & pedicle*   *3. Transverse process & the body*

A. 1 only
B. 2 only

C. 3 only
D. 1, 2 & 3

47. The bony ridge formed by the junction of the two upper portions of the sternum is called the:

A. Sternal notch
B. Sternal angle

C. Xiphoid ridge
D. Costal ridge

Referring to the diagram of the lateral aspect of the lumbar vertebra, place the number that corresponds to the following structures for questions 48-53.

48. _____ Vertebral Body
49. _____ Lamina
50. _____ Superior articulating process
51. _____ Pedicle
52. _____ Spinous process
53. _____ Inferior articulating process

54. The articulation between the first and second cervical vertebrae is called the:

A. Axio-occipital joint
B. Atlanto-epistrophyseal joint
C. Densa-occipital joint
D. Axio-epistrophyseal joint

55. The destruction of a vertebra, which can lead to spinal cord and nerve involvement, is termed:

A. Lordosis
B. Spondylolisthesis
C. Spondylitis
D. Spondylolysis

56. The abnormal lateral curvature of the spine results in a condition called:

A. Scoliosis
B. Lordosis
C. Subluxation
D. Ankylosis

57. Protrusion of the central portion of an intervertebral disc on the spinal cord leads to a painful condition known as (a):

A. Hunchback
B. Thoracic spine
C. Sciatica
D. Herniated disc

58. Kyphosis is a relatively common condition of the elderly that results in a pronounced forward curvature of the:

A. Cervical spine
B. Thoracic spine
C. Lumbar spine
D. Sacrum

59. A congenital deformity resulting in the incomplete formation of the spinal canal is called:

A. Spondylolisthesis
B. Lordosis
C. Spina bifida
D. Spondylitis

60. The total number of sacral foramina found in the adult human sacrum is:

A. Four
B. Eight
C. Twelve
D. Sixteen

1. B   The vertebral column of the adult is separated into five divisions containing 26 bones: 7 cervical, 12 thoracic, 5 lumbar, a single sacrum and coccyx bone. The adult sacrum develops from the fusion of 5 rudimentary vertebrae. The terminal segment of the spine, the coccyx, is formed by the fusion of the last 4 vertebral segments.

2. D   A unique characteristic of the cervical vertebra is a large oval foramen in each transverse process which protects the vertebral arteries and veins in their journey toward the brain.

3. B   Each intervertebral disc is composed of an outer fibrous ring, the annulus fibrosus, and the pulpy elastic center called nucleus pulposus.

4. A   The first 7 pairs are classified as true ribs because of their direct costal cartilage attachment to the sternum. The 3 pairs of false ribs have only an indirect sternal connection and the last 2 pairs are termed floating because they have no sternal attachment.

5. B   The tooth-like odontoid process found on the second cervical vertebra most likely represents the remnants of the body of first cervical vertebra which is no longer present in modern man.

6. C   The notches found on the inferior surface of the upper vertebra and superior surface of an inferior vertebra form a new structure called the intervertebral foramen. This foramen allows the passage of spinal nerves in the cervical, thoracic, and lumbar regions to exit from the spinal canal.

7. D   Because it supports the head in a similar manner to the Greek god Atlas holding the world, first cervical vertebra shares this name. The junction between first cervical vertebra and the occipital bone is called the atlantooccipital joint.

8. A   The greatest flexibility is found in the cervical region of the spine followed by that of the lumbar region. The thoracic spine permits only a small amount of motion while the sacrum and coccyx permit virtually no movement.

9. B   The first 10 pairs of ribs articulate with a thoracic vertebra at two separate locations: the head of the rib joins with the body across the costovertebral joint while the tubercle of the rib attaches to the transverse process at the costotransverse joint.

10. D   The vertebral or sacral canal extends through the five sacral vertebrae and ends in a small opening at the level of fourth sacral vertebra called the sacral hiatus.

11. C   The adult sacrum develops from the fusion of 5 rudimentary vertebrae. The terminal segment of the spine, the coccyx, is formed by the fusion of the last 4 vertebral segments.

12. 6   13. 4   14. 5   15. 7   16. 2

17. 3   18. 1

19. A   The large prominence on the anterosuperior surface of the first sacral vertebrae is termed the sacral promontory.

20. D   The ear-shaped auricular surfaces on the upper lateral border of the sacrum articulate with the auricular surface on the medial portion of the ilium.

21. C   The most inferior portion of the sternum, the xiphoid process is normally located at the same level as the 10th thoracic vertebra and shaft of the 6th rib.

22. B   The body is normally the most anterior portion of the vertebra. The body is anterior to the spinous process, the lamina transverse process and the vertebral foreman.

23. A   The vertebral foramen which protects the spinal cord is bounded anteriorly by the body, laterally by the pedicles, and posteriorly by the lamina.

24. A   At birth, the vertebral column contains only the two kyphotic curves, those in the thoracic and sacral regions. With the beginning of the weight bearing erect posture, the lordotic curves in the cervical and lumbar region develop.

25. B   The principal characteristic of the thoracic or dorsal vertebra is the presence of the articulating processes for the ribs called rib facets.

26. C   The transitional vertebra between the cervical and thoracic spines, the seventh cervical vertebra is known for its long horizontal spine called the vertebral prominence.

27. B   The 18" or 45 cm long spinal cord terminates at the lower border of the first lumbar vertebra. Below this point the cord is tapered onto the structure called the conus medullaris.

28. C   The atlantoaxial joint between the first and second cervical vertebra is responsible for the lateral rotation of the head.

29. B   The small, slender 11th and 12th ribs, which have no sternal connection, are called the floating ribs.

30. C   The first cervical vertebra is characterized by a lack of a spinous process and vertebral body. In addition, it has 2 specialized superior and inferior articulating processes that join the condyles of the occipital bone. This joint called the atlantooccipital joint permits the nodding of the head.

31. C   The large sternal or jugular notch on the superior surface of the manubrium is normally found at the same vertebral level as the body of first or second thoracic vertebra.

32. D   The body of a vertebra is the most anterior portion of a normal vertebra. It articulates with the structures of the neural arch.

33. B   The vertebral foramen or canal is formed anteriorly by the vertebral body, laterally by the pedicles, and posteriorly by the lamina.

34. A   The upper segment of the sternum, the manubrium, articulates with the cartilages from the first 2 pairs of ribs.

35. A   The highly resilient semi-liquid center of an intervertebral disk, the nucleus pulposus, aids in its action as a shock absorber for the spinal column.

36. D   The direct sternal attachments to the costal cartilages of the first 7 pairs of ribs leads to their classification as true ribs.

37. C   By definition, an apophysis is a bony outgrowth lacking its own ossification center. Examples include the superior and inferior articulating processes of the vertebra.

38. 5   39. 2   40. 4   41. 6   42. 3

43. D   The deep notches above and below each pedicle creates an oval foramen formed when two vertebrae are articulated called the intervertebral foramen. These structures surround and protect the pair of spinal nerves which exit at each vertebral level.

44. C   The junction between the atlas (CI) and the occipital bone permits the forward and backward bending of the head called flexion and extension.

45. A   The posterior portion of the vertebral foramen called the neural arch is formed by the two lamina and the spinous process which projects posteriorly from their junction.

46. C   To increase the stability of the posterior connection of the ribs with a thoracic vertebra two points of articulation are present. The head of the rib connects with the body of the vertebrae and the tubercles connect with the transverse process.

47. B   The manubrium, the upper portion of the sternum, is fused to the body or gladiolus at a raised ridge (sternal angle) at the level of the third thoracic vertebra.

48. 6   49. 3   50. 1   51. 5   52. 2   53. 4

54. B   The atlantoaxial or atlanto epistrophyseal joint refers to the articulation between CI (atlas) and C2 (axis).

55. D   Spondylolysis is defined as the destruction of a vertebra. Spondylolisthesis is the forward slippage of one vertebra over another. Spondylitis is inflammation of a vertebra.

56. A   If the spine has an abnormal curve (scoliosis), the body will attempt to keep the head centered over the spine. In most individuals these compensatory actions worsen the condition leading to the progressive nature of the untreated scoliosis.

57. D   The rupture (herniation) of the nucleus pulposus and subsequent leakage and pressure on the spinal cord causes pain and muscle weakness on the effected side.

58. B   The exaggerated curve of the dorsal spine gives rise to a condition called humpback, hunchback, or kyphosis.

59. C   Rachischisis, or spina bifida, is a serious condition in which the lamina of the neural arch fail to fuse. The spinal cord and its membranes are pushed through this opening leading to brain and/or spinal cord damage.

60. D   The sacrum has 8 ventral and 8 dorsal foramen which enable the passage the spinal nerves from the vertebral canal. These nerves innervate the anterior and posterior structures within the pelvis.

1. The adult cranial vault or calvaria that houses the brain is formed by all or part of _____ cranial bones.

   A. Four
   B. Eight
   C. Fourteen
   D. Twenty-eight

2. The large prominent ridges on the anterior surface of the frontal bone (superciliary ridges) are closely related to the surface structure termed the:

   A. Acanthion
   B. Bregma
   C. Inner canthus
   D. Eyebrows

3. The pterygoid processes that form part of the posterior nasal cavity are the inferior extensions of the:

   A. Sphenoid bone
   B. Temporal bone
   C. Maxillary bone
   D. Mandible

4. The roof of the mouth (hard palate) is formed by the inferior surfaces of the:

   1. Maxillary bone   *2. Palatine bones*   *3. Malar bones*

   A. 1 & 2 only
   B. 1 & 3 only
   C. 2 & 3 only
   D. 1, 2 & 3

5. The meninges, the membranes that help to protect the brain, have an attachment on a small triangular process of the ethmoid bone called the:

   A. Falx cerebri
   B. Clinoid process
   C. Crista galli
   D. Styloid hamulus

6. The mastoid air cells are located posterior to the ear within a large, bony prominence of the:

   A. Temporal bone
   B. Sphenoid bone
   C. Malar bone
   D. Occipital bone

7. The majority of the sides and top of the cranial vault are formed by a pair of bones called the:

   A. Parietal bones
   B. Frontal bones
   C. Ethmoid bones
   D. Palatine bones

8. The large foramen that passes through the occipital bone, which allows for the passage of the spinal cord and brain stem into the calvaria, is called:

   A. Foramen ovale
   B. Foramen magnum
   C. Foramen rotundum
   D. Foramen lacerum

9. The majority of the middle cranial fossa are formed by the body of the:

   A. Occipital bone
   B. Maxillary bone
   C. Temporal bone
   D. Sphenoid bone

10. The nasal cavities have a continuous membrane with a series of cavities within the cranial bones called the:

    A. Paranasal sinuses
    B. Infraorbital fissure
    C. Mastoid air cells
    D. Olfactory foramina

11. The junction of the nasal bones and the frontal bones corresponds to a useful surface landmark called the:

    A. Inion
    B. Gonion
    C. Nasion
    D. Acanthion

12. The sagittal suture is located between the:

A. Two parietal bones
B. Frontal & parietal bones
C. Occipital & temporal bones
D. Occipital & parietal bones

13. The zygomatic arch is formed by portions of the:

A. Zygomatic & temporal bones
B. Temporal & occipital bones
C. Sphenoid & temporal bones
D. Frontal & malar bones

14. Nerves, arteries and veins pass in and out of the floor of the cranium by way of the:

A. Auditory canals
B. Eustachian tubes
C. Basal foramina
D. Paranasal sinuses

Referring to the diagram of the skull, place the number that corresponds to the following structures for questions 15-24.

15. _____ Occipital bone
16. _____ Zygomatic (malar) bone
17. _____ Parietal bone
18. _____ Styloid process
19. _____ Nasal bones
20. _____ Frontal bone
21. _____ Anterior nasal spine
22. _____ Coronal structure
23. _____ Maxillary bone
24. _____ Temporal bone

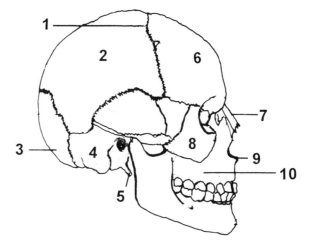

25. The flattened, horizontal portion on the superior aspect of the ethmoid bone is called the:

A. Crista galli
B. Perpendicular plate
C. Superior conchae
D. Cribriform plate

26. In the adult skull, the immovable articulation between the parietal and temporal bones is called the:

A. Lambdoidal suture
B. Squamosal suture
C. Sagittal suture
D. Coronal suture

27. The bony orbit is formed by all of the following bones EXCEPT:

A. The lacrimal bone
B. The ethmoid bone
C. The temporal bone
D. The sphenoid bone

28. The lateral border of the eyelids meets at a point termed the:

A. Acanthion
B. Nasion
C. Outer canthus
D. Optic fissure

29. The widest transverse measurement of the skull is found between the two:

A. Parietal bones
B. Temporal bones
C. Malar bones
D. Lacrimal bones

30. The condyles for the articulation with the first cervical vertebra are located on the inferior surface of the:

A. Malar bone
B. Occipital bone
C. Vomer bone
D. Ethmoid bone

31. The most superior paranasal sinus is the:

A. Ethmoidal sinus
B. Maxillary sinus
C. Frontal sinus
D. Zygomatic sinus

32. The thick, spongy ridge of bone for the reception of the roots of the teeth is termed the:

A. Antrum of Highmore
B. Condyloid process
C. Parietal eminences
D. Alveolar process

33. The only movable bones in the normal adult skull are the:

A. Zygoma
B. Mandible
C. Nasal bone
D. Inferior conchae

34. Which bone does NOT form any portion of the lateral cranial wall?

A. Occipital bone
B. Sphenoid bone
C. Frontal bone
D. Ethmoid bone

35. The lacrimal canal is a bony groove through the lacrimal bone that serves to connect the:

A. Oral and nasal cavities
B. Auditory and nasal cavities
C. Orbital and oral cavities
D. Nasal and orbital cavities

36. In the adult skull, the immovable articulation between the parietal and occipital bones is called the:

A. Sagittal suture
B. Coronal suture
C. Lambdoidal suture
D. Squamosal suture

37. The maxillary sinus is often referred to as the:

A. Pyloric antrum
B. Venous sinus
C. Antrum of Highmore
D. Hypoglossal antrum

38. The surface areas of the nasal cavities are divided into two more or less equal sides by a bony separation called the:

A. Alveolar process
B. Nasal septum
C. Horizontal sinus
D. Charismatic groove

39. The hard, dense bony process of the temporal bone that houses the structures of the middle and inner ear is called the:

A. Mastoid process
B. Carotid canal
C. Petrous pyramid
D. Tympanic ridge

40. The coronal suture is located between the:

A. Frontal and the parietal bones
B. Parietal and the temporal bones
C. Frontal and the temporal bones
D. Occipital and the parietal bones

41. The localizing point that is found between the eyebrows over the frontal bone is called the:

A. Nuchel point
B. Articular point
C. Glabella
D. Lambda

Referring to the diagram of the mandible, place the number that corresponds to the following structures for questions 42-48.

42. _____ Mental foramen
43. _____ Gonion (angle)
44. _____ Coronoid process
45. _____ Body
46. _____ Ramus
47. _____ Mental protuberance
48. _____ Condyloid process

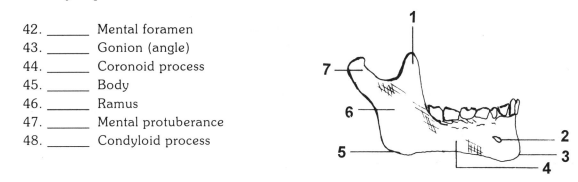

49. The majority of the lateral orbital wall and the cheek is formed by the:

A. Malar bone
B. Maxillary bone

C. Parietal bone
D. Temporal bone

50. Which of the following cranial bones contains a paranasal sinus?

   1. Frontal bone   *2. Temporal bone*   *3. Sphenoid bone*

A. 1 & 2 only
B. 1 & 3 only

C. 2 & 3 only
D. 1, 2 & 3

51. The rounded articulating condyles of the mandible that help form the temporomandibular joint are called the:

A. Coronoid processes
B. Coracoid processes

C. Condyloid processes
D. Alveolar processes

52. The greatest number of cranial foramina are located on the:

A. Temporal bone
B. Sphenoid bone

C. Occipital bone
D. Frontal bone

53. The external occipital protuberance is often referred to as the:

A. Inion
B. Gonion

C. Pterion
D. Asterion

54. Extending anteriorly from the dorsum sella are two bony processes that help in the formation of the lesser wing of the sphenoid These are termed the:

A. Petrous ridges
B. Pterygoid processes

C. Anterior clinoid processes
D. Anterior condyloid processes

55. In the adult, the junction between the sagittal and lambdoidal sutures is known as the:

A. Bregma
B. Pterion

C. Inion
D. Lambda

56. The superior and middle nasal conchae that serve to increase the surface area of the nasal cavities are found on the:

A. Sphenoid bone
B. Ethmoid bone

C. Frontal bone
D. Nasal bones

57. What are the two L-shaped facial bones that form part of the posterior hard palate and nasal septum known as?

A. Maxillae
B. Palatine bones

C. Vomer bones
D. Inferior conchae

58. In the lateral projection of the mandible, the condyloid process is:

A. Superior to the coronoid process
B. Anterior to the coronoid process

C. Posterior to the coronoid process
D. More than one, but not all of the above

59. All of the following structures are located on the sphenoid bone EXCEPT:

1. Foramen magnum   2. Optic foramen   3. Sella turcica

A. 1 only
B. 2 only

C. 3 only
D. 1, 2 & 3

Referring to the diagram of the skull, place the number that corresponds to the following structures for questions 60-67.

60. _____ External auditory canal
61. _____ Ethmoid bone
62. _____ Occipital bone
63. _____ Parietal bone
64. _____ Sphenoid sinus
65. _____ Pterygoid process
66. _____ Frontal sinus
67. _____ Alveolar process

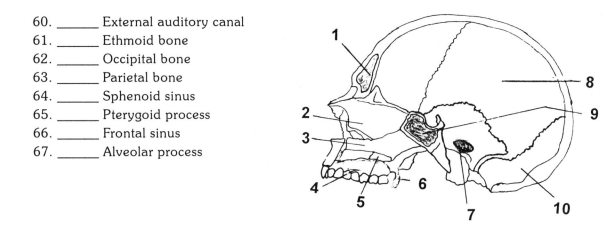

68. The long, sliver-like processes that extend inferiorly from each temporal bone, serving for the attachment of the muscles of speech, are called:

A. Styloid processes
B. Zygomatic processes

C. Sphenoid struts
D. Coronoid processes

69. The small, U-shaped hyoid bone situated below the floor of the mouth aids in the act of:

A. Phonation
B. Swallowing

C. Both A & B
D. Neither A nor B

70. The floor of the cranial vault is formed by all or part of the following EXCEPT:

A. Occipital bone
B. Frontal bone

C. Parietal bones
D. Sphenoid bone

71. The majority of the upper portion of the nasal septum is formed by the:

A. Perpendicular plate of the ethmoid bone
B. Cribriform plate of the lacrimal bone

C. Lateral masses of the frontal bone
D. Crista galli of the sphenoid bone

72. The sphenoid sinus is located anterior and inferior to what other structure of the sphenoid bone?

A. Optic foramen
B. Styloid process

C. Greater wing
D. Sella turcica

73. The surface landmark associated with the anterior nasal spine at the junction of the upper lip and nose is called the:

A. Glabella
B. Acanthion

C. Gonion
D. External occipital protuberance

74. Which bone does NOT form any portion of the nasal cavity or septum?

A. Vomer bone
B. Palatine bones

C. Ethmoid bone
D. Malar bone

75. The most posterior paranasal sinus is the:

A. Frontal sinus
B. Sphenoid sinus

C. Ethmoid sinus
D. Maxillary sinus

Explanations

1. B   The cranium or cranial vault is formed by 8 separate bones; namely, the frontal bones, the 2 parietal bones, the 2 temporal bones, the occipital bone, the sphenoid bone and ethmoid bone. Collectively these bones house the brain and it's covering membranes, the meninges.

2. D   The raised bony prominence on the frontal bone above the orbit is called the superciliary ridges. These form the under-structure for the eyebrows.

3. A   The slender feet-like extensions on either side of the posterior nasal cavity are called the pterygoid processes of the sphenoid bone. The appearance of these structures is often compared to the feet of a bat. These serve at the point of attachment from some of the muscles of the mastication.

4. A   The arched roof of the mouth (hard palate) is composed of the palatine processes of the maxillae and the horizontal portions of the palatine bones.

5. C   The falx cerebri, a specialized layer of dura mater occupying the cerebral fissure is attached to the crista galli of the ethmoid bone to increase stability of the brain.

6. A   The air cells which provide a resonance chamber for sound waves are located within a bony (mastoid) process located at the inferolateral portion of the temporal bone.

7. A   A majority of the superior and lateral portions of the cranium are formed by a pair of large flat bones called the parietals.

8. B   The lower central portion of the skull contains a large foramen, the foramen magnum, that enables the passage of the brainstem (spinal cord), vertebral and spinal arteries into and out of the skull.

9. D   Two deep depressions known as the greater wings, found on the base and sides of the sphenoid bone comprise the majority of the floor of the middle cranial fossa.

10. A   The mucus membranes that line the four paranasal sinuses of the frontal, ethmoid, sphenoid and maxillary bones are continuous with those of the nasal cavity.

11. C   The midpoint of the frontonasal suture at the roof of the nose is termed the nasion.

12. A   The sagittal plane which divides the body into right and left sides derives its name from the cranial suture that separates the two parietal bones.

13. A   The prominence of the cheek is due in large part to the underlying zygomatic arch. This is formed by the temporal process of the malar bone and the zygomatic process of the temporal bone.

14. G   The numerous pairs of foramen found on the floor of the cranium are collectively termed the basal foramina. These consist of the jugular foramen, the foramen lacerum, the foramen spinosum, the foramen ovale and the foramen rotundum.

15.	3	16.	8	17.	2	18.	5	19.	7	20.	6
21.	9	22.	1	23.	10	24.	4				

25. D   The horizontal or cribriform plate of the ethmoid bone contains numerous perforations that allow the passage of the olfactory nerves. These small bare nerve endings detect others that are carried through the roof of the nasal passages toward the temporal lobe of the brain.

26. B   The cranial sutures, the fixed junctions between the bones of the skull, are normally named for the articulating bones. One exception is the connection between the temporal and parietal bones, called the squamosal suture.

27. C   The bony orbit consists of seven bones including the frontal bone, the malar bone, the maxillary bone, the sphenoid bone, the lacrimal bone, the ethmoid bone and the palatine bones.

28. C   The two corners of the eye where the eyelids meet are called the inner and outer canthi.

29. A   The widest point in the adult skull is between the two parietal eminences of the parietal bones.

30. B   Just lateral to the foramen magnum are the 2 oval-shaped occipital condyles. These are for the articulation with the superior articular process of the first cervical vertebra.

31. C   The skull contains four paranasal sinuses which link the mucus membranes. The frontal sinus most superior, the maxillary sinus most inferior, and the sphenoid sinus most posterior. The ethmoid sits between the orbits.

32. D   From the Latin for cavity, the alveolar process containing the tooth sockets are found on the inferior border of the maxillae and superior surface of the body of the mandible.

33. B   The only movable bone of the adult skull is the lower jaw bone or mandible.

34. D   The lateral cranial wall is formed by the frontal bone, the temporal bone, the sphenoid bone, the parietal bones, and occipital bone.

35. D   The nasolacrimal duct passes through the lacrimal bone on the middle surface of the orbits. This conveys tears from the eye into the nasal cavity.

36. C   The most posterior suture of the adult skull formed at the junction of the parietal and occipital bones is called the lambdoidal suture.

37. C   The largest and most inferior paranasal sinus located within the maxillary bone is often referred to as the antrum of Highmore.

38. B   The two halves of the nasal cavity are separated by a bony septum this structure formed by portions of the ethmoid bone, the palatine bones, the maxillae bone, and the vomer bones.

39. C   The air-filled middle ear and the fluid-filled bony and membranous labyrinths are housed in hollowed cavities within the petrous ridge of the temporal bone.

40. A   The fixed articulation which passes laterally across the skull between the frontal and parietal bone is called the frontal or coronal suture.

41. C   A useful landmark in skull positioning is a flattened area between the superciliary ridges called the glabella.

42. 2   43. 5   44. 1   45. 4   46. 6   47. 3

48. 7

49. A   The majority of the lateral orbit, body of the cheek and anterior portion of the zygomatic arch are formed by a superficial facial bone called the zygoma or malar bone.

50. B   The mucus membrane, the lining of the nasal passages are continuous with those of the paranasal sinuses. These include the frontal bone, the sphenoid bone, the ethmoid bone and the maxillary bones.

51. C   The junction between the capitulum on the superior surface of the condyloid process of the mandible and the mandibular fossa on the temporal bone form an articulation called the temporomandibular (TMJ) joint. This joint allows for the opening and the closing of the mouth.

52. B   In addition to the large foramen magnum of the occipital bone, the floor of the cranium contains five paired foramina; 4 are associated with the sphenoid bone: foramen ovale, lacerum, spinosum and rotundum, while the fifth, jugular foramen, is found between the temporal and occipital bones.

53. A   The most prominent structure on the posterior aspect of the occipital bone is the external occipital protuberance (EOP) or the inion.

54. C   The deep saddle-shaped depression on the superior surface of the sphenoid bone, the sella turcica, is bounded anteriorly by the anterior clinoid process on the lesser wing of the sphenoid and dorsally by the posterior clinoid process.

55. D   The sagittal and lambdoidal sutures converge at the back of the head in an area named for its similarity to the Greek letter λ (lambda).

56. B   The two upper scroll-like processes of the ethmoid bone are called the nasal conchae. The third set or inferior conchae are extensions of a deep facial bones called the turbinates.

57. B   The posterior one-third of the hard palate is formed by the horizontal portions of the palatine bones, with the remaining two-thirds consisting of the palatine portion of the maxillae.

58. C   The ramus of the mandible divides superiorly into an anterior coronoid process and the more posterior condylar process which articulates with the temporal bone.

59. A   The structures of the sphenoid bone include the sella turcica, a pair of clinoid processes, two wings, two pterygoid processes and a sphenoid sinus and the optic foramen. The foramen magnum is located on the occipital bone.

60. 7   61. 2   62. 10   63. 8   64. 9   65. 6

66. 1   67. 4

68. A   The styloid processes of the petrous portion of the temporal bone project inferoanterior and serve to attach the stylohyoid, stylopharyngeus and styloglossus muscles used to aid in speaking.

69. C   The hyoid bone has numerous muscular attachments to the mandible, styloid process, and tongue which help in phonation and deglutition.

70. C   The floor of the cranial vault is formed by portions of the frontal, ethmoid, temporal, sphenoid, and occipital bones.

71. A   The majority of the superior aspect of the nasal septum is formed by a thin bony process called the cribriform or horizontal plate of the ethmoid bone.

72. D   The air-filled sphenoid sinus is normally found under the sella turcica just posterior to the back of the nasal passages.

73. B   The fusion of the maxillae shortly before birth leaves a small ridge and a sharp spine at the base of the nasal aperture called the anterior nasal spine.

74. D   The nasal septum is formed by portions of the ethmoid, palatine, maxillae and vomer. Other bones contributing to the nasal cavity are the inferior conchae, sphenoid, nasal, lacrimal, and frontal bones.

75. B   The most posterior of the paranasal sinuses is the sphenoid sinus which extends to the back of the sella turcica.

1. The entire process by which the body exchanges gasses with the atmosphere is called:

   A. Osmosis
   B. Hypoxia
   C. Respiration
   D. Transferral

2. Some of the dust and other particles in the air that enter the nose are filtered by a series of:

   A. Internal nasal hairs
   B. Mucus membranes
   C. Lymphatic ducts
   D. Nasal foramina

3. The separation of the nasal cavity into right and left portions is accomplished by a vertical partition called the:

   A. Turbinates
   B. Conchae
   C. Nostrils
   D. Septum

4. Which of the following is the correct sequence of air passage in the upper respiratory tract?

   A. Nose, larynx, pharynx, trachea
   B. Mouth, pharynx, larynx, trachea
   C. Nose, trachea, pharynx, larynx
   D. Mouth, larynx, trachea, pharynx

5. The vocal cords are housed in a structure formed by a number of singular and paired cartilages called the:

   A. Pharynx
   B. Trachea
   C. Larynx
   D. Bronchi

6. Due to its large diameter and more vertical placement, the most common location of foreign body blockages is the:

   A. Right primary bronchus
   B. Right pleural cavity
   C. Left primary bronchus
   D. Left pleural cavity

7. The bifurcation of the trachea into the mainstem bronchi occurs at the vertebral level of the _____ thoracic vertebra.

   A. First or second
   B. Fourth or fifth
   C. Seventh or eighth
   D. Ninth or tenth

8. The principal muscle involved with pulmonary ventilation is a dome-shaped muscle called the:

   A. Intercostal muscle
   B. Psoas muscle
   C. Diaphragm
   D. Pleura

9. The upper, bluntly pointed portion of the lung sitting just above the clavicles is called the:

   A. Lingula
   B. Hilum
   C. Superior fissure
   D. Apex

10. The numerous U-shaped cartilages which form the anterior and lateral borders of the trachea serve to:

    A. Separate the esophagus and trachea
    B. Aid in the swallowing of food
    C. Prevent collapse during respiration
    D. Produce the sounds of speech

11. The small sac at the end of a respiratory bronchiole that serves in the exchange of respiratory gasses is called the:

    A. Alveolus
    B. Glottis
    C. Bronchiolus
    D. Larynx

12. The left lung of the normal adult contains:.

    A. A single distinct lobe
    B. Two distinct lobes
    C. Three distinct lobes
    D. Five distinct lobes

13. The large, shield-shaped cartilage that forms the anterior portion of the larynx is called the:

A. Hyoid cartilage
B. Thyroid cartilage

C. Cricoid cartilage
D. Arytenoid cartilage

14. The inner layer of pleura that attaches directly to the lung tissue is termed the:

A. Visceral pleura
B. Parietal pleura

C. Axillary pleura
D. Alveolar pleura

15. The nose functions to:

1. Moisten incoming air   2. Warm incoming air   3. Filter incoming air

A. 1 & 2 only
B. 1 & 3 only

C. 2 & 3 only
D. 1, 2 & 3

Referring to the diagram of the bronchial tree, place the number that corresponds to the following structures for questions 16-21.

16. _____ Cricoid cartilage
17. _____ Right primary bronchus
18. _____ Thyroid cartilage
19. _____ Left primary bronchus
20. _____ Secondary bronchus
21. _____ Trachea

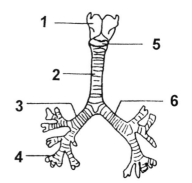

22. During normal respiration, the diaphragm will reach the lowest point of its excursion during:

A. Supine inspiration
B. Erect inspiration

C. Supine exhalation
D. Erect exhalation

23. The openings of the mainstem bronchi at their bifurcation are separated by a ridge of cartilage called the:

A. Bronchial fissure
B. Bronchial septum

C. Carina
D. Vocal fold

24. Each individual alveolar sac is surrounded by a:

A. Mucus membrane
B. Pleural membrane

C. Cartilaginous ring
D. Capillary network

25. The medial border of each lung has a deep depression for entry of the pulmonary vessels and bronchi termed the:

A. Mediastinum
B. Apex

C. Hilus
D. Diaphragmatic recess

26. In the lower respiratory tract, dust is filtered and moved by the action of mucus and small, hairlike structures called:

A. Cilia
B. Villi

C. Flagella
D. Cornu

27. The lungs are covered by a transparent, double-layered serous membrane called the:

A. Dura
B. Pleura

C. Meninges
D. Lingula

28. The mediastinum, the potential space between the lungs, contains all of the following EXCEPT the:

A. Heart
B. Thyroid gland

C. Esophagus
D. Thymus gland

29. The thin, serous fluid that exists between the layers of the pleura serves to:

A. Remove dust particles
B. Prevent collapse

C. Reduce friction
D. Prevent infection

30. The blood supply to the lungs is accomplished by the:

A. Systemic arteries
B. Portal arteries

C. Aorta
D. Pulmonary arteries

31. The normal amount of air that enters the lungs during quiet inspiration is about 500cc and is called the:

A. Tidal volume
B. Vital capacity

C. Inspiratory reserve
D. Residual volume

32. The slit-like opening between the vocal folds is called the:

A. Glossus
B. Glottis

C. Carina
D. Septum

33. The upper portion of the left lower lobe of the lung lies superior to the lower portion of the left upper lobe of the lung.

A. True
B. False

34. The separation of the lobes of the lungs is accomplished by long, narrow grooves called:

A. Faci
B. Lobules

C. Fissures
D. Vagi

35. Contractions of the diaphragm will result in an action of the lung tissue called:

A. Exhalation
B. Inhalation

C. Segmentation
D. Coarctation

36. The collection of fluids or gasses in the pleural space may lead to a lung collapse. This condition may be referred to as a:

   1. Pneumothorax   *2. Hydrothorax*   *3. Hemothorax*

A. 1 only
B. 2 only

C. 3 only
D. 1, 2 & 3

37. If any portion of the upper respiratory tract becomes swollen or blocked, an artificial opening can be made to improve breathing which is called a:

A. Laryngotomy
B. Rhinoplasty

C. Bronchography
D. Tracheostomy

38. Many air movements NOT related to breathing are termed non-respiratory movements, which are reflex in nature. A common reflex for clearing the lower respiratory tract is called a:

A. Sneeze
B. Cough

C. Hiccup
D. Laugh

39. A common degenerative disease characterized by the destruction of the alveolar walls is termed:

A. Asthma
B. Pneumonia

C. Emphysema
D. Dyspnea

40. The collapse of part or all of the lung, which often results from bronchial obstruction, is termed:

A. Bronchiectasis
B. Atelectasis

C. Pneumitis
D. Pleurisy

41. The insertion of a needle into the pleural cavity to remove air or fluid is called a:

A. Tracheotomy
B. Bronchiectasis

C. Thoracoplasty
D. Thoracocentesis

42. The allergic reaction to a foreign substance, such as pollen, may cause the constriction of the bronchial passages. This condition is termed:

A. Bronchial asthma
B. Bronchiectasis

C. Bronchopneumonia
D. Bronchoesophageal fistula

43. The production of free fluid in the chest cavity is often seen on an erect chest radiograph in the lower portion of the lung called the:

A. Mediastinum
B. Apical region

C. Costophrenic angle
D. Axillary border

44. A condition resulting from a marked deficiency of oxygen and an excess of carbon dioxide in the blood from a respiratory obstruction is termed:

A. Bradypnea
B. Asphyxia

C. Eupnea
D. Pulmonary edema

45. In the normal adult, the resting inspiration rate is about:

A. 70 respirations/minute
B. 40 respirations/minute

C. 15 respirations/minute
D. 10 respirations/minute

Match the following for questions 46-50.

46. _____ Apnea
47. _____ Dyspnea
48. _____ Eupnea
49. _____ Anoxia
50. _____ Bradypnea

A. Abnormally fast breathing
B. Normal breathing
C. Difficult breathing
D. Temporary absence of breathing
E. Low oxygen in tissues
F. Abnormally slow breathing

Explanations

1. C Respiration or the exchange of gases may take one of two forms; internal respiration, which involves the gaseous transfer within the tissues of the body, and external respiration, that involves the movement of air in and out of the lungs and the pulmonary exchange of gases.

2. A The buildup of foreign particles in the lungs is prevented by the filtering effects of coarse nasal hairs and ciliated membrane of the bronchial passages. Any trapped particles are carried by the muscles to the oropharynx for their elimination by the gastrointestinal tract.

3. D The two sides of the nasal cavity are separated by a bony partition called the nasal septum, that is formed by portions of the ethmoid bone, palatine bone and vomer bones and the maxillae.

4. B Air from the outside air passes through the mouth or nose into the pharynx, larynx, trachea, main stem bronchi, primary bronchiole, terminal bronchiole to its termination point in the alveolar sacs.

5. C The voice box or larynx that houses the true and false vocal cords and the glottis is formed by nine individual cartilages; namely, the thyroid, epiglottic, cricoid and the 3 paired cartilages: the arytenoid, corniculate and cuneiform.

6. A The right primary bronchus is shorter, wider and more vertical than the left and is therefore more readily accessed by foreign objects or infectious agents.

7. B The last tracheal cartilage is expanded inferiorly forming a hook-like process at the level of the 4th or 5th thoracic vertebrae called the carina.

8. C The large dome shaped diaphragm which has extensive muscular attachments to the xiphoid process, ribs, costal cartilages and vertebrae is the principal muscle of respiration.

9. D The narrow superior portion of the lung located behind the first three pairs of ribs and clavicle is called the apex of the lung.

10. C The solid ring of C-shaped cartilages which form the larynx, trachea and bronchiole provide the rigid support to prevent the inward collapse of these structures during respiration.

11. A At the terminal end of the respiratory passages are a series of thin-walled membranous sacs called alveoli. These are surrounded by a capillary network that enables the free exchange of oxygen and carbon dioxide between the lungs and the red blood cells.

12. B The left lung has two distinct lobes called the upper and lower lobes. The right lung contains an upper, middle, and lower lobe. Surprisingly both lungs have approximately the same surface area and capacity.

13. B The characteristic shape of the larynx is due in a large part to the fused triangular shaped thyroid cartilage that forms the anterior prominence or Adam's apple.

14. A The lungs are protected and separated by a double layered serous membrane called the pleura. The parietal or outer layer of pleura is attached to the wall of the thoracic cavity while the inner visceral layer forms a direct covering of the lung tissue. Between these is a thin fluid-filled cavity called the pleural space.

15. D The nose is lined with coarse nasal hairs and mucus membranes to filter out dust particles. The scroll shaped nasal conchae create the turbulence which circulates and warms the air. Tears from the orbital cavity pass into the nasal cavity to humidify the air prior to its passage into the lungs.

16. 5   17. 3   18. 1   19. 6   20. 4   21. 2

22. B On inspiration, the intrathoracic pressure is lowered by the contraction of the diaphragm and the resulting expansion of the lung. The further dropping of the abdominal organs in the erect position aids in maximizing the volume lung tissue which can be visualized radiographically.

23. C   The prominent last tracheal cartilage at the bifurcation of the trachea forms a ridge called the carina.

24. D   The exchange of respiratory gasses in the lungs occurs across the thin walls of the alveolus and the surrounding capillary vessels.

25. C   The centrally located bronchi, pulmonary arteries, veins and lymphatics extend laterally from their origins causing an indentation as they enter the lung called the hilum.

26. A   The trachea and bronchi are lined with a specialized membrane containing goblet cells and ciliated columnar cells. The goblet cells produce the mucous that work with the hair-like cilia to trap many of the larger inhaled particles before they reach the lungs.

27. B   The lungs are protected and separated by a double layered serous membrane called the pleura. The parietal or outer layer of pleura is attached to the wall of the thoracic cavity while the inner visceral layer forms a direct covering of the lung tissue. Between these is a thin fluid-filled cavity called the pleural space.

28. B   The two lungs are separated by a space called the mediastinum which contains the heart, great vessels (aorta and vena cava), esophagus, pulmonary vessels, trachea, thymus gland and the vagus, phrenic and cardiac nerves.

29. C   The potential space between the two layers of pleura is filled with an oil-like lubricating fluid that helps to reduce the friction occurring during respiration.

30. D   The blood supply to the lung is provided by the pulmonary artery originating at the right ventricle of the heart.

31. A   The normal volume of air taken into the lung from quiet breathing is about 500 cc and is called the tidal volume. A forced deep breath (inspiratory reserve volume) can pull in about 3600 cc of air.

32. B   The glottis or opening between the vocal cords is covered during swallowing by a cartilaginous tissue called the epiglottis.

33. A   The lobes of the lung are not separated segmentally by a straight axial plane but instead are divided by a superior to inferior oblique fissure which provide a great deal of lobular overlap when viewed radiographically.

34. C   The lobes of each lung are divided by fissures. The oblique fissure of the left lung separates the upper and lower lobes. In the right lung an additional horizontal fissure separates the middle and inferior lobes.

35. B   The contraction of the diaphragm increases the volume of the chest cavity and lowers the intrathoracic pressure enabling air to flow into the lungs during the act of inspiration or inhalation.

36. D   The presence of excessive fluids in the pleural cavity increases the intrathoracic pressure which collapses the lung. The resulting conditions are often named for the type of agent present in the pleural spaces: blood (hemothorax), fluid (hydrothorax), or air (pneumothorax).

37. D   A blockage of the upper respiratory tract can be alleviated by an artificial surgical opening into the trachea just below the cricoid cartilage called a tracheostomy.

38. B   There are a number of modified respiratory movements effecting lung and respiratory passages; namely, the cough, which expels unwanted substances from the lower respiratory passages; sneezing, for the clearing of the upper respiratory passages and the spasmodic contraction of the diaphragm; the hiccup, whose function is still undetermined.

39. C   The partial or near complete destruction of the alveolar sacs results in a chronic degenerative process effecting the degree of air exchange between the lungs and blood called emphysema.

40. B   The inward recoil of the alveoli and subsequent collapse of the sacs may result in segmental collapse of the lung called atelectasis.

41. D   The relief of a pneumothorax or hydrothorax will often involve the removal of these substances from the pleural cavity by a technique called thoracocentesis. This involves the placement of a chest drainage tube into the cavity and the use of a low negative air pressure (suction) device.

42. A   The spasmodic contraction of the smooth muscles in the bronchial walls in response to an allergic reaction to pollen, fumes, or physiological stress, may result in an asthmatic attack.

43. C   The domed shape of the diaphragm creates two recesses in the lower lung near the junction with the ribs. In the erect position, free fluids in the chest cavity will tend to collect in these costophrenic angles at the bottom of the lungs.

44. B   Insufficient intake of oxygen (asphyxia) may lead to symptoms including cyanosis, tachycardia, mental impairment and in severe cases, convulsions and death.

45. C   A complete respiration, including one inspiration and one expiration normally occurs at a rate of 12-16 respirations per minute.

46. D   The prefix a, meaning without, combined with pnea for breathing, apnea (without breathing).

47. C   The prefix dys, meaning difficult, dyspnea (difficult breathing).

48. B   The prefix eu, meaning average or normal, eupnea (normal breathing).

49. E   The prefix an, meaning without, and oxia for oxygen, anoxia (without oxygen).

50. F   The prefix brady, meaning slow, bradypnea (slow breathing).

1. The principal function of the urinary system is the elimination of the body's major metabolic byproduct which is:

    A. Carbon dioxide
    B. Urea

    C. Hydrochloric acid
    D. Thrombin

2. The medial border of each kidney has a deep longitudinal fissure which vessels and nerves enter and leave the kidney called the:

    A. Calyx
    B. Medulla

    C. Hilus
    D. Cortex

3. The kidneys lie on either side of the vertebral column resting on the psoas and quadratus lumborum muscle:

    A. In front of the peritoneum
    B. Inside of the Glisson's capsule

    C. Between the layers of pleura
    D. Behind the peritoneum

4. The upper expanded end of each ureter enters the kidney at an area termed the:

    A. Renal pelvis
    B. Renal pyramid

    C. Renal column
    D. Renal cortex

5. The kidneys are held in the upper abdominal position by a connective tissue called:

    A. Renal columns
    B. Renal papillae

    C. Renal fascia
    D. Convoluted tubules

6. The principal function of the ureters is to actively aid the passage of urine from the:

    A. Renal pelvis to the urinary bladder
    B. Urinary bladder to the urethra

    C. Urethra to the urinary bladder
    D. Urinary bladder to the kidney

7. The term applied to the expelling of urine from the urinary bladder is:

    A. Mastication
    B. Catheterization

    C. Tubular secretion
    D. Micturition

8. In the normal adult female, the urethra functions as a/an:

    1. Reproductive organ   *2. Excretory organ*   *3. Endocrine organ*

    A. 1 only
    B. 2 only

    C. 3 only
    D. 1, 2 & 3

9. The muscular coat of the urinary bladder that plays an important part in expelling urine is called the:

    A. Trigone
    B. Psoas muscle

    C. Detrusor muscle
    D. External urinary sphincter

10. The renal pyramids have a striated appearance due to a number of _____ which are directed into the renal papillae.

    1. Collecting ducts   *2. Calyces*   *3. Renal arteries*

    A. 1 only
    B. 2 only

    C. 3 only
    D. 1, 2 & 3

11. The glandular appearance of the cortex of the kidney is due to the arrangement of the functional units of the kidney called the:

    A. Renal sinuses
    B. Nephrons

    C. Peritubular capillaries
    D. Afferent arterioles

12. In the normal adult's urinary system, the ureters will attach to the urinary bladder at its:

A. Anterior lateral surface
B. Posterior lateral surface

C. Anterior inferior surface
D. Posterior inferior surface

13. Due to the effect of other abdominal structures, the right kidney is normally located above the left kidney.

A. True
B. False

14. The renal corpuscle consists of a thin-walled, saclike structure surrounding a cluster of capillary beds called the:

A. Proximal convoluted tubule
B. Glomerulus

C. Afferent arteriole
D. Collecting duct

Referring to the diagram of the kidney, place the number that corresponds to the following structures for questions 15-20.

15. _____ Renal pyramid
16. _____ Major calyx
17. _____ Ureter
18. _____ Renal pelvis
19. _____ Minor calyx
20. _____ Cortex

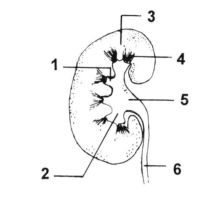

21. In the normal adult female, urine passes from the urinary bladder to the outside in a short muscular tube called the:

A. Urethra
B. Ureter

C. Trigone
D. Renin

22. The filtration of the organic salts, fluids and nitrogenous compounds from the blood occurs across a membrane called the:

A. Collecting ducts
B. Hypothalamus

C. Bowman's capsule
D. Major calyx

23. In the male adult, a mechanical urinary bladder obstruction may result from a/an:

A. Ureteral reflux
B. Renal colic

C. Enlarged prostate gland
D. Paralyzed external urethra sphincter

24. Blood is carried to the glomerulus by which of the following?

A. Vasa recta
B. Loop of Henle

C. Afferent arteriole
D. Efferent arteriole

25. The urinary bladder is situated:

1. Below the peritoneum   *2. Anterior to the rectum*   *3. Anterior to the uterus*

A. 1 only
B. 2 only

C. 3 only
D. 1, 2 & 3

26. The concentration of urine and tubular reabsorption occurs within the:

1. Proximal convoluted tubules   **2. Distal convoluted tubules**   **3. Loop of Henle**

A. 1 only
B. 2 only

C. 3 only
D. 1, 2 & 3

27. The normal adult's urinary bladder can hold a maximum of about:

A. 100cc of fluid
B. 300cc of fluid

C. 600cc of fluid
D. 1500cc of fluid

28. In the normal adult male, which is NOT one of the divisions of the male urethra?

A. Prostatic urethra
B. Membranous urethra

C. Erectile urethra
D. Penile urethra

Referring to the diagram of the nephron, place the number that corresponds to the following structures for questions 29-36.

29. _____ Proximal convoluted tubule
30. _____ Afferent arteriole
31. _____ Loop of Henle
32. _____ Bowman's capsule
33. _____ Distal convoluted tubule
34. _____ Efferent arteriole
35. _____ Collecting duct
36. _____ Glomerulus

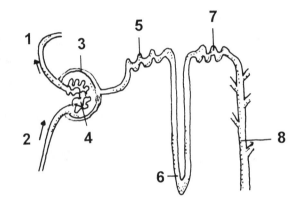

37. An individual who has lost all kidney function must have the blood cleaned out. This is normally accomplished by some type of renal:

A. Micturition
B. Diuresis

C. Urinalysis
D. Dialysis

38. Which of the following is NOT normally a function of the urinary system?

A. Elimination of urea
B. Maintenance of electrolyte balance

C. Regulation of fluid balance
D. Production of erythrocytes

39. Many substances, such as alcohol, caffeine and digitalis, may cause an excessive urine output termed:

A. Pyuria
B. Diuresis

C. Uremia
D. Hematuria

40. The obstruction of the blood flow to the kidney is one of the common causes of:

A. Hypertension
B. Hypotension

C. Diabetes
D. Renal colic

41. A blockage of the ureter by a calculus or other mechanical obstruction will often lead to the dilatation of the renal pelvis and calyces called:

A. Ptosis
B. Hydronephrosis

C. Polyuria
D. Agenesis

42. Water and fluid waste products can be eliminated by way of the:

1. Skin   2. Kidneys   3. Liver

A. 1 only
B. 2 only

C. 3 only
D. 1, 2 & 3

43. The term employed for a kidney NOT found in its normal location is:

A. Ectopic kidney
B. Staghorn kidney

C. Horseshoe kidney
D. Micro-kidney

44. The buildup of poisonous waste products in blood is commonly caused by poor kidney function which is called:

A. Ptosis
B. Anuria

C. Uremia
D. Renal reflux

45. Many types of trauma may lead to a kidney falling from its original location. This is termed (a):

A. Pyelitis
B. Horseshoe kidney

C. Extravasation
D. Nephroptosis

Explanations

1. B   The breakdown of food (proteins) by the body is associated with a number of toxic waste products such as ammonia and urea which are normally removed by the urinary system.

2. C   The notch on the concave medial border of the kidney for the passage of the ureters, blood and lymphatic vessels into the kidney is called the hilum.

3. D   The kidney rests between the level of T12-L3 behind the peritoneum resting on the psoas muscles posterior abdominal wall.

4. A   The expanded proximal end of the ureter (pelvis) is located in the renal sinus and receives urine from the major and minor calyces of the kidney.

5. C   Both the surrounding adipose capsule and the fibrous layer called the renal fascia of the kidney help in anchoring this organ to the adjacent structures on the posterior abdominal wall.

6. A   The muscular walls of the ureters which contain both longitudinal and circular muscle fibers enable the peristaltic motion of this tubular structure and encourage the movement of urine toward the bladder.

7. D   Also called urination or voiding, micturition is the act by which urine is passed from the urinary bladder into the urethra for its expulsion from the body.

8. B   Unlike the male urethra, which has dual urinary and reproduction functions, the female urethra, which is attached to the anterior bladder wall between the clitoris and vaginal opening, serves only to discharge urine.

9. C   The outer coat of the bladder consists of a muscular layer, the detrusor, which can contract to actively force urine from the urinary bladder.

10. A   The appearance of the collecting ducts and the associated blood vessels give the pyramid its striated appearance when the kidney is seen in cross-section.

11. B   The tubular components, with their associated vasculature, make up the essential components or functional units of the kidney, the nephrons.

12. B   The ureters descend from the renal pelvis through the posterior abdomen to their final resting point near the lateromedial portion of the posterior bladder.

13. B   The large space occupied by the liver on the right side of the abdomen displaces the right kidney inferiorly and the right diaphragm superiorly.

14. B   The renal (Bowman's) capsule and the vascular glomerulus provide the membrane over which wastes are filtered from the blood stream and directed into the collecting system of the kidney.

15. 4   16. 2   17. 6   18. 5   19. 1   20. 3

21. A   The female urethra with its inner mucosa, middle venous coat and outer muscular coat directs urine from the urinary bladder to the outside of the body.

22. C   The first step in urine production glomerular filtration occurs as the blood pressure forces water and dissolved blood particles through the capillary wall and basement membrane of the Bowman's capsule.

23. C   Since the male urethra descends vertically through the prostate gland before its junction with the penis, any enlargements of this gland may be associated with a lower urinary tract obstruction.

24. C   The 1200 ml of blood that passes through the kidney each minute is distributed to the renal cortex by the afferent arterioles and the interlobular arterial branches.

25. D The midline urinary bladder is bordered anteriorly by the vagina and uterus in the female, posteriorly by the rectum and peritoneum in both males and females.

26. D The glomerular filtration rate which provides about 180 liters of filtrate each day is largely reabsorbed in the epithelial layers of the proximal and distal convoluted tubules and the loop of Henle.

27. C Even though the maximum capacity of the urinary bladder is about 6-800 ml, specialized stretch receptors in the wall will trigger a desire to urinate when the bladder reaches one half of its maximum capacity.

28. C The three main divisions of the male urethra are the prostatic, membranous and penile or spongy portions.

29. 5   30. 2   31. 6   32. 3   33. 7   34. 1

35. 8   36. 4

37. D Kidney failure results in a dangerous buildup in toxic wastes in the blood. Hemodialysis (artificial kidney) is used in these patients to clean blood and maintain the appropriate electrolyte balance.

38. D The kidney aids in the removal of the nitrogenous wastes, ammonia and urea from the blood stream, and maintains the electrolyte balance by its control of sodium, chloride, sulfate and phosphate ions in addition to regulating the volume of fluid in the tissues and blood.

39. B Any drug which stimulates the flow of urine is classified as a diuretic (Lasix) and may be prescribed to assist patients in the control of edema, congestive heart failure and hypertension.

40. A Any reduction in the blood flow to the kidney or the obstruction of urine passage to the bladder may cause a painful spasm of the kidney called renal colic.

41. B The failure of urine to pass from the kidney into the bladder increases the pressure in the calyces causing their distension (hydronephrosis) and eventual atrophy if the condition is chronic in nature.

42. D Though the urinary tract is the principal system involved in the maintenance of the fluid balance, the lungs, skin, biliary and GI tract all contribute to normal fluid homeostasis.

43. A Ectopic kidneys or those located in other than the usual areas are most commonly found in the pelvis and only rarely cause significant problems for the patient.

44. C Renal failure and the resulting toxic buildup causes uremia, which is diagnosed by increased levels in a blood urea nitrogen (BUN) or creatinine test.

45. D Damage to the supporting tissues of the kidney may result in its dropping to an abnormal abdominal location called nephroptosis.

1. The process by which food particles are mechanically reduced into smaller particles in the mouth is termed:

   A. Absorption
   B. Peristalsis
   C. Mastication
   D. Deglutition

2. Projecting from the middle arch of the soft palate is a small soft-tissue structure concerned with the gag reflex. This is called the:

   A. Vulva
   B. Uvula
   C. Macula
   D. Sebum

3. Small circular bands of smooth muscle fibers that serve to restrict the flow of liquids and solids within the digestive tract are termed:

   A. Sphincter muscles
   B. Cardiac muscles
   C. Diaphragmatic muscle
   D. Erectile muscles

4. The small raised structures of the tongue, papillae, contain the receptors for the sensation of:

   1. Smell   2. Speech   3. Taste

   A. 1 only
   B. 2 only
   C. 3 only
   D. 1, 2 & 3

5. The underside of the tongue is attached to the floor of the mouth by a membranous fold called the:

   A. Omentum
   B. Uvula
   C. Gingiva
   D. Frenulum

6. The neural intervention that controls much of the digestive function results primarily from the action of the 10th cranial nerve called the:

   A. Vagus nerve
   B. Trigeminal nerve
   C. Hypoglossal nerve
   D. Buccal nerve

7. The act of swallowing involves both voluntary and involuntary actions by the muscles of the:

   1. Tongue   2. Pharynx   3. Upper esophagus

   A. 1 only
   B. 2 only
   C. 3 only
   D. 1, 2 & 3

8. The wavelike muscular contractions that move the food bolus through the digestive system is termed:

   A. Fauces
   B. Micturition
   C. Peristalsis
   D. Mastication

9. The teeth that fit into the alveolar processes of the jaw bones are held in place by the substance:

   A. Dentin
   B. Cementum
   C. Enamel
   D. Pulp

10. The first liquid secretion (saliva) encountered by food during the digestive process is formed by the:

    A. Tonsils
    B. Salivary glands
    C. Glottis
    D. Eustachian glands

11. The process by which food is actively transported from the mouth into the esophagus is termed:

    A. Pepination
    B. Salivation
    C. Deglutition
    D. Phonation

12. The different types of teeth are concerned with specific functions. The grinding of food is normally accomplished by the larger, flatter teeth called:

A. Molars
B. Incisors
C. Canines
D. Milk teeth

13. The nerves and blood vessels of the teeth are located in the centralized cavity called the:

A. Crown
B. Dentin
C. Enamel
D. Pulp cavity

14. The Wharton's ducts that open under the tongue are concerned with the passage of saliva from the:

A. Sublingual gland
B. Submandibular gland
C. Parotid gland
D. Tyroid gland

15. The first set of temporary teeth, which appear in early childhood, are referred to as:

A. Permanent teeth
B. Deciduous teeth
C. Wisdom teeth
D. Partition teeth

16. The bulk of a mature tooth lying beneath the outer enamel covering is composed of:

A. Cementum
B. Periosteum
C. Caries
D. Dentin

17. The largest set of salivary glands sitting in front and below each ear are the:

A. Parotid glands
B. Sublingual glands
C. Submaxillary glands
D. Submandibular glands

18. The portion of the tooth that extends above the gum line is termed the:

A. Root
B. Capitellum
C. Crown
D. Stalk

19. Saliva from the parotid gland passes within the _____ to its opening onto the inner surface of the cheek.

A. Wharton's duct
B. Stensen's duct
C. Ducts of Wirsung
D. Ducts of Rivinus

20. In the normal adult, the esophagus is located:

A. Posterior to the trachea
B. Anterior to the voicebox
C. Medial to the aorta
D. Lateral to the epiglottis

21. All of the following structures contribute to the mechanical or chemical digestive process EXCEPT:

A. Larynx
B. Stomach
C. Mouth
D. Small intestine

22. The protection of the intestinal tract from the action of digestive acids and enzymes is primarily accomplished by the secretion of:

A. Maltese
B. Cholecystokinin
C. Sodium bicarbonate
D. Mucus

23. The highest portion of the normal adult stomach sitting just below the left diaphragm is the:

A. Pylorus
B. Cardiac orifice
C. Fundus
D. Greater curvature

24. The stomach is involved with the:

A. Chemical digestion of food
B. Production of growth hormone
C. Elimination of liquid wastes
D. Production of red blood cells

25. The semifluid, partially digested food found in the upper gastrointestinal tract is referred to as:

A. Chyme
B. Meconium

C. Flora
D. Bolus

26. The terminal segment of the normal human stomach is called the:

A. Duodenum
B. Pylorus

C. Fundus
D. Trigone

27. The periodontal membrane is a close-fitting protective membrane associated with the structure of the:

A. Stomach
B. Intestines

C. Teeth
D. Peritoneum

28. In the stomach, the production of hydrochloric acid aids the digestive process. This is produced by specialized cells called:

A. Acini cells
B. Goblet cells

C. Parietal cells
D. Zymogenic cells

29. Most of the capillary beds concerned with the absorption of food are located within numerous finger-like soft tissue extensions of the small intestines called the:

A. Cilia
B. Villi

C. Papillae
D. Taenia coli

30. The empty stomach is thrown into numerous mucosal folds, that give it its characteristic appearance, called:

A. Rugae
B. Haustra

C. Epiploicae
D. Voluli

31. The name of the last portion of the small intestines is the:

A. Ilium
B. Jejunum

C. Duodenum
D. Ileum

32. The junction of the transverse and descending colons takes place at the:

A. Angular notch
B. Splenic flexure

C. Hepatic flexure
D. Rectosigmoid junction

Referring to the diagram of the stomach, place the number that corresponds to the following structures for questions 33-41.

33. _____ Pyloric sphincter
34. _____ Cardiac orifice
35. _____ Duodenum
36. _____ Esophagus
37. _____ Fundus of the stomach
38. _____ Lesser curvature of the stomach
39. _____ Body of the stomach
40. _____ Greater curvature of the stomach
41. _____ Pylorus

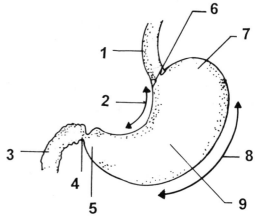

42. Food entering the stomach is prevented from exiting into the small intestines by the action ring of smooth muscles called the:

A. Duodenal sweep
B. Ileocecal valve

C. Cardiac sphincter
D. Pyloric sphincter

43. The majority of blood vessels and nerves, Meissner's plexus, are located in the:

A. Serosa layer of the omentum
B. Submucosa layer of the walls of the intestines

C. Muscular layer of the esophagus
D. Mucosa layer of the oral cavity

44. The digestive tract organ most responsible for the absorption of water is the:

A. Large intestine
B. Stomach

C. Esophagus
D. Duodenum

45. The most distal portion of the large intestine is called the:

A. Rectum
B. Sigmoid colon

C. Ileum
D. Cecum

46. The folds of mucosa that help to increase the surface area of the small intestines are termed:

A. Rugae
B. Circular folds

C. Angulations
D. Maculi

47. Which of the following organs are concerned with the chemical digestion of food?

1. Colon   *2. Small intestines*   *3. Salivary gland*

A. 1 & 2 only
B. 1 & 3 only

C. 2 & 3 only
D. 1, 2 & 3

48. The small intestines receive large amounts of sodium bicarbonate from pancreatic secretions to help neutralize the:

A. Fatty acids present in the bile
B. Digestive enzymes in the small intestine

C. Hydrogen chloride passing from the stomach
D. Bile salts passing from the liver

49. The walls of the intestinal tract are composed of:

A. Two distinct tissue layers
B. Four distinct tissue layers

C. Six distinct tissue layer
D. Eight distinct tissue layer

50. The undigested food of the small intestines will make its way into the colon through the:

A. Ileocecal valve
B. Pyloric canal

C. Cardiac orifice
D. Sphincter of Oddi

51. The majority of the organs in the gastrointestinal tract are contained within the:

A. Omentum
B. Mesentery

C. Peritoneum
D. Pleural cavity

52. The longest segment of the normal human intestinal tract is the:

A. Esophagus
B. Jejunum

C. Ileum
D. Cecum

53. The external layer of the digestive organs are composed of fibrous membranes called the:

A. Serosa
B. Mucosa

C. Endothelium
D. Mesentery

54. The first portion of the small intestines, which is about 25cm in length, is called the:

A. Pylorus   C. Ileum
B. Duodenum   D. Rectum

55. The vermiform appendix is a small, blind pouch which extends from the inferior surface of the:

A. Jejunum   C. Cecum
B. Duodenum   D. Rectum

56. Which of the following organs is concerned with the absorption of digested foods?

    1. Small bowel   **2. Colon**   **3. Esophagus**

A. 1 only   C. 3 only
B. 2 only   D. 1, 2 & 3

57. The fibrous structure that serves to attach the intestines to the posterior abdominal wall is termed the:

    1. Omentum   **2. Peritoneum**   **3. Mesentery**

A. 1 only   C. 3 only
B. 2 only   D. 1, 2 & 3

58. A small slap of cartilage on the pharynx that helps to prevent solid or liquid foods from entering the trachea is called the:

A. Frenulum   C. Epiglottis
B. Uvula   D. Adenoids

59. The digestive enzymes are produced in specialized cells of the small intestines called:

A. Chief cells   C. Islets of Langerhans
B. Crypts of Lieberkuhn   D. Brunner's gland

60. The haustra are the folds that give the _____ its characteristic appearance.

A. Esophagus   C. Stomach
B. Small intestines   D. Colon

    Referring to the diagram of the large intestine, place the number that corresponds to the *following structures for questions 61-72.*

61. _____ Rectum
62. _____ Splenic flexure
63. _____ Sigmoid colon
64. _____ Ileum
65. _____ Transverse colon
66. _____ Cecum
67. _____ Taeniae coli
68. _____ Descending colon
69. _____ Hepatic flexure
70. _____ Appendix
71. _____ Ileocecal valve
72. _____ Ascending colon

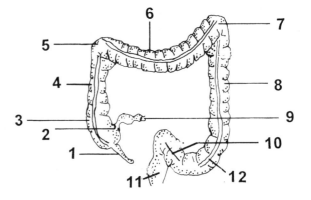

73. The ascending and transverse colon join at an area under the right lobe of the liver called the:

A. Splenic flexure
B. Gastric flexure
C. Hepatic flexure
D. Duodenal flexure

74. Digestive enzymes are manufactured by the body in which of the following organs?

1. Stomach   2. Pancreas   3. Salivary glands

A. 1 & 2 only
B. 1 & 3 only
C. 2 & 3 only
D. 1, 2 & 3

75. The appearance of haustral folds is due in large part to the action of three flat bands of muscle fibers on the exterior colon called the:

A. E-coli
B. Taenia coli
C. Epiploicae
D. Gyri

76. The formation of a stalked tumor that protrudes into the intestinal tract is called a:

A. Volvulus
B. Fistula
C. Hemorrhoid
D. Polyp

77. The telescoping of one part of the bowel into another is termed:

A. Herniation
B. Varices
C. Intussusception
D. Atresia

78. An outpouching or weakening of the intestinal wall most commonly seen in the large intestine is termed a/an:

A. Diverticulum
B. Aneurism
C. Calculus
D. Fistula

79. Bright red blood in the stool is a common indication of bleeding in the:

1. Colon   2. Stomach   3. Small intestines

A. 1 only
B. 2 only
C. 3 only
D. 1, 2 & 3

80. Valsalva's maneuver is used to increase abdominal cavity pressure and is useful for the demonstration of:

A. Diverticuli
B. Polyps
C. Esophageal varices
D. Ascites

81. An abnormal dilation of the veins near the anal canal leads to a condition termed:

A. Colitis
B. Hemorrhoids
C. Cholelithiasis
D. Rectal abscess

82. All of the following gastrointestinal structures are located within the peritoneum EXCEPT:

A. Stomach
B. Pancreas
C. Splenic flexure
D. Jejunum

83. The loss of the mucosa of the stomach and/or duodenum will often lead to development of (an):

A. Appendicitis
B. Jaundice
C. Intussusception
D. Ulceration

84. The only major secretion of the large intestines is:

A. Digestive enzymes
B. Hormones
C. Mucus
D. Water

85. The abnormal protrusion of the stomach through the diaphragm is called:

   A. Hiatal hernia
   B. Pyloric stenosis

   C. Esophageal atresia
   D. Meckel's diverticulum

86. In the normal adult male, the rectum is located:

   A. Anterior to the urinary bladder
   B. Posterior to the urinary bladder

   C. Superior to the diaphragm
   D. Posterior to the peritoneum

87. In the normal adult, the normal sequence for the passage of food is:

   A. Esophagus, Stomach, SI, Colon
   B. Stomach, Colon, Esophagus, SI

   C. Colon, SI, Stomach, Esophagus
   D. SI, Esophagus, Stomach, Colon

88. A common abnormality occurring more frequently in male infants resulting from a persistent spasm of the terminal portion of the stomach is called:

   A. Pyloric stenosis
   B. Umbilical hernia

   C. Gastritis
   D. Hypoglycemia

89. An abnormal collection of serous fluids in the peritoneal cavity is termed:

   A. Peritonitis
   B. Ascites

   C. Varices
   D. Dehydration

90. The abnormal twisting of a loop of intestine around itself can lead to an abnormal condition called (a/an):

   A. Imperforate colon
   B. Ileitis

   C. Volvulus
   D. Thrombosis

1. C   Of the various movements that aid in the digestive process the most important for the pulverization of food in the mouth is chewing or mastication. Deglutition is the process by which food is swallowed.

2. B   The uvula and surrounding fauces when irritated trigger gagging and vomiting which serves to clear objects obstructing the pharynx.

3. A   In its normal contracted state, a circular sphincter muscle serves to close an orifice. The relaxation of a sphincter muscle enables the passage of materials into more distal sections of the organ or tract.

4. C   The four primary tastes, sweet, sour, salt and bitter are detected on the receptors (taste buds) located on the fungiform papillae of the tongue.

5. D   In order to prevent the accidental swallowing of the tongue and the resulting asphyxiation, the tongue is attached to the floor of the mouth by the short flap of tissue called the frenulum.

6. A   The wide distribution of the vagus nerve to the muscles of the pharynx, larynx, respiratory passages, lungs, heart, esophagus, stomach, small intestine and colon make it critical for normal respiratory and digestive functions.

7. D   The act of swallowing which can be initiated voluntarily ends in a reflex action involving the muscles of the tongue, pharynx, larynx and upper esophagus which push the food into the proximal esophagus.

8. C   In organs possessing both longitudinal and circular layers of smooth muscle fibers, a wavelike constriction called peristalsis assists in the movement of substances within the tract.

9. B   A bone-like substance which covers the dentin on the roots of the tooth that serves for its attachment to the periodontal membrane of the alveolar process is called cementum.

10. B   Saliva is a watery substance which assists in the swallowing of food, and triggers the digestive process. This slightly acidic fluid contains both bacteriolytic and digestive enzymes produced in the three sets of salivary glands.

11. C   The act in which the tongue, pharynx and esophagus move the food toward the stomach is termed deglutition or swallowing.

12. A   A study of teeth indicates a relationship exists between their shape and function. The chisel shaped incisors are adapted to cutting food, the pointed canines tear and shred, while the larger molars crush and grind the food.

13. D   The outer dentin encloses the hollowed pulp cavity and its slender root canals. This cavity houses the blood vessels, nerves and lymphatic vessels needed to nourish the tooth.

14. B   Saliva passes from the submandibular gland beneath the base of the tongue to its openings just behind the central incisors through the Wharton's ducts.

15. B   The normal human goes through two dentitions during their lives: the first ending between 6 and 12 years of age consists of 20 deciduous or baby teeth, the second is the permanent dentition consisting of the 32 adult teeth.

16. D   The majority of a normal tooth is composed of a bone-like substance called dentin which gives the tooth its characteristic shape.

17. A   The largest and most posterior set of salivary glands, the parotids, are located in front of the ears on top of the masseter muscle.

18. C   The portion of the tooth extending above the gum (crown) is covered with the hardest material of the body, enamel.

19. B   A small opening in the cheek near the second molar provides the entry point for Stensen's ducts which pass anteriorly from the parotid gland to carry saliva into the mouth.

20. A   The esophagus is a midline structure which connects the oropharynx and the stomach . It is located anterior to the cervical and dorsal spines and posterior to the trachea within the mediastinum.

21. A   Chemical digestion begins in the mouth with the secretion of saliva and continues in the stomach and small intestine. The mechanical digestive process begins in the mouth with chewing and is completed by the churning action of the stomach and small intestine. The esophagus has no active chemical or mechanical digestive function.

22. D   The acidic conditions which aid the digestive process in the stomach would, without protection, digest the stomach itself. The gastric mucosa and its associated mucus-producing glands secrete a protective mucus for the stomach. Mucus-producing cells are also present in the esophagus, small intestine, and large intestine.

23. C   The esophagus descends within the mediastinum of the thoracic cavity to its junction with the stomach (cardia). The fundus, the highest portion of the stomach, extends superiorly to a space under the left hemidiaphragm.

24. C   The digestive enzymes and acids produced in the gastric walls and the churning motion provided by the muscles of the stomach assist in both the chemical and mechanical digestive process.

25. A   The food entering the stomach is subjected to rippling peristaltic movement (mixing waves) which macerate the food into a thin partially digested liquid called chyme.

26. B   The body of the stomach narrows at its distal end in an area called the pylorus just proximal to its junction with the duodenum.

27. C   The roots of the teeth and their articulating surfaces of the alveolar processes or tooth sockets are lined with a periodontal membrane or ligament.

28. C   The gastric mucosa contains three types of secreting cells: the goblet cell that produces mucus, the chief or zymogenic cells for the digestive enzyme pepsinogen, and the parietal cells which secrete hydrochloric acid.

29. B   The mucosa of the small intestine contains 4-5 million, 1 mm high projections called villi which greatly increase the surface area of the epithelium over which the absorption of the digested food takes place.

30. A   The interior of the empty stomach is not smooth and contains a number of large deep folds of mucosa called rugae. When fully distended these folds stretch and are no longer visible.

31. D   The small intestine is divided into three segments based on the appearance of the mucosa. The first and shortest portion, the duodenum, is only about a foot long. The eight foot long jejunum extends to the 12 foot long final portion called the ileum.

32. B   The ascending and descending portions of the colon have abrupt turns at their junction with the transverse colon. The hepatic (right colic) flexure is named for its proximity to the liver. The higher splenic (left colic) flexure derives its name from its association to the spleen.

33. 4I   34. 6   35. 3   36. 1   37. 7   38. 2

39. 9   40. 8   41. 5

42. D   The narrowed end of the stomach, the pylorus, is separated from the duodenum by a thick circular sphincter muscle called the pyloric sphincter.

43. B   The intestines consist of four distinct tissue layers. The innermost layer of the mucosa contains the crypts of Lieberkuhn which produce digestive enzymes. The vascular submucosa possesses the nerves and vessels of the Meissner's plexus. The third layer consists of smooth muscle tissue which aids in peristalsis and the outer covering membrane called the serosa or adventitia.

44. A   The absorption of water is an important function of both the small intestine and large intestine.

45. A   From proximal to distal, the large intestine consists of a cecum, ascending colon, hepatic flexure, transverse colon, splenic flexure, descending colon, sigmoid colon, rectum, and anus.

46. B   The plicae circularis or circular folds are the true mucosa of the walls in the small intestine. They act to agitate the chyme and provide additional surface area for the absorption of food.

47. C   The chemical digestion of food begins in the mouth by the action of saliva which is augmented by the stomach and small intestine.

48. C   The hormonal secretions from the intestinal mucosa trigger the release of the pancreatic juice which is rich in bicarbonates to assist in the neutralization of the acidic chyme entering the small intestine.

49. B   The 4 layers of the digestive tract are the mucosa, submucosa, muscular, and serosa or adventitia.

50. A   The small intestine and large intestine are joined across a muscular sphincter called the ileocecal valve.

51. C   The largest serous membrane of the body that covers most of the organs in the gastrointestinal tract is called the peritoneum.

52. C   At a length of nearly 25 feet, the small intestine is by far the longest segment of the GI tract. The colon has a length of about 8 feet.

53. A   The serosa or outer layer for most structures in the gastrointestinal tract is a double-layered membrane known as the visceral peritoneum.

54. B   The first and shortest 10-12 inch long segment of the small intestine is called the duodenum.

55. C   Derived from the word vermis, for worm, the coiled tube called the vermiform appendix attaches to the lower margin of the cecum. The appendix, once thought to be functional, no longer plays any detectable role in the digestive process.

56. A   The structural variation including the circular fold, villi, microvilli, and the extended length of the small intestine, make it well adapted for the absorption of food and water.

57. C   The peritoneum has two major extensions: the mesentery, which attaches the small intestine to the posterior abdomen wall, and the omentum (fatty apron) which surrounds the transverse colon and help prevent the spread of infections.

58. C   In response to deglutition, the larynx moves upward shifting the epiglottis into a position which seals the glottis and larynx.

59. B   The intestinal glands or crypts of Lieberkuhn of the small intestine produce sucrose, lactase and maltase which break down the named sugars into glucose.

60. D   Unlike other portions of the gastrointestinal tract, the longitudinal muscle layer of the colon is gathered into 3 flat bands called the taeniae coli that rest on the external surface. Under the normal tonic conditions a series of pouches called haustral are formed.

| 61. 11 | 62. 7 | 63. 10 | 64. 9 | 65. 6 | 66. 3 |
| 67. 12 | 68. 8 | 69. 5 | 70. 1 | 71. 2 | 72. 4 |

73. C   The colon ascends from the ileocecal valve in the right lower quadrant into the right upper quadrant before turning just under the liver to become the right colic or hepatic flexure.

74. D   Digestive enzymes are produced in many areas including the salivary glands, stomach, small intestine, pancreas and liver (bile).

75. B   The contraction of the taenia coli gives the large intestine its characteristic puckered (haustral) appearance.

76. D   Both benign and malignant tumors may occur in the lower GI tract. A class of stalked tumors called polyps are among the most common, though histological examinations are required for a definite diagnosis, short-stemmed polyps with irregular borders are generally cancerous while a longer stemmed smooth bordered polyp is most often benign.

77. C   Most commonly seen in the ileocecal region in children, an intussusception involves the slippage of one part of the intestine into another. The result is often an obstruction which is alleviated by abdominal surgery or in some cases by a therapeutic barium enema.

78. A   A weakening in the lining of the walls of the colon may result in the formation of small blind pouches (diverticulum). When inflamed (diverticulitis), these may bleed causing cramp-like pains due to the muscle spasms of the colon.

79. A   The examination of the stool can be helpful in distinguishing the region of gastrointestinal bleeding. Dark tar-colored stools indicate bleeding in the stomach or small intestine. Fresh blood mixed with the stool is associated with lesions in the large intestine, rectum, and anus.

80. C   The increased intra-abdominal pressure associated with the bearing down action of the valsalva maneuver is helpful in the evaluation of esophageal varices, stress incontinence, laryngopharyngography and nasopharyngography.

81. B   Tortuous veins that have developed in the anorectal region can result in a painful swelling and inflammation called hemorrhoids or piles.

82. B   The double serous membrane, the peritoneum, encapsulates all or part of the stomach, jejunum, ilium and the majority of the colon. The duodenum, pancreas, liver, spleen, and urinary tract organs are located outside of this covering.

83. D   The body's own protective mucosa is not a perfect barrier to the effects of gastric and intestinal fluids. Its failure may result in the erosion of the mucosal lining and the formation of an ulcer.

84. C   The large intestine's main function is the absorption of water. Its only secretion is the protective mucus needed to avoid irritation of its walls.

85. A   The protrusion of all or part of an organ through an adjoining membrane or cavity is called herniation. Three types of hernias are commonly seen in the abdominal cavity. The diaphragmatic or hiatal hernia, involves the protrusion of the stomach or intestine into the thoracic cavity; the umbilical hernia, in which the abdominal organs pass through a weakening near the umbilicus, and the inguinal hernia in which a loop of intestine passes through a weakness of the inguinal ring of the groin.

86. B   In a normal male, the rectum is located anterior to the sacrum and coccyx and posterior to the urinary bladder and peritoneum.

87. A   The correct sequence in the gastrointestinal tract is the mouth, pharynx, esophagus, stomach, small intestine, and colon.

88. A   The narrowing of the pyloric valve (pyloric stenosis) by an enlargement of the circular muscle fibers and persistent spasm of the pylorus (pylorospasm) are two common conditions which occur more frequently in the stomach of male infants.

89. B   Interference with the venous return from the abdomen, can cause lymphatic obstructions or cirrhosis of the liver, and may result in an excess fluid accumulation in the peritoneum called ascites.

90. C   Obstruction in the intestinal tract caused by a twisting of the bowel (volvulus) is most commonly seen in the sigmoid and ileocecal regions.

1. Which of the following are among the 100 known functions of the liver?

   1. Formation of blood cells   2. Storage of glycogen   3. Detoxification of toxins

   A. 1 & 2 only
   B. 1 & 3 only

   C. 2 & 3 only
   D. 1, 2 & 3

2. The bulk of the pancreas is made up of pancreatic acinar cells that serve to secrete a number of enzymes which aid in the:

   A. Emulsification process
   B. Digestion of food products

   C. Hemopoietic process
   D. Detoxification process

3. The spleen is concerned with both the formation and breakdown of blood cells and is normally located under the:

   A. Right lobe of the liver
   B. Left lobe of the liver

   C. Right diaphragm
   D. Left diaphragm

4. The two major lobes of the liver are separated by folds of visceral peritoneum called the:

   A. Caudate ligament
   B. Broad ligament

   C. Quadrate ligament
   D. Falciform ligament

5. A musculomembranous sac that serves to store and concentrate bile between meals is called the:

   A. Gallbladder
   B. Alveolar sac

   C. Biliary node
   D. Urinary bladder

6. The bilary ducts are principally concerned with the transport of bile and digestive enzymes into the:

   A. Colon
   B. Left lobe of the liver

   C. Small intestines
   D. Duodenum

7. The gallbladder is normally found by its attachment to the ventral surface of the:

   A. Right lobe of the liver
   B. Left lobe of the liver

   C. Spleen
   D. Duodenum

8. The duct formed at the junction of the cystic and the common hepatic ducts is termed the:

   A. Ampulla of Vater
   B. Common bile duct

   C. Duct of Santorini
   D. Common cystic duct

9. The head of the pancreas will normally be located within the:

   A. Lesser curvature of the stomach
   B. Splenic flexure

   C. Sweep of the duodenum
   D. Falciform ligament

10. The largest single organ contained within the abdominal cavity is the:

    A. Spleen
    B. Pancreas

    C. Small intestines
    D. Liver

11. The breakdown of the red blood cells' hemoglobin into bilirubin is principally the function of the:

    A. Spleen
    B. Pancreas

    C. Liver
    D. Kidney

12. The exocrine secretions of the pancreas are carried to the common bile duct by way of the:

    A. Duct of Santorini
    B. Stenson's duct

    C. Pancreatic duct
    D. Wharton's duct

13. The absorption of water is an important factor in the maintenance of homeostasis. Which of the following organs are concerned with this function?

1. Esophagus   2. Kidney   3. Colon

A. 1 & 2 only

B. 1 & 3 only

C. 2 & 3 only

D. 1, 2 & 3

14. The control of glucose utilization is directly associated with a hormonal secretion of the pancreas called:

A. Urea

B. Insulin

C. Bile

D. Bilirubin

Referring to the diagram of the biliary system, place the number that corresponds to the following structures for questions 15-25.

15. _____ Gallbladder

16. _____ Sphincter of Oddi

17. _____ Duodenum

18. _____ Left hepatic duct

19. _____ Pancreatic duct

20. _____ Common hepatic duct

21. _____ Duct of Santorini

22. _____ Common bile duct

23. _____ Right hepatic duct

24. _____ Cystic duct

25. _____ Sphincter of Boyden

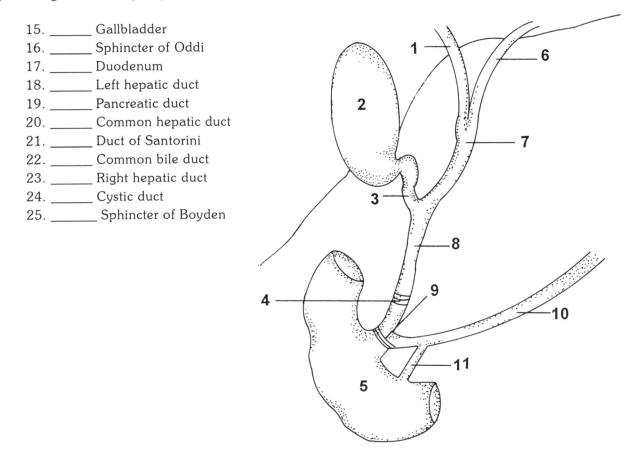

26. The release of bile by the gallbladder is controlled by the hormonal secretion of the small intestines called:

A. Cholografin

B. Cholecystokinin

C. Ferritin

D. Bilirubin

27. The tail of the pancreas is attached to what other abdominal cavity structure:

A. Kidney

B. Duodenum

C. Liver

D. Spleen

28. The gallbladder is filled by the reverse flow of bile caused by this strong sphincter muscle located at the distal end of the common bile duct called the:

A. Sphincter of Boyden
B. Sphincter of Wirsung

C. Sphincter of Rivinus
D. Sphincter of Kupffer

29. The formation of bile from the bile salts and bilirubin is one of the functions associated with the:

1. Spleen   *2. Liver*   *3. Kidney*

A. 1 only
B. 2 only

C. 3 only
D. 1, 2 & 3

30. The stored bile from the gallbladder passes into the biliary duct system by way of the:

A. Cystic duct
B. Ampulla of Vater

C. Hepatic duct
D. Papillary duct

31. Which of the following is NOT a common function of the liver?

A. Storage of blood
B. Storage of Vitamins A and D

C. Deamination of amino acids
D. Secretion of digestive enzymes

32. The junction of the biliary collecting system and the duodenum is guarded by a sphincter muscle called the:

A. Sphincter of Lieberkuhn
B. Sphincter of Oddi

C. Sphincter of Beta
D. Sphincter of Pauli

33. The major function of bile in the gastrointestinal tract is the:

A. Emulsification of fats
B. Production of amino acids

C. Emulsification of carbohydrates
D. Absorption of vitamins

34. The process by which the liver manufactures sugars from non-carbohydrate materials is termed:

A. Glycolysis
B. Metabolism

C. Gluconeogenesis
D. Maturation

35. The release of bile by the gallbladder is triggered by the presence of _____ in the stomach.

1. Carbohydrates   *2. Lear Liquids*   *3. Fats*

A. 1 only
B. 2 only

C. 3 only
D. 1, 2 & 3

36. The intake of fecal contaminated food or water may lead to an inflammation of the liver called:

A. Hepatitis
B. Cirrhosis

C. Colitis
D. Jaundice

37. Cholelithiasis is a common condition of the biliary tract and is commonly referred to as:

A. Duodenal ulcers
B. Arteriosclerosis

C. Gallstones
D. Jaundice

38. A serious condition that results from the hypofunction of the islets of Langerhans in the pancreas is termed:

A. Hemophilia
B. Diabetes mellitus

C. Hydrophobia
D. Gastroenteritis

39. The failure of the liver to absorb bilirubin will often lead to a yellowish discoloration of the skin and eyes called:

A. Cyanosis
B. Yanthosis

C. Jaundice
D. Chlorosis

40. Cirrhosis of the liver is often associated with the excessive intake of toxic material known as:

A. Cholesterol
B. Heparin

C. Renin
D. Alcohol

Explanations

1. C   The liver is one of only a few organs that is essential to life. Among its important functions are the manufacture of heparin, prothrombin, fibrinogen, and albumin. It also detoxifies toxins such as nitrogen wastes and alcohol, stores whole blood, glycogen and many other vitamins and essential metals as well as producing bile from bilirubin.

2. B   The exocrine secretion of the pancreas (pancreatic juice) is a mixture of digestive enzymes including trypsin, chymotrypsin, pancreatic amylase and lipase which are produced in the specialized acinar cells of the gland.

3. D   The largest mass of lymphatic tissue is contained within a 5 inch oval organ called the spleen located on the posterolateral surface of the stomach just below the left hemidiaphragm.

4. D   The liver maintains its position in the large part due to a fold of the peritoneum, the falciform ligament, which attaches this organ to the anterior abdominal wall.

5. A   The gall bladder serves as a reservoir for bile excreted into the biliary collecting system by the liver. In response to hormonal signals from the intestinal mucosa in the presence of fats its muscular wall contracts to expel the bile due to the duct network.

6. C   The digestive juices of the pancreas and bile from the liver and gall bladder, empty into the duodenum by way of the duodenal sphincter or the sphincter of Oddi.

7. A   The gall bladder is a muscular sac which normally contains bile and is located along the under surface of the right lobe of the liver.

8. B   The cystic duct provides a passageway for the flow of bile from the gall bladder into the common hepatic duct just above its junction with the larger common bile duct.

9. C   The 6" oblong pancreas has a broad head that is situated within the C-shaped curve of the duodenum. Its tapered tail extends laterally to its attachment on the medial surface of the spleen.

10. D   Weighing in at about 4 pounds, the liver is the largest single organ in the body. It occupies most of the right hypochondrium and part of the epigastric regions.

11. A   The phagocytic bacteria present in the spleen, break down worn out red blood cells and platelets converting the released hemoglobin into a green-colored fluid called bilirubin.

12. C   The pancreatic juices pass into the biliary collecting system by way of a long duct running the length of the pancreas known as the pancreatic duct or duct of Wirsung.

13. C   To maintain the proper fluid balance in the body, a number of systems aid in this process including the kidneys, small intestine and large intestine, lungs and gall bladder.

14. B   The islets of Langerhans in the pancreas contain two types of cells, the alpha cells which secrete glucagon that tend to increase blood glucose levels, and the insulin-producing beta cells which serve to lower blood glucose levels.

15. 2   16. 9   17. 5   18. 6   19. 10   20. 7

21. 11   22. 8   23. 1   24. 3   25. 4

26. B   Stimulated by fats in the stomach or duodenum, the intestinal mucosa secretes cholecystokinin which triggers the release of bile and pancreatic juices.

27. D   The upper end of the pancreas extends laterally across the abdomen to its attachment on the outer surface of the spleen.

28. A   The filling of the gall bladder with bile is principally due to the action of the sphincter of Boyden, located on the common bile duct just above its junction with the pancreatic duct.

29. B   The principal secretion of the liver, bile, is composed of a number of compounds including bilirubin, cholesterol, fatty acids, and bile salts.

30. A   The gall bladder is connected to the common hepatic duct by way of a thin hollow tube called the cystic duct.

31. D   The liver is one of only a few organs that is essential to life. Among its important functions are the manufacture of heparin, prothrombin, fibrinogen, and albumin. It also detoxifies toxins such as nitrogen wastes and alcohol, stores whole blood, glycogen and many other vitamins and essential metals as well as producing bile from bilirubin.

32. B   The weak sphincter at the end of the ampulla of Vater which guards the opening into the duodenum is called the sphincter of Oddi.

33. A   Bile and more specifically the bile salt component of bile formed during the breakdown of cholesterol is the active constituent which serves to emulsify fats and aid in the absorption of fatty acids from the intestine.

34. C   During periods of low carbohydrate intake (starvation) the liver can form glycogen from amino and fatty acids.

35. C   The release of cholecystokinin by the intestinal mucosa is triggered by the presence of fats or fatty acids in the stomach and small intestine.

36. A   A liver inflammation called hepatitis may have many possible causes including exposure to viruses, bacteria, protozoa or other chemical agents such as carbon tetrachloride, anesthetics or alcohol.

37. C   The presence of stones in the biliary tract sometimes associated with pain and obstruction develops in two main forms: the low density cholesterol stones which tend to float in the bile and the more dense calculi made of calcium salts.

38. B   The main clinical symptoms of insufficient insulin production resulting from diabetes mellitus are hyperglycemia and excretion of glucose in the urine. The inability to reabsorb water, dehydration and electrolytic imbalance are the cause of the life-threatening consequences seen with diabetes.

39. C   The yellow-green fluid bilirubin produced in the spleen following the breakdown of hemoglobin can not be removed by a poorly functioning liver, therefore it remains in circulation causing the yellow staining of the skin and eyes called jaundice.

40. D   Cirrhosis is a chronic disease of the liver in which fibrous connective tissue replaces the normal cells. This condition can occur from hepatitis, parasitic infection, or chronic exposure to toxic chemicals such as alcohol.

1. The ability to detect the external environment is accomplished by a series of specialized structures called:

   A. Motor end plates
   B. Adaptors

   C. Sensory receptors
   D. Accommodators

2. The sense of smell is related to the first cranial nerve, which passes through the ethmoid bone into the nasal cavity, called the:

   A. Acoustic nerve
   B. Olfactory nerve

   C. Optic nerve
   D. Cochlear nerve

3. Which of the following are among the number of cutaneous (exteroceptive) senses found in the normal adult?

   1. Pain   *2. Pressure*   *3. Heat*

   A. 1 & 2 only
   B. 1 & 3 only

   C. 2 & 3 only
   D. 1, 2 & 3

4. The detection of certain low energy forms of electromagnetic radiation can be accomplished by photoreceptors located in the:

   A. Eye
   B. Ear

   C. Nose
   D. Mouth

5. The degree of distention in some muscular organs, such as the rectum and urinary bladder, is accomplished by:

   A. Thermoreceptors
   B. Chemoreceptors

   C. Acousto-receptors
   D. Stretch receptors

6. The transparent outer portion of the eye that allows passage of light into the eye is called the:

   A. Cornea
   B. Pupil

   C. Optic disc
   D. Retina

7. The tympanic cavity is a small, irregular space within the temporal bone that serves to house the structures associated with:

   A. Taste
   B. Hearing

   C. Smell
   D. Sight

8. The specialized organs on the tongue for the detection of taste are called:

   A. Gustatory glands
   B. Auditory receptors

   C. Taste buds
   D. Ossicles

9. The external funnel-like structure of the ear that helps to direct sound vibrations to the ear is called the:

   A. Tragus
   B. Auricle

   C. Ceruminous
   D. Eustachian cartilage

10. The tympanic membrane is a thin membrane which separates the structures of the:

    A. Chambers of the eye
    B. Middle and inner ear

    C. External and middle ear
    D. Oral and nasal cavities

11. Pressure differences in the middle ear must be maintained to prevent hearing impairment. The structure associated with the function is the:

    A. Eustachian tube
    B. Nasolacrimal duct

    C. Semicircular canal
    D. Fallopian tube

12. The small bones of the middle ear that serve to transmit sound vibrations to the inner ear are the:

   1. Cochlea　　*2. Stapes*　　*3. Malleus*

   A. 1 & 2 only　　　　　　　　　　C. 2 & 3 only
   B. 1 & 3 only　　　　　　　　　　D. 1, 2 & 3

13. The sense of equilibrium (balance) is controlled by a series of small fluid canals called the:

   A. Cochlear canals　　　　　　　C. Canaliculi
   B. Semicircular canals　　　　　　D. Lacrimal canals

14. The specialized receptor for hearing located on the basilar membrane of the cochlea is called the:

   A. Utricle　　　　　　　　　　　C. Organ of Corti
   B. Conjunctiva　　　　　　　　　D. Ciliary body

Referring to the diagram of the membranous labyrinth, place the number that corresponds to the following structures for questions 15-20.

15. _____ Stapes at oval window
16. _____ Cochlear nerve
17. _____ Vestibule
18. _____ Semicircular canals
19. _____ Cochlea
20. _____ Vestibular portion of
　　　　　　the acoustic nerve

21. The membranous labyrinth of the inner ear is surrounded by a protective fluid called:

   A. Perilymph　　　　　　　　　　C. Vitreous humor
   B. Plasma　　　　　　　　　　　　D. Aqueous humor

22. The action of otoliths on the hairlike extensions of the macula in response to gravity is interpreted by the brain to help control:

   A. Positional orientation　　　　　C. Tympanic reflex
   B. Visual reflexes　　　　　　　　D. Sensory adaptation

23. The tongue is able to detect four primary taste sensations. Among these are:

   1. Sweet　　*2. Sour*　　*3. Salt*

   A. 1 & 2 only　　　　　　　　　　C. 2 & 3 only
   B. 1 & 3 only　　　　　　　　　　D. 1, 2 & 3

24. Visceral pain impulses may be felt at body parts other than where the stimulus is occurring. This phenomenon is called:

   A. Adaptation　　　　　　　　　　C. Referred pain
   D. Discrimination　　　　　　　　D. Anosmia

25. The upper eyelid that forms a protective covering for the eye also houses a small gland that produces tears called the:

   A. Lacrimal gland　　　　　　　　C. Rectus gland
   B. Conjunctiva　　　　　　　　　D. Basilar gland

26. The production of tears helps to moisten and protect the eye from bacterial infection and will drain into the nasal cavity through the:

A. Iris
B. Vestibule
C. Eustachian tube
D. Nasolacrimal duct

27. The visual receptor cells are located on the back of the eye in a cellular layer called the:

A. Sclera
B. Vitreous humor
C. Retina
D. Ciliary body

28. The amount of light passing through the pupil of the eye is controlled by a set of circular muscles called the:

A. Lens
B. Iris
C. Retus muscle
D. Obicularis oculi

29. The majority of the eyeball is composed of a white, fibrous protective covering called the:

A. Sclera
B. Cornea
C. Retina
D. Macula

30. The impulses for visual sensation will pass into the occipital lobe of the brain by way of the second cranial nerve, which is also called the:

A. Olfactory nerve
B. Oculomotor nerve
C. Optic nerve
D. Ophthalmic nerve

31. The action of the ciliary bodies controls the shape of a transparent oval body called the _____ which serves to focus light on the back of the eye.

A. Lens
B. Cornea
C. Retina
D. Iris

32. Eye color is due principally to the appearance of a circular muscle called the:

A. Lens
B. Iris
C. Suspensory ligament
D. Optic disc

33. The retina is composed of two distinct types of cells called:

A. Actin and myosin
B. Fovea and stroma
C. Cones and rod
D. Stroma and incus

34. The point on the retinal surface with the greatest amount of visual acuity is the:

A. Fovea centralis
B. Fovea femoris
C. Fovea lateralis
D. Fovea rectus

Referring to the diagram of the eye, place the number that corresponds to the following structures for questions 35-44.

35. _____ Vitreous humor
36. _____ Cornea
37. _____ Lens
38. _____ Optic nerve
39. _____ Retina
40. _____ Sclera
41. _____ Iris
42. _____ Ciliary body
43. _____ Pupil opening
44. _____ Optic disc

45. The normal value for the interpupillary measurement is:

   A. 1.5" (3-4cm)   C. 3.5" (9-10cm)
   B. 2.5" (6-7cm)   D. 5.0" (12-13cm)

46. High interocular pressure most commonly results from excessive amounts of:

   A. Perilymph in the vestibule   C. Vitreous humor in the posterior chamber
   B. Mucus in the lacrimal duct   D. Aqueous humor in the anterior chamber

47. There are no photoreceptors in the area where the optic nerve joins the posterior portion of the eye leaving a blind spot called the:

   A. Fovea centralis   C. Optic disc
   B. Choroid plexus   D. Focal point

48. The clouding of the lens of the eye is a common cause of blurred vision. This condition is called (a):

   A. Glaucoma   C. Conjunctivitis
   B. Night blindness   D. Cataract

49. The specialized cell of the retina that is primarily responsible for color vision is the:

   A. Fovea   C. Cone
   B. Stroma   D. Myosin

50. The principal group of muscles responsible for eye movements are the:

   A. Rectus muscles   C. Oculi muscles
   B. Palpebrae muscles   D. Ciliary muscles

1. C   The body receives information about the external environment through a number of specialized sensory receptors or exteroceptors.

2. B   The olfactory receptors, for the sense of smell, are found on the membranous lining of the upper nasal cavity. Sensory signals pass along the olfactory nerves into the temporal lobe of the brain for interpretation.

3. D   Each of five cutaneous (skin) sensations, light touch, pressure, pain, cold, and heat are associated with a distinct exteroceptor.

4. A   The exteroceptor for vision lies on the retina, the nervous coat, covering the posterior and lateral surfaces of the eyeball.

5. D   The visceroreceptors are a class of sensory receptors that provide the body with information concerning the status of the internal organs including hunger, pain, pressure, and distention. These are collectively referred to as stretch receptors.

6. A   The anterior portion of the eye is composed of a clear connective tissue, the cornea, which enables the unimpeded flow of light into the eye.

7. B   The middle ear, or tympanic cavity, is an air-filled space housing the auditory ossicles or middle ear bones.

8. C   The gustatory, or taste receptors, called the taste buds, are located on the raised elevations of the tongue, the papillae.

9. B   The skin covered cartilage commonly called the ear is also known as the pinna or auricle.

10. C   At the medial end of the external auditory canal sits the tympanic membrane or eardrum which separates external and middle ears.

11. A   The pressure in the air-filled middle ear is equalized with that of the outside environment to prevent its rupture. The eustachian or auditory tubes which connect the middle ear and nasopharynx serve to balance this pressure.

12. C   The auditory ossicle from external to internal are the malleus (hammer), incus (anvil), and stapes (stirrup).

13. B   The inner ear houses both the organs of hearing and balance. The organs associated with equilibrium and the positional senses are the saccule, utricle, and semicircular canals.

14. C   The true exteroceptors for hearing, the organs of Corti, contain hair cells and support structures which generate an electronic signal that passes into the temporal lobe of the brain.

15. 3   16. 5   17. 2   18. 1   19. 6   20. 4

21. A   A series of sacs and tubes having the same general form as the bony labyrinth, the membranous labyrinth, are protected by a fluid called perilymph.

22. A   Static and positional equilibrium are provided in the utricle, saccule, and semicircular canals. These fluid-filled cavities are lined with sensory hairs that respond to small grains of calcium carbonate (otolith) which trigger the hairs during the changing orientation of the head.

23. D   The 4 primary taste sensations are salt and sweet, detected on the tip and sides of the tongue, and sour on the side, and bitter which is detected on the back of the tongue.

24. C   Because many afferent nerve (pain) fibers have a wide cutaneous distribution, the brain is often unable to pinpoint the exact location of a stimulus. The pain that is felt may be referred to an area other than the effected site; e.g., heart pain may radiate down the left arm, kidney pain may be felt on the sides of the upper leg.

25. A   The superolateral portion of each upper eye lid contains an almond sized lacrimal gland which bathes the eyes with tears through 6-12 small lacrimal duct openings under the lid.

26. D   Tears, which contain both mucus and bactericidal enzymes, spread over the eye before passing to the nasolacrimal duct near the inner canthus. Tears eventually are dispersed within the nasal cavity to help humidify the air passing into the lungs.

27. C   The exteroceptors for vision, the cone and rod cells, are dispersed on the back and sides of the nervous coat or retina of the eye.

28. B   Surrounding the pupil is a doughnut shaped muscle, the iris, which contains the pigments for eye color.

29. A   The tough outer fibrous covering of the eye which both protects and gives shape to the eyeball is formed by the clear cornea and white sclera.

30. C   The retina on which the visual image is formed is continuous with the optic nerve which carries the visual impulses to the occipital lobe of the brain.

31. A   Immediately behind the pupil and iris, the clear oval lens is suspended by the ciliary muscles. Changes in the shape of the lens focus the incoming light providing for both near and far vision.

32. B   The circular muscle, the iris, surrounding the pupil, contains the protective pigment, melanin. The distribution and scattering of the incoming light by the iris provides the various shades of eye color. Blue eyes which lack this pigment tend to be more sensitive to the potential damage caused by ultraviolet radiation.

33. C   The nervous coat of the eye, the retina, contains two separate cell types; the cone cells, which respond to bright and colored light and the more sensitive rod cells, used in dim light conditions which are only capable of black and white vision.

34. A   The distribution of the cells for vision is uneven with the cones being densely packed on the center of the eye (fovea centralis) and the rods predominating on the sides of the retina.

35. 6   36. 2   37. 4   38. 10   39. 5   40. 9

41. 3   42. 7   43. 1   44. 8

45. B   The stereoscopic vision that humans possess is largely due to the 2.5" separation of the eyes and the independent images each sends to the brain.

46. D   The interior of the eye is divided into anterior and posterior chambers by the lens of the eye. The anterior chamber contains a watery fluid called aqueous humor which maintains the desired intraocular pressure. Excessive pressure in this chamber may cause a condition called glaucoma resulting in circulatory disturbances, retinal damage, and eventual blindness. In the posterior chamber, a jelly-like vitreous humor helps maintain the characteristic shape of the eyeball.

47. C   The junction of the optic nerve ganglions at the retina forms a small circular area which lacks visual receptors known as the optic disc or blind spot.

48. D   The loss of transparency of the lens of the eye as is seen with cataracts may be accelerated by exposure to x rays and ultraviolet (UV) radiation.

49. C   The larger of the two cells for vision is the color sensitive cone cell. The cones, which function only at high light levels (daytime) possess a higher level of visual acuity than the smaller rod cells.

50. A   The movements of the eye are controlled by 6 extrinsic eye muscles namely the superior, inferior, medial and lateral rectus muscles and the superior and inferior oblique muscles.

1. The principal hormone that is responsible for both proper function and development of the male reproductive organ is:

   A. Acrosome
   B. Progesterone

   C. Estrogen
   D. Testosterone

2. In the male fetus, the testes originate and develop behind the peritoneum near the:

   A. Prostate gland
   B. Kidneys

   C. Urinary bladder
   D. Symphysis pubis

3. The male sex cell (spermatozoa) is produced by a specialized lining in tiny coil passages of the testes called the:

   A. Epididymis
   B. Vas deferens

   C. Prostate gland
   D. Seminiferous tubules

4. Before birth, the testes descend into a pouchlike sac below the abdominal cavity called the:

   A. Prepuce
   B. Acrosome

   C. Vas deferens
   D. Scrotum

5. Which of the following structures are concerned with the formation of seminal fluids?

   ### 1. Prostate gland   2. Prepuce   3. Seminal vesicle

   A. 1 & 2 only
   B. 1 & 3 only

   C. 2 & 3 only
   D. 1, 2 & 3

6. The male germ cells undergo a special type of cell division that reduces the number of chromosomes by half. This is termed:

   A. Osmosis
   B. Mitosis

   C. Meiosis
   D. Chromosis

7. In the adult male, urinary tract problems often result from enlargement of the_____ gland which surrounds the urethra.

   A. Seminiferous
   B. Scrotal

   C. Inguinal
   D. Prostate

8. After spermatogenesis has occurred, the immature sperm passes into the _____ for their final maturation.

   A. Epididymis
   B. Urethra canal

   C. Scrotal sac
   D. Ejaculatory duct

9. The proper temperature required in the formation of the male sex cell is:

   A. Above normal body temperature
   B. Below normal body temperature

   C. The same as normal body temperature

Referring to the diagram of the male reproductive system, place the number that corresponds to the following structures for questions 10-19.

10. _____ Epididymis
11. _____ Urinary bladder
12. _____ Ejaculatory duct
13. _____ Seminiferous tubules
14. _____ Scrotal sac
15. _____ Prostate gland
16. _____ Urethra
17. _____ Vas deferens
18. _____ Seminal vesicle
19. _____ Head of the penis

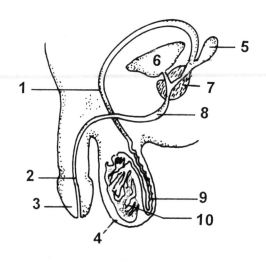

20. The normal mature spermatozoa will contain _____ autosomes and _____ sex chromosome(s).

   A. 44, two   C. 21, one
   B. 45, one   D. 22, one

21. The body of the penis is composed of three columns of specialized tissue that allows for the stiffening necessary during intercourse. This is called:

   A. Seminal tissue   C. Nocturnal tissue
   B. Erectile tissue   D. Pubescent tissue

22. A loose fold of skin covering the glans penis (prepuce) is often removed shortly after birth by a procedure called:

   A. Caversionization   C. Cryptorectomy
   B. Circumcision   D. Vasectomy

23. The average male ejaculation will normally contain about:

   A. Ten thousand spermatozoa   C. One million spermatozoa
   B. Fifty thousand spermatozoa   D. One hundred million spermatozoa

24. A vasectomy is a relatively simple procedure for male sterilization. This involves the cutting or tying off of a portion of the:

   A. Urethra   C. Vas deferens
   B. Seminal duct   D. Ejaculatory duct

25. Possible causes of male infertility are:

   1. Cryptorchidism   2. Low sperm count   3. Insufficient sperm motility

   A. 1 & 2 only   C. 2 & 3 only
   B. 1 & 3 only   D. 1, 2 & 3

Explanations

1. D In response to the hormones of the anterior pituitary gland, the testes secrete testosterone which stimulates bone and muscle growth, aids in the maturation of sperm and stimulates the development of the secondary sex characteristics.

2. B In fetal life, the testes develop near the kidneys in the upper abdomen. Their descent out of the abdominal cavity and into the scrotum is normally completed about 2 months prior to birth.

3. D The immature sperm cells (spermatogonia) are produced in a set of tightly coiled tubules of the testes called the seminiferous tubules.

4. D The testes are supported in an external pouch, the scrotum, which helps to maintain their temperature at about 3 degrees below that of the body's temperature.

5. B The slightly alkaline semen or seminal fluid not only provides the sperm with a transportation and nutrient medium it also serves to neutralize the acid environment of the vagina. It is produced in the bulbourethral (Cowper's gland), prostate gland and a pair of seminal vesicles.

6. C The formation of the male sperm and female ova involves a type of haploid-type division called meiosis.

7. D The doughnut shaped prostate gland sitting just inferior to the urinary bladder surrounds the first portion of the urethra. In older men the enlargement of the prostate may compress the urethra and obstruct the flow of urine from the bladder.

8. A The immature sperm produced in the seminiferous tubules pass into the epididymis where maturation is completed. During ejaculation, peristaltic contractions of the smooth muscular walls of the ejaculatory ducts system propel sperm toward the urethra.

9. B To increase the production and survival of the sperm, they are produced at a temperature below that of the body. The external location of the testes within the scrotal sac help provide this desired difference in temperature.

10. 9   11. 6   12. 8   13. 10   14. 4   15. 7

16. 2   17. 1   18. 5   19. 3

20. A Unlike the majority of human cells that contain 44 autosomes and 2 sex chromosomes, the spermatozoa is haploid, possessing only 22 autosomes and a single sex chromosome. Since males have both x and y sex chromosomes, a sperm has either an X (female sperm) or Y (male sperm) genetic potential.

21. B The sponge-like (erectile) tissue of the penis possesses blood sinuses that dilate during sexual excitement or stimulation.

22. B The covering of the glands by the foreskin helps protect this sensitive structure from irritation. The removal of this tissue by circumcision soon after birth is common in some religious faiths and in many medical systems such as our own.

23. D The average volume of ejaculate of about 2-4 cc of semen may contain as many as 100 million sperm per cc of fluid.

24. C The main permanent form of male sterilization is a procedure involving the cutting or tieing of the vas deferens. The interruption of the ducts closer to the urethra are not employed because of the potential negative effects to the urinary tract functions.

25. D There are numerous physiologic causes of male infertility with both physiological and psychological roots. Among the more common are inflammation of the testes, undescended testicles and low sperm count or sperm morbidity.

1. The process by which a secondary oocyte is released from a mature ovarian follicle is termed:

   A. Oogenesis
   B. Implantation

   C. Ovulation
   D. Fertilization

2. After its release from the ovary, the egg is directed into the uterine tube by a series of fingerlike extensions of the infundibulum called the:

   A. Fimbriae
   B. Villi

   C. Cilia
   D. Fornices

3. The normal site for implantation of a fertilized egg is the wall of the:

   A. Vagina
   B. Uterus

   C. Oviduct
   D. Ovary

4. During intercourse, the fibromuscular tube that serves as the depository for the spermatozoa is called the:

   A. Clitoris
   B. Mons pubis

   C. Labia major
   D. Vagina

5. If fertilization of the egg cell does NOT occur, menstruation will begin. Which of the following layers of the uterine wall disintegrate and slough away?

   1. Endometrium   2. Myometrium   3. Perimetrium

   A. 1 only
   B. 2 only

   C. 3 only
   D. 1, 2 & 3

6. Shortly after giving birth, the alveolar glands within the mammary tissue begin producing a special nutrient fluid called:

   A. Muconium
   B. Colostrum

   C. Androgen
   D. Chorion

7. In the normal female, the development of the mature oocyte (egg cell) is accomplished within the:

   A. Vagina
   B. Ovary

   C. Uterus
   D. Oviducts

8. The lower constricted portion of the thick-walled muscular sac (uterus) is called the:

   A. Myometrium
   B. Labia Major

   C. Clitoris
   D. Cervix

9. The maintenance of the uterus, ovaries and oviducts position within the abdominal cavity is accomplished by the:

   1. Ligament of Treitz   2. Broad ligament   3. Ovarian ligament

   A. 1 & 2 only
   B. 1 & 3 only

   C. 2 & 3 only
   D. 1, 2 & 3

10. The penetration of a secondary oocyte by a spermatozoan is termed:

    A. Implantation
    B. Parturition

    C. Fertilization
    D. Ovulation

11. The small collection of erectile tissue in the female that corresponds in structure and origin to the male penis is called the:

    A. Hymen
    B. Areola

    C. Clitoris
    D. Prepuce

12. The secondary oocyte passes from its origin in the ovary into the uterus by way of the:

A. Cervix
B. Vaginal orifice

C. Urethral tubes
D. Fallopian tubes

Referring to the diagram of the female reproductive system, place the number that corresponds to the following structures for questions 13-20.

13. _____ Fundus of the uterus
14. _____ Ovary
15. _____ Suspensory ligament
16. _____ Cervix
17. _____ Myometrium
18. _____ Vagina
19. _____ Fimbriae
20. _____ Oviduct

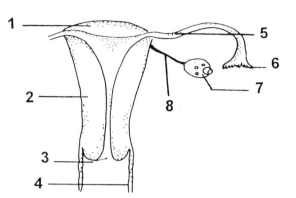

21. During fetal development, the highly vascular structure that serves as respiratory, nutritive and excretory organs of the fetus is called the:

A. Placenta
B. Zygote

C. Amnion
D. Mammary glands

22. In the female, the hormone principally responsible for the development of the secondary sexual characteristics is:

A. Testosterone
B. Oxytocin

C. Estrogen
D. Lactogen

23. In the female, puberty begins around the thirteenth year and is marked by the first menstrual flow called:

A. Menarche
B. Phimosis

C. Menopause
D. Parturition

24. Many routing radiographic procedures should be performed during the first third of the menstrual cycle. This cycle begins on the:

A. First day of menses
B. Last day of menses

C. Day of ovulation

25. After fertilization of the female egg cell by a single male spermatozoa, a single cell containing 46 chromosomes is formed called a:

A. Zona
B. Zygote

C. Polar body
D. Primary follicle

26. Upon entering the oviducts, the egg cell is actively transported toward the uterus by:

1. Whip-like motions of the egg's tail   **2. Ciliary action**   **3. Peristaltic contractions**

A. 1 & 2 only
B. 1 & 3 only

C. 2 & 3 only
D. 1, 2 & 3

27. Implantation of the blastocyst into the endometrium will generally occur within about _____ after fertilization.

   A. 2 hours
   B. 24 hours

   C. 2 days
   D. 1 week

28. At the present time, the most effective form of temporary contraception for females is:

   A. Coitus interruptus
   B. The rhythm method

   C. An intrauterine device
   D. An oral contraceptive

29. Pregnancy normally continues for about 40 weeks or 9 calendar months. This development stage is referred to as:

   A. Conization
   B. Gestation

   C. Menopause
   D. Parturition

30. A surgical method for permanent sterilization of the female, in which the oviducts are cut and tied, is called a/an:

   A. Curettage
   B. Eclampsia

   C. Abortion
   D. Tubal ligation

Explanations

1. C   The ovum is released from the Graafian or ovarian follicle of the ovary. The maturation process which takes approximately 56 days alternates between the two ovaries resulting in the more familiar 28-day cycle.

2. A   The ovum is released into the pelvic cavity but is directed by the finger-like fimbriae into the uterine (Fallopian) tube.

3. B   Fertilization which normally takes place in the fallopian (uterine) tubes causes a small change in the electrical potential of the egg which encourages its implantation into the uterine wall.

4. D   The muscular vagina serves as the passageway for the menstrual flow to leave the body. It also forms the lower portion of the birth canal and the receptacle for the penis during coitus.

5. A   During a non-reproductive menstrual cycle, the outer layer of the endometrium, the stratum functionalis, is shed during menses. This layer is re-formed during middle portion of the cycle each month.

6. B   During pregnancy, the mammary glands reach their final development stage under the influence of estrogen and progesterone. These specialized tissues began with the production of the immune boosting colostrum for 2-5 days after birth. After this time, true mother's milk replaces this fluid. Recent evidence shows that breast-fed babies may have better physical and mental advantages compared to formula-fed babies.

7. B   The ovarian cycle consists of development stages by which the primary follicle matures into the secondary follicle and Graafian follicle.

8. D   The uterus is shaped like an inverted pear with the fundus at the expanded upper end and the narrowed cervical portion on the lower aspect opening into the vagina.

9. C   The majority of the female reproductive organs are held in place by a series of supporting ligaments including the broad, uterosacral, cardinal, round ovarian, and suspensory ligaments.

10. C   Fertilization refers to the union of the nuclear material of the sperm and ovum into a new single cell called the zygote.

11. C   The small cylindrical mass of erectile tissue, the clitoris, is located anterior to the labia minor just above the vaginal orifice.

12. D   After its release from the ovary, the ovum or oocyte enters the fallopian tube and is carried by ciliated cells and muscular waves towards the body of the uterus.

13. 1   14.   7   15.   8   16.   3   17.   2   18.   4

19. 4   20.   5

21. A   The fetus is physically connected to the mother by way of a highly vascular organ called the placenta. In the developing infant, the lack of fully functional lungs, liver, GI tract and kidneys necessitates the exchange of food, oxygen and wastes. The placenta provides this essential exchange enabling proper growth and development of the fetus.

22. C   Unlike males, who have only one major sex hormone, females possess two. Estrogen which is responsible for the development of the primary and secondary sex characteristics such as hair distribution and breast development, and progesterone which works in conjunction with estrogen to prepare the uterus and egg for implantation and development of mother's milk.

23. A   The normal monthly menstrual cycle provides the desired conditions for the development of a viable ovum and the preparation for implantation in the event that fertilization occurs. Beginning with menarche at about 11 years of age, the cycle continues until menopause in the fifth decade of life.

24. A   The menstrual cycle progresses in 3 distinct phases. The first, lasting for the first 5 days, is the period of menstruation characterized by the loss of the outer layer of endometrium. In the middle 7-10 days, or pre-ovulatory phase, the endometrium thickens with increased vascularity. During the last week, or ovulatory phase, the egg is released and upon fertilization the egg may attach to the fully developed endometrium.

25. B   During fertilization, the 23 chromosomes of the male and female gametes (egg and sperm) are fused to form a single zygote containing 44 autosomes and 2 sex chromosomes. A male zygote possesses XY pattern and the female zygote XX pattern of sex chromosomes.

26. C   The ciliary action and muscular contraction of the fallopian tubes provide the necessary force to move the egg cell toward the uterus.

27. D   Implantation or the actual attachment of the dividing zygote (blastocyst) and the uterine wall occurs about a week after fertilization takes place.

28. D   Temporary contraceptive methods for females take many forms, the diaphragm, sponge and female condom which are designed to prevent the passage of sperm into the uterus. The IUD (intra uterine device) prevents implantation of the egg and the pill acts to prevent ovulation by the hormonal disruption of the menstrual cycle. All have their potential benefits and disadvantages. The most effective at preventing pregnancy at this time is the pill.

29. B   The length of time the embryo-fetus is carried by the mother is termed gestation and normally lasts about 40 weeks. The last step in this process is labor which results in the actual birth of the fetus called parturition.

30. D   The permanent form of female birth control is termed a tubal ligation, in which the fallopian tubes are cut or tied. Though sometimes reversible with micro-surgery it should only be considered when no other pregnancies are contemplated.

1. The normal adult nervous system is subdivided into two major divisions called the:

   A. Central & peripheral nervous systems
   B. Meningeal & cutaneous nervous systems
   C. Thalmic & spinal nervous systems
   D. Skeletal & vascular nervous systems

2. The sensory areas of the brain associated with hearing, sight and smell are located in the:

   A. Cerebrum
   B. Cerebellum
   C. Midbrain
   D. Brain stem

3. In a cross-section cut of the spinal cord, the gray appearance of the central portion is due in large part to the:

   A. Neuro-fibrils
   B. Myelinated fibers
   C. Nissl bodies
   D. Nerve cell bodies

4. The activity of the pituitary gland is controlled by the action of the:

   A. Telencephalon
   B. Cerebellum
   C. Hypothalamus
   D. Midbrain

5. The coordination of fine muscular movement of the extremities is primarily accomplished by the:

   A. Cerebrum
   B. Cerebellum
   C. Limbic system
   D. Diencephalon

6. The folds of the brain giving the appearance of ridges and valleys are called the:

   A. Gyri and sulci
   B. Villi and cilia
   C. Ganglia and fissures
   D. Axons and dendrites

7. The lateral ventricles are the largest of the fluid-filled cavities of the brain and are located within the mass of the:

   A. Brain stem
   B. Cisterna magnum
   C. Cerebral hemispheres
   D. Falx cerebri

8. The brain and spinal cord are protected by three membranous layers collectively known as the:

   A. Diploë
   B. Meninges
   C. Sinuses
   D. Fissures

9. The cranial venous sinuses that drain the blood from the brain are formed by the two layers of the:

   A. Periosteum
   B. Dura mater
   C. Lateral ventricles
   D. Arachnoid villi

10. The brain and spinal cord develop from a tube-like nerve bundle called the:

    A. Neural tube
    B. Cerebral nuclei
    C. Commissural fibers
    D. Neural synapse

11. The dorsal root of the spinal nerve is composed of:

    1. Motor fibers   **2. Sensory fibers**   **3. Afferent fibers**

    A. 1 & 2 only
    B. 1 & 3 only
    C. 2 & 3 only
    D. 1, 2 & 3

12. The medulla oblongata exerts a great deal of control over:

    1. Respiration rate   **2. Heart rate**   **3. Blood pressure**

    A. 1 only
    B. 2 only
    C. 3 only
    D. 1, 2 & 3

13. The largest portion of the normal adult brain is the:

A. Cerebellum   C. Cerebrum
B. Pons   D. Diencephalon

14. The normal connection between the third and fourth ventrals is accomplished by a narrow canal called the:

A. Cerebral aqueduct   C. Foramen of Monro
B. Choroid plexus   D. Falx cerebri

15. The innermost layer of the membrane covering the brain is termed the:

A. Arachnoid mater   C. Subarachnoid mater
B. Dura mater   D. Pia mater

16. The cerebrospinal fluid that is secreted by masses of specialized capillary beds in the walls of the ventricles serves for the:

A. Regulation of hormones C. Filtration of wastes
B. Protection of the brain   D. Production of antibodies

17. How many distinct pairs of spinal nerves originate from the spinal cord?

A. Twelve   C. Twenty-four
B. Twenty-one   D. Thirty-one

18. The subarachnoid space that contains the cerebrospinal fluid is formed by the:

A. Periosteum & dura mater C. Arachnoid & pia mater
B. Dura & pia mater   D. Transverse & sigmoid sinuses

19. The junction of two or more nerve cell extensions occurs at a conduction pathway called the:

A. Syntaxis   C. Synapse
B. Ganglion   D. Reflex arc

20. The conduction of electrical impulses from the central nervous system to the peripheral nerves occurs over the:

1. Efferent fibers *2. Sensory fibers* *3. Motor fibers*

A. 1 & 2 only   C. 2 & 3 only
B. 1 & 3 only   D. 1, 2 & 3

21. The interventricular foramen permits the passage of fluids between the _____ ventricles.

1. Third and fourth *2. Fourth and lateral* *3. Lateral and third*

A. 1 only   C. 3 only
B. 2 only   D. 1, 2 & 3

22. The phrenic nerve that innervates the primary muscle of respiration, the diaphragm, is a branch off of the:

A. Cervical plexus   C. Lumbar plexus
B. Brachial plexus   D. Sacral plexus

23. A neuron consists of a cell's body and two thin, branching extensions called the:

A. Actin and myosin   C. Synapse and ganglion
B. Axon and dendrite   D. Myelin and neurolemma

24. The fluid-filled cavities of the brain and spinal canal communicate by way of the foramina of:

A. Sylvius and Moore   C. Monro and Rinivus
B. Schwan and Ranvier   D. Luschka and Magendie

25. Severe pain in the facial region, called tic douloureux, is caused by mechanical irritation of the:

A. Oculomotor nerve
B. Vagus nerve

C. Facial nerve
D. Trigeminal nerve

26. Drop foot, due to the loss of the dorsiflexor muscles of the foot, often results from damage to the:

A. Musculocutaneous nerve
B. Obturator nerve

C. Median nerve
D. Peroneal nerve

27. The cutaneous areas supplied by a single dorsal root is called a:

A. Dermatome
B. Ganglia

C. Spinothalamic tract
D. Nerve plexus

28. The regeneration of peripheral nerve fibers is in large part due to a special nerve covering called the:

A. Nodes of Ranvier
B. Myelin sheath

C. Neurolemma
D. Synaptic cleft

29. The conduction pathway between the right and left sides of the cerebrum is accomplished by a large band of white fibers called the:

A. Corpus callosum
B. Basal ganglia

C. Cerebral aqueduct
D. Islets of Reil

Referring to the diagram of the superior aspect ventricles, place the number that corresponds to the following structures for questions 30-34.

30. _____ Anterior horn (lateral ventricle)
31. _____ Fourth ventricle
32. _____ Cerebral aqueduct
33. _____ Posterior horn (lateral ventricle)
34. _____ Third ventricle

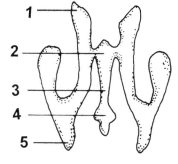

35. The sensory and motor impulses to and from the brain for the head and neck region are carried by the _____ pairs of cranial nerves.

A. Six
B. Twelve

C. Eighteen
D. Thirty-one

Explanations

1. A   The brain and spinal cord are collectively referred to as the central nervous system. The remaining nerves which carry signals between the receptors, muscles, and glands are called the peripheral nerves.

2. A   The vast majority of the senses: sight, sound, taste, smell, and feeling are interpreted in the area known as the cerebrum or cerebral cortex of the brain.

3. D   Neural tissues are grouped by their relative gray or white coloration. White matter's appearance is derived from the covered axons extending from the nerve cell body. The gray matter contains a large number of nerve cell bodies and its short associated dendrites.

4. C   The pituitary gland consists of two divisions: the adenohypophysis and neurohypophysis which produces the 9 major hormones. Its physical and functional connection to the hypothalamus of the brain are unique to this gland.

5. B   Though gross muscle movements are initiated by the motor areas of the cerebrum, the ability to co-ordinate fine muscle movement is accomplished in the cerebellum.

6. A   The convolutions of the brain are divided into raised areas called gyri and depressions known as sulci.

7. C   In addition to the protection provided by the cranial bones, the cerebrospinal fluid circulating around the brain and spinal cord provides additional protection from injury. This fluid is contained within the subarachnoid spaces and within the cavities of the cerebral hemispheres called ventricles.

8. B   Cerebral spinal fluid is contained between two layers of the meningeal coverings in the subarachnoid space; the arachnoid and pia mater.

9. B   The vascular structures of the brain including arteries, capillaries, veins, and specialized cavities are formed by the layers of dura mater called venous sinuses.

10. A   In early embryonic life, the central nervous system begins to form from a bundle of nerves called the neural tube. Tissues of the brain and spinal cord continue to grow until about the age of 18.

11. C   Each spinal nerve arises by 2 roots; the anterior (ventral) root contains the motor or efferent fibers that conduct impulses away from the central nervous system, the afferent or sensory fibers which conduct impulses toward the central nervous system are found in the posterior (dorsal) root.

12. D   The upper portion of the spinal column and lower portion of the brain is called the medulla oblongata. The medulla is responsible for cardiac and respiratory functions and constitutes the majority of the primitive brain found in most animals.

13. C   The largest portion of the brain is the cerebrum, which is responsible for most of the sensory, gross motor, and all of the creative thought functions.

14. A   Cerebral spinal fluid passes between the 3rd and 4th ventricles by way of the sylvian aqueduct or cerebral aqueduct.

15. D   The pia mater is the name given to the close-fitting membrane which covers the brain and spinal cord.

16. B   The clear watery fluid, the cerebral spinal fluid, fills the subarachnoid space. This fluid contains salts, sugars, and proteins for the exchange of nutrients as well as its physical protection of the brain and spinal cord.

17. D   As the spinal cord descends from the brain it gives rise to 31 pairs of spinal nerves which exit the vertebral canal by way of the intervertebral foramina.

18. C   The subarachnoid space which contains the cerebral spinal fluid is formed between pia and arachnoid mater layers.

19. C   The nerve cell has 2 cytoplasmic extensions called the axon and dendrites. These connect with the other nerve extensions across a specialized junction called a synapse.

20. B   Electrical signals generated in the brain and spinal cord are directed along efferent or motor fibers to the peripheral structures such as muscles or glands.

21. C   The 2 lateral ventricles are connected by the right and left intervertebral foramina that enables the passage of cerebrospinal fluid between the lateral ventricles.

22. A   The motor fibers which innervate the diaphragm are formed by the ventral rami of C3, C4 and C5 of the cervical plexus. Spinal cord injuries above this point are extremely dangerous because they may result in paralysis of the diaphragm.

23. B   The two cytoplasmic extensions of a nerve fiber are the axon and dendrite.

24. D   The subarachnoid space of the spinal cord and ventricles of the brain communicate at an area at the back of the head called the cisterna magna. Fluids pass from one area to another through the two foramina of Luscka and the foramen of Magendie.

25. D   A painful spasm of the face due to irritation or pressure on the trigeminal nerve is called trigeminal neuralgia or tic douloureux.

26. D   The principal muscles of the anterior lower leg responsible for the lifting of the foot are innervated by the peroneal nerve.

27. A   The extensive pattern of superficial nerve fiber distribution for a single cutaneous nerve is called a dermatome.

28. C   The regeneration of a peripheral nerve is possible if its covering sheath, the neurilemma formed by the Schwann cells, remains viable. The nerve cells of the brain and spinal cord lacking this membrane are incapable of repair.

29. A   Although both sides of brain function independently, they must communicate and share information through a pathway of fibers called the corpus callosum.

30. 1   31. 4   32. 3   33. 5   34. 2

35. B   Most of the structures of the face and neck area are enervated by branches from the 12 pairs of cranial nerves. The 12 nerves and their general functions are summarized below:

    I.   Olfactory - sense of smell

    II.   Optic - vision pathway

    III.   Oculomotor - movements of the eye and eyelid

    IV.   Trochlear - movement of the eye

    V.   Trigeminal - jaw movement and facial sensations

    VI.   Abducens - eye movement

    VII.   Facial - muscles of the face

    VIII.   Vestibulocochlear - hearing and balance

    IX.   Glossopharyngeal - muscles of swallowing

    X.   Vagus - muscles of digestion

    XI.   Accessory - muscles of neck and shoulder

    XII.   Hypoglossal - muscles of the tongue

1. The movement of blood to the lungs and other tissues of the body is accomplished by a specialized, muscular organ called the heart. The normal human heart consists of:

   A. Two chambers
   B. Four chambers

   C. Six chambers
   D. Eight chambers

2. The movement of blood away from the heart occurs in thick-walled elastic muscular tubes called:

   A. Arteries
   B. Veins

   C. Capillaries
   D. Lymphatic vessels

3. The right and left sides of the heart are separated by a fibrous membrane called the:

   A. Chordae tendineae
   B. Bundle of His

   C. Septum
   C. Mitral valve

4. Blood is returned to the heart by way of two major trunk veins that empty into the:

   A. Right atrium
   B. Right ventricle

   C. Left atrium
   D. Left ventricle

5. The majority of the heart is composed of a specialized type of cardiac muscle called the:

   A. Pericardium
   B. Endocardium

   C. Epicardium
   D. Myocardium

6. The prevention of the back flow of blood as it passes through the heart is the principal function of the:

   A. Septum
   B. Valves

   C. Trunk vessels
   D. Coronary sinus

7. The heart is covered by a serous, double-walled fibrous membrane called the:

   A. Pericardium
   B. Papillary muscle

   C. Purkinje fibers
   D. Sino-atrial nodes

8. The largest, most muscular chamber of the heart concerned with the movement of blood to the tissues of the body is the:

   A. Right atrium
   B. Right ventricle

   C. Left atrium
   D. Left ventricle

9. The right side of the heart is primarily concerned with the passage of deoxygenated blood into the tissues of the:

   A. Liver
   B. Kidney

   C. Lungs
   D. Intestines

10. The muscle of the heart derives its blood supply by way of the:

    A. Pulmonary vessels
    B. Coronary vessels

    C. Pericardial vessels
    D. Atriventricular vessels

Referring to the diagram of the heart, place the number that corresponds to the following structures for questions 11-17.

11. _____ Tricuspid valve
12. _____ Superior vena cava
13. _____ Left atrium
14. _____ Septum
15. _____ Right ventricle
16. _____ Right atrium
17. _____ Left ventricle

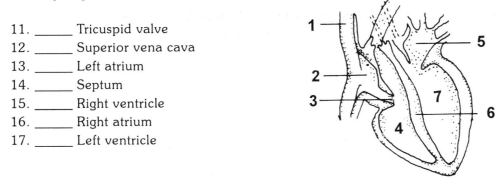

18. The cardiac conduction system that both initiates and distributes impulses to the heart muscles consists of:

   1. Sino-atrial node   2. Atrioventricular node   3. Purkinje fibers

   A. 1 & 2 only
   B. 1 & 3 only
   C. 2 & 3 only
   D. 1, 2 & 3

19. The exchange of oxygen, carbon dioxide and other nutrients and wastes to the tissues of the body occurs across the walls of the:

   A. Arteries
   B. Veins
   C. Lymphatic vessels
   D. Capillaries

20. The highest blood pressure measurements are most often recorded in the:

   1. Arteries   2. Veins   3. Capillaries

   A. 1 only
   B. 2 only
   C. 3 only
   D. 1, 2 & 3

21. The celiac axis is the first branch of the abdominal aorta which serves to supply arterial blood to the:

   1. Spleen   2. Liver   3. Pancreas

   A. 1 & 2 only
   B. 1 & 3 only
   C. 2 & 3 only
   D. 1, 2 & 3

22. The external division of the common carotid artery supplies many of the structures located in the:

   A. Anterior cranium
   B. Posterior cranium
   C. Face and neck
   D. Shoulder and clavicle

23. The diaphragm is supplied with arterial blood by branches of both the thoracic and abdominal aorta called the inferior and superior:

   A. Phrenic arteries
   B. Intercostal arteries
   C. Apical arteries
   D. Coronary arteries

24. The femoral artery is the continuation of the _____ below the inguinal ring.

   A. Popliteal artery
   B. Subclavian artery
   C. Internal iliac artery
   D. External iliac artery

25. Which is NOT a paired branch of the abdominal aorta?

   A. Renal arteries
   B. Mesenteric arteries
   C. Spermatic arteries
   D. Lumbar arteries

26. The arterial blood supply to the brain is accomplished by the:

   1. Internal carotid arteries   2. Vertebral arteries   3. Jugular arteries

   A. 1 & 2 only   C. 2 & 3 only
   B. 1 & 3 only   D. 1, 2 & 3

27. The first major division of the aorta is termed the:

   A. Ascending aorta   C. Descending aorta
   B. Aortic arch   D. Pulmonary aorta

28. The vessels principally involved with the circulation of blood through the lungs are the:

   A. Cardiac vessels   C. Pulmonary vessels
   B. Portal vessels   D. Azygos vessels

Referring to the diagram of the aortic arch, place the number that corresponds to the following arteries for questions 29-34.

29. _____ Right vertebral
30. _____ Left common carotid
31. _____ Left subclavian
32. _____ Innominate
33. _____ Right common carotid
34. _____ Right subclavian

35. The bifurcation of the abdominal aorta normally occurs at the level of the:

   A. 10th thoracic vertebra   C. 4th lumbar vertebra
   B. 1st lumbar vertebra   D. 2nd sacral vertebra

36. The largest portion of the intestinal tract is supplied with arterial blood by way of the:

   A. Celiac axis   C. Inferior mesenteric artery
   B. Superior mesenteric artery   D. Hepatic artery

37. The popliteal artery is named for its location in the posterior fossa of the:

   A. Ankle   C. Armpit
   B. Elbow   D. Knee

38. The ovarian arteries are a branch-off of the:

   A. Internal iliac artery   C. Abdominal aorta
   B. External iliac artery   D. Obturator artery

39. A specialized arterial structure in the brain that helps in the prevention of irreparable damage in cases of arterial blockage is called the:

   A. Circle of Willis   C. Plantar arch
   B. Loop of Henle   D. Cerebral sinuses

40. The urinary bladder, reproductive organs and gluteal muscles derive the majority of blood from the:

   A. External iliac arteries   C. Mesenteric arteries
   B. Internal iliac arteries   D. Femoral arteries

41. From the aortic arch to the hand, describe the correct sequence of arterial branches.

    A. Axillary, subclavian, brachial, radial   C. Brachial, subclavian, axillary, radial
    B. Subclavian, brachial, axillary, ulnar   D. Subclavian, axillary, brachial, ulnar

42. The bifurcation of the common carotid arteries occurs at the level of the:

    A. Manubrial notch   C. Thyroid cartilage
    B. External auditory meatus   D. Inion

43. The largest artery in the normal human body is the:

    A. Aorta   C. Portal artery
    B. Vena cava   D. Pulmonary artery

44. Arterial blood is delivered to the liver primarily by the:

    A. Hepatic artery   C. Basilic artery
    B. Portal artery   D. Renal artery

45. The longest blood vessel of the body is the:

    A. Femoral artery   C. Aorta
    B. Saphenous vein   D. Vena cava

46. The jugular veins are primarily concerned with the drainage of blood from the:

    A. Feet   C. Lungs
    B. Arms   D. Brain

47. The most important deep vein of the upper arm is the:

    A. Cephalic   C. Antebrachial
    B. Brachial   D. Basilar

Referring to the diagram of the arteries of the lower abdomen and pelvis, place the number that corresponds to the following vessels for questions 48-52.

48. _____ Left common iliac artery
49. _____ Right external iliac artery
50. _____ Abdominal aorta
51. _____ Left internal iliac artery
52. _____ Right femoral artery

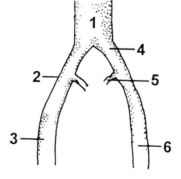

53. The largest veins of the normal human that are primarily concerned with the return of blood to the heart are the superior and inferior:

    A. Portal veins   B. Vena cava
    B. Azygos veins   D. Saphenous veins

54. The majority of the blood from the small intestines will be drained and returned to the portal circulation by the:

    A. Celiac vein   C. Brachiocephalic veins
    B. Axillary vein   D. Superior mesenteric vein

55. The individual layer of the walls of the arteries will receive their own nutrients by way of vessels termed:

A. Sinusoids
B. Varicosities

C. Vaso vasorum
D. Capillaries

56. Which characteristics are generally associated with the venous circulation?

1. Low blood pressure   2. Low oxygen content blood   3. High waste content blood

A. 1 & 2 only
B. 1 & 3 only

C. 2 & 3 only
D. 1, 2 & 3

57. In the fetus, shunting of blood between the two atria is accomplished by an opening called the:

A. Ductus arteriosus
B. Ductus venosus

C. Foramen lacerum
D. Foramen ovale

58. The umbilical arteries of the fetus which return deoxygenated blood to the mother arise from the:

A. Abdominal aorta
B. Pulmonary arteries

C. Celiac axis
D. Internal iliac arteries

59. The highly vascular capillary membrane that permits the exchange of nutrients and oxygen between the fetus and mother is termed the:

A. Placenta
B. Lingual

C. Ovary
D. Macula

60. The white blood cells of the lymphatic system enter the systemic circulation at the:

A. Cisterna chyli
B. Subclavian veins

C. Cisterna magna
D. Splenic artery

61. In many large veins, back flow of blood is prevented by the presence of:

A. Sphincter muscles
B. Sinusoids

C. Venules
D. Valves

62. Because of its high oxygen content, arterial blood is characterized by its _____ appearance.

A. Bright red
B. Dark red

C. Straw-colored
D. Blue-colored

63. The fluid medium in which the blood cells circulate is called:

A. Serum
B. Plasma

C. Hemocrit
D. Lymph

64. The most common type of blood cell has no nucleus and has a characteristic biconcaval shape. This is called a/an:

A. Leukocyte
B. Lymphocyte

C. Erythrocyte
D. Thrombocyte

65. The major function of platelets (thrombocytes) is:

A. Destruction of bacteria
B. Transport of oxygen

C. Production of antibodies
D. Clotting of blood

66. A condition in which there is a pronounced deficiency of red blood cells is called:

A. Cyanosis
B. Hemophilia

C. Anemia
D. Leukemia

67. Insufficiency or regurgitation of the heart valves, or septal defects, are often diagnosed by an abnormal sound called a:

A. Branch block
B. Heart murmur
C. Fibrillation
D. Hypertrophy

68. A partial loss of blood flow to the heart muscle with subsequent tissue damage is termed:

A. Mitral stenosis
B. Pericarditis
C. Myocardial infarction
D. Pericardial effusion

69. A localized dilation of the walls of an artery, due to a pre-existing weakening, leads to a defect called a/an:

A. Aneurysm
B. Anastomosis
C. Atherosclerosis
D. Phlebitis

70. Varicose veins that sometimes occur in the rectum are termed:

A. Emboli
B. Thrombi
C. Hemorrhoids
D. Infarcts

Explanations

1. B The interior of the heart contains four distinct chambers, separated into two sides by the septum, and into upper (atria) and lower ventricles by the atrioventricular valves.

2. A Blood is contained within 3 types of blood vessels: the arteries, capillaries, and veins. The arteries, which transport blood away from the heart, possess thick, elastic walls which aid in the maintenance of the blood flow.

3. C Between the two sides of the heart is a fibrous partition called the interatrial and interventricular septum.

4. A Blood for the general circulation is returned to the right atria by way of 2 large veins, the superior and inferior vena cava.

5. D The 3 types of muscle tissue differ in both their appearance and function. Though smooth muscle can provide only weak contractions, they can be sustained for indefinite periods of time. Skeletal muscles are extremely strong but tire easily and cannot provide the repetitive contraction required of the heart. The cardiac muscles possess moderate strength and short recovery time required for repetitive function.

6. B Blood which normally travels through the heart in a single direction is assisted by the two atrioventricular valves between the upper and lower chambers and the semilunar valves which sit on top of each ventricle. All of these one-way valves help to prevent the backward flow of blood through the heart.

7. A The heart is encapsulated by a layer of fibrous pericardium which is connected to the mediastinum and helps prevent over distension. Two additional tissue layers called the serous pericardium consisting of visceral and parietal layers and the pericardial space found between the layers reduce the friction associated with the beating of the heart.

8. D The ventricles are the pumping chambers of the heart. The smaller right ventricle pumps blood into the lungs while the stronger left ventricle provides the force required to distribute the blood to all of the other tissues of the body.

9. C The poorly oxygenated blood from the vena cava is emptied into the right aorta. As this chamber contracts, blood flows into the right ventricle for its transit into the pulmonary arteries and lungs.

10. B The myocardium or heart muscle is supplied with oxygen-rich blood by way of the first branch of the ascending aorta, the coronary artery.

11. 3   12. 1   13. 5   14. 6   15. 4   16. 2

17. 7

18. D The efficient contraction of the chambers of the heart requires a precise timing and the efficient spread of the electrical impulses. Specialized muscle tissue found in the sinoatrial (SA) nodes, atrioventricular (AV) nodes, and bundle of His, generate the electrical signals that pass through the Purkinje fibers to the desired area of the heart muscle.

19. D The thin single-cell thick walls of the capillaries are ideally suited for the exchange of respiratory gasses and nutrients between the blood cells and the individual cells of the body.

20. A The proximity of arteries to the heart will result in high blood pressure measurements not apparent in the capillaries or veins.

21. D After the abdominal aorta descends through the diaphragm its first main branch is the celiac axis which provides oxygen-rich blood to the liver (hepatic artery), spleen (splenic artery), and the stomach, gall bladder, and pancreas (gastric artery).

22. C   The external and internal divisions of the common carotid artery divide at the level of the angle of the mandible (gonion). The external branches distribute oxygenated blood to the tongue, face, pharynx, scalp, and thyroid gland.

23. A   Due to its importance in respiration, the diaphragm derives its blood supply from phrenic arteries originating from both the thoracic and abdominal aorta.

24. D   After the external iliac artery descends through the pelvic cavity, at the inguinal ring, it is renamed the femoral artery.

25. B   The five paired branches of the abdominal aorta are the renal, suprarenal, testicular or ovarian, lumbar, and inferior phrenic arteries.

26. A   The four main vessels which supply oxygenated blood to the brain are the right and left internal carotids and the right and left vertebral arteries.

27. A   The major divisions of the aorta are the ascending, arch, thoracic, and abdominal. The ascending aorta comes from the left ventricle and ends in the curved aortic arch. As it descends through the chest cavity it becomes the thoracic aorta, and it is renamed the abdominal aorta below the diaphragm.

28. C   The pulmonary trunk which arises from the right ventricle bifurcates into the right and left pulmonary arteries which penetrate the lung at its hilus. The pulmonary veins return blood to the left atrium of the heart.

29. 1   30. 5   31. 7   32. 3   33. 4   34. 2

35. C   The abdominal aorta passes along the lumbar spine to an area just in front of the 4th lumbar vertebrae where it bifurcates into the right and left common iliac arteries.

36. B   The second major branch-off of the abdominal aorta, the superior mesenteric artery, supplies oxygenated blood to the pancreas, duodenum, jejunum, ileum, ascending, and transverse colon.

37. D   The femoral artery descends down the medial and posterior side of the femur to the back of the knee where it is renamed the popliteal artery.

38. C   Because both the ovaries and testes develop near the kidney, the branches of the abdominal aorta which supply these structures will also follow their descent into the pelvic cavity.

39. A   The brain derives its blood from an anastomosis near the base of the skull called the circle of Willis. This structure is formed by right and left internal carotids and basilar artery which give rise to the anterior and posterior cerebral and communicating arteries.

40. B   The majority of structures and muscle groups in the mid and lower pelvis receive much of their blood supply by way of the internal iliac arteries.

41. D   Blood travels to the hand in the following sequence of arteries: subclavian, axillary, brachial, ulnar, and radial, palmer arch.

42. C   The bifurcation of the common carotid artery into its internal and external branches occurs at the same level as the thyroid cartilage of the larynx.

43. A   The large vessel originating at the left ventricle of the heart is the ascending aorta.

44. A   The majority of the arterial blood entering the liver is supplied by the common hepatic artery, one of the main branches off of the celiac axis.

45. B   The longest blood vessel of the body which begins in the foot and extends into the lower pelvis is called the great saphenous vein.

46. D   The internal jugular vein drains venous blood from all of the cranial venous sinuses; namely, the superior and inferior sagittal, the straight, transverse, and sigmoid sinuses.

47. B   The largest deep vein of the upper arm is the brachial vein. The 2 other major veins of the arm, the cephalic and basilic veins, are classified as superficial.

48. 4   49. 2   50. 1   51. 5   52. 3

53. C   The largest veins of the body, the vena cava, collect blood from the upper and lower portions of the body and return it to the right atria.

54. D   To prevent high levels of sugars or proteins from gaining access to the general circulation, venous blood from the intestines (mesenteric vein), drain directly into the portal circulation associated with the liver.

55. C   The blood vessels which provide oxygen and nutrients to their own walls are called the vaso vasorum.

56. D   Venous blood which is being returned to the heart is normally associated with low blood pressure and low oxygen content and a high level of metabolic waste products.

57. D   Because the fetal lungs are not functional until birth, blood is directed away from the lungs through an opening between the atria called the foramen ovale.

58. D   The umbilical arteries located in the umbilical cord which direct fetal blood to the placenta arise from the internal iliac arteries of the pelvis.

59. A   The exchange of gasses, nutrients and wastes between the fetal and maternal circulatory system occur across the membranes of the placenta.

60. B   The low blood pressure in the subclavian veins provides a safe entry point for the thoracic ducts of the lymphatic system.

61. D   The majority of the veins in the lower and upper limbs contain one-way valvular structures to help ensure that blood continues to flow back towards the heart.

62. A   Venous and arterial blood differs in both content and appearance. The presence of oxygen in arterial blood gives this fluid a bright red tint. The lower oxygen content of venous blood is characterized by a deep red color.

63. B   The intercellular matrix for the blood cells is a straw-colored fluid containing many of the blood proteins called plasma.

64. C   The most numerous type of blood cells with a count of between 3-5 million per cc are the erythrocytes or red blood cells.

65. D   The platelets or thrombocytes play an essential role in repairing damaged vessels and preventing excessive blood loss through clotting and have a normal count of 300,000 per/cc.

66. C   The reduction in the number of erythrocytes or hemoglobin may lead to a condition called anemia. Among the more common causes are poor nutrition, low vitamin B12 levels, bleeding disorders, and deficient bone marrow production.

67. B   Many types of heart problems associated with the heart valves or defects in the septum, create an abnormal sound called a heart murmur.

68. C   One of the most common causes of a heart attack is a blocked coronary artery which reduces the blood supply to the heart muscle called a myocardial infarction.

69. A   The abnormal dilatation of an artery due to a weakening in the wall of the vessel is termed an aneurysm.

70. C   Tortuous veins resulting from excessive pressure, which are commonly found in the legs, are called varicose veins. Related conditions occurring in the gastrointestinal tract are esophageal varices and hemorrhoids or piles of the rectum and anus.

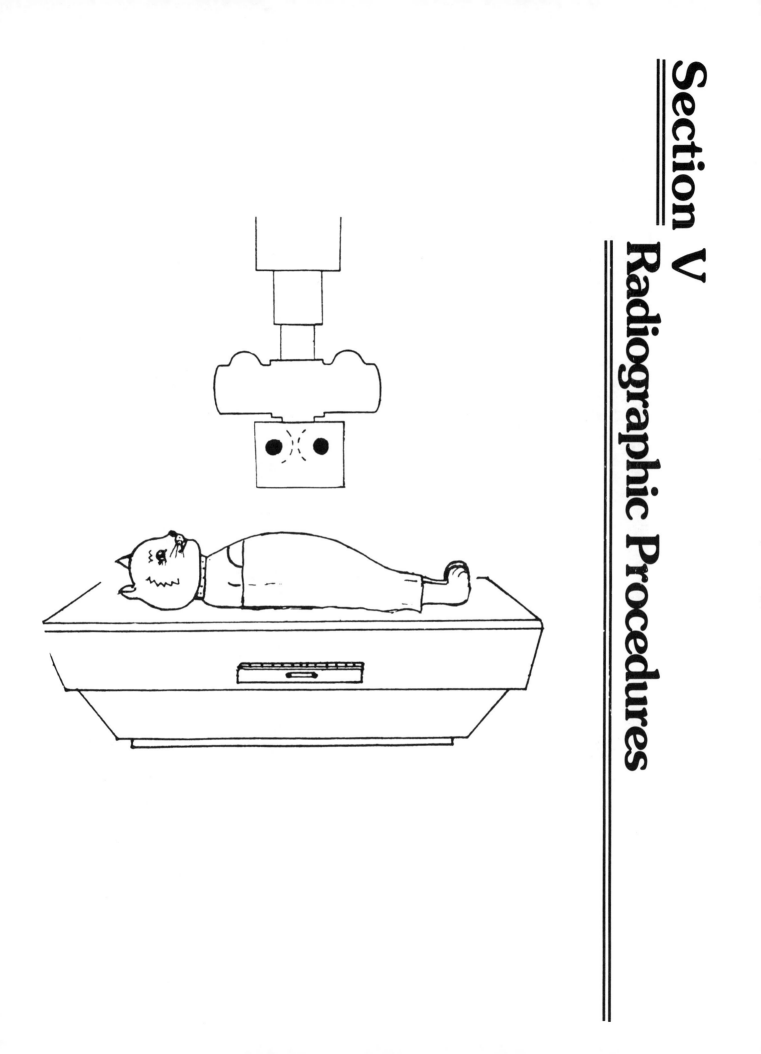

Radiographic Graffiti IV

Can you identify what caused these common mistakes.   Answers appear at the bottom of the page.

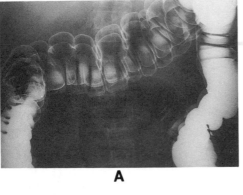

A

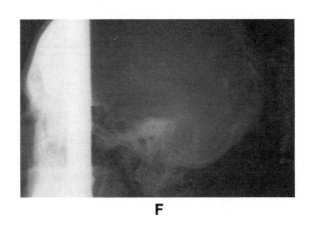

B

C

D

E

F

1. A plane that divides the body into upper and lower proportions is called:

   A. Sagittal plane
   B. Decubitus plane
   C. Coronal plane
   D. Transverse plane

2. The most common body habitus consisting of nearly 50% of the population is called the:

   A. Hyposthenic type
   B. Sthenic type
   C. Hypersthenic type
   D. Asthenic type

3. Which of the following terms refer to a body part or a central ray angulation directed toward the head?

   1. Cranial   **_2. Cephalic_**   **_3. Caudad_**

   A. 1 & 2 only
   B. 1 & 3 only
   C. 2 & 3 only
   D. 1, 2 & 3

4. The body position defined as a person standing erect with palms facing forward is termed:

   A. Fetal position
   B. Recumbent position
   C. Habital position
   D. Anatomical position

5. A longitudinal angulation of the central ray with the long axis of the body describes a/an:

   A. Axial projection
   B. Transverse projection
   C. Tangential projection
   D. Horizontal projection

6. The ventral decubitus position is most similar to the _____ position.

   A. Prone
   B. Supine
   C. Oblique
   D. Lateral

7. The general size, shape, and position of the internal organs can be, to a large degree, predicted by the classifications of a person's:

   A. Addison's types
   B. Bodily habitus
   C. Body cavities
   D. Primary lobules

8. A vertical plane that divides the body into anterior and posterior is called a:

   A. Transverse plane
   B. Sagittal plane
   C. Coronal plane
   D. Squamosal

9. The term used to designate a structure away from its source is:

   A. Visceral
   B. Medial
   C. Proximal
   D. Distal

10. The turning point of a body part inward is termed:

    A. Inversion
    B. Eversion
    C. Extension
    D. Flexion

11. Which of the following terms refers to the front surface of the body?

    1. Anterior   **_2. Posterior_**   **_3. Ventral_**

    A. 1 only
    B. 3 only
    C. 1 & 2 only
    D. 1 & 3 only

12. A movement of a joint that increases the angle between the articulating bones is called:

    A. Adduction
    B. Abduction
    C. Flexion
    D. Extension

13. Which of the following organs is NOT located in the pelvic cavity?

A. Spleen
B. Rectum

C. Vagina
D. Urinary bladder

14. A projection in which visualization of a structure is improved by directing the central ray to skim the profiled body part is termed a/an:

A. Tangential projection
B. Axial projection

C. Decubitus projection
D. Oblique projection

15. The gonion or angle of the mandible is found at the level of:

A. First cervical vertebra
B. Third cervical vertebra

C. Fifth cervical vertebra
D. Seventh cervical vertebra

16. The portions of the body that are located away from the midline of the body are referred to as:

A. Caudal
B. Lateral

C. Cranial
D. Ventral

17. A very thin person will often have a low, vertical stomach and gallbladder. This person would be classified as having the _____ body type.

A. Sthenic
B. Asthenic

C. Hypersthenic
D. Optesthetic

18. A physician has requested a tightly collimated image of the tenth thoracic vertebra. The external landmark that would be most helpful is the:

A. Costal margin
B. Xiphoid tip

C. Iliac crest
D. Sternal angle

19. During a ventral decubitus examination of the abdomen using a horizontal beam, free fluid will accumulate on the patient's:

A. Right side
B. Left side

C. Anterior surface
D. Posterior surface

20. The term 'projection' describes:

A. The image as seen from the film through the patient
B. The path of the radiation from the x-ray tube
C. Both of the above
D. Neither of the above

21. The movement of a structure toward the central axis of the body is termed:

A. Abduction
B. Adduction

C. Pronation
D. Supination

22. A right lateral decubitus radiograph is taken of a patient. This is an example of a/an:

A. Lateral projection
B. Oblique projection

C. Frontal projection
D. Axial projection

23. The dorsum of the foot refers to the:

A. Superior aspect
B. Inferior aspect

C. Medial aspect
D. Lateral aspect

24. The most superior aspect of the iliac crest corresponds to the level of:

A. T-12
B. L-2

C. L-4
D. Sacral promontory

25. When the hand is rotated from an anatomic position for a posteroanterior projection, this is termed:

A. Extension   C. Pronation
B. Flexion   D. Supination

26. A person with a large, high peripherally located colon, and high transverse gallbladder and stomach would be classified as:

A. Hypersthenic   C. Asthenic
B. Hyposthenic   D. Sthenic

27. The right anterior oblique position (RAO) is described as a/an:

A. Anteroposterior (AP) oblique position   C. Decubitus position
B. Lateral position   D. Posteroanterior (PA) oblique position

28. The body plane that passes vertically through the midline of the body and divides it into equal right and left portions is termed the:

A. Mid-coronal plane   C. Mid-transverse plane
B. Mid-sagittal plane   D. Mid-lambdoidal plane

29. The turning of the foot outward at the ankle would be an example of:

A. Pronation   C. Eversion
B. Extension   D. Inversion

30. Which of the following structures are found in the same transverse plane?

1. Ischial tuberosity   *2. Symphysis*   *3. Greater trochanters*

A. 1 & 2 only   C. 2 & 3 only
B. 1 & 3 only   D. 1, 2 & 3

31. The anterior superior iliac spine (ASIS) is found at the same level as:

A. L-4   C. S-1
B. L-5   D. S-3

32. The division of the abdomen into four major quadrants involves a transverse and midsagittal plane which intersects at the:

A. Umbilicus   C. Xiphoid tip
B. Sternal notch   D. Inguinal fold

33. Which of the following can be used to describe a patient lying on his or her back?

1. Supine   *2. Dorsal decubitus*   *3. Ventral decubitus*

A. 1 & 2 only   C. 2 & 3 only
B. 1 & 3 only   D. 1, 2 & 3

34. The torso is divided into two major subdivisions called the:

A. Epi and hypogastric regions   C. Thoracic and abdominal cavities
B. Hypochondrial and iliac regions   D. Pleural and pelvic cavities

35. In the posteroanterior (PA) projection of the abdomen, the gallbladder of a sthenic patient is normally located at a level near the:

A. Sacral promontory   C. Ninth costal cartilage
B. Dome of the diaphragm   D. Iliac crest

36. The term applied to a position in which the body is rotated so that it is NOT in a frontal or lateral position is called a/an:

A. Axial position
B. Tangential position

C. Decubitus position
D. Oblique position

37. A club-shaped process usually located at the end of a bone is termed a:

A. Fissure
B. Facet

C. Malleolus
D. Sulcus

38. Which of the following is NOT located within the pelvic cavity?

A. Bladder
B. Duodenum

C. Rectum
D. Reproductive organs

39. The right PA oblique position will correspond to a patient in the:

A. Right anterior oblique position (RAO)
B. Left anterior oblique position (LAO)

C. Right posterior oblique position (RPO)
D. Left posterior oblique position (LPO)

40. A hole in a bone through which vessels and nerves pass is called a:

A. Fovea
B. Fossa

C. Fistula
D. Foramen

Explanations

1. D   A transverse plane also known as the axial plane passes crosswise through the body at right angles to both the midsagittal and midcoronal planes. This will divide the body into superior and inferior sections.

2. B   Bodily habitus refers to the general form of the body and can be used to help predict the size, shape and position of the internal organs. individuals are classified in one of four types: (1) Sthenic type: the most common or average type consisting of 50% of the population; (2) Hypersthenic type: (5%) A massive build characterized by a high laterally placed gallbladder, high transverse stomach and large laterally positioned colon; (3) Hyposthenic type: (35%) Smaller than average individual found to have long vertical stomach, low vertical gallbladder and more centralized smaller colon; (4) Asthenic type (10%): An extremely slender person characterized by organs which are smaller and lower than in a hyposthenic individual.

3. A   Cranial and cephalic refer to skull, and parts toward the head end of the body. Caudad or caudal refer to parts away from the head, or toward the feet. Dorsal refers to the posterior or back side of the part; i.e., the dorsum of the hand is the back of the hand. Ventral refers to the anterior or front surface of the body.

4. D   In order to provide a single reference for describing a patient position, the concept of anatomic position was developed. This refers to a person facing forward, standing erect, feet together with arms at the sides, palms facing forward. The radiograph placed on a view box or illuminated viewer should follow this convention.

5. A   The term axial or semi-axial normally refers to a central ray angulation following the long axis of the body or part; i.e., AP axial (Towne) skull. The definition also includes extension and flexion movements on longitudinal planes which can also provide axial projections. In a tangential projection, the central ray skims a rounded body segment in an attempt to profile and eliminate superimposition with adjacent structures; i.e., Settegast (Sunrise) method for the patella.

6. A   When a patient is placed recumbent or lying down and the central ray is directed horizontal to the floor, a decubitus position has been obtained. These positions are especially useful in the diagnosis of air-fluid levels especially in the chest, abdomen and skull. A lateral decubitus position is defined as a patient lying on the affected side, with the anterior or posterior body surface closest to the image receptor. Because of the horizontal central ray, a frontal (PA/AP) image will be demonstrated. In a ventral decubitus position, the patient is lying on the ventral (anterior) surface of the body as in the prone position. In a dorsal decubitus position, the patient is lying on the dorsal (posterior) surface of the body facing-up or supine. The resulting radiograph for the dorsal and ventral decubitus positions provide a lateral image of the anatomic region.

7. B   Bodily habitus refers to the general form of the body and can be used to help predict the size, shape and position of the internal organs. Individuals are classified in one of four types: (1) Sthenic type: the most common or average type consisting of 50% of the population; (2) Hypersthenic type: (5%) A massive build characterized by a high laterally placed gallbladder, high transverse stomach and large laterally positioned colon; (3) Hyposthenic type: (35%) Smaller than average individual found to have long vertical stomach, low vertical gallbladder and more centralized smaller colon; (4) Asthenic type (10%): An extremely slender person characterized by organs which are smaller and lower than in a hyposthenic individual.

8. C   The midcoronal, or midfrontal plane is named for the coronal suture, and passes vertically through the midaxillary region of the body, at right angles to the midsagittal plane. It divides the body into anterior (ventral) and posterior (dorsal) portions. Any vertical plane passing parallel to the midcoronal plane is called a coronal plane.

9. D   The term distal designates a structure farthest from the center or midline of the body. Proximal refers to a structure or part closest to the center or midline of the body; i.e., the foot is distal to the ankle; the radial head is located at the proximal end of the radius.

10. A   Inversion refers to turning of a body part inward. Eversion refers to turning of a body part outward; i.e., movement of the foot by turning it outward at the ankle joint.

11. D   Anterior and ventral are both terms that describe the front surface of the body. Posterior and dorsal describe the back surface of the body.

12. D   Extension, the opposite of flexion, increases the angle of the articulating bones by straightening the joint and stretching the anatomic structure. Flexion is a bending movement where the angle of the bones of the joint is diminished. In the vertebral column a backward bending movement is called extension, as opposed to forward bending movement known as flexion.

13. A   The torso of the body is divided into two main sections, called the thoracic and abdominal cavities, which are separated by a dome shaped muscle called the diaphragm. The respiratory and cardiovascular structures associated with the heart occupy the thoracic cavity while most of the major organs of digestion and the urinary tract occupy the abdominal cavity. The pelvis is a subdivision of the lower portion of the abdominal cavity and contains the urinary bladder, the rectum and female reproductive organs.

14. A   In a tangential projection, the central ray skims a rounded body segment in an attempt to profile and eliminate superimposition with adjacent structures; i.e., Settegast (Sunrise) method for the patella.

15. B   Since most anatomical structures cannot be directly visualized, external indicators must be used to accurately position the patient. The gonion (angle of the mandible) is an external landmark that is located at the level of the 2nd and 3rd cervical vertebrae.

16. B   Portions of the body located away from the midline or middle of an anatomic structure are referred to as lateral. The opposite, medial, refers to anatomic structures located toward the middle or median plane of the body; i.e., the pisiform is located on the medial side of the wrist.

17. B   Asthenic type (10%): An extremely slender person characterized by organs which are smaller and lower than in a hyposthenic individual.

18. B   The tenth thoracic vertebra is located in the same axial plane as the xiphoid tip on the inferior aspect of the sternum.

19. C   The ventral decubitus position places the patient prone. Due to gravity, free fluid will drain toward the patient's anterior surface. The lighter free air will rise and accumulate on the patient's posterior surface.

20. B   Projection is a term applied to recording of an image resulting from the passage of the x rays through the patient. It is defined by the surfaces passed by the beam; for example, an AP projection indicates that the beam enters on the anterior surface and exits on the posterior surface of the patient.

21. B   Adduction refers to the movement of an anatomic structure toward the central axis of the body. Abduction refers to the movement of an anatomic structure away from the central axis of the body. Example: Moving the extended arm from the anatomical position "down to the side" is called adduction, lifting the arm away from the body is called abduction.

22. C   When a patient is placed recumbent or lying down and the central ray is directed horizontal to the floor, a decubitus position has been obtained. These positions are especially useful in the diagnosis of air-fluid levels especially in the chest, abdomen and skull. A lateral decubitus position is defined as a patient lying on the affected side, with the anterior or posterior body surface closest to the image receptor.

23. A   Dorsum usually refers to the posterior aspect or the back of a part or organ. When referring to the foot, the dorsum is the superior aspect, while the inferior surface is called the plantar aspect.

24. C   The most superior aspect of the palpable iliac crest is at the same level as the body of the fourth lumbar vertebra. This external landmark is widely used when radiographing the abdomen, spine, kidneys (IVU), colon (BE), stomach (GI), and pelvis.

25. C   Pronation refers to the rotation of the hand so that the palm faces the back of the body; supination, the opposite of pronation, refers to the rotation of the hand into true anatomical position.

26. A   Hypersthenic type: (5%) A massive build characterized by a high laterally placed gallbladder, high transverse stomach and large laterally positioned colon.

27. D   The term position most often describes position of the body or anatomic structure as seen from the imaging system when entry and exit points are not specified. In an oblique position, the patient is rotated between the frontal and lateral position; producing an oblique image of the anatomic structure. In the right anterior oblique (RAO) position, the patient's right anterior surface is closest to the imaging system, and is also described as a right PA oblique. By convention, it is always the side of the body closest to the film that defines position. A right posterior oblique (RPO) position has the patient with the right posterior surface in contact with the imaging system and is also described as a right AP oblique. Describing oblique positions is sometimes confusing. The majority of obliques describe the body surface that is closest to the image receptor (RAO/right anterior oblique) rather than the projection description (right PA oblique). The abbreviations are being used with the PA/AP oblique description parenthetically listed as follows: RAO/right anterior oblique (right PA oblique); RPO/right posterior oblique (right AP oblique); LAO/left anterior oblique (left PA oblique); and LPO/left posterior oblique (left AP oblique). Example: The patient is rotated 45 degrees from the PA so that the right anterior chest wall is placed closest to the image receptor (RAO). The central ray enters the posterior chest and exits the anterior chest wall with the right side closest to the image receptor (right PA oblique). This will produce an oblique image of the chest cavity.

28. B   The transverse, or axial, plane passes crosswise through the body at right angles to the midsagittal and coronal planes, dividing it into superior and inferior portions.

29. C   Inversion refers to turning of a body part inward. Eversion refers to turning of a body part outward; i.e., movement of the foot by turning it outward at the ankle joint.

30. C   The greater trochanter is located on the proximal lateral aspect of the femur at the level of the symphysis pubis and the coccyx. The ischial tuberosity is located inferior to these structures. The greater trochanter is sometimes easier to palpate than the symphysis and can be used to find the lower portion of the abdomen or the hip joint.

31. C   The anterior superior iliac spine (ASIS) is an external landmark located at the level of the first sacral segment and is commonly used to locate the sacroiliac and hip joint.

32. A   The abdomen is divided into four quadrants by the transverse and midsagittal planes which intersect at the umbilicus. The four quadrants, located between the diaphragm and pelvic inlet, are named as follows: right upper quadrant (RUQ), left upper quadrant (LUQ), right lower quadrant (RLQ), and left lower quadrant (LLQ).

33. A   In a dorsal decubitus position, the patient is lying on the dorsal (posterior) surface of the body facing-up or supine. The resulting radiograph for the dorsal and ventral decubitus positions provide a lateral image of the anatomic region.

34. C   The torso of the body is divided into two main sections, called the thoracic and abdominal cavities, which are separated by a dome-shaped muscle called the diaphragm.

35. C   The gall bladder of a sthenic or "average" patient is normally located at the level of the ninth costal cartilage. Depending on the body habitus, the gallbladder can vary in position by as much as 8 inches both vertically and laterally from the level of the 8th rib down to well within the iliac fossa and anywhere from the midsagittal plane to the lateral wall of the abdomen. For example, for a hypersthenic patient, an initial radiograph may be centered 2" higher than the sthenic; for an asthenic patient, 2" lower. See question number 2.

36. D   In an oblique position, the patient is rotated between the frontal and lateral position, producing an oblique image of the anatomic structure.

37. C   A process is an anatomic term used to describe a projection that extends or juts out from the main body of a structure. Malleolus is an example of a club-shaped process found on the medial and lateral aspect of the distal end of the tibia and fibula (ankle).

38. B   The pelvis is a subdivision of the lower portion of the abdominal cavity and contains the urinary bladder, the rectum and female reproductive organs.

39. A   By convention, it is always the side of the body closest to the film that defines position. A right posterior oblique (RPO) position has the patient with the right posterior surface in contact with the imaging system and is also described as a right AP oblique.

40. D   A foramen is an anatomic term used to describe a hole in a bone for the transmission of blood vessels and nerves. For example, the foramen magnum allows the spinal cord (medulla oblongata) to pass into the cranial cavity; the jugular foramen allows passage of the jugular vein out of the cranium and into the neck.

1. In a radiographic image of the hand in the PA oblique projection, the hand should be placed at an angle of:

   A. 15 degrees to the plane of the image receptor
   B. 25 degrees to the plane of the image receptor
   C. 45 degrees to the plane of the image receptor
   D. 60 degrees to the plane of the image receptor

2. During radiography of the clavicle, which projection is used to reduce the amount of focal spot blur?

   A. A PA projection with the patient erect
   B. A PA oblique projection with the patient supine
   C. A lateral projection with the patient erect
   D. An AP oblique projection with the patient supine

3. Which radiographic projection of the shoulder will demonstrate the humerus in true anatomic position?

   A. The transthoracic lateral projection
   B. The anteroposterior (AP) projection with internal rotation
   C. The anteroposterior (AP) projection with external rotation
   D. The anteroposterior (AP) projection with neutral rotation

4. To insure proper position of the elbow during a radiographic image taken in the lateral projection the:

   1. Epicondyles should be placed perpendicular to the plane of the image receptor
   2. Radial and ulnar styloids should be superimposed
   3. Forearm, elbow, and humerus should be placed in the same plane

   A. 1 & 2 only
   B. 1 & 3 only
   C. 2 & 3 only
   D. 1, 2 & 3

5. In a radiographic image of the carpal bones obtained in the PA projection, the fingers are flexed in order to:

   A. Decrease rotation of the carpal bones
   B. Decrease the part-to-image receptor distance
   C. Open the carpal interspaces
   D. Superimpose the lateral carpal bones

6. When bilateral hands or wrists are requested, it is important to radiograph each side separately to prevent:

   A. Excessive patient exposures
   B. The distortion of the phalanges
   C. The distortion of the joint spaces
   D. The magnification of the wrist bones

7. Radiographic images of the acromioclavicular articulations are normally performed:

   1. In the decubitus position   *2. With the patient holding weights*   *3. Bilaterally*

   A. 1 & 2 only
   B. 1 & 3 only
   C. 2 & 3 only
   D. 1, 2 & 3

8. A radiographic image of the shoulder is obtained in the AP projection with a medial rotation. This will place the humerus in the:

   A. Anatomic position
   B. 25-degree oblique position
   C. Lateral position
   D. 45-degree oblique position

9. A radiographic image of the wrist is obtained in the AP oblique projection with a medial rotation of about 45 degrees. This is normally used to demonstrate:

   A. The trapezium and trapezoid bones
   B. The lunate and pisiform bones
   C. The scaphoid and hamate bones
   D. The ulnar and radial styloid processes

10. To avoid crossing the radius and ulna during the radiographic evaluation of the forearm in the AP projection, the hand should be:

   A. Hyperextended to its full extent
   B. Fully pronated
   C. Fully supinated
   D. Placed in the true lateral

11. Which of the following radiographic projections will best demonstrate the carpal interspaces?

A. The AP projection
B. A 45 degree medial oblique projection

C. The PA projection
D. A 35 degree lateral oblique projection

Referring to the radiograph of the hand, place the letter that corresponds to the following structures for questions 12-20.

12._____ Scaphoid (navicular) bone
13._____ Proximal interphalangeal joint
14._____ Head of the metacarpal bone
15._____ Styloid process of the ulna
16._____ Base of the proximal phalanx
17._____ Metacarpophalangeal joint
18._____ Middle phalanx
19._____ Distal interphalangeal joint
20._____ Head of the proximal phalanx

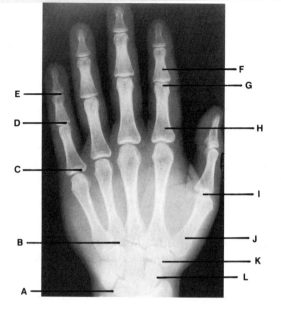

21. A radiographic image of the finger is obtained in the PA projection. The central ray should be directed perpendicular to the:

A. Third metacarpophalangeal joint
B. Middle interphalangeal joint

C. Proximal interphalangeal joint
D. Distal interphalangeal joint

22. In evaluating a radiographic image of the scapula in the true lateral projection, which of the following structures should be superimposed?

A. The vertebral and axillary borders
B. The acromion and coracoid process

C. The scapular body and axillary ribs
D. The humeral head and glenoid fossa

23. When radiographing the proximal humerus using the transthoracic (Lawrence) projection, the plane of the epicondyles should be:

A. Parallel to the plane of the image receptor
B. Perpendicular to the plane of the image receptor

C. 45 degrees to the plane of the image receptor
D. 70 degrees to the plane of the image receptor

24. A bony outgrowth at the base of the third metacarpocarpal joint (carpe bossu) is best demonstrated in the:

A. Lateral projection of the hand in acute dorsiflexion flexion
B. PA projection of the hand in ulnar deviation
C. PA projection of the wrist in acute palmar flexion
D. PA projection of the wrist in radial deviation

25. Which of the following radiographic projections is most commonly used to provide a profile image of the olecranon process?

A. The anteroposterior projection of the elbow
B. An AP oblique projection of the elbow with medial rotation
C. An AP oblique projection of the forearm with lateral rotation
D. A lateral projection of the elbow

26. When performing a radiographic image of the hand in the PA projection, which of the following surfaces should be in contact with the image receptor?

   1. Dorsal surface   **2. Palmar surface**   **3. Ventral surface**

   A. 1 & 2 only   C. 2 & 3 only
   B. 1 & 3 only   D. 1, 2 & 3

27. A radiographic image is taken on a supine patient with their arm abducted to a right angle and the elbow flexed. This best describes:

   A. An AP projection of the scapula   C. A tangential projection of the elbow
   B. An PA oblique projection of the scapula   D. A lateral projection of the elbow

28. During the radiographic evaluation of the hand for the localization of foreign bodies, the hand should be placed in the:

   A. Lateral position with fingers flexed   C. Oblique position with the fingers flexed
   B. Lateral position with the fingers extended D. Oblique position with the fingers extended

29. During a radiographic evaluation of the scapula in the AP projection, respiration should be:

   A. Suspended on full inspiration   C. Suspended at the end of forced inspiration
   B. Suspended on full expiration   D. Continued with quiet breathing

30. In order to obtain a radiographic image of the humerus (non-trauma) in the lateral projection, the arm should be placed in:

   A. Neutral rotation   C. Internal rotation
   B. Counter rotation   D. External rotation

31. A radiographic image is obtained with the palmar surface of the wrist in contact with the image receptor, and dorsiflexed with a slight radial shift. If the central ray is directed 25 degrees to the long axis of the hand to a point 2 centimeters distal to the base of the third metacarpal bone, what structures are clearly demonstrated?

   A. The anterior wrist bones and the carpal canal   C. The posterior wrist bones and the carpal bridge
   B. A defect known as carpe bossu   D. The scaphoid bone without superimposition

In the radiographic images of the elbow, place the letter that corresponds to the following structures for questions 32-37.

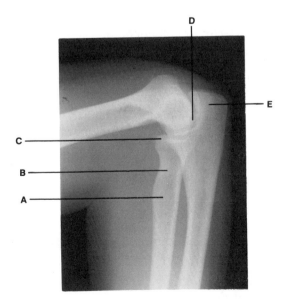
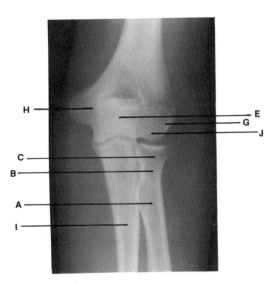

32._____ Radial head   35._____ Capitulum

33._____ Olecranon process   36._____ Semilunar notch

34._____ Medial epicondyle   37._____ Ulna

38. A radiographic image of the elbow is obtained in the lateral projection. In order to improve the visualization of the soft tissue structures, the elbow should be flexed to an angle of about:

   A. 45 degrees   C. 60 degrees
   B. 90 degrees   D. 35 degrees

39. In order to obtain a radiographic image of the first finger in the AP projection, the hand should be placed in the:

   A. Anteroposterior position   C. 45-degree oblique position
   B. Posteroanterior oblique position   D. Lateromedial position

40. A radiographic image of the scapula is obtained in the lateral projection. The upper portion of the scapula is demonstrated as the letter Y. The superior aspect of the letter Y is formed by:

   A. The glenoid and humeral heads   C. The acromion and coracoid processes
   B. The scapular spine and vertebral border   D. The vertebral and axillary borders

41. A radiographic image of the humerus is obtained in the anteroposterior (AP) projection. This will require that a plane passing between the epicondyles be placed parallel to the plane of the image receptor in order to obtain a profile view of the:

   A. Acromion   C. Glenohumeral joint
   B. Lesser tuberosity   D. Greater tuberosity

42. A radiographic image which best demonstrates the carpal interspaces on the medial side of the wrist should be taken in the:

   A. PA projection with ulnar deviation   C. Lateral projection with radial deviation
   B. AP projection with radial deviation   D. PA projection with radial deviation

43. A radiographic image of the elbow is obtained in the AP oblique projection for the demonstration of the radial head. This will require that the elbow be:

   A. Medially rotated 15 degrees   C. Laterally rotated 45 degrees
   B. Medially rotated 35 degrees   D. Laterally rotated 15 degrees

44. A radiographic image of the hand is obtained in the PA projection. The central ray should be directed perpendicular to the:

   A. Third metacarpocarpal joint   C. Third metacarpophalangeal joint
   B. Third proximal interphalangeal joint   D. Midcarpal region

45. In order to obtain a radiographic image of the clavicle in the PA axial projection, the central ray should be directed at a:

   A. Caudal angulation of 15-30 degrees   C. Caudal angulation of 35-45 degrees
   B. Cephalic angulation of 20-35 degrees   D. Cephalic angulation of 35-45 degrees

46. A radiographic image of the shoulder is obtained in the AP projection. If the patient's arm is medially rotated this will provide a:

   A. Profile view of the humeral epicondyles   C. Profile view of the glenoid fossa
   B. Profile view of the greater tuberosity   D. Profile view of the lesser tuberosity

47. Fractures of the dorsal aspect of the carpal bones are well demonstrated on a radiographic image taken in the tangential projection. This is also refereed to as the:

   A. Extension method   C. Carpal canal method
   B. Stecher method   D. Carpal bridge method

48. A radiographic image of the elbow is obtained in the AP oblique projection with medial rotation. This projection is most often used to provide an unobstructed view of the:

A. Coronoid process   C. Olecranon process
B. Radial head   D. Radial styloid process

49. A radiographic image of the first digit is obtained in the AP projection with the central ray directed at an angle of 45 degrees to the elbow. This projection is most often used for the demonstration of the:

A. Bones of the dorsum of the wrist   C. Base of the fifth metacarpal bone
B. First carpometacarpal joint   D. Proximal phalanges

50. A radiographic image of the scapula is obtained in the tangential projection. The patient is placed supine with the central ray directed through the posterosuperior region of the shoulder at a caudal angle of 45 degrees. Which of the following structures is best demonstrated?

A. The acromioclavicular articulation   C. The coracoid process
B. The glenoid fossa   D. The scapular spine

Referring to the radiograph of the shoulder, place the letter that corresponds to the following structures for questions 51-60.

51._____ Humeral head
52._____ Inferior angle of the scapula
53._____ Lesser tuberosity
54._____ Acromion process
55._____ Surgical neck of the humerus
56._____ Axillary border of the scapula
57._____ Clavicle
58._____ Greater tuberosity
59._____ Glenoid fossa
60._____ Coracoid process

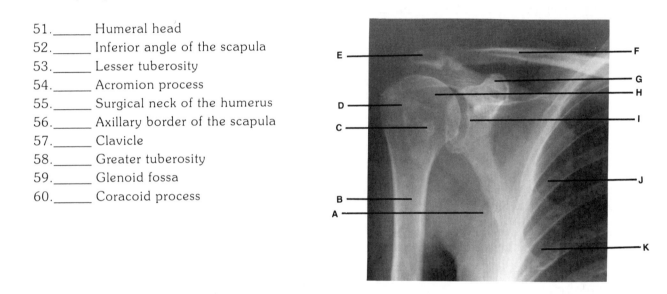

61. In order to demonstrate the coracoid process on a radiographic image obtained in the AP axial projection, the central ray should be directed:

A. Perpendicular to the shoulder joint   C. 15-45 degrees caudal to the humeral head
B. 15-45 degrees cephalic to the coracoid process   D. 65-70 degrees caudal to the glenoid fossa

62. A radiographic image of the wrist is obtained in the tangential (Gaynor Hart) projection (inferosuperior position). The only carpal bone demonstrated free of all bony superimposition is the:

A. Navicular bone   C. Hamate bone
B. Capitate bone   D. Pisiform bone

63. For the demonstration of the entire circumference of the radial head, how many lateral radiographic projections are required?

A. Four   C. Two
B. Three   D. One

64. The early changes of rheumatoid arthritis can best be demonstrated on radiographic images of the hands in the:

A. AP oblique projection   C. AP projection
B. PA projection   D. Lateral projection with fingers flexed

65. A radiographic image of the scapula is obtained in the PA oblique (Lorenz and Lilienfeld) projection. This projection is most commonly used to demonstrate the:

A. Sternoclavicular joints and body of the sternum   C. Space between scapula and the clavicle
B. Scapula free of the ribs   D. Fractures of the anterior scapula

66. In order to visualize the glenoid fossa on radiographic image of the humerus in the AP oblique (Grashey) projection, the patient is medially rotated:

A. 10-20 degrees   C. 35-45 degrees
B. 20-30 degrees   D. 60-70 degrees

67. All of the following projections may be used for the radiographic evaluation of the scaphoid or navicular bone EXCEPT:

A. A PA projection of the wrist with the fingers flexed
B. A PA projection of the wrist with ulnar deviation
C. A PA axial projection of the wrist in acute flexion

68. A radiographic image of the forearm is obtained in the AP projection with the arm in acute flexion. The central ray is directed perpendicular to enter 5 centimeters superior to the olecranon process. This image is most often used for the demonstration of the:

A. Radial notch   C. Semi-lunar notch
B. Radial head   D. Olecranon process

69. A radiographic image of the shoulder is obtained in the inferosuperior axial (Lawrence) projection. This projection requires that the humerus be:

A. Abducted 90 degrees and medially rotated   C. Fully adducted and laterally rotated
B. Fully adducted and medially rotated   D. Abducted 90 degrees and laterally rotated

70. The radiographic projection of the hand that will demonstrate the greatest amount of distortion of the phalanges and the interphalangeal joints is:

A. The PA projection   C. The PA oblique projection
B. The AP projection   D. The lateral projection

71. A radiographic image of the shoulder in the PA oblique (scapular Y) projection is obtained to rule out a dislocation of the humeral head. The patient should be rotated so that the midcoronal plane forms an angle of:

A. 5-10 degrees to the plane of the image receptor
B. 15-25 degrees to the plane of the image receptor
C. 55-75 degrees to the plane of the image receptor
D. 45-60 degrees to the plane of the image receptor

72. The maximum relaxation of the fat pads of the elbow is seen with the joint flexed to an angle of about:

A. 15 degrees   C. 90 degrees
B. 35 degrees   D. 180 degrees

73. A radiographic image for the bicipital groove is obtained in the tangential projection. If the patient is placed supine on the imaging table, the central ray should be directed about:

A. 20 degrees anterior to the horizontal to skim the anterior aspect of the humerus
B. 15 degrees posterior to the horizontal to skim the anterior aspect of the humerus
C. 25 degrees laterally to the glenohumeral joint
D. 45 degrees medially to the coracoid process

74. A radiographic image of the hand is obtained in the PA projection. This will provide a:

A. Lateral projection of the thumb
B. Tangential projection of the thumb
C. AP projection of the thumb
D. Oblique projection of the thumb

75. In order to obtain a radiographic image of the shoulder joint in the inferosuperior axial (West Point) projection, the central ray should be directed:

A. 25 degrees anteriorly and 25 degrees medially
B. 10 degrees anteriorly and 45 degrees medially
C. 30 degrees posteriorly, 25 degrees laterally
D. 45 degrees posteriorly and 10 degrees medially

Referring to the radiographic image, answer questions 76 & 77.

76. This tangential projection of the wrist is most commonly referred to as the:

A. Carpal canal
B. Carpal bridge
C. Radial flexion
D. Ulnar flexion

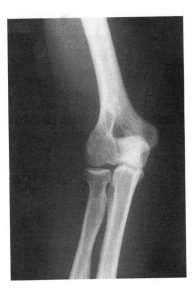

77. The structures which are best shown in this projection are the:

1. Pisiform   2. Styloid process   3. Hook of the hamate

A. 1 & 2 only
B. 1 & 3 only
C. 2 & 3 only
D. 1, 2 & 3

Referring to the radiographic image, answer questions 78 & 79.

78. This projection was obtained with the elbow placed in the:

A. Medial oblique position
B. Lateral oblique position
C. Acute flexion
D. Forced extension

79. The structure that is best demonstrated free of superimposition, is the:

A. Semilunar notch
B. Olecranon process
C. Trochlea
D. Radial head

Referring to the radiographic image, answer questions 80 & 81.

80. This projection is most often used for the demonstration of the:

**1. Acromioclavicular articulations
2. Sternoclavicular articulations
3. Costovertebral articulations**

A. 1 only
B. 2 only
C. 3 only
D. 1, 2 & 3

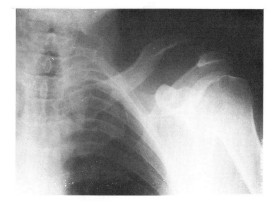

81. In order to maximize visualization of the dislocation shown, the radiograph was taken with the patient:

A. Supine without stress
B. Erect without stress

C. Supine with weights
D. Erect with weights

Referring to the radiographic image, answer questions 82 & 83.

82. The radiographic image of the shoulder in the AP projection was taken with the arm in:

A. Internal rotation
B. Neutral rotation
C. External rotation
D. Full abduction

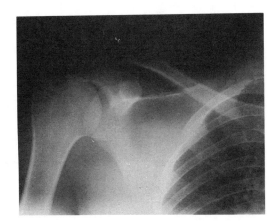

83. This projection is most commonly employed to show the _____ in profile.

A. Scapular spine
B. Coracoid process
C. Coronoid process
D. Greater tuberosity

Referring to the radiographic image, answer questions 84 & 85.

84. This projection is principally used in the evaluation of the:

A. Axillary border of the ribs
B. Glenohumeral joint
C. Coracoid process
D. Scapula

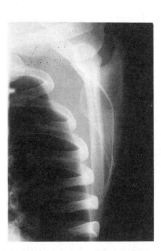

85. This projection demonstrates the scapula in the:

A. AP position
B. Lateral position
C. Oblique position
D. Axial position

Explanations

1. C   A radiographic image of the hand is obtained in the PA oblique projection. This projection is taken with the palmar surface of the hand rotated laterally 45 degrees to the image receptor and the central ray directed perpendicular to the third metacarpal phalangeal joint. This image is most commonly taken to open the metacarpal phalangeal and interphalangeal joint spaces. A properly positioned oblique will allow for a slight overlapping of the base and head of the metacarpal.

2. A   The clavicle is directed obliquely between the manubrium and the acromion process of the scapula. Due to its location on the anterior surface of the body, a shorter object-to-image receptor distance is obtained in the PA projection. This will reduce the amount of focal spot in the image. The erect position is often preferred to make the patient more comfortable. Radiographic images in the recumbent PA projection should be avoided in cases of severe trauma to eliminate the possibility of secondary injury and pain. This projection is obtained by directing the central ray perpendicular to the center of the clavicle.

3. C   A radiographic image of the shoulder in the AP projection with external rotation will place the humeral head in true anatomic position. This projection will demonstrate a profile view of the greater tuberosity and the soft tissue structures of the shoulder. In a properly positioned external rotation position, a plane passing through the humeral epicondyles should be parallel to the plane of the image receptor. This image is obtained with the patient supine or erect and directing the central ray perpendicular to a point 2 centimeters below the coracoid process.

4. D   A radiographic image of the elbow in the lateral projection is obtained to evaluate the joint space, semilunar notch and olecranon process. The proper positioning is accomplished by placing the forearm, elbow and humerus in the same plane and adjusting a plane passing through the epicondyles perpendicular to the image receptor. The hand is placed in the lateral position and the elbow is flexed 90 degrees to superimpose the styloid processes of the ulna and radius. The central ray is directed perpendicular to the elbow joint.

5. B   The magnification on a radiographic image of the wrist obtained in the PA projection can be reduced by flexing the finger, which places the wrist closer to the image receptor. This projection is obtained by placing the palmar surface of the hand at the top of the image receptor and directing the central ray perpendicular to the midcarpal area. This will also reduce the amount of focal spot blur and improve the definition of the joint spaces. This image provides a PA projection of the carpal bones, proximal metacarpal bones and the distal ulna and radius.

6. C   Bilateral views of both hands on a single radiographic image are discouraged because the centered ray must be directed to the midpoint of the image receptor and not the anatomical structure. This causes the divergent rays to obscure the narrow joint spaces found between phalanges and the metacarpal bones.

7. C   A radiographic image of the acromioclavicular (AC) joints is obtained in the AP projection. The patient is placed erect with the central ray directed to the midline of the body at the level of the AC joint. The patient is asked to hold weights to maximize the degree of separation occurring at the joint. Bilateral views of both AC joints is recommended so that the joints can be compared. An AP axial (Alexander) projection can be obtained of the AC joints by directing the central ray at a cephalic angulation of 15 degrees.

8. C   In order to obtain a radiographic image of the shoulder in the anatomical position, the humerus should be externally rotated until a plane passing between the humeral epicondyles is placed parallel to the image receptor. This projection will provide a profile view of the greater tuberosity. In the AP projection of the shoulder in internal rotation, a plane passing through the humeral epicondyles should be placed perpendicular to the image receptor. This projection will provide a profile view of the lesser tuberosity. Both projections require that the central ray be directed perpendicular to the image receptor to a point 2 centimeters below the coracoid process.

9. B   A radiographic image of the wrist in the AP oblique projection with a medial rotation of about 45 degrees is employed to give an improved visualization of the bones on the medial side of the wrist. This projection will reduce the amount of a superimposition of the pisiform bone and improve visualization of the lunate and triquetrum bones.

10. C   A radiographic image of the forearm is obtained in the AP projection. This projection requires that the arm be placed in an anatomic position with the hand supinated and the central ray directed perpendicular to the mid-forearm. This demonstrates the two bones of the forearm, the ulna and radius. The supination of the hand shows these bones in an uncrossed position.

11. A   The carpal interspaces, which extend obliquely, are most closely aligned with the divergent primary beam in a radiographic image of the wrist obtained in the AP projection.

12. K   13. D   14. I   15. A   16. H   17. C

18. F   19. E   20. G

21. C   In order to include all of the phalanges and the distal portion of the metacarpal bones on a radiographic image of the finger, the central ray is directed perpendicular to a point at the level of the proximal interphalangeal joint.

22. A   A radiographic image of the scapula, obtained in a lateral projection, demonstrates the vertebral and axillary borders superimposed on each other. This is best accomplished by placing the patient in the anterior oblique position with the coronal plane forming about a 60 degree angle with the plane of the image receptor. In the lateral projection, the affected scapula is placed closest to the image receptor. The dependent arm is placed at an angle of about 45 degrees to the body and the central ray is directed perpendicular to enter the mid-scapula. The position of the arm should be adjusted to avoid being imaged in the area of the primary injury.

23. B   A true lateral projection of the humerus can be made without internal rotation of the humerus by obtaining a radiographic image in a transthoracic (Lawrence) lateral projection. In cases of severe trauma, this provides the only method by which a right angle projection of an unstabilized shoulder or humerus can be made. When possible, the Lawrence method requires that a plane passing through the humeral epicondyles be placed perpendicular to the image receptor. The central ray should be directed perpendicular to the image receptor entering the midcoronal plane at the level of the neck of the humerus.

24. C   A radiographic image of the wrist obtained in the lateral projection with acute palmar flexion improves the visualization of the carpe bossu. This bony lesion first described by Foille is located on the posterior surface of the third metacarpocarpal joint .

25. D   The bony process at the tip of the elbow located on the expanded proximal end of the ulna is called the olecranon process. This structure is best demonstrated on a radiographic image taken in true lateral projection of the elbow. When the bony structures are of primary interest, the elbow should be flexed to an angle of 90 degrees; if the soft tissue structures are of primary concern, the elbow should be flexed to an angle of 30 degrees.

26. C   Palmar and ventral refer to the anterior surface of the hand. In order to perform a radiographic image of the hand in the PA projection, the palmar surface should be placed against the image receptor.

27. A   A radiographic image of the scapula is obtained in the AP projection. The patient is placed supine with the dependent arm abducted to a 90 degree angle with the elbow flexed to a right angle. This helps to draw the scapula laterally away from the chest wall and to prevent the superimposition of the scapula with the shadows of the lungs and rib cage. The central ray is directed perpendicular to the midscapular area at a point 5 centimeters inferior to the coracoid process.

28. B   In a radiographic image of the hand in the lateral projection, if the fingers are of primary interest they should be fanned to eliminate superimposition of the phalanges. In cases of foreign body localizations and metacarpal fractures, the lateral projection should be obtained with the fingers in extension. This projection requires that the central ray be directed perpendicular to the metacarpal phalangeal joint of the second digit when the hand is in the lateromedial position.

29. D   A quiet breathing technique is used for radiographic images of the scapula obtained in the AP projection. This serves to blur the shadows caused by the overlying lung tissues and ribs.

30. C   A radiographic image of the shoulder obtained in the AP projection in internal rotation, is used to provide a profile view of the lesser tuberosity. The patient's arm is adjusted so a plane passing through the humeral epicondyles is perpendicular to the image receptor. The central ray should be directed perpendicular to the image receptor to a point 2 centimeters below the coracoid process.

31. A   A radiographic image of the wrist in the tangential (carpal canal) projection is used to improve the visualization of the palmar aspect of the proximal row of carpal bones including unobstructed images of the hook of the hamate and pisiform bone. This projection is obtained with the hand hyperextended and having the central ray directed inferosuperiorly at an angle of 25-30 degrees to the center of the wrist.

32. C   33. E   34. H   35. J   36. D   37. I

38. D   A radiographic image of the elbow in partial flexion of about 35 degrees is used instead of the right angle flexion to reduce stretching and compression of the soft tissues of the elbow.

39. D   A radiographic image of the thumb is obtained in the AP projection. In a normal adult or child, the degree of counter rotation required to place the thumb into the AP position will result in the hand being placed in the lateral positioning. This projection requires that the central ray should be directed perpendicular to the metacarpal phalangeal joint.

40. C   The lower portion of the Y, which appears in a lateral projection of the scapula, is formed by the superimposed body, vertebral and axillary borders. The portion of the upper Y closest to the ribs is formed by the coracoid process, the lateral arm of the Y is formed by the acromion process.

41. D   In a radiographic image of the shoulder to profile the greater tuberosity, the humerus must be laterally rotated until a plane passing through the epicondyles is parallel to the plane of the image receptor. This will place the humerus in the true anatomic position.

42. A   Radiographic images of the wrist in the PA projection taken with radial flexion or ulnar deviation or AP oblique projection, are used to improve the visualization of the carpal bones on the medial aspect of the wrist. A radiographic image in the PA projection with radial deviation or ulnar flexion, is used to improve the visualization of the scaphoid (navicular) bone.

43. C   A radiographic image of the elbow is obtained in the AP oblique projection with lateral (external) rotation. This projection is obtained with the arm rotated laterally about 45 degrees from the AP position with the hand supinated. The central ray is directed perpendicular to the elbow joint to provide an unobstructed image of the radial head without superimposition with the ulna.

44. C   The approximate center of the hand is over the third metacarpophalangeal joint. All radiographic images should be obtained by directing the central ray perpendicular to this structure.

45. A   In a radiographic image of the clavicle taken without a tube angle, will result in the superimposition with the apical region of the lungs and first rib. The PA axial projection is often taken to avoid this superimposition. With the patient prone, the central ray should be directed at a caudal angulation of a 15-30 degrees to direct the clavicles above the shadow of the apical region. The AP axial projection can provide a similar view. With the patient supine, a 15-40 degree cephalic angulation of the central ray will direct the shadow of the clavicles above the apical region of the chest.

46. D   A radiographic image of the humerus obtained in the AP projection, will require an internal rotation in order to place the humerus in the lateral position. This is accomplished by rotating the humerus until a plane passing through the epicondyles is perpendicular to the plane of the image receptor. This projection will give a lateral image of the humerus which profiles the lesser tuberosity.

47. D   A radiographic image of the dorsal surface of the wrist in the tangential projection (carpal bridge) requires that the hand be flexed to form a 90 degree angle with the forearm, resting on the dorsal surface of the hand. The central ray is directed tangentially at an angle of about 45 degrees to the dorsum of the wrist. This projection is recommended for visualization of lunate dislocations, navicular fractures and foreign bodies or calcifications of the posterior aspect of the wrist.

48. A   A radiographic image of the elbow obtained in the medial (internal) oblique projection requires a 45 degree rotation to free the coronoid process superimposition.

49. B   A radiographic image of the first carpometacarpal joint in the AP projection can be obtained using a modified radial tilt of the carpal canal projection. This projection provides a magnified view of the first metacarpocarpal joint and the trapezium or greater multangular bone. The central should directed at a 45 degree angle to the elbow.

50. D   A radiographic image of the scapular spine, which projects posteriorly from the body of the scapula, can be visualized in the tangential projection. If the patent is supine, the central ray should be directed at an angle of 45 degrees caudal, entering at the posterosuperior aspect of the shoulder. If the patient is prone, the tangential projection can be obtained by directing the central ray at a posteroinferior angle of 45 degrees to the lower border of the spine. These projections show a profile view of the scapular spine free of superimposition with the body of the scapula.

51. H   52. K   53. C   54. E   55. B   56. A

57. F   58. D   59. I   60. G

61. B   A radiographic image in the AP axial projection is obtained to demonstrate the coracoid process free of superimposition. This image is obtained with the patient supine and having the central ray directed at a cephalic angle of between 15-45 degrees. The degree of angulation depends on the roundness of the patient's shoulder and back.

62. D   A radiographic image of the wrist in the tangential (carpal canal) projection, is used to improve the visualization of the palmar aspect of the proximal row of carpal bones, including unobstructed images of the hook of the hamate and pisiform bone. This projection is obtained with the hand hyperextended and having the central ray directed inferosuperiorly at an angle of 25-30 degrees to the center of the wrist.

63. A   In order to provide radiographic images that will demonstrate the entire circumference of the radial head, four lateral projections are required. The lateral projections of the elbow are obtained with the hand supinated as much as possible, with the hand pronated, with the hand in lateral extension and with the hand in extreme internal rotation.

64. A   A radiographic image of the hands in the AP oblique (Norgaard) projection or ball catcher's positions is most often used to diagnose early changes associated with rheumatoid arthritis. The physician will look for changes on borders of the distal phalanges and slight demineralization of the bones of the fingers. This position requires that the hands be placed in the half-supinate position at a 45 degree angle.

65. B   Radiographic images of the scapula obtained in the PA oblique projections, described by Lorenz and Lilienfeld, require the patient to be placed in the lateral position with the affected scapula closest to the image receptor. The dependent arm is placed at a right angle to the body and the central ray is directed perpendicular to enter the mid-scapula. These projections are used to provide an oblique view of the scapula projected free of the rib cage.

66. C   A radiographic image of the glenohumeral joint space is obtained in the AP oblique (Grashey) projection. The image is obtained with the patient RPO or LPO with the affected joint closest to the image receptor. The patient is rotated 35-45 degrees and the central ray is directed perpendicular to a point 5 centimeters medial and 5 centimeters inferior to the superolateral border of the shoulder This projection is used to open the joint space between the humeral head and the glenoid cavity.

67. A   A radiographic image of the scaphoid (navicular) bone is obtained in the PA axial (Stecher) projection. The scaphoid bone, which is directed obliquely to the bones of the forearm, can be placed at a right angle by placing the wrist on a 20 degree incline toward the elbow and directing the central ray perpendicular to the wrist. A PA axial (Stecher) projection can also be obtained when the wrist is horizontal by directing the central ray at an angle of 20 degrees toward the elbow. Other variations for imaging the scaphoid bone include PA projections with ulnar deviation (flexion) and by directing the central rays at various cephalic angulations of 10, 20 and 30 degrees.

68. D   A radiographic image of the distal humerus is obtained in the AP projection in acute flexion. This projection, which provides a profile view of the olecranon process, is obtained by directing the central ray perpendicular to the image receptor at a point on the humerus 5 centimeters superior to the olecranon process. The same position can also be used to provide a PA projection of the proximal forearm by directing the central ray perpendicular to the bones of the forearm. This modification is used to demonstrate the proximal portions of the ulna and radius.

69. D   A radiographic image of the shoulder joint is obtained in the inferosuperior axial (Lawrence method) projection. With the patient supine, the arm is abducted to a right angle and externally rotated, the central ray is directed horizontally through the axilla, with a medial angle of between 15-30 degrees. This projection is used in the visualization of the proximal humerus, the scapulohumeral and acromioclavicular joints. This can also be used to demonstrate the sites of insertion for the subscapular tendon on the lesser tubercle of the humerus and the teres minor tendon on the greater tubercle.

70. C   A radiographic image of the hand obtained in the PA oblique projection with lateral rotation, will often show a great deal of distortion between the phalanges due to the forward tilting of the fingers. The use of a 45 degree stepped-wedge sponge should be used to reduce this effect.

71. D   A radiographic image of the shoulder joint is obtained in the PA oblique (scapular Y) projection. With the patient in a RAO or LAO position, the patient is rotated so that the midcoronal plane forms an angle of 45-60 degrees to the plane of the image receptor. The central ray is directed perpendicular to the scapulohumeral joint. This projection demonstrates an oblique image of the shoulder superimposed over the lateral scapula and is used to help in differentiating between anterior and posterior dislocation of the humeral head.

72. C   Partial or complete flexion can elevate the fat pads and distort the appearance of the joint. It is recommended that the elbow be flexed 90 degrees to relax these pads.

73. B   A radiographic image is obtained of the bicipital (intertubular) groove in the tangential (Fisk) projection. If the patient is supine, the central ray is angled 15 degrees posterior to the horizontal at a point on the anterior surface of the shoulder. This projection shows the bicipital groove free of any bony superimposition.

74. D   A radiographic image of the hand is obtained in the PA projection. The hand should be placed prone on the image receptor, with the central ray directed perpendicular to the third metacarpal joint. This will demonstrate a PA projection of the 2nd-5th phalanges, the metacarpal and carpal bones and a PA oblique projection of the thumb.

75. A   A radiographic image of the shoulder joint is obtained in the inferosuperior axial (West Point) projection. If the patient is prone and the arm abduced to a right angle, the central ray is directed at a dual angle of 25 degrees anteriorly from the horizontal and 25 degrees medially to the center glenohumeral joint. This projection is used to show bony defects of the rim of the glenoid fossa and for patients with chronic instability of the shoulder joint.

76. A   This radiographic image of the wrist in the tangential (carpal canal) projection is used to improve the visualization of the palmar aspect of the proximal row of carpal bones including unobstructed images of the hook of the hamate and pisiform bone. This projection is obtained with the hand hyperextended and having the central ray directed inferosuperiorly at an angle of 25-30 degrees to the center of the wrist.

77. B   See the explanation above.

78. B   This radiographic image of the elbow was obtained in the AP oblique projection with lateral (external) rotation. This projection was taken with the arm rotated laterally about 45 degrees from the AP position with the hand supinated. The central ray was directed perpendicular to the elbow joint. This projection is used to provide an unobstructed image of the radial head without superimposition of the ulna.

79. D   See the explanation above.

80. A   This radiographic image of the acromioclavicular (AC) joint was obtained in the AP projection to rule out a separation. The image was obtained with the patient in the erect position with the patient holding weights, to maximize the dislocation of the joint.

81. D   See the explanation above.

82. C   With the body AP, the external rotation position of the arm places the humeral head in true anatomic position. This position shows a profile view of the greater tuberosity and is accomplished by placing the humeral epicondyles parallel to the image receptor.

83. D   See the explanation above.

84. D   In the true lateral position of the scapula, the vertebral and axillary borders should be superimposed. This is best accomplished by placing the patient in the PA oblique position, with the coronal plane forming about a 60 degree angle with the plane of the image receptor. In the PA projection, the affected scapula is placed closest to the image receptor.

85. C   See the explanation above.

1. A radiographic image of the foot is to be obtained in the true lateral projection. This is more consistently obtained when the:

   A. Medial surface is in contact with the image receptor
   B. Lateral surface is in contact with the image receptor
   C. Plantar surface is in contact with the image receptor
   D. Dorsal surface is in contact with the image receptor

2. During radiography of the knee, in order to relax the muscles and demonstrate the maximum volume of the knee joint a:

   A. 20-30 degrees of flexion is required
   B. 40-45 degrees of flexion is required
   C. 55-60 degrees of flexion is required
   D. 70-90 degrees of flexion is required

3. A radiographic image of the pelvis is obtained in the anteroposterior (AP) projection. Which of the following can be employed to reduce the normal anteversion of the femoral neck?

   A. A 15 degree caudal angle of the central ray
   B. A 15 degree cephalic angle of the central ray
   C. A 15 degree medial rotation of the legs
   D. A 12 degree lateral rotation of the legs

4. A radiographic image of the lower leg is obtained in the AP projection. In order to prevent the joints of the lower leg from being projected off the image due to the divergence of the x-ray beam, the image receptor should extend:

   A. 13 centimeters beyond each joint
   B. 8 centimeters beyond each joint
   C. 4 centimeters beyond each joint

5. A radiographic image of the calcaneus is obtained in the lateral projection. The central ray is directed to enter perpendicular to a point:

   A. 2.5 centimeters proximal to the medial malleolus
   B. 2.5 centimeters distal to the medial malleolus
   C. 2.5 centimeters posterior to the medial malleolus
   D. 2.5 centimeters anterior to the medial malleolus

6. Which of the following radiographic projections should NOT be used to evaluate the hip of a patient with a suspected fracture?

   1. Axiolateral Danelius-Miller   *2. AP axiolateral Cleaves*   *3. Lateral Lauenstein-Hickey*

   A. 1 & 2 only
   B. 1 & 3 only
   C. 2 & 3 only
   D. 1, 2 & 3

7. A radiographic image of the foot is obtained in the AP oblique projection. The foot is medially rotated so the plantar surface forms a 30 degree angle with the plane of the image receptor. This will improve the visualization of which of the following structures?

   1. First and second metatarsal bases   *2. Fourth and fifth metatarsal bases*   *3. Cuboid bone*

   A. 1 & 2 only
   B. 1 & 3 only
   C. 2 & 3 only
   D. 1, 2 & 3

8. A radiographic image of the knee is obtained in the posteroanterior (PA) projection on a sthenic patient. Due to the slightly inclined position of the lower leg, the central ray should be directed at a:

   A. Caudal angle of 5 degrees to the image receptor
   B. Cephalic angle of 7 degrees to the image receptor
   C. Caudal angle of 15 degrees to the image receptor
   D. Cephalic angle of 90 degrees to the image receptor

9. A radiographic image of the femoral neck is obtained in the AP oblique (Cleaves) projection. In order to provide optimum visualization of the neck, the patient's thighs should be abducted:

   A. 15 degrees from the vertical plane
   B. 40 degrees from the vertical plane
   C. 60 degrees from the vertical plane
   D. 90 degrees from the vertical plane

10. A radiographic image of the ankle is obtained in the mediolateral lateral projection. Dorsiflexion of the foot is required to help prevent:

A. The toes from obscuring the ankle joint   C. Lateral rotation of the ankle joint
B. The cuboid bone from obscuring the ankle joint   D. Additional stress fractures at the joint

11. A radiographic image of the patella is obtained in the tangential (Settegast) projection. This is often used for the evaluation of:

A. Vertical fractures of the patella   C. Subtalar dislocations
B. Joint mice   D. Tears of cruciate ligaments

12. A radiographic image of the pelvis is obtained in the axial (Chassard-Lapiné) projection. This projection is taken to evaluate:

1. Defects of the pelvic outlet
2. Defects in the urinary bladder
3. Defects of the rectosigmoid region

A. 1 & 2 only   C. 2 & 3 only
B. 1 & 3 only   D. 1, 2 & 3

Referring to the radiographic image of the ankle, place the letter that corresponds to the following structures for questions 13-15.

13. _____Lateral malleolus
14. _____Talus (Astragalus) bone
15. _____Tibia

16. A radiographic image of the foot is obtained in the AP oblique projection. If the foot is laterally rotated so the plantar surface forms an angle of 30 degrees with the plane of the image receptor, which structure is best demonstrated?

A. The interspaces between the 1st and 2nd metacarpal and the cuneiform bones
B. The sesamoid bones free of superimposition
C. The cuboid bone free of superimposition
D. The base of the 5th metatarsal

17. A radiographic image of the knee is obtained in the lateral projection. Which of the following will prevent the medial femoral condyle from obscuring the joint space?

A. A 30-45 degree flexion of the knee joint
B. A 20-30 degree medial rotation of the leg
C. A 5-7 degree cephalic angulation of the central ray
D. A 10-15 degree lateral rotation of the leg

18. A radiographic image of the hip is obtained in the axiolateral (shoot-through) projection. If a stationary grid is used to reduce the amount of scattered radiation, it is important to place the grid strips:

1. *Parallel to the neck of the femur*
2. *Perpendicular to the neck of the femur*
3. *Perpendicular to the plane of the table*

A. 1 only
B. 2 only
C. 3 only
D. 1, 2 & 3

19. A radiographic image of the ankle is obtained in the AP oblique projection. In order to maximize the opening of the tibiofibular joint space the leg should be rotated:

A. 45 degrees medially
B. 25 degrees medially
C. 45 degrees laterally
D. 25 degrees laterally

20. A radiographic image of the calcaneus is obtained in the axial (plantodorsal) projection. In order to demonstrate the subtalar joint the central ray should be directed at an angle of:

A. 15 degrees to the long axis of the foot
B. 25 degrees to the long axis of the leg
C. 40 degrees to the long axis of the foot
D. 60 degrees to the long axis of the leg

21. A radiographic image obtained in the anteroposterior (AP) oblique (Cleaves) projection is most often used for the evaluation of the:

A. Pelvic inlet and outlet
B. Symphysis pubis
C. Femoral head and neck
D. Sacroiliac joints

22. Which of the following radiographic projections of the foot should be obtained to best demonstrate the cuboid bone and its articulations?

A. The AP axial projection
B. The axial plantodorsal projection
C. The AP oblique projection with medial rotation
D. The AP oblique position with lateral rotation

23. A radiographic image of the patella in the tangential projection, should NOT be attempted until a lateral image has been obtained to rule out:

A. Damage to the collateral ligaments
B. Transverse fractures of the patella
C. Severe arthritis of the patellofemoral joint space
D. Damage to the patellofemoral bursa

24. A radiographic image of the hip is obtained in the anteroposterior (AP) projection. The central ray is directed about 6 centimeters distal to a line drawn perpendicular to the midpoint between the:

A. Iliac crest and the symphysis pubis
B. Iliac crest and the greater trochanter
C. Symphysis pubis and the ASIS
D. Lesser trochanter and the symphysis pubis

25. When a radiographic image is obtained of the ankle in the lateral projection, which position of the ankle will place the joint closest to the image receptor?

A. Lateromedial position
B. Anteroposterior position
C. Mediolateral position
D. posteroanterior position

26. A radiographic image of the third toe is obtained in the AP projection. The central ray should be directed perpendicular to the:

A. Second tarsometatarsal joint
B. Third tarsometatarsal joint
C. Third metacarpophalangeal joint
D. Third proximal interphalangeal joint

27. A patient is brought into the radiography department to rule out a trimalleolar fracture. The radiographic projection that is most commonly obtained to demonstrate this condition is an:

A. AP projection of the proximal femur   C. Axial projection of the distal femur
B. AP projection of the ankle joint   D. Axial projection of the calcaneus

28. A radiographic image of the foot is obtained in the AP oblique projection. The foot should be rotated medially until the plantar surface forms an angle of:

A. 15 degrees to the plane of the image receptor   C. 45 degrees to the plane of the image receptor
B. 30 degrees to the plane of the image receptor   D. 60 degrees to the plane of the image receptor

29. A radiographic image obtained in the posteroanterior (PA) axial (Camp-Coventry) projection may be used to demonstrate:

A. Joint mice in the intercondyloid fossa   C. Dislocations of the talofibular articulation
B. Vascular malformation in the popliteal fossa   D. Dislocations of patellofemoral articulation

Referring to the radiograph of the foot, place the letter that corresponds to the following structures for questions 30-36.

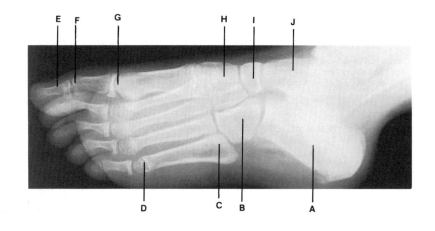

30._____Navicular (scaphoid) bone   34._____Interphalangeal joint
31._____Metacarpophalangeal joint   35._____Cuboid bone
32._____Head of the metatarsal bone   36._____Talus (astragalus)
33._____Calcaneus

37. The dome of the acetabular cavity can be localized by finding the mid-point of a line drawn between the symphysis pubis and the:

A. Greater trochanter   C. Ischial tuberosity
B. Sacral promontory   D. Anterior superior iliac spine (ASIS)

38. A radiographic image of the os calcis or calcaneus is obtained in the axial dorsoplantar projection. The central ray should be directed at a caudal angle of 40 degrees to enter at the dorsal surface of the ankle parallel with the:

A. Tuberosity of the os calcis   C. Tibia tuberosity
B. Base of the third metatarsal bone   D. Head of the third metatarsal

39. A radiographic image of the ankle is obtained in the AP oblique projection. In order to maximize the opening of the mortise joint, the ankle and leg should be rotated:

A. 5-12 degrees laterally   C. 35-45 degrees laterally
B. 15-20 degrees medially   D. 45-55 degrees medially

40. The magnification on a radiographic image of the anterior pelvic bones can be reduced by obtaining images in the:

A. AP axial projection
B. PA projection
C. AP oblique axial projection
D. Lateral projection

41. A radiographic image of the anterior pelvic bones is obtained in the AP axial (Taylor) projection on a female patient. The central ray should be directed to enter a point 5 centimeters distal to the pubic symphysis at a:

A. Cephalic angle of 30-45 degrees
B. Caudal angle of 20-35 degrees
C. Cephalic angle of 10-12 degrees
D. Caudal angle of 15-20 degrees

42. A radiographic image of the foot is obtained in the anteroposterior (AP) axial projection. The central ray should be directed:

A. 10 degrees toward the heel to pass through the base of the third metatarsal
B. 25 degrees toward the heel to pass through the head of the third metacarpophalangeal joint
C. 10 degrees toward the toes to pass through the head of the third metatarsal
D. 25 degrees toward the toes to pass through the third tarsometatarsal joint

43. A radiographic image of the knee is obtained in the anteroposterior (AP) projection on a patient measuring 22 centimeters from the ASIS to the table top. The central ray should be directed perpendicular to a point:

A. 1.5 centimeters distal to the medial border of the patella
B. 1.5 centimeters distal to the lateral border of the patella
C. 1.5 centimeters distal to the base of the patella
D. 1.5 centimeters distal to the apex of the patella

44. A radiographic image of the femur is obtained in the anteroposterior (AP) projection. In order to place the femoral neck parallel with the plane of the image receptor the:

A. Central ray should be angled 15 degrees toward the midsagittal plane
B. Central ray should be angled 15 degrees toward the head
C. Lower limb should be medially rotated 15-20 degrees
D. Lower limb should be laterally rotated 15-20 degrees

45. A radiographic image of the ankle is obtained in the AP projection with inversion stress. This image is most commonly employed to evaluate a rupture of the:

A. Medial collateral ligament
B. Lateral collateral ligament
C. Medial meniscus
D. Lateral meniscus

46. In order to improve the delineation of the joint spaces on a radiographic image of the toes in the AP axial projection, the central ray should be directed to the third metacarpophalangeal joint at a:

A. Cephalic angulation of 15 degrees
B. Caudal angulation of 15 degrees
C. Cephalic angulation of 45 degrees
D. Caudal angulation of 45 degrees

47. When a patient has bilateral hip fractures and a radiographic image cannot be obtained in the axiolateral (shoot-through) projection, an axiolateral (Clements-Nakayama) projection may be required. With the patient supine, the central ray is directed to a vertical image receptor at an angle of:

A. 15 degrees posteriorly
B. 35 degrees posteriorly
C. 25 degrees anteriorly
D. 45 degrees anteriorly

48. In order to obtain a radiographic image of the foot in the AP axial projection to demonstrate the tarsal articulations, the central ray is directed to enter the base of the third metatarsal at an angle of:

A. 10 degrees to the heel
B. 10 degrees to the toes
C. 35 degrees to the heel
D. 35 degrees to the toes

49. A radiographic image of the patellofemoral joint is obtained in the tangential (Hughston) projection. If the knee is flexed to form a 55 degree angle with the image receptor, the central ray should be directed through the joint space at a cephalic angle of:

A. 5 degrees
B. 15 degrees

C. 30 degrees
D. 45 degrees

50. A radiographic image of the hip is obtained in the axiolateral (shoot-through) projection. On the resultant image, which of the following the structures will be seen directly posterior to the femoral neck and acetabulum?

A. The ischial tuberosity
B. The anterior superior iliac spine (ASIS)

C. The symphysis pubis
D. The lesser trochanter

51. A radiographic image of the foot is obtained in the tangential (Lewis and Causton) projection. This projection is most commonly employed for the evaluation of:

A. Sesamoid bones
B. Subtalar joint

C. Sustentaculum tali
D. Longitudinal arch

52. A patient is supine with their leg resting in a 18 x 43 centimeter image receptor with the foot in the vertical position. The central ray is directed perpendicular to the midtibia. This projection is best described as:

A. An AP axial projection of the leg
B. A PA oblique projection of the leg

C. A lateral projection of the leg
D. An AP projection of the leg

53. A radiographic image of the knee is obtained in the AP oblique projection. In order to prevent overlapping of the proximal tibiofibular articulation the leg should be rotated:

A. Laterally 45 degrees
B. Medially 45 degrees

C. Laterally 25 degrees
D. Medially 25 degrees

54. A radiographic image of a postoperative fractured hip is to be obtained. Which of the following would be TRUE?

1. The AP axiolateral (Cleaves) projection is recommended to obtain a lateral projection
2. The entire prosthesis must be included on the image
3. Only an AP projection is required for a follow-up examination

A. 1 only
B. 2 only

C. 3 only
D. 1, 2 & 3

Referring to the radiography of the knee, place the letter that corresponds to the following structures for questions 55-64.

55._____Head of the fibula
56._____Medial condyle of the femur
57._____Tibial tuberosity
58._____Intercondyloid spaces
59._____Lateral condyle of the femur
60._____Intercondyloid eminences (tibial spaces)
61._____Medial tibial plateau
62._____Lateral epicondyle of femur
63._____Lateral tibial plateau
64._____Medial epicondyle of the femur

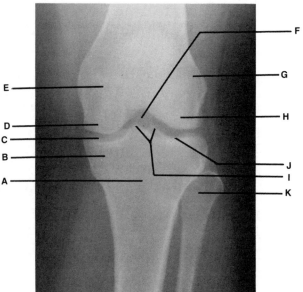

65. A radiographic image of the knee is obtained in a standing AP projection to evaluate arthritis. The central ray is directed:

A. At a cephalic angulation of 5-7 degrees to the knee joint
B. At a caudal angulation of a 5-7 degrees to the knee joint
C. Perpendicular to the knee joint
D. Parallel to the patellofemoral joint

66. Radiographic images of the feet of a child are obtained in AP and lateral (Kite) projections. These images are most often obtained to evaluate:

A. Juvenile arthritis
B. Talipes equinovarus
C. Coxa valga
D. Osgood Slatter's syndrome

67. A radiographic image of the femoral necks is obtained in the axiolateral (original Cleaves) projection. In order to direct the central ray parallel to the shafts of the femur the tube should be angled:

A. 25-45 degrees caudal
B. 15 degrees cephalic
C. 15 degrees caudal
D. 25-45 degrees cephalic

68. A radiographic image of the ankle is obtained in the anteroposterior (AP) projection. The central ray should be directed:

A. 15 degrees cephalic to the midpoint between the tibial plateaus
B. Vertically to the midpoint between the malleoli
C. 15 degrees cephalic to the midpoint between the malleoli
D. Vertically to the midpoint between the metatarsal bones

69. Radiographic images of the foot are to be obtained to evaluate the subtalar joint. Which of the following projections would most commonly be requested?

A. AP axial oblique projections of the ankle
B. Lateral projections of the foot
C. Tangential projections of the toes
D. AP projections of the ankle

70. A radiographic image of the knee is obtained in the anteroposterior (AP) axial (Beclere) projection to demonstrate the intercondylar fossa. The central ray should be directed to enter the knee joint:

A. Parallel to long axis of the femur
B. Perpendicular to long axis of the femur
C. Perpendicular to long axis of the patella
D. Perpendicular to long axis of the tibia

71. A radiographic image of the ilium is obtained to evaluate the broad surface wing without rotation. The AP oblique projection that is used for this purpose will require that the:

A. Unaffected side be elevated 15 degrees
B. Affected side be elevated 25 degrees
C. Unaffected side be elevated 40 degrees
D. Affected side be elevated 70 degrees

72. A radiographic image of the intercondyloid fossa is obtained in the posteroanterior (PA) axial (Holmblad) projection. If the long axis of the femur forms an angle of 70 degrees to the plane of the image receptor, the central ray should be directed through the knee joint:

A. At a caudal angulation of 20 degrees
B. At a cephalic angulation of 20 degrees
C. At a cephalic angulation of 45 degrees
D. Vertical to the image receptor

73. A radiographic image of the foot is obtained in the lateral (weight-bearing) projection. This is most commonly employed to evaluate the status of the:

A. Longitudinal arch
B. Transverse arch
C. Subtalar articulation
D. Sesamoid bones

74. A radiographic image of the foot is obtained in the AP axial projection. This composite view requires that two separate exposures are made to produce a single image. This method requires the central ray to be directed:

A. 15 degrees posterior and 45 degrees posterior
B. 25 degrees anterior and 45 degrees anterior
C. 25 degrees posterior and 45 degrees anterior
D. 15 degrees posterior and 25 degrees anterior

75. A series of radiographic images are made for the measurement of the length of the lower leg. In order to insure the proper orthoroentgenographic measurements are obtained it is important to place the legs in an anatomic position and:

A. Make one exposure of the entire leg
B. Move the image receptor and the leg in opposite directions
C. Make three exposures over the joints without moving the patient
D. Make three exposures as the leg is moved

Referring to the radiograph of the pelvis, place the letter that corresponds to the following structures for questions 76-85.

76.____ Femoral head
77.____ Sacrum
78.____ Ischial tuberosity
79.____ Anterior superior iliac spine (ASIS)
80.____ Greater trochanter
81.____ Sacroiliac joint
82.____ Pubic symphysis
83.____ Iliac crest
84.____ Lesser trochanter
85.____ Obturator foramen

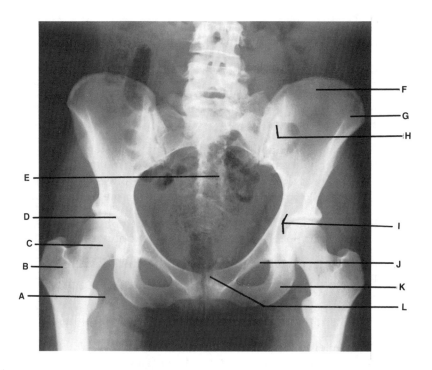

Referring to the radiographic image, answer questions 86 & 87.

86. This radiographic image of the knee was obtained in the:

A. PA axial projection
B. Tangential projection
C. AP oblique projection
D. AP projection with stress

87. The principal structure of interest in this position is/are the:

1. Intercondylar fossa
2. Fibular head
3. Tibial eminences

A. 1 only
B. 2 only
C. 1 & 2 only
D. 1 & 3 only

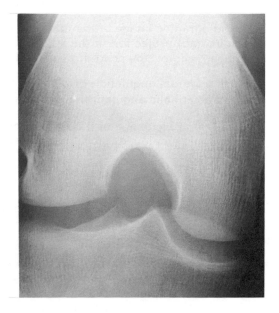

Referring to the radiographic image, answer questions 88 & 89.

88. In order to accomplish this tangential projection of the patella the central ray is directed perpendicular to the:

A. Femur
B. Joint space
C. Tibia
D. Fibula

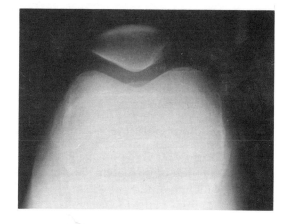

89. In addition to providing a profile view of the patella this projection is often used to evaluate the:

A. Intercondylar fossa
B. Cruciate ligaments
C. Femoropatellar articulation
D. Menisci

Referring to the radiographic image, answer questions 90 & 91.

90. The radiographic image shown is best described as a/an:

A. Superoinferior projection of the femoral necks
B. Lateral projection of the femoral necks
C. AP axial projection of the femoral necks
D. Inferosuperior projection of the femoral necks

91. This image demonstrates a profile view of the:

1. *Symphysis pubis*
2. *Greater trochanter*
3. *Lesser trochanter*

A. 1 only   C. 3 only
B. 2 only   D. 1, 2 & 3

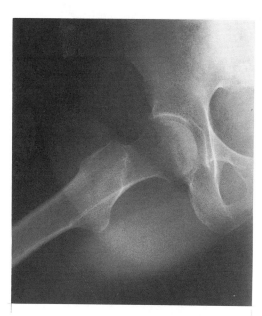

Referring to the radiographic image, answer questions 92 & 93.

92. In an AP projection of the knee, improved visualization of the joint space can be accomplished by:

A. Acute flexion of the knee
B. Partial flexion of the knee
C. Medial rotation of the knee
D. Increasing the mAs value

93. The radiolucent lines seen proximal and distal to the knee joint are termed:

A. Chondral folds
B. Epiphysis
C. Collateral ligaments
D. Pseudo joints

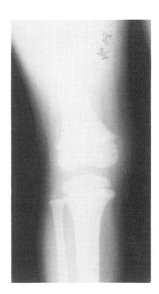

94. This weight-bearing AP projection of the knee was taken with the:

    A. Patient standing
    B. Patient seated
    C. Patient supine
    D. Patient recumbent

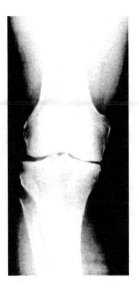

95. The radiographic image indicates that the patient has a:

    A. Narrowing of the medial joint space
    B. Rupture of the patellar bursa
    C. Fracture of the patella
    D. Rupture of the fovea capitis

1. A A radiographic image of the foot is obtained in the lateral projection with the medial surface of the foot against the image receptor. The foot should be dorsiflexed to form a right angle with the lower leg. The central ray should be directed perpendicular to the base of the third metatarsal. This projection is performed to the lateromedial position because it places the foot into a more true lateral position. The lateromedial position is more commonly used because it can be obtained with more ease. Both lateral projections are used to provide a profile view of the foot and ankle and the distal ends of the tibia and fibula.

2. A A radiographic image of the knee is obtained in the lateral projection. The patient is turned on to the affected side to place the knee in a lateral position. The knee should be flexed to 20-30 degrees to relax the muscles and to demonstrate the maximum volume of the joint space. The central ray is directed to the knee joint 2.5 centimeters distal to the medial epicondyle at a cephalic angulation of 5-7 degrees.

3. C A radiographic image of the pelvis is obtained in an AP projection. The patient is placed supine with the top of a crosswise 35 x 43 centimeter image receptor positioned just above the iliac crests. The central ray is directed perpendicular to the midpoint of the image receptor. Prior to making the exposure, the legs of the patient should be medially rotated 15-20 degrees to rotate the femoral necks to more closely parallel the plane of the image receptor. In the relaxed AP position of the legs, the femoral neck is tilted obliquely from the pelvis (anteversion). This projection provides an AP projection of the bony pelvis and the proximal portions of the femur.

4. C A radiographic image of the lower leg is obtained in the AP projection. The patient is supine with their leg resting in a 18 x 43 centimeter image receptor with the foot in the vertical position. The central ray is directed perpendicular to the midtibia. The relatively long distance between the ankle and knee joint, combined with the use of the peripheral portion of the beam, requires that the image receptor extend about 4 centimeters beyond the actual location of the joints if they are to be included on the image.

5. B A radiographic image of the calcaneus (os calcis) is obtained in the lateral projection. The foot is placed on the lateral surface with the foot centered to the image receptor. The central ray is directed perpendicular to a point about 2.5 centimeters distal to the medial malleolus. This projection will show the calcaneus just below the talus bone and lower portion of the ankle joint.

6. C In order to prevent excessive movement in a patient with a suspected fracture of the hip, the AP oblique and axiolateral (Cleaves) projections should be used. The lateral (Lauenstein) projection may put the patient at risk for further injury and should only be attempted after fractures have been ruled out. The axiolateral (Danelius-Miller) projection, or a shoot-through projection, may also be used without a considerable risk to the patient.

7. C A radiographic image of the foot is obtained in the AP oblique projection. The foot is rotated medially until the plantar surface of the foot forms a 30 degree angle with the plane of the image receptor. The central ray should be directed perpendicular to the base of the third metatarsal. This projection is used to demonstrate the interspaces between the cuboid bone, calcaneus and fourth and fifth metatarsal bones and the lateral conforms as well as the articulations between the navicular and talus, navicular and cuneiform, and the sustentaculum tali and talus bone. This projection often provides an unobstructed view of the entire cuboid bone.

8. D A radiographic image of the knee is obtained in the posteroanterior (PA) projection. The leg is extended with the knee joint centered to a 24 x 30 centimeter image receptor. The central ray is directed perpendicular to a point 1.5 centimeters inferior to the apex of the patella. This projection is used to evaluate bony and soft tissue structures of the knee. In the AP position, the inclination of the knee joint is dependent on the patient's bodily habitus. In thinner hyposthenic patients, a 5 degree caudal angle is preferred to maximize the joint space. No angle is recommended for the sthenic patient and a 5 degree cephalic angulation is suggested on individuals that are hypersthenic.

9. B   A radiographic image of the femoral neck is obtained in the anteroposterior (AP) oblique (Cleaves) projection. With the patient supine, the hips and knees are abducted with the soles of the feet together to place the femoral necks as close to a 45 degree angle as possible. This serves to place them parallel to the image receptor. The top of a crosswise 35 x 43 centimeter image receptor is positioned just above the iliac crests. The central ray is directed perpendicular to the imaging system for the modified AP oblique (Cleaves) projection. In the axiolateral (original Cleaves) projection, a 25-45 degrees cephalic angulation of the central ray is employed. Both projections are used to demonstrate an oblique projection of the femoral heads, necks and trochanteric regions.

10. C   A radiographic image of the ankle is obtained in the mediolateral lateral projection. The foot is placed so the lateral surface is in contact with the 18 x 24 centimeter image receptor. The toes are dorsiflexed to eliminate lateral rotation of the ankle. The central ray is directed perpendicular and enters at the medial malleolus. This projection shows a lateral projection of the lower tibia and fibula and the talus bone.

11. A   A radiographic image of the patella is obtained in the tangential (Settegast) projection. This projection, also known as the sunrise view, is taken with the knee flexed to a right angle with the central ray directed to the knee joint parallel to the patellofemoral joint. This projection is often used to provide a profile view of the patella for the evaluation of vertical fractures and the patellofemoral articulation.

12. D   A radiographic image of the pelvic region is obtained in the axial (Chassard-Lapiné) projection. The patient is seated near the side of the table and told to lean forward as much as possible. The central ray is directed perpendicular through the lumbosacral region at the level of the greater trochanters. This projection provides an axial projection of the pelvis that can be used to measure the transverse diameter of the pelvis or demonstrate the relationship of the femoral heads and the acetabula. This projection can also be used in conjunction with a barium enema to evaluate rectosigmoid area and with urography to visualize the junction between the ureters and the posterior portion of the urinary bladder.

13. A   14.   F   15.   D

16. A   A radiographic image of the foot is obtained in the AP oblique projection. The patient is placed with the affected foot on top of a 24 x 30 centimeter image receptor The foot is laterally rotated until the plantar surface of the foot forms an angle of 30 degrees with the plane of the image receptor. The central ray should be directed perpendicular to the base of the third metatarsal. Because of the natural curve of the anterior aspect of the foot, a lateral oblique projection improves the visualization of the articulations between the medial and intermediate cuneiforms and the first and second metatarsal bones.

17. C   A radiographic image of the knee is obtained in the lateral projection. The patient is turned on to the affected side to place the knee in a lateral position. The knee should be flexed to 20-30 degrees to relax the muscles and to demonstrate the maximum volume of the of the joint space. The central ray is directed to the knee joint 2.5 centimeters distal to the medial epicondyle at an angle of 5-7 degrees cephalic to reduce the magnification of the more distal femoral condyle and improve the visualization of the joint space.

18. A   A radiographic image of the hip is obtained in the axiolateral (Danelius-Miller) projection. This projection is also called shoot-through hip. This projection is performed using a stationary grid that is placed vertically parallel to the femoral neck. With the patient supine, the unaffected leg is raised to enable the central ray to be directed horizontal with the table and perpendicular to the femoral neck. This provides lateral projection of the upper femur and the acetabulum. When a grid is employed, its visualization is improved by placing the grid strips parallel with the plane of the femoral neck.

19. A   A radiographic image of the ankle is obtained in the AP oblique projection. The ankle is rotated medially by about 45 degrees and the toes are dorsiflexed. The central ray is directed perpendicular and enters at the midpoint of the malleoli. This projection is used to maximize the opening of the tibiofibular articulation.

20. C   A radiographic image of the calcaneus is obtained in the axial (plantodorsal) projection. The foot is placed vertically on the image receptor with toes in dorsiflexion. With the foot vertical, the central ray is directed to enter at the level of the base of the third metatarsal with a 40 degree angulation to the long axis, exiting proximal to the posterior ankle joint. This projection is used to obtain a frontal projection of the calcaneus and the subtalar joint.

21. C   A radiographic image of the femoral neck is obtained in the anteroposterior (AP) oblique (Cleaves) projection. With the patient supine, the hips and knees are abducted with the soles of the feet together as much as is possible to place the femoral necks parallel to the image receptor. The top of a crosswise 35 x 43 centimeter image receptor is positioned just above the iliac crests. The central ray is directed perpendicular to the imaging system for the modified AP oblique (Cleaves) projection. In the axiolateral (original Cleaves) projection, a 25-45 degrees cephalic angulation of the central ray is employed. Both projections are used to demonstrate an oblique projection of the femoral heads, necks and trochanteric regions.

22. C   A radiographic image of the foot is obtained in the AP oblique projection. The foot is rotated medially until the plantar surface of the foot forms a 30 degree angle with the plane of the image receptor. The central ray should be directed perpendicular to the base of the third metatarsal. This projection is used to demonstrate the interspaces between the cuboid bone, calcaneus and fourth and fifth metatarsal bones and the lateral conforms as well as the articulations between the navicular and talus, navicular and cuneiform, and the sustentaculum tali and talus bone. This projection often provides an unobstructed view of the entire cuboid bone.

23. B   Since all of these radiographic images of the patella are obtained in the tangential projection, they will require the knee to be placed in acute flexion. Therefore, transverse fractures of the patella which can be displaced by this motion, must be ruled out prior to attempting these projections.

24. C   A radiographic image of the hip is obtained in the anteroposterior (AP) projection. The central ray is directed to enter a point about 6 centimeters distal to a line drawn perpendicular to the midpoint between the symphysis pubis and the anterior superior iliac spine(ASIS).

25. A   A radiographic image of the ankle is obtained in the lateromedial lateral projection. The foot is placed so the medial surface is in contact with 18 x 24 centimeter image receptor. The toes are dorsiflexed to eliminate lateral rotation of the ankle. The central ray is directed perpendicular and enters at the medial malleolus. Because the medial side of the ankle is less prominent, this position enables the ankle joint to be closer to the image receptor than in the mediolateral lateral position.

26. C   A radiographic image of the toe is obtained in the AP projection. The toes are placed on the image receptor with the central ray directed perpendicular to the center of the third digit at the third metacarpophalangeal joint. When there is a need to open the joint spaces, the AP axial projection with 15 degree posterior angulation should be employed.

27. B   A trimalleolar fracture is associated with the injury to the three malleoli at the ankle; namely, the lateral malleolus of the fibula, the medial malleolus and the lateromedial malleolus of the tibia. A radiographic image of the ankle obtained in the AP projection should clearly show this injury.

28. B   A radiographic image of the foot is obtained in the AP dorsoplantar oblique projection. In the projection, the leg is rotated 45 degrees medially to form a 30 degree angle between the plantar surface of the foot and the plane of the image receptor.

29. A   There are three methods for visualizing the intercondylar fossa of the knee; the PA axial (Camp Coventry) projection in which the patient lays prone with the lower leg supported at an angle of about 40 degrees, the PA axial (Holmblad) projection in which the patient rests on the knee with the lower leg resting on the table, and the AP axial (Beclere) where the knee rests on a curved image receptor. In all of these projections it is essential to direct the central ray parallel to the tibial plateau and perpendicular to the long axis of the tibia. These projections are often used to demonstrate calcifications of the joint called joint mice.

30. I   31. G   32. D   33. A   34. F   35. B

36. J

37. D   The dome of the acetabulum is located at a point about 6 centimeters distal to a line drawn perpendicular to the midpoint between the symphysis pubis and the anterior superior iliac spine (ASIS).

38. B   A radiographic image of the calcaneus is obtained in the axial (plantodorsal) projection. The foot is placed vertically on the image receptor with toes in dorsiflexion. The central ray is directed axially to a vertical foot. It is directed to enter the foot at the level of the base of the third metatarsal at a cephalic angulation of 40 degrees, exiting proximal to the posterior ankle joint. This projection is used to obtain a frontal projection of the calcaneus and the subtalar joint.

39. B   Because of the relationship of the malleoli, a radiographic image of the ankle is obtained in the AP oblique projection with a medial rotation of the leg and foot together of about 15-20 degrees. This projection will place the two malleoli nearly parallel to the image receptor and improve the visualization of the ankle mortise.

40. B   Radiographic images of the anterior pelvic bones obtained in the PA projection or a PA axial projection, are generally preferred over AP projections, because the patient is placed in the prone position resulting in a reduced part-to-image receptor distance and reduction in the amount of magnification occurring in the image.

41. A   A radiographic image of the anterior pelvic bones is obtained in the AP axial (Taylor) projection on a female patient. The central ray should be directed at a cephalic angulation of 30-45 degrees. In a male patient, the degree of angulation required is reduced to 20-30 degrees cephalad. In either situation, the central ray is directed to enter at a point 5 centimeters distal to the pubic symphysis. This projection is employed to provide a magnified view of the ischial and pubic rami projected over the sacrum and coccyx and images of the obturator foramen.

42. A   A radiographic image of the foot is obtained in the anteroposterior (AP) axial projection. The foot should be placed flat on a 24 x 30 centimeter lengthwise image receptor. The central ray should be directed perpendicular to the base of the third metatarsal. A perpendicular central ray is used for general survey of the foot and foreign body localizations. A 10 degree angle toward the heel may be used to improve the visualization of the bony articulations in the middle region of the foot.

43. D   A radiographic image of the knee is obtained in the anteroposterior (AP) projection on a patient measuring 22 centimeters from the ASIS to the table top. The central ray should be directed perpendicular to a point 1.5 centimeters distal to the apex of the patella to pass through the joint space of the knee. If the patient measures less than 18 centimeters from the ASIS to the table top, the central ray should be angled 5 degrees caudad. In patients measuring more than 25 centimeters, the central ray should be angled 5 degrees cephalad.

44. D   A radiographic image of the hip is obtained in the anteroposterior (AP) projection. In the relaxed AP position of the leg, the femoral neck is tilted obliquely from the pelvis (anteversion). If the lower limb is inverted 15-20 degrees, the femoral neck will more closely parallel the plane of the image receptor, thus improving the visualization of both the femoral neck and greater trochanter.

45. B   A radiographic image of the ankle is obtained in the AP projection with inversion stress. In this projection, the appearance of a widened joint space is nearly always related to tears in the collateral ligaments. Inversion stress is used to indicate damage to the lateral collateral ligament. Eversion stress is used to demonstrate tears of the medial collateral ligament. Stress studies are normally performed after the acute pain of the injury has subsided or following the administration of a local anesthetic.

46. A   A radiographic image of the toe is obtained in the AP projection. The toes are placed on the image receptor with the central ray directed perpendicular to the center of the third digit at the third metacarpophalangeal joint. When there is a need to open the joint spaces, the AP axial projection using a 15 degree posterior angulation should be employed. Since the joint spaces between the toes are directed at about a 15 degree angle to the horizontal, a 15 degree cephalic central ray angulation is required. This can also be accomplished by placing the toes on a 15 degree wedge and directing the central ray perpendicular to the image receptor.

47. A   When the unaffected leg cannot be raised to enable a shoot-through position of the hip, a lateral view may be obtained using the Clements Nakayama method. With the patient supine, a cassette is placed vertically alongside the affected hip, parallel to the femoral neck and perpendicular to the table. The central ray is directed at a posterior angle of 15 degrees to pass through the femoral neck.

48. A   A radiographic image of the foot is obtained in the AP axial projection. The foot should be placed flat on a 24 x 30 centimeter lengthwise image receptor. The central ray should be directed perpendicular to the base of the third metatarsal. A perpendicular central ray is used for general survey of the foot and foreign body localizations. A 10 degree angulation of the central ray toward the heel can be used to improve the visualization of the bony articulations in the middle region of the foot.

49. D   A radiographic image of the patellofemoral joint is obtained in the tangential (Hughston) projection. If the knee is flexed to form a 55 degree angle with the image receptor, the central ray should be directed through the joint space at a cephalic angulation of 45 degrees through the patellofemoral joint.

50. A   A radiographic image of the hip is obtained in the axiolateral (Danelius-Miller) projection. This projection is also called shoot-through hip. This projection is performed using a stationary grid that is placed vertically parallel to the femoral neck. With the patient supine, the unaffected leg is raised to enable the central ray to be directed horizontal with the table and perpendicular to the femoral neck. This projection provides lateral projection of the upper femur and the acetabulum. When a grid is employed, its visualization is improved by placing the grid strips parallel with the plane of the femoral neck. On the resultant image, the ischial tuberosity lies posterior to the dome of the acetabulum, femoral head and neck. The greater trochanter is seen just distal to the neck projecting posteroanteriorly with the lesser trochanter seen toward the posterior surface of the body.

51. A   A radiographic image of the metacarpophalangeal sesamoid bones is obtained in the tangential (Lewis and Causton) projection. The foot is placed vertically over the image receptor with the toes dorsiflexed. The central ray is directed perpendicular to the ball of the foot superoinferiorly to pass through the joint. Other methods for demonstrating the sesamoid bones are the inferosuperior tangential projection and the lateromedial tangential (Causton) projection.

52. D   A radiographic image of the lower leg is obtained in the AP projection. The patient is supine with their leg resting in an 18 x 43 centimeter image receptor with the foot in the vertical position. The central ray is directed perpendicular to the midtibia. This projection shows the tibia and fibula in the AP position and should include both the ankle and knee joints when possible.

53. B   A radiographic image of the knee is obtained in the AP oblique projection. In order to prevent overlapping of the proximal tibiofibular articulation, the leg should be medially rotated 45 degrees toward the affected side. Because the head of the fibula is located more laterally and distal to the knee joint, a medial rotation will improve the demonstration of the tibiofibular articulation.

54. B   A radiographic image of a postoperative fractured hip is to be obtained for both the axiolateral and AP projections. It is important to include the entire prosthesis on the radiographic images to ensure its alignment and position within the shaft of the femur.

55. K   56. D   57. A   58. F   59. H   60. I

61. C   62. G   63. J   64. E

65. C   A radiographic image of the knees is obtained in standing AP projection to evaluate arthritis. With the patient standing in front of the image receptor, the central ray is directed horizontal to the knee joints. This projection is also used in the evaluation of the degree of valgus (outward turning) or varus (inward turning) deformities of the hip and knees.

66. B   Radiographic images of the feet of a child are taken in the AP and lateral (Kite and Kandel) projections. These are most often used to evaluate talipes equinovarus or club feet. Kite described the appropriate dorsoplantar AP and lateral projections to best show this condition. Kandel described the dorsoplantar axial projection with the infant standing and the central ray angled 35, 45, and 55 degrees caudal, entering the posterior lower leg just superior to the calcaneus and exiting the plantar portion of the foot.

67. D   A radiographic image is obtained of the femoral necks in the axiolateral (original Cleaves) projection. With the patient supine, the hips and knees are abducted with the soles of the feet together as much as is possible to place the femoral necks parallel to the image receptor. The top of a crosswise 35 x 43 centimeter image receptor is positioned just above the iliac crests. In the axiolateral (original Cleaves) projection, a 25-45 degrees cephalic angulation of the central ray is employed. These projections are used to demonstrate an oblique projection of the femoral heads, necks and trochanteric regions.

68. B   A radiographic image of the ankle is obtained in the anteroposterior (AP) projection. Since the ankle joint is located in the same plane as the malleoli, a vertical central ray directed between the malleoli will best demonstrate the joint.

69. A   Radiographic images of the subtalar joint are normally obtained in the AP axial oblique (Broden) projections. With the foot vertical and medially rotated 45 degrees, the central ray is directed to the subtalar joint at angles of 10, 20, 30, and 40 degrees cephalad. The subtalar (talocalcaneal) joint which may be compromised in comminuted fractures of the foot, is also demonstrated in any number of axiolateral or oblique axial positions as described by Broden, Isherwood, and Feist-Mankin.

70. D   There are three methods for visualizing the intercondylar fossa of the knee: the PA axial (Camp Coventry) projection in which the patient lays prone with the lower leg supported at an angle of about 40 degrees, the PA axial (Holmblad) in which the patient rests on the knee with the lower leg resting on the table, and the AP axial (Beclere) where the knee rests on a curved cassette. In all positions it is essential to direct the central ray parallel to the tibial plateau and perpendicular to the long axis of the tibia.

71. C   A radiographic image of the ilium is obtained using an AP oblique projection to evaluate the broad surface (wing) without rotation. Because the wing of the ilium is projected obliquely at an angle of about 40 degrees, an AP oblique projection with the unaffected side elevated 40 degrees will place the affected ilium parallel to the image receptor. The central ray is directed perpendicular to enter the middle of the ilium. This projection is used to demonstrate the entire ilium and the structures of the hip and upper femur.

72. D   There are three methods for visualizing the intercondylar fossa of the knee; the PA axial (Camp Coventry) projection in which the patient lays prone with the lower leg supported at an angle of about 40 degrees, the PA axial (Holmblad) in which the patient rests on the knee with the lower leg resting on the table, and the AP axial (Beclere) where the knee rests on a curved cassette. In all positions it is essential to direct the central ray parallel to the tibial plateau and perpendicular to the long axis of the tibia. Since all intercondylar projections require that the central ray be directed perpendicular to the tibial plateaus and long axis of the lower leg, no tube angulation is required in the PA axial (Holmblad) projection.

73. A   A radiographic image of the foot is obtained in the lateral (weight-bearing) projection. With the patient standing next to a vertical image receptor, the central ray is directed horizontally to a point just above the base of the third metatarsal. This projection is taken to show the longitudinal arch which extends from the metatarsal bones along the plantar surface of the foot to the calcaneus. The flattening of the arch, pes planus, (flat feet) is most clearly seen when weight-bearing lateral images are obtained.

74. D   A radiographic image of the foot in the composite view is obtained in the AP axial projection. This composite view requires that two separate exposures are made to produce a single image. This method requires the use of two axial projections; the first requires that the central ray to be directed anteriorly at a 25 degrees angulation, the second with the central ray directed posteriorly at an angle of 15 degrees. The resulting image provides a unique composite image of the weight-bearing foot without the shadows of the bones of the lower leg.

75. C   A series of radiographic images are made to measure the length of the lower leg. In order to insure the proper orthoroentgenographic measurements are obtained it is important to place the legs in an anatomic position and place a metallic ruler next to the legs. The central ray is directed over each joint in sequence without any movement of the patient. Recent advances in CT has allowed accurate measurements of the extremities at reduced exposures by use of computed tomographic scanography.

76. D   77. E   78. K   79. G   80. B   81. H

82. L   83. F   84. A   85. J

86. A   There are three methods for visualizing the intercondylar fossa of the knee; the PA axial (Camp Coventry) position in which the patient lays prone with the lower leg supported at an angle of about 40 degrees, the PA axial (Holmblad) in which the patient rests on the knee with the lower leg resting on the table, and the AP axial (Beclere) where the knee rests on a curved cassette. In all positions it is essential to direct the central ray parallel to the tibial plateau and perpendicular to the long axis of the tibia.

87. D   See the explanation above.

88. B   A radiographic image of the patella is obtained in the tangential (Settegast) projection. This projection, also known as the sunrise view, is taken with the knee flexed to a right angle with the central ray directed to the knee joint parallel to the patellofemoral joint. This projection is often used to provide a profile view of the patella for the evaluation of vertical fractures and the patellofemoral articulation.

89. C   See the explanation above.

90. C   A radiographic image of the femoral neck is obtained in the anteroposterior (AP) oblique (modified Cleaves) projection. With the patient supine, the hips and knees are abducted with the soles of the feet together as much as is possible to place the femoral necks parallel to the image receptor. The top of a crosswise 35 x 43 centimeter image receptor is placed positioned just above the iliac crests. The central ray is directed perpendicular to the imaging system for the modified AP oblique (Cleaves) projection. In the axiolateral (original Cleaves) projection, a 25-45 degree cephalic angulation of the central ray is employed. Both projections are used to demonstrate an oblique projection of the femoral heads, necks and trochanteric regions.

91. C   See the explanation above.

92. D   This radiographic image of the knee of a child was obtained in the AP projection. The technical factors selected by the radiographer were insufficient to provide the appropriate optical density on the image. The appropriate correction would require a doubling of the mAs value or a 15% increase in the kVp setting.

93. B   The radiolucent lines represent the cartilage at the growth plates called the epiphyseal plate.

94. A   This radiographic image of the knee was obtained in the AP projection weight-bearing position. In this standing projection, this technique is used to show changes in the joint space and help demonstrate varus and valgus deformities. This image shows the narrowing of the joint space on the medial surface of the tibial plateau.

95. A   See the explanation above.

1. The best projection for the evaluation of the lumbar vertebral bodies and the intervertebral disc spaces is a radiographic image obtained in the:

    A. Anteroposterior (AP) projection
    B. AP oblique projection
    C. PA oblique projection
    D. Lateral projection

2. In order to obtain a lateral projection of the cervical spine in flexion, the patient is placed in the true lateral position with the:

    A. Chin lowered toward the chest
    B. Chin raised as much as possible
    C. Body of the mandible parallel to the floor
    D. Ramus of the mandible parallel to the floor

3. A lead strip, which is placed just behind the posterior surface of a patient for a radiographic image of the lumbar spine in the lateral projection, serves to:

    A. Absorb scattered radiation produced in the patient
    B. Filter radiation from the x-ray beam
    C. Increase the amount of divergence of the beam
    D. Eliminate the need for a radiographic grid

4. A radiographic image of the thoracic spine is obtained in the anteroposterior (AP) projection. With the patient supine, the normal degree of kyphosis can be reduced by:

    A. Flexing the knees
    B. Fully extending the legs
    C. Inverting the toes by 15 degrees
    D. Raising the arms above the head

5. In a radiographic image of the lumbar spine, which of the following projections will best demonstrate the transverse processes of the lumbar vertebrae?

    A. Anteroposterior (AP) projection
    B. AP oblique projection
    C. Axiolateral projection
    D. Lateral projection

6. If a patient is in traction for a cervical spine injury, the radiographic image should normally be obtained:

    A. After removing the patient from the traction apparatus
    B. After removing the weights from the traction apparatus
    C. With the patient in acute flexion and extension
    D. Without adjusting the traction apparatus

7. Which of the following radiographic images are most often obtained to evaluate a patient with lateral scoliosis of the thoracic spine?

    A. An AP projection with the patient supine
    B. A PA projection with the patient erect
    C. A lateral projection with the patient recumbent
    D. A lateral projection with the patient erect

8. A radiographic image is obtained in the lateral (Twining) projection or swimmer's position. This projection is most commonly employed for the evaluation of the:

    A. Atlanto-occipital region of the spine
    B. Atlanto-axial region of the spine
    C. Cervicothoracic region of the spine
    D. Lumbosacral region of the spine

9. In a radiographic image of the lumbar spine in the AP oblique projection, the nose of the "Scotty dog" corresponds to the:

    A. Pedicle of the vertebra
    B. Lamina of the vertebra
    C. Body of the vertebra
    D. Transverse process of the vertebra

10. A radiographic image of the cervical spine is obtained in the AP axial oblique projection to demonstrate open intervertebral foramina. This projection requires the cervical spine be rotated:

    A. 20 degrees away from the image receptor
    B. 45 degrees away from the image receptor
    C. 60 degrees away from the image receptor
    D. 90 degrees away from the image receptor

11. A radiographic image of the sternum is obtained in the right posteroanterior oblique (RAO) projection. This projection will project the image of the sternum:

A. Away from the shadow of the heart
B. Over the shadow of the heart
C. Over the shadow of the thoracic vertebrae
D. Away from the shadow of the posterior ribs

12. Which of the following radiographic projections of the cervical spine should be obtained to demonstrate articulations between the superior and inferior articulating processes of the cervical vertebrae?

A. Anteroposterior axial projection
B. AP axial (Pillar) projection
C. AP axial oblique projection
D. Lateral projection

13. A radiographic study of the lumbar spine is ordered to determine motion in the area of a spinal fusion. Which of the following projections and positions are indicated?

1. AP projection taken with the legs flexed and extended
2. AP projection taken with the patient in the right and left bending positions
3. Lateral projections taken with the patent in flexion and extension

A. 1 & 2 only
B. 1 & 3 only
C. 2 & 3 only
D. 1, 2 & 3

14. If a radiographic image of the sacrum is obtained in the lateral projection, the central ray should be directed perpendicular to enter at a point:

A. 13 centimeters anterior to sacral prominence
B. 9 centimeters posterior to the ASIS
C. 2 centimeters above the iliac crest
D. 5 centimeters above the pubic symphysis

15. A radiographic image of the vertebral arch of the cervical region is obtained in the AP axial projection. The central ray should be directed to enter the neck at the level of the thyroid cartiledge at a:

A. Cephalic angle of 10-20 degrees
B. Caudal angle of 20-30 degrees
C. Caudal angle of 10-15 degrees
D. Cephalic angle of 45-55 degrees

16. A radiographic image of the upper thoracic and lower cervical vertebrae is obtained in the lateral (Twining) projection. The central ray is directed 5 degrees caudal with the arm closest to the image receptor:

A. Raised above the head
B. Lowered to the side
C. Extended at a right angle

17. A radiographic image of the sternum is obtained in the lateral projection. In order to improve contrast in the resulting image, which of the following could be employed?

A. A 180 centimeter (72") source-to-image receptor distance
B. A radiographic exposure that is made on full inspiration
C. A radiographic exposure that is made on full expiration
D. Positioning the patent with their shoulders rotated posteriorly

18. A radiographic image of the lumbar spine is obtained in the anteroposterior oblique projection. This projection is most commonly employed for the evaluation of the:

A. Dependent zygapophyseal articulations
B. Independent zygapophyseal articulations
C. Dependent intervertebral foramina
D. Independent intervertebral foramina

19. A radiographic image of the sternoclavicular joints is obtained in the PA oblique projection. The proper degree of obliquity to best evaluate an affected SC joint is a:

A. 10-15 degree rotation
B. 25-35 degree rotation
C. 45-55 degrees rotation
D. 60-70 degree rotation

20. A radiographic image of the thoracic spine is obtained in the lateral projection. What action can be taken to prevent the ribs from obscuring the intervertebral foramina?

A. Directing the central ray 15 degrees caudal
B. Adjusting the arms to right angles
C. Taking the exposure on full inhalation
D. Directing the central ray 5 degrees caudal

21. A radiographic image of the cervical spine is obtained in the AP axial oblique projection. This projection is commonly used for the evaluation of the:

A. Pedicles closest to the image receptor
B. Pedicles farthest from the image receptor
C. Spinus processes closest to the image receptor
D. Spinus processes farthest the image receptor

22. A patient comes from the emergency room with a possible cervical spine fracture. The preliminary radiographic image that should be obtained for this purpose is:

A. A lateral projection in the dorsal decubitus position
B. A lateral projection in extension with the patient prone
C. A lateral projection in flexion with the patient erect
D. A lateral projection in the erect position holding weights

23. In a radiographic image of the lumbar spine in the AP oblique projection, the "Scotty dogs" are seen when the patient is properly positioned. The front feet of the " dog" corresponds to the:

A. Lamina of the vertebra
B. Pedicle of the vertebra
C. Superior articular process of the vertebra
D. Inferior articular process of the vertebra

24. Which of the following radiographic projections of the thoracic spine is most commonly employed to demonstrate the intervertebral disc spaces?

A. An anteroposterior projection
B. A 45 degree AP oblique projection
C. A 15 degree PA oblique projection
D. A lateral projection

25. A radiographic image of the right sacroiliac (SI) joint is to be obtained in the AP oblique projection. The patient should be placed in the:

A. LAO position with the affected joint farthest from the image receptor
B. RAO position with the affected joint farthest from the image receptor
C. LAO position with the affected joint closest to the image receptor
D. RAO position with the affected joint closest to the image receptor

26. A radiographic image of the thoracic spine is obtained in the AP projection. In order to maintain a uniform optical density, which of the following conditions should be employed?

1. The anode should be placed over the region having the greatest thickness
2. The anode should be placed over the Bucky side of the table
3. The cathode should be placed over the region having the greatest thickness

A. 1 only
B. 2 only
C. 3 only
D. 1, 2 & 3

27. Which of the following radiographic projections should be employed to determine the presence or absence of cervical ribs?

A. An AP oblique projection
B. A lateral projection
C. A PA oblique projection
D. An AP axial projection

28. A radiographic image of the cervical spine is obtained in the lateral projection. A source-to-image receptor distance of 180 (72") centimeters should be employed to:

A. Reduce exposure received by the patient
B. Increase the magnification of the vertebrae
C. Decrease the magnification of the vertebrae
D. Improve radiographic contrast on the image

29. A radiographic image of the coccyx is obtained in the PA axial projection. The central ray is directed to tip of the coccyx at a:

A. Cephalic angulation of 10 degrees
B. Caudal angulation of 10 degrees
C. Cephalic angulation of 25 degrees
D. Caudal angulation of 25 degrees

30. Which projection of the spinal column will most likely demonstrate a lordotic curve of the spine?

A. An AP axial projection of the cervical spine
B. A lateral projection of the thoracic spine
C. An AP projection of the lumbar spine
D. A lateral projection of the cervical spine

31. Which of the following projections of the lumbar spine is most commonly employed to demonstrate the zygapophyseal articulations between the lumbar vertebrae?

A. An AP projection
B. An AP 30 degree oblique projection
C. An AP 45 degree oblique projection
D. A lateral projection

32. A radiographic image of the odontoid process is obtained in the AP projection open-mouth position. The patient should open the mouth as wide as possible with a line drawn from the lower edge of the upper incisor to the:

A. Mastoid tip placed perpendicular to the image receptor
B. Mental process placed perpendicular to the image receptor
C. External auditory meatus placed perpendicular to the image receptor
D. Inion (EOP) placed perpendicular to the image receptor

33. A radiographic image of the sacrum is obtained in the AP axial projection. In order to compensate for the normal degree of kyphosis of the sacrum, the central ray should be directed at a:

A. Cephalic angulation of 5 degrees
B. Caudal angulation of 5 degrees
C. Cephalic angulation of 15 degrees
D. Caudal angulation of 25 degrees

34. A radiographic image of the thoracic spine is obtained in the lateral projection. The patient's arms should be elevated no more than 90 degrees from the body. Any more than this will cause:

1. The correction of an abnormal amount of kyphosis
2. The scapulae to obscure the upper thoracic vertebrae
3. Unwanted discomfort to the patient

A. 1 only
B. 2 only
C. 3 only
D. 1, 2 & 3

35. A radiographic image of the cervical spine is obtained in the PA axial oblique projection. This projection is commonly employed to demonstrate the:

A. Vertebral foramina closest to the image receptor
B. Transverse foramina closest to the image receptor
C. Zygapophyseal articulations closest to the image receptor
D. Intervertebral foramina closest to the image receptor

36. A radiographic image of the thoracic spine is obtained in the AP oblique projection. In order to demonstrate the zygapophyseal articulations between the thoracic vertebrae, the patient should be rotated so the midcoronal plane forms an angle of:

A. 15 degrees with the plane of the image receptor
B. 25 degrees with the plane of the image receptor
C. 45 degrees with the plane of the image receptor
D. 70 degrees with the plane of the image receptor

37. A radiographic image of the odontoid process (dens) is obtained in the AP projection open mouth position. The resultant radiograph demonstrated the occipital bone obscuring the dens. In order to correct for this positioning error, the radiographer should:

A. Tuck the patient's chin closer to the chest
B. Raise the patient's chin farther away from the chest
C. Oblique the patient's head by 10 degrees
D. Place the head in the lateral position

38. A radiographic image of the right sacroiliac (SI) joint is to be obtained in the AP oblique projection. The joint space is best visualized when the patient is rotated:

A. 15 degrees to the plane of the image receptor   C. 45 degrees to the plane of the image receptor
B. 25 degrees to the plane of the image receptor   D. 70 degrees to the plane of the image receptor

39. A radiographic image of the thoracic spine is obtained in the AP projection. The central ray should be directed perpendicular to the:

A. Second thoracic vertebra   C. Seventh thoracic vertebra
B. Fourth thoracic vertebra   D. Tenth thoracic vertebra

40. A radiographic image of the L5-SI interspace is obtained in the AP axial projection. The central ray should be directed through the lumbosacral joint with a:

A. Caudal angulation of 15-25 degrees   C. Cephalic angulation of 30-35 degrees
B. Cephalic angulation of 10-15 degrees   D. Caudal angulation of 45-55 degrees

41. Which radiographic projection of the cervical spine will most clearly demonstrate the spinous processes of the cervical vertebrae?

A. AP axial projection   C. Axiolateral projection
B. PA oblique projection   D. Lateral projection

42. Which of the following positions should be employed to obtain a radiographic image on a patient that has injured their anterior 5th rib?

1. Erect PA position   **2. An AP position on inspiration**   **3. Prone position on expiration**

A. 1 & 2 only   C. 2 & 3 only
B. 1 & 3 only   D. 1, 2 & 3

43. A radiographic image of the cervical spine is obtained in the AP axial oblique projection. This projection is commonly used for the evaluation of the:

A. Vertebral foramina farthest from the image receptor
B. Transverse foramina farthest from the image receptor
C. Intervertebral foramina farthest from the image receptor
D. Zygapophyseal articulations farthest from the image receptor

44. A radiographic image of the lumbar spine is obtained in the lateral projection. This projection is most often requested to demonstrate the:

1. Transverse processes of the vertebrae
2. Spinous processes of the vertebrae
3. Height of the vertebral disc spaces

A. 1 & 2 only   C. 2 & 3 only
B. 1 & 3 only   D. 1, 2 & 3

45. A radiographic image of the thoracic spine is obtained in the lateral projection. If the lower vertebral column is not elevated to the horizontal position, the central ray should be directed to the level of T7 at a:

A. Cephalic angulation of 10-15 degrees   C. Caudal angulation of 10-15 degrees
B. Caudal angulation of 25-35 degrees   D. Cephalic angulation of 25-30 degrees

46. A radiographic image of the odontoid process is obtained in the AP (Fuchs) projection. The chin should be extended until:

A. The infra-orbital margin is parallel to the image receptor
B. The acanthion and glabella are vertical
C. The lower incisors are perpendicular to the image receptor
D. The tip of the chin and mastoid process at vertical

47. A radiographic image of the costal joints is obtained in the AP axial projection for the demonstration of rheumatoid spondylitis. The central ray is directed to enter the midsagittal plane 5 centimeters above the xiphoid process at a:

A. Caudal angulation of 50 degrees
B. Cephalic angulation of 45 degrees
C. Caudal angulation of 30 degree
D. Cephalic angulation of 20 degrees

48. A radiographic image of the sternoclavicular joints is obtained in the axiolateral (Kurz Bauer) projection. The patient is placed in the lateral recumbent position on their affected side with the arm raised. The central ray is directed through the sternoclavicular joint closest to the image receptor at a:

A. Caudal angulation of 35 degrees
B. Caudal angulation of 15 degrees
C. Cephalic angulation of 25 degrees
D. Cephalic angulation of 12 degrees

49. A radiographic image of the cervical spine is obtained in the lateral projection. The use of a radiographic grid is generally NOT required because of the scatter reduction that occurs:

A. From the use of the anode heel effect
B. From the use of a cylinder or cone
C. From the use of the natural air gap
D. From the use of a high kVp technique

50. A radiographic image of the cervical spine is obtained in the AP axial projection. A 15-20 degree cephalic angulation of the central ray should be employed to better define the:

A. Cervical rib facets
B. Cartiledge of the larynx
C. Atlanto-occipital joint
D. Intervertebral disk spaces

51. A radiographic image of the thoracic spine is obtained in the PA oblique projection. The patient is placed in a right lateral recumbent position and rotated anteriorly 20 degrees. The central ray is directed through T7. The structures best demonstrated would be the:

A. Right zygapophyseal articulations
B. Left zygapophyseal articulations
C. Right intervertebral foramen
D. Left intervertebral foramina

52. A radiographic image obtained in the PA axial oblique (Kovacs) projection is most often taken to demonstrate the:

A. Last lumbar zygapophyseal articulation
B. Last lumbar spinous process
C. L5-S1 interspace
D. Last lumbar intervertebral foramen

53. A radiographic image of the cervical spine is obtained in the AP (Ottonello) projection while the patient is asked to open and close their mouth during the exposure. This maneuver is employed in the evaluation of:

A. Basilar fractures
B. Cervical instability
C. Cervical tumors
D. The entire cervical spine

54. A radiographic image of the symphysis pubis is taken in the PA (Chamberlain) projection with the patient erect. This weight-bearing position is used to evaluate abnormal slippage of the:

A. Sacrococcygeal region
B. Sacroiliac joints
C. L5-S1 junction
D. Humeral head

55. A radiographic image of the cervical spine is obtained in the AP axial projection to demonstrate the lower cervical vertebrae. The central ray should be directed 15 degrees cephalic to enter at the:

A. Mental point
B. Gonion
C. Thyroid cartilage
D. Hyoid bone

56. Which of the following projections is most commonly used to more closely align the intervertebral disk spaces of the lumbar spine with the divergence of the x-ray beam?

A. The AP projection
B. The left lateral projection
C. The right lateral projection
D. The PA projection

57. A radiographic image of the entire spine is obtained to evaluate a patient with lateral scoliosis. Which of the following projections can be used to reduce the radiation dosage to the thyroid, gonads and breasts of the patient?

A. The AP projection
B. The PA oblique projection
C. The PA projection
D. The lateral projection

58. A radiographic image of the atlas and axis is obtained in the AP (open mouth) projection. In order to keep the tongue in the floor of the mouth the patient is instructed to:

A. Breath rapidly during the exposure
B. Softly phonate "ah" during the exposure
C. Slowly expire during the exposure
D. Suspend respiration during the exposure

59. A radiographic image of the lumbar spine is obtained on a female patient in the lateral projection. If the spine is not adjusted to the horizontal plane for the exposure, the central ray should be directed:

A. Vertically to the image receptor
B. At a cephalic angulation of 5 degrees
C. At a caudal angulation of 8 degrees
D. At a caudal angulation of 12 degrees

60. A radiographic image of the axillary portion of the right ribs can be best demonstrated in the:

A. AP oblique (LPO) projection
B. PA oblique (LAO) projection
C. Right lateral projection
D. Apical lordotic projection

Referring to the radiographic image, answer questions 61 & 62.

61. This image was obtained with the patient positioned for the:

A. AP projection
B. AP 20 degrees oblique projection
C. AP 45 degrees oblique projection
D. Lateral projection

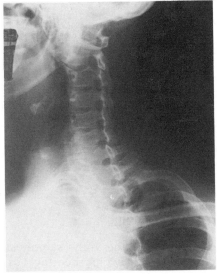

62. This projection is most frequently used for the evaluation of the:

1. Intervertebral foramina
2. Odontoid process
3. Zygapophyseal joints

A. 1 only
B. 2 only
C. 3 only
D. 1, 2 & 3

Referring to the radiographic image, answer questions 63 & 64.

63. This radiographic image was obtained with the patient positioned for the:

A. AP projection
B. PA 20 degree oblique projection
C. AP 45 degree oblique projection
D. Lateral projection

64. This projection is most frequently used for the evaluation of the:

    1. Intervertebral foramina
    2. Odontoid process
    3. Zygapophyseal joints

    A. 1 only   C. 3 only
    B. 2 only   D. 1, 2 & 3

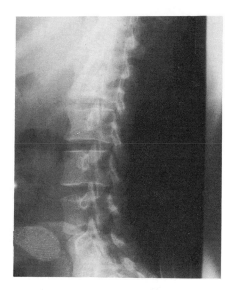

Referring to the radiographic image, answer question 65.

65. The low optical density (Harrington) rods seen on this PA projection of the spine are used in the treatment of:

    A. Severe scoliosis
    B. Whip lash injuries
    C. Disk slippage
    D. Sacroiliac dislocation

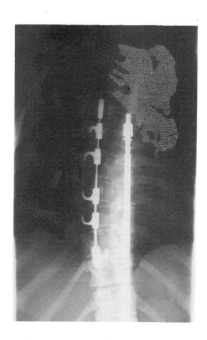

Explanations

1. D   A radiographic image of the lumbar spine is obtained in the lateral projection. The patient is placed on their side with the hips and knees flexed to make the patient comfortable. If the spine of the patient is adjusted to the horizontal plane, the central ray is directed perpendicular to enter at the level of the iliac crest. If the spine is not adjusted to the horizontal plane, in the male patient, the central ray should be directed with a caudal angulation of 5 degrees, in a female patient, the CR is directed with a caudal angulation of 8 degrees. This projection is used to evaluate the disk spaces between the vertebral bodies and spinous processes of the lumbar vertebrae.

2. A   A radiographic image of the cervical spine is obtained in the lateral projection. The patient is placed erect and asked to depress the shoulder to improve the visualization of the lower vertebrae. The central ray is directed horizontally to enter at the level of the 4th cervical vertebra. A 180 centimeter source-to-image receptor distance is employed to reduce the magnification of the spine. This projection is used to demonstrate the vertebral bodies and their interspaces, the zygapophyseal joints and spinous processes of the cervical spine After cervical fractures have been ruled out, lateral projections in hyperflexion and extension may be ordered to determine the degree of anterior/posterior motion of the spine. The appropriate extension images are obtained with the chin raised so the body of the mandible is parallel to the floor and fully flexed when the body of the mandible is nearly perpendicular to the floor.

3. A   A radiographic image of the lumbar spine is obtained in the lateral projection. Because a large percentage of scatter radiation may be produced in the patient, the use of a leaded shield behind the patient may help absorb some of the scattered radiation that would otherwise reach the image receptor. This will serve to improve the radiographic contrast of the resulting image.

4. A   A radiographic image of the thoracic spine is obtained in the anteroposterior (AP) projection. The patient is placed supine, with the central ray directed perpendicular to enter between the jugular notch and the xiphoid process at C7. The normal kyphosis of the dorsal spine can be reduced by flexing the knee and hips to place the back in contact with the table. This projection is used to evaluate the vertebral bodies, the intervertebral disk spaces, the transverse processes and the costovertebral articulations.

5. A   Any radiographic image of the spinal column obtained in the AP or PA projections will demonstrate the transverse processes of the vertebra, the vertebral bodies, the intervertebral disk spaces.

6. D   During a radiographic evaluation of a patient with a possible injury to the cervical spine, cervical collars or other types of support devices placed on the patient by a physician should remain in place during imaging procedure to reduce further injury due to subluxation or complications due to movement of bone fragments. When a patient is in traction, all images should be obtained without having any modification to the traction apparatus.

7. B   A radiographic image of the thoracic and lumbar spines is obtained in the PA projection to evaluate an abnormal lateral curvature of the spine due to the scoliosis. The patient should be placed erect with the central ray directed to enter at the midpoint of the scoliosis image receptor. The elevation of the hip on the concave side of the curve can be used to help distinguish between compensatory and deforming curves. The PA projection is employed during the examination to reduce the exposure to the thyroid glands, ovaries and breasts in female patients.

8. C   A radiographic image is obtained in the lateral (Twining) projection or swimmer's position. The patient is placed in the lateral position with the arm closest to the image receptor raised and the other arm depressed. The central ray is directed perpendicular to the C7-T1 interspace. This projection is most commonly employed for the evaluation of the lower cervical and upper dorsal vertebra when the overlaying structures of the shoulder joints are superimposed over the vertebra. This will project the C7-T2 vertebrae between the shoulders, minimizing the degree of bony superimposition.

9. D   A radiographic image of the lumbar spine is obtained in the AP oblique projection with 45 degree rotation of the patient. With the patient in the RPO position, the central ray is directed perpendicular to enter 5 centimeters medial to the ASIS at a level of the iliac crest. This shows an oblique projection of the lumbar vertebrae and zygapophyseal joints closest to the image receptor. These are the joints between the superior and inferior articulating processes of the vertebra. Radiographically, this image of the area should resemble an image of a "Scotty dog". The nose is formed by the transverse process, the eye is from the pedicle and the body from the lamina of the vertebra.

10. B   A radiographic image of the cervical spine is obtained in the AP axial oblique projection to demonstrate open intervertebral foramina. The patient is rotated obliquely to form a 45 degree angle with the midsagittal plane. The central ray is directed to C4 at a cephalic angulation of 15-20 degrees. The AP axial oblique projections with the patient RPO and LPO demonstrate the foramina and pedicles furthest from the image receptor and open views intervertebral disk spaces of the cervical spine.

11. B   A radiographic image of the sternum is obtained in the right posteroanterior oblique (RAO) projection. The patient is placed in the RAO position with a 15-20 degree rotation. The central ray is directed perpendicular to enter the midsternum at the level of T7. Due to the low contrast between the sternum and overlaying shadows of the ribs, vertebra and lungs, the right PA 15-20 degree oblique RAO position will superimpose the heart and sternum, thus improving the visualization of the sternum. The use of a short source-to-image receptor distance of 15" (38 cm) is also recommended to magnify and blur the posterior ribs. Breathing techniques which tend to blur the lung tissues are also recommended by some authors.

12. D   A radiographic image of the cervical spine obtained in the lateral projection will best demonstrate the zygapophyseal joints. These joints are formed between the superior and inferior articulating processes of the 6 lower cervical vertebrae. The lateral projection is also used to demonstrate the cervical bodies, cervical interspaces and the spinous processes.

13. C   Spinal motion prior to or following a lumbar laminectomy or other lower spinal surgeries is best visualized in lateral projections taken with the patient in flexion and extension and AP projections taken with the patient bending laterally

14. B   A radiographic image of the sacrum and coccyx is obtained in the lateral projection. The patient is placed in the lateral position with the hips and knee in partial flexion. The central ray is directed perpendicular to a point 9 centimeters posterior to the ASIS for the sacrum or 5 centimeters lower for the coccyx. This projection will demonstrate the profile view of the sacrum and coccyx.

15. B   A radiographic image of the vertebral arch of the cervical region is obtained in the AP axial projection. The patient is placed supine with the central ray directed to the thyroid cartiledge at a caudal angulation of 20-30 degrees to place it parallel with the plane of the articular facets. This projection will demonstrate the structures of the vertebral arch including the laminae, the zygapophyseal elements and spinous processes.

16. A   A radiographic image is obtained in the lateral (Twining) projection or swimmer's position. The patient is placed in the lateral position with the arm closest to the image receptor raised and the other arm depressed. The central ray is directed perpendicular or with a 5 degree caudal angle to the C7-T1 interspace. This projection is most commonly employed for the evaluation of the lower cervical and upper dorsal vertebra when the overlaying structures of the shoulder joints are superimposed over the vertebra. This will project the C7-T2 vertebrae between the shoulders minimizing the degree of bony superimposition.

17. B   A radiographic image of the sternum is obtained in the lateral projection. The patient is placed in the lateral position with the shoulders rotated posteriorly. The central ray is directed perpendicular to enter the midsternum at the level of T7. The patient should suspend their breath on full inspiration to fill the lung with more air and provide a higher contrast between the lung and sternum.

18. A   A radiographic image of the lumbar spine is obtained in the AP oblique projection with 45 degree rotation of the patient. With the patient in the RPO or LPO position, the central ray is directed perpendicular to enter 5 centimeters medial to the ASIS at a level of the iliac crest. This shows an oblique projection of the lumbar vertebrae and zygapophyseal joints closest (dependent) to the image receptor. These are the joints between the superior and inferior articulating processes of the vertebra. Radiographically, this image of the area should resemble an image of a "Scotty dog". The nose is formed by the transverse process, the eye is from the pedicle and the body from the lamina of the vertebra.

19. A   A radiographic image of the sternoclavicular joints is obtained in the PA oblique projection. The patient is placed prone with their head turned to the joint of interest. Because the AC joints are located near the midline of the body, their visualization will require that the central ray is directed 10-15 degrees medially toward the midsagittal plane to enter at a the level of the T2-T3. Each AC joint should be examined separately.

20. B   A radiographic image of the thoracic spine is obtained in the lateral projection. The patient is placed in the lateral position with the arms extended to a right angle to the long axis of the body. If the patient is adjusted to the horizontal plane, the central ray is directed perpendicular to enter at a point at the level of T7 just posterior to the midcoronal plane. If the spine is not adjusted to the horizontal, the CR is directed at a cephalic angulation of 10-15 degrees. This projection is used to demonstrate the space between the vertebral bodies and their interspaces and spinous process of the lower thoracic vertebrae. It may be necessary to use a swimmer's projection to improve the demonstration of the upper vertebrae.

21. B   A radiographic image of the cervical spine is obtained in the AP axial oblique projection to demonstrate open intervertebral foramina of the cervical region. In the cervical region, the intervertebral foramina are directed obliquely from the spinal canal at about a 45 degree angle laterally and 15-20 degrees inferiorly. This requires that the patient is rotated into the 45 degree LPO or RPO position. The central ray is directed to C4 at a cephalic angulation of 15-20 degrees. The AP axial oblique projection with the patient RPO and LPO, demonstrates the foramina and pedicles farthest from the image receptor and open views of the intervertebral disk. If images are obtained in the PA axial oblique projection, the patient is placed in the LAO or RAO position with the central ray directed caudally 15-20 degrees. This will demonstrate the foramina and pedicles closest to the image receptor.

22. A   The single best position for the evaluation of cervical spine fractures is a lateral projection in the dorsal decubitus position. A shoot-through lateral can be performed without moving the patient thus minimizing the risk of additional spinal cord injury. In this projection, the central ray is directed horizontally to the fourth cervical vertebra with the image receptor being vertical to the table.

23. D   A radiographic image of the lumbar spine is obtained in the AP oblique projection with 45 degree rotation of the patient. With the patient in the RPO or LPO position, the central ray is directed perpendicular to enter 5 centimeters medial to the ASIS at a level of the iliac crest. This projection shows an oblique projection of the lumbar vertebrae and zygapophyseal joints closest to the image receptor. These are the joints between the superior and inferior articulating processes of the vertebra. Radiographically, this image of the area should resemble an image of a "Scotty dog". The nose is formed by the transverse process, the eye is from the pedicle and the body from the lamina of the vertebra. The superior and inferior articulating processes of the vertebra forms the ears and front foot of the "dog".

24. D   Any radiographic image of the cervical, thoracic or lumbar spine in the lateral projection can be used to demonstrate the height of the intervertebral disk spaces.

25. A   A radiographic image of the right sacroiliac (SI) joint is to be obtained in the AP oblique projection. The patient is placed in the LAO position with the affected joint farthest from the image receptor. The central ray is directed perpendicular to the joint entering 2.5 centimeters medial to the ASIS. The RPO position is used for the demonstration of the left SI joint. This joint is formed by the junction of the sacrum and the two innominate bones at their medial surface. Since the SI joint projects obliquely at an angle of 25-30 degrees from the midsagittal plane, this is the desired degree of obliquity to place the independent joint perpendicular to the image receptor.

26. C   A radiographic image of the thoracic spine obtained in either the AP or lateral projection should be taken with the x-ray tube directed to best utilize the anode heel effect. Since the angle of the anode reduces the amount of exposure at the anode side of the tube, a greater amount of radiation is directed toward the cathode side of the tube. In order to take advantage of this effect, the radiographer should place the cathode over the thicker or lower parts of the body and the anode toward the head of the table.

27. D   A radiographic image of the cervical spine is obtained in the AP axial projection. With the patient supine or erect, the central ray is directed at a cephalic angulation of 15-20 degrees to enter just inferior to the thyroid cartiledge. This projection is used to show the bodies and intervertebral disk spaces of the lower 5 cervical vertebrae. It is also used to demonstrate the presence of cervical ribs which project anteriorly from the lateral surface of C7.

28. C   A radiographic image of the cervical spine is obtained in the lateral projection. The patient is placed erect and asked to depress the shoulder to improve the visualization of the lower vertebrae. The central ray is directed horizontally to enter at the level of the 4th cervical vertebra. A 180 centimeter source-to-image receptor distance is employed to reduce the magnification of the spine. This projection is used to demonstrate the vertebral bodies and their interspaces, the zygapophyseal joints and spinous processes of the cervical spine.

29. A   A radiographic image of the coccyx is obtained in the PA axial projection. Due to the kyphotic curve of the coccyx, a 10 degree cephalic angulation directed to enter at the coccyx of a patient in the prone position will tend to reduce the foreshortening that would occur without this angulation. In an AP axial projection of the coccyx, the central ray is directed at a caudal angulation of 10 degrees to enter a point about 5 centimeters above the pubic symphysis.

30. B   A radiographic image of the thoracic spine is obtained in the lateral projection. The patient is placed in the lateral position with the arms extended to a right angle to the long axis of the body. If the patient is adjusted to the horizontal plane, the central ray is directed perpendicular to enter at a point at the level of T7 just posterior to the midcoronal plane. If the spine is not adjusted to the horizontal, the CR is directed at a cephalic angulation of 10-15 degrees. This projection is also used to demonstrate the normal anterior (kyphotic) or posterior lordotic curves of the spinal column. Scoliosis or the lateral curvature of the spine, is best seen in the AP or PA projections.

31. C   A radiographic image of the lumbar spine is obtained in the AP oblique projection with 45 degrees rotation of the patient. With the patient in the RPO or LPO position, the central ray is directed perpendicular to enter 5 centimeters medial to the ASIS at a level of the iliac crest. This projection shows an oblique projection of the lumbar vertebrae and zygapophyseal joints closest to the image receptor. These are the joints between the superior and inferior articulating processes of the vertebra. In the PA oblique projection, the zygapophyseal joints farthest from the image receptor are visualized.

32. A   A radiographic image of the odontoid process is obtained in the AP projection open mouth position. The patient is placed supine with the mouth opened as wide as possible. The head is adjusted so that a line drawn from the lower edge of the upper incisor to the tip of the mastoid process is perpendicular to the image receptor. The central ray is directed perpendicular to the midpoint of the mouth. Shadows from the back of the head and teeth should be eliminated from the odontoid process in an open mouth AP projection. The phonation of "ah" during the exposure helps to fix the tongue to the bottom of the mouth.

33. C   A radiographic image of the sacrum is obtained in the AP axial projection. Due to the kyphotic curve of the sacrum, the central ray is directed at a cephalic angulation of 15 degrees to enter at a point 5 centimeters superior to the pubic symphysis with the patient supine. This angulation is used to reduce the foreshortening that would occur without this angulation. In a PA axial projection of the sacrum, the central ray is directed at a caudal angulation of 15 degrees to enter at the sacral curve.

34. B   A radiographic image of the thoracic spine is obtained in the lateral projection. The patient is placed in the lateral position with the arms extended to a right angle to the long axis of the body. An increase of the 90 degree arm position will result in the scapula being projected over the upper thoracic vertebrae.

35. D A radiographic image of the cervical spine is obtained in the PA axial oblique projection to demonstrate the open intervertebral foramina. The patient is rotated obliquely to a 45 degree angle with the midsagittal plane. The central ray is directed to C4 at a caudal angulation of 15-20 degrees. The PA 45 degree axial oblique with the patient RAO and LAO demonstrates the foramina and pedicles closest to the image receptor and an open view of the intervertebral disk spaces.

36. D A radiographic image of the thoracic spine is obtained in the AP oblique projection. Since the articulation between the inferior and superior articulating processes (zygapophyseal joints) in the thoracic spine forms an angle of 70 degrees with the image receptor, the AP 70 degree RPO or LPO position is required to demonstrate these articulations. This projection shows the articulations farthest from the image receptor. In the PA 70 degree LAO or RAO position, the joints closest to the image receptor are demonstrated.

37. A A radiographic image of the odontoid process is obtained in the AP projection open mouth position. The patient is placed supine with the mouth opened as wide as possible. The head is adjusted so that a line drawn from the lower edge of the upper incisor to the tip of the mastoid process is perpendicular to the image receptor. If the odontoid process is superimposed on the teeth in an open mouth position, the head must be extended or directed away from the chest. The base of the skull will appear over the odontoid when the head is not tucked close enough to the chest wall.

38. B A radiographic image of the right sacroiliac (SI) joint is to be obtained in the AP oblique projection. The patient is placed in the LAO position with the affected joint farthest from the image receptor. The central ray is directed perpendicular to the joint entering 2.5 centimeters medial to the ASIS. The RPO position is used for the demonstration of the left SI joint. This joint is formed by the junction of the sacrum and the two innominate bones at their medial surface. Since the SI joint projects obliquely at an angle of 25-30 degrees from the midsagittal plane, this is the desired degree of obliquity to place the independent joint perpendicular to the image receptor.

39. C A radiographic image of the thoracic spine is obtained in the AP oblique projection. Since the articulation between the inferior and superior articulating processes (zygapophyseal joints) in the thoracic spine forms an angle of 70 degrees with the image receptor, the AP 70 degree RPO or LPO position is required to demonstrate these articulations. This projection shows the articulations farthest from the image receptor. Because the upper thoracic vertebrae are smaller than those of the lower thoracic spine, the appropriate centering point is at level T7 instead of T6.

40. C A radiographic image of the L5-SI interspace and the sacroiliac joints is obtained in the AP axial projection. The patient is placed supine, with the central ray directed through the lumbosacral joint at a cephalic angulation of 30 degrees in male patients and an angulation of 35 degrees in a female patient. The angle of the central ray should be reversed in the PA axial projection.

41. D A radiographic image of the cervical spine is obtained in the lateral projection. The patient is placed erect and asked to depress the shoulder to improve the visualization of the lower vertebrae. The central ray is directed horizontally to enter at the level of the 4th cervical vertebra. A 180 centimeter source-to-image receptor distance is employed to reduce the magnification of the spine. This projection is used to demonstrate the vertebral bodies and their interspaces, the zygapophyseal joints and spinous processes of the cervical spine.

42. A An anterior rib injury would be best examined in the PA projection with the patient erect to eliminate potential complications due to lung involvement while maintaining the best definition and true image size. With the patient in the PA projection, the central ray should be directed horizontal to enter at the level of T7 for the upper ribs. Images of the upper ribs are taken with respirations suspended at the end of full inspiration. Images of the lower ribs may be improved by directing the central ray at a caudal angulation of 10-15 degrees and obtaining the exposure at the end of expiration.

43. C   A radiographic image of the cervical spine is obtained in the AP axial oblique projection to demonstrate open intervertebral foramina. In the cervical region, the intervertebral foramina are directed obliquely from the spinal canal at about a 45 degree angle laterally and 15-20 degrees inferiorly. This requires that the patient be rotated into the 45 degree LPO or RPO position. The central ray is directed to C4 at a cephalic angulation of 15-20 degrees. The AP axial oblique projection with the patient RPO and LPO, will demonstrate the foramina and pedicles furthest from the image receptor and open view of the intervertebral disk spaces of the cervical spine.

44. C   A radiographic image of the lumbar spine is obtained in the lateral projection. The patient is placed on their side with the hips and knees flexed to make the patient comfortable. If the spine of the patient is adjusted to the horizontal plane, the central ray is directed perpendicular to enter at the level of the iliac crest. This projection is used in the evaluation of the disk spaces between the vertebral bodies and spinous processes of the lumbar vertebra and the lateral aspect of the lumbar vertebrae, the intervertebral foreman and spinous processes.

45. A   A radiographic image of the thoracic spine is obtained in the lateral projection. The patient is placed in the lateral position with the arms extended to a right angle to the long axis of the body. If the patient is adjusted to the horizontal plane, the central is directed perpendicular to enter at a point at the level of T7 just posterior to the midcoronal plane. If the spine is not adjusted to the horizontal, the CR is directed at a cephalic angulation of 10-15 degrees. This projection is used to demonstrate the space between the vertebral bodies and their interspaces and spinous process of the lower thoracic vertebrae. It may be necessary to use a swimmer's projection to improve the demonstration of the upper vertebrae.

46. D   A radiographic image of the odontoid process is obtained in the AP (Fuchs) projection. The patient is placed supine with the chin extended until tip of the chin and the mastoid tip are vertical to the image receptor. The central ray is directed perpendicular to enter a point just distal to the chin. This projection is used to demonstrate the odontoid process lying within the shadow of the foramen magnum.

47. D   A radiographic image of the costal joints is obtained in the AP axial projection for the demonstration of rheumatoid spondylitis. The patient is placed supine, with the central ray directed to enter on the midsagittal plane 5 centimeters above the xiphoid process at a cephalic angulation of 20 degrees. This projection shows the open costovertebral and costotransverse joints. A 5-10 degree increase in this angle is recommended for patients with noticeable kyphosis.

48. B   The radiographic image of the sternoclavicular joints is obtained in the axiolateral (Kurz Bauer) projection. The patient is placed in the lateral recumbent position on their affected side with the arm raised similar to the swimmer's position. The central ray is directed at a caudal angulation 15 degrees to enter the lower SC joint.

49. C   A radiographic image of the cervical spine is obtained in the lateral projection. The patient is placed erect and asked to depress the shoulder to improve the visualization of the lower vertebrae. The central ray is directed horizontally to enter at the level of the 4th cervical vertebra. A 180 centimeter source-to-image receptor distance is employed to reduce the magnification of the spine from the long object-to-image distance between the cervical spine and the image receptor. Because of the air gap created, most of the divergent scattered radiation produced in the patient will miss the image receptor thus eliminating the need to use a radiographic grid while improving the contrast of the image.

50. D   A radiographic image of the cervical spine is obtained in the AP axial projection. With the patient supine or erect, the central ray is directed at a cephalic angulation of 15-20 degrees to enter just inferior to the thyroid cartiledge. This projection is used to show the bodies and intervertebral disk spaces of the lower 5 cervical vertebrae. It is also used to demonstrate the presence of cervical ribs which project anteriorly from the lateral surface of C7.

51. A   A radiographic image of the thoracic spine is obtained in the AP oblique projection. Since the articulation between the inferior and superior articulating processes (zygapophyseal joints) in the thoracic spine forms an angle of 70 degrees with the image receptor, the AP 70 degree RPO or LPO position is required to demonstrate these articulations. This projection shows the articulations farthest from the image receptor.

52. C   The radiographic image obtained in the PA axial oblique (Kovacs) projection is most often taken to demonstrate the intervertebral foramina formed by L5 and SI which is not well visualized in the lateral projection. This modification requires an additional forward rotation of 30 degrees of the pelvis and a central ray angulation of 15-30 degrees caudad.

53. D   A radiographic image of the cervical spine is obtained in the AP (Ottonello) projection while the patient is asked to open and close their mouth during the exposure. This maneuver is employed in the evaluation of the entire cervical spine by preventing the shadow of the mandible from obscuring C1 and C2. The AP Ottonello projection (wagging jaw), uses a long exposure time and the movement of the jaw to blur its image on the receptor, thus providing an image of the entire cervical spine on a single radiograph.

54. B   A radiographic image of the symphysis pubis is taken in the PA (Chamberlain) projection with the patient erect. This weight-bearing position is used to evaluate abnormal slippage of the SI joints. A slipped SI joint is most readily seen by the relationship of the pubic bones at the symphysis pubis where the images are centered.

55. C   A radiographic image of the cervical spine is obtained in the AP axial projection. With the patient supine or erect, the central ray is directed at a cephalic angulation of 15-20 degrees to enter just inferior to the thyroid cartiledge. This projection is used to show the bodies and intervertebral disk spaces of the lower 5 cervical vertebrae. It is also used to demonstrate the presence of cervical ribs which project anteriorly from the lateral surface of C7.

56. D   A radiographic image of the lumbar spine is obtained in the AP projection. Because of the lordotic curve in this region, the intervertebral spaces are improperly aligned with the divergent rays of the x-ray beam. A radiographic image taken in the PA projection is more aligned with the divergence of the x-ray beam. The main disadvantage of the PA projection, is an increased magnification of the vertebrae due to a longer object-to-image receptor distance.

57. C   A radiographic image of the thoracic and lumbar spines is obtained in the PA projection to evaluate an abnormal lateral curvature of the spine due to scoliosis. The PA projection is employed to reduce the radiation exposure to the thyroid gland, ovaries and breasts in female patients.

58. B   A radiographic image of the odontoid process is obtained in the AP projection open mouth position. The patient is placed supine with the mouth opened as wide as possible. The head is adjusted so that a line drawn from the lower edge of the upper incisor to the tip of the mastoid process is perpendicular to the image receptor. The phonation of "ah" during the exposure helps to fix the tongue to the bottom of the mouth.

59. C   During lateral positions of the spine, the vertebral column should be adjusted to be parallel with the image receptor. When recumbent, the spine must be supported through the use of radiolucent sponges placed under the lateral thorax or lumbar region to achieve this goal. When unsupported, a 10 degree cephalic angle for females and a 15 degree angle for males will compensate for this curve in the thoracic region. In the lumbar region, a 5 degree caudal angle for males and an 8 degree caudal angle for females, is required on the unsupported spine.

60. B   A radiographic image of the axillary portion of the right ribs can be best demonstrated in the AP oblique projection. The patient is placed in the RPO or LPO position with a 45 degree rotation. The central ray should be directed perpendicular toT7 for the upper ribs and T1 for the lower ribs. In the AP projection, the affected side is placed closest to the image receptor. In the PA obliques projection with the patient in the LAO or RAO positions, the affected side is rotated away from image receptor.

61. C   This radiographic image of the cervical spine was obtained in the AP axial oblique projection to demonstrate open intervertebral foramina. In the cervical region, the intervertebral foramina are directed obliquely from the spinal canal at about a 45 degree angle laterally and 15-20 degrees inferiorly. This requires that the patient be rotated into the 45 degree LPO or RPO position. The central ray is directed to C4 at a cephalic angulation of 15-20 degrees. The AP 45 degree axial oblique with the patient RPO and LPO, demonstrate the foramina and pedicles furthest from the image receptor and open view of the intervertebral disk spaces of the cervical spine.

62. A   See the explanation above.

63. C A radiographic image of the lumbar spine is obtained in the AP oblique projection with 45 degrees rotation of the patient. With the patient in the RPO or LPO position, the central ray is directed perpendicular to enter 5 centimeters medial to the ASIS at a level of the iliac crest. This projection shows an oblique projection of the lumbar vertebrae and zygapophyseal joints closest to the image receptor. These are the joints between the superior and inferior articulating processes of the vertebra.

64. B See explanation above.

65. A The low optical density (Harrington) rods seen on this PA projection of the spine are used in the treatment of scoliosis.

1. The line extending from the smooth prominence between the superciliary arches to the external auditory meatus is called the:

   A. Acanthiomeatal line
   B. Orbitomeatal line
   C. Supraorbitomeatal line
   D. Glabellomeatal line

2. The projection that will best demonstrate all of the paranasal sinuses on a single radiographic image is the:

   A. Submentovertical projection
   B. Parietoacanthial projection
   C. Lateral projection
   D. Posteroanterior projection

3. The central ray for a radiographic image of the nasal bones in the lateral projection should be directed:

   A. Perpendicular to the image receptor at a point 2 centimeters distal to the glabella
   B. Perpendicular to the image receptor at a point 5 centimeters distal to the anterior nasal spine
   C. Perpendicular to the image receptor at a point 2 centimeters distal to the nasion
   D. Perpendicular to the image receptor at a point 5 centimeters distal supraorbital ridge

4. Which of the following projections would be most useful for the radiographic evaluation of a single zygomatic arch?

   A. Parietoacanthial projection
   B. Tangential (submentovertical) projection
   C. Posteroanterior axial projection
   D. Axial (intraoral) projection

5. The cranial type in which the petrous ridges projects anteriorly and medially from the midsagittal plane at an angle of 54 degrees is called a:

   A. Mesocephalic skull
   B. Hypercephalic skull
   C. Brachycephalic skull
   D. Dolichocephalic skull

6. The most superior portion of the skull is referred to as:

   A. Lambda
   B. Vertex
   C. Corona
   D. Mentum

7. A radiographic image obtained in the parietoacanthial (Waters) projection is useful for the evaluation of the:

   A. Foramen rotundum
   B. Foramen spinosum
   C. Foramen magnum
   D. Foramen Magendie

8. When obtaining radiographic images of the temporomandibular joints (TMJ), it is important to include all of the following EXCEPT:

   A. Open mouth views of the TMJ
   B. Closed mouth views of the TMJ
   C. Images of both TMJs
   D. Wagging jaws of the TMJs

9. The radiographic line that extends from the anterior nasal spine to the external auditory meatus (EAM) is known as:

   A. Glabellomeatal line
   B. Nasomeatal line
   C. Acanthiomeatal line
   D. Infraorbitomeatal line

10. The mesocephalic cranium measures approximately:

    A. 15 centimeters between the parietal eminences
    B. 18 centimeters between the zygomatic arches
    C. 25 centimeters between the orbital margins
    D. 25 centimeters between the parietal eminences

11. In a radiographic image of the sella turcica in the lateral projection, the central ray is directed perpendicular to enter a point that is:

A. 2 centimeters anterior and superior to the external auditory meatus (EAM)
B. 5 centimeters posterior and inferior to the external auditory meatus (EAM)
C. 2 cm anterior and inferior to the outer canthus
D. 5 cm anterior and inferior to the temporomandibular joint

12. In a radiographic image of the petromastoid region in the axiolateral oblique (Stenvers) projection, the midsagittal plane is adjusted to form an angle of:

A. 12 degrees to the plane of the image receptor   C. 45 degrees to the plane of the image receptor
B. 27 degrees to the plane of the image receptor   D. 53 degrees to the plane of the image receptor

13. A radiographic image of the cranium is to be obtained in the verticosubmental position. The central ray should be directed to enter at the point near the:

A. Vertex   C. Pterion
B. Nasion   D. Foramen magnum

Referring to the radiograph of the AP axial projection of the skull, place the letter that corresponds to the following structures for questions 14-20.

14. _____ Occipital bone
15. _____ Mastoid air cells
16. _____ Petrous ridge
17. _____ Dorsum sellae
18. _____ Parietal bone
19. _____ Condylar process of mandible
20. _____ Foramen magnum

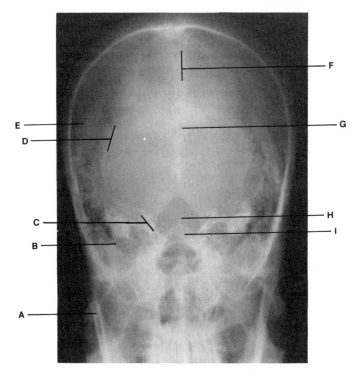

21. The difference between the orbitomeatal line and the infraorbitomeatal line is approximately:

A. 3 degrees   C. 12 degrees
B. 7 degrees   D. 17 degrees

22. A radiographic image of the optic foramen in the orbitoparietal oblique projection requires that the midsagittal plane form an angle of:

A. 15 degrees to the plane of the image receptor   C. 40 degrees to the plane of the image receptor
B. 30 degrees to the plane of the image receptor   D. 53 degrees to the plane of the image receptor

23. In order to remove overlying shadows when radiographic images are obtained of the mastoid process in the axiolateral projection, it is recommended that:

A. The mouth be opened during the exposure
B. The eyes are closed during the exposure

C. The auricles be taped forward for the exposure
D. The jaw is moved during the exposure

24. A radiographic image in the full basal or verticosubmental (Schuller) projection, is commonly used in the evaluation of all of the following EXCEPT:

A. The foramen ovale
B. The mastoid processes

C. The mandibular condyles
D. The sagittal suture

25. The radiographic line that passes from the outer canthus of the eye to the external auditory meatus (EAM) is described as the:

A. Orbitomeatal line
B. Canthomeatal line

C. Acanthiomeatal line
D. Interpupillary line

26. A radiographic image obtained in submentovertical (Schuller) projection, is often employed for the evaluation of:

1. Sphenoid sinuses   2. Zygomatic arches   3. Foramen magnum

A. 1 & 2 only
B. 1 & 3 only

C. 2 & 3 only
D. 1, 2 & 3

27. The average measurement of a skull from the vertex to the submental area is:

A. 15 centimeters
B. 22 centimeters

C. 12 centimeters
D. 28 centimeters

28. The radiographic line that bisects the pupils of the eye is termed the:

A. Intercanthial line
B. Interorbital line

C. Interpupillary line
D. Orbitomeatal line

29. In evaluating the proper positioning of a radiographic image of the skull obtained in the submentovertical projection, which of the following is normally located anterior to the petrous ridges?

A. The odontoid process
B. The mandibular condyles

C. The clivus
D. The foramen magnum

30. All of the following are clearly demonstrated on a radiographic image of the paranasal sinuses in the posteroanterior axial (Caldwell) projection EXCEPT:

A. The anterior ethmoidal air cells
B. The maxillary sinuses

C. The frontal sinuses
D. The sphenoidal sinuses

31. The usual difference between the orbitomeatal line and the glabellomeatal line is about:

A. 5 degrees
B. 8 degrees

C. 12 degrees
D. 17 degrees

32. In a radiographic image of the optical canal in the parieto-orbital oblique (Rhese) projection, the head is positioned so that the:

A. Glabella, nose and superciliary ridge are in contact with the table
B. Zygoma, nose and chin are in contact with the table
C. External auditory meatus, chin and parietal eminence are in contact with the table
D. Rami of the mandibles are in contact with the table

33. A radiographic image of the temporomandibular joints is obtained in the anteroposterior axial projection. In the closed mouth position the mandibular condyles can be moved out of the mandibular fossae by having the patient:

A. Occlude the incisors rather than the front teeth
B. Occlude the molars rather than the incisors

C. Shift the mandible laterally
D. Extend the mandible anteriorly

34. A radiographic image is obtained with the patient placed prone, and the head is adjusted so that the acanthiomeatal line is perpendicular to the plane of the image receptor. If the midsagittal plane forms an angle of 37 degrees from the perpendicular and the central ray is directed vertically to pass through the orbit, this projection was most likely taken to evaluate the:

A. Maxillary sinus
B. Eustachian tube

C. Internal auditory canal
D. Optic foramen

35. The radiographic line that passes from the inferior border of the orbits to the auricular point is called the:

A. Cranial base line
B. Orbitomeatal line

C. Acanthiomeatal line
D. Infraorbitomeatal line

36. In a normal adult male, the orbit and orbital foramina project anteriorly from the midsagittal plane at an angle of about:

A. 23 degrees
B. 30 degrees

C. 37 degrees
D. 62 degrees

37. Which of the following radiographic projections is most commonly used to evaluate mediolateral displacement fractures of the mandibular ramus?

A. Axiolateral projection
B. AP axial projection

C. Axiolateral oblique projection
D. PA axial projection

38. In a radiographic image of the petromastoid region in the axiolateral oblique (Stenvers) projection, the central ray is directed at a:

A. 10 degree caudal angle to the petrous pyramid farthest from the image receptor
B. 45 degree caudal angle to the petrous pyramid farthest from the image receptor
C. 12 degree cephalic angle to the petrous pyramid closest to the image receptor
D. 53 degree cephalic angle to the petrous pyramid closest to the image receptor

39. Which of the following lines will be perpendicular to the image receptor for a radiographic image of the cranium in the lateral projection?

A. The orbitomeatal line
B. The interpupillary line

C. The infraorbitomeatal line
D. The glabelloalveolar line

40. When performing a radiographic image of the facial bones in the parietoacanthial (Waters) projection, the failure to extend the chin sufficiently will result in the:

A. Frontal sinuses obscuring the optic foramen
B. Zygomatic arches obscuring nasal septum

C. Teeth obscuring the sphenoid sinuses
D. Petrous ridges obscuring the maxillary sinuses

Referring to the radiographic image of the lateral skull, place the letter that corresponds to the following structures for questions 41-48.

41. _____ Parietal bone
42. _____ Frontal sinus
43. _____ Posterior clinoid process
44. _____ Anterior nasal spine
45. _____ Sphenoid sinus
46. _____ Coronal suture
47. _____ Maxillary sinus
48. _____ Occipital bone

49. When a radiographic image of the skull is obtained in the lateral projection which of the following is adjusted so that it is parallel to the plane of the image receptor?

A. The midsagittal plane
B. The interpupillary line
C. The coronal plane
D. The lambdoidal plane

50. A radiographic image of the petromastoid region is obtained in the axiolateral oblique (Arcelin) projection. The petrous ridge that is demonstrated lies:

A. Perpendicular and farthest from the image receptor
B. Perpendicular and closest to the image receptor
C. Parallel and farthest from the image receptor
D. Parallel and closest to the image receptor

51. A radiographic image of the skull obtained in the anteroposterior axial (Towne) projection, is useful for the demonstration of all of the following EXCEPT:

A. Zygomatic arches
B. Dorsum sella
C. Pars petrosa
D. Frontal sinuses

52. A radiographic image of the mastoid process is obtained in the axiolateral oblique (Arcelin) projection. The central ray should be directed at a 10 degree caudal angle to enter at a point:

A. 5 centimeters anterior and 5 centimeters above the EAM
B. 5 centimeters anterior and 5 centimeters below the EAM
C. 2 centimeters anterior and 2 centimeters above the EAM
D. 2 centimeters anterior and 2.5 centimeters below the EAM

53. The patient is supine with the infraorbitomeatal line parallel to the plane of the image receptor. The central ray enters vertically at a coronal plane 2 cm anterior to the EAM. This describes a:

A. True lateral projection
B. Submentovertical projection
C. Anteroposterior projection
D. Verticosubmental projection

54. Which of the following projections of the facial bones would be most helpful in diagnosing a "blowout" fracture of the floor of the orbit?

A. Parieto-orbital oblique projection
B. Orbitoparietal oblique projection
C. Parietoacanthial projection
D. Posteroanterior projection

55. A radiographic image is obtained in the parieto-orbital projection. If the central ray is directed perpendicular to the image receptor, this projection is used for the demonstration of the:

A. Optic foramen within the orbital shadow
B. Internal auditory canals
C. Antra of Highmore
D. Jugular foramen

56. A radiographic image of the occipital region is obtained in the PA axial (Haas) projection. This projection is often employed to provide a comparable image to one taken in the:

A. Parietoacanthial projection
B. Orbitoparietal oblique projection
C. Anteroposterior axial projection
D. Submentovertical projection

57. A radiographic image of the zygomatic arch is obtained in the oblique tangential (May) projection. The patient is placed prone with the infraorbitomeatal line parallel and the midsagittal plane perpendicular to the image receptor. Which of the following would be TRUE?

A. The midsagittal plane should be rotated 15 degrees away from the affected side
B. The midsagittal plane should be rotated 15 degrees toward the affected side
C. The central ray should be directed 15 degrees caudal
D. The central ray should be directed 15 degrees cephalic

58. A radiographic image of the auditory ossicles is obtained in the PA axiolateral oblique (Mayer) projection. The patient is placed prone with the midsagittal plane rotated 45 degrees toward the affected side. If the infraorbital line is parallel to the transverse axis of the image receptor, the central ray should be directed at a:

A. Cephalic angulation of 12 degrees
B. Caudal angulation of 23 degrees
C. Cephalic angulation of 30 degrees
D. Caudal angulation of 45 degrees

59. In order to obtain an AP axial (Towne) projection on an infant or young child, the central ray should be directed through the foramen magnum at a caudal angulation of:

A. Less than 15 degrees to enter at the nasion
B. More than 37 degrees to enter at the superciliary ridge
C. Less than 30 degrees to enter 6 centimeters below the glabella
D. More than 45 degrees 6 centimeters below the nasion

60. A radiographic image of the optic canal is obtained in the parieto-orbital oblique (Rhese) projection. The head is positioned so that the affected orbit is over the image receptor. Which of the following lines should be perpendicular?

A. The acanthiomeatal line
B. The infraorbitomeatal line
C. The mentomeatal line
D. The interpupillary line

61. When a radiographic image is obtained in the axiolateral (Stenvers) projection, the patient's head is rotated from the prone position to place the midsagittal plane at an angle of 45 degrees with the plane of the image receptor. The central ray is directed at a cephalic angulation of 12 degrees to enter about 10 centimeters posterior and 2 centimeters inferior to the external auditory meatus. The projection is most commonly used to:

A. Demonstrate the auditory canal farthest from the image receptor
B. Demonstrate the optic foramen within the bony orbit
C. Profile view of the styloid process closest to the image receptor
D. Profile view of the petromastoid portion closest to the image receptor

62. In the PA projection of the cranium for a general survey examination, the orbitomeatal line is perpendicular and the central ray is directed at a:

A. At a caudal angulation 23 degrees to exit the skull at the acanthion
B. At a cephalic angulation of 15 degrees to exit the skull at the metal point
C. Perpendicular to exit the skull at the nasion
D. At a cephalic angulation of 23 degrees to exit the skull at the glabella

63. A radiographic image of the petromastoid portion is obtained in the axiolateral oblique (Modified Law) projection. The patient's head is placed in the relaxed lateral position with the central ray directed at a caudal angulation of 15 degrees. This projection is used in the evaluation of the demonstration of the:

A. Mastoid air cells closest to the image receptor
B. Orbital rim closest to the image receptor
C. Maxillary sinus closest to the image receptor
D. Anterior clinoid process closest to the image receptor

64. A radiographic image is to be obtained in the PA axial projection. If the head is adjusted so that the orbitomeatal line and midsagittal plane are perpendicular to the image receptor in order to demonstrate the frontal bone, the central ray should be directed to the nasion at a:

A. Caudal angulation of 15 degrees
B. Caudal angulation of 35 degrees
C. Cephalic angulation of 23 degrees
D. Cephalic angulation of 37 degrees

65. A radiographic image of the inferior orbital fissure and orbital floor is obtained in the PA axial (Bertel) projection. If the patient's head is placed with the infraorbitomeatal line perpendicular to the image receptor, the central ray should be directed to a point 8 centimeters below the external occipital protuberance at a:

A. Caudal angulation of 30-35 degrees   C. Cephalic angulation of 20-25 degrees
B. Caudal angulation of 10-15 degrees   D. Cephalic angulation of 5-10 degrees

66. A radiographic image of the optic foramen is obtained in the parieto-orbital oblique (Rhese) projection. This projection will project the optic foramen at the end of the sphenoid ridge into the shadow of the:

A. Superomedial quadrant of the orbit   C. Inferolateral quadrant of the orbit
B. Inferior mandibular ramus   D. Greater wing of the temporal bone

67. When radiographic images are obtained of the paranasal sinuses, which of the following gives the best indication the proper exposure factors have been employed?

   1. Adequate penetration of the sella turcica is demonstrated
   2. The optical density of the upper and lower orbits is nearly the same
   3. The optical density of the sinuses is similar to that of the zygomatic arches

A. 1 only   C. 3 only
B. 2 only   D. 1, 2 & 3

68. A radiographic image is obtained in the PA axial (Cahoon) projection. If the patient's head is adjusted so that the orbitomeatal line is perpendicular to the image receptor and the central ray is directed to pass through the nasion at a cephalic angle of 25 degrees, this projection is most commonly used for the evaluation of the:

A. Temporal styloid processes   C. External occipital protuberance
B. Greater wing of the sphenoid bone   D. Hypoglossal canals

69. A radiographic image is obtained in the PA axial projection. The patient's head is adjusted so that the orbitomeatal line and midsagittal plane are perpendicular to the plane of the image receptor. The central ray is directed 25 degrees caudal to pass through the inferior margin of the orbit. This projection is used to demonstrate:

A. Supraorbital fissure   C. External auditory meatus
B. Foramen magnum   D. Hard palate

70. A radiographic image of the zygomatic arch is obtained in tangential (May) projection. The patient is placed with the neck hyperextended to place the infraorbitomeatal line parallel to the plane of the image receptor. The central ray is directed perpendicular IOML to a point 4 centimeters posterior to the outer canthus. The head should be rotated:

A. 15 degrees away from affected side   C. 35 degrees toward the affected side
B. 25 degrees toward the affected side   D. 30 degrees away from the affected side

71. A radiographic image of the frontal bone is obtained in the PA projection. With the patient prone, the central ray is directed perpendicular to exit at the:

A. Pterion   C. Acanthion
B. Inion   D. Nasion

72. Which of the following projections will demonstrate the least angular distortion on images taken of both petrous ridges?

A. The PA axial (Valdini) projection   C. The axiolateral oblique (Laws) projection
B. The axiolateral (Haas) projection   D. The AP axial (Towne) projection

73. A radiographic image is obtained in the AP axial (Towne) projection. With the patient supine, the head is adjusted to place the infraorbitomeatal line perpendicular to the image receptor. If the central ray is directed through the foramen magnum at a cephalic angulation of 30 degrees, all of the following structures will be demonstrated EXCEPT:

A. The occipital bone
B. The posterior clinoid process
C. The dorsum sella
D. The optic canal

74. A radiographic image of the skull is obtained to determine the presence of effusion in the sphenoid sinus. Which of the following projections will best demonstrate this condition?

A. A cross-table lateral projection
B. An axiolateral projection with the patient erect
C. An AP projection with the patient erect
D. A submentovertical projection with the patient supine

75. A radiographic image of the facial bones is obtained in the acanthioparietal (reverse Waters) projection. With the patient supine, the head is adjusted so the mentomeatal line is perpendicular to the image receptor. The central ray should be directed:

A. At a caudal angle of 30 degrees
B. At a cephalic angle of 30 degrees
C. At a medial angle of 15 degrees
D. Perpendicular to the image receptor

76. A radiograph of a parieto-orbital oblique is taken. The radiograph demonstrates the optic foramen in the upper outer quadrant of the orbit. In order to improve the positioning, the technologist should:

A. Rotate the affected orbit medially
B. Extend the patient's chin
C. Rotate the affected orbit laterally
D. Tuck the patient's chin

77. It is recommended that radiographs of the paranasal sinuses be taken erect in order to demonstrate the existence of:

A. Fracture fragments
B. Layering of stones
C. Fluid levels
D. Lacrimal duct obstructions

78. A radiographic image of the petromastoid region is obtained in the axiolateral oblique lateral (original Law) projection. The central ray is directed to enter 5 centimeters posterior and 5 centimeters superior to the external auditory meatus with a double tube angulation of:

A. 15 degrees anteriorly and 15 degrees caudal
B. 25 degrees anteriorly and 15 degrees cephalic
C. 15 degrees anteriorly and 45 degrees caudal
D. 20 degrees anteriorly and 45 degrees caudal

79. A radiographic image of the mandible is obtained in the PA axial projection. The central ray is directed at the level of the TMJs at a cephalic angulation of 20-25 degrees. This projection is used for the demonstration of the:

A. Mandibular body
B. Superior alveolar border
C. Mandibular symphysis
D. Condylar processes

80. A radiographic image is obtained in the parieto-orbital (open mouth Waters) projection. The central ray is directed perpendicular to the image receptor to exit at the acanthion. This projection is used to demonstrate the:

A. Styloid process through the open mouth
B. Sphenoid sinus through the open mouth
C. Hypoglossal canal through the open mouth
D. Frontal sinus through the open mouth

81. A radiographic image is obtained in the submentovertical (Schuller) projection. The patient extends their head to place the infraorbitomeatal line parallel to the plane of the image receptor. The central ray enters at a right angle with the image receptor to enter a point just above the thyroid cartiledge. This position is used for the demonstration of structures in the:

A. Anterior cranial vault
B. Sagittal suture
C. Base of the cranium
D. Posterior cranial vault

82. A radiographic image of the petrous pyramids is obtained with the patient prone with the orbitomeatal line perpendicular to the plane of the image receptor. The central ray is directed at a cephalic angulation of 25 degrees to enter 4 centimeters below the external occipital protuberance. This describes a/an:

A. Axiolateral projection of the cranium
B. Verticosubmental projection of the cranium
C. Axiolateral oblique projection of the cranium
D. PA axial (Haas) projection of the cranium

83. A radiographic image is obtained in the PA axial (Valdini) projection. The patient is prone to place the infraorbitomeatal line at a 50 degree angle to the plane of the image receptor.. The central ray is directed perpendicular to enter at a point just above the external auditory meatus. This projection is used for the demonstration of the:

A. Internal auditory canals and the labyrinths
B. Nasal septum and nasopharynx
C. Optic foramen and optic canals
D. Anterior nasal spine and oropharynx

84. A radiographic image of the facial bones is obtained in the parietoacanthial (Waters) projection. The central ray is directed perpendicular to the image receptor to exit at the acanthion. In this projection, the orbitomeatal line in the average adult will form an angle of:

A. 15 degrees with the plane of the image receptor
B. 23 degrees with the plane of the image receptor
C. 37 degrees with the plane of the image receptor
D. 48 degrees with the plane of the image receptor

85. A series of radiographic images of the paranasal sinuses are to be obtained for preoperative measurements. It is normally recommended that these projections be taken at a:

A. 28 centimeter source-to-image receptor distance
B. 40 centimeter source-to-image receptor distance
C. 100 centimeter source-to-image receptor distance
D. 180 centimeter source-to-image receptor distance

86. A radiographic image is obtained of the cranium in the PA projection with the patient prone and the orbitomeatal line perpendicular to the image receptor. The central ray is directed perpendicular to exit at the nasion. This projection will normally demonstrate the petrous ridges:

A. In the lower third of the orbital shadows
B. Below the maxillary sinuses
C. Within the orbital shadows
D. Below the orbital shadows

87. Radiographic images for the cranial base are to be obtained in the submentovertical (SMV) or verticosubmental (VSM) projections. Which baseline should be placed parallel to the plane of the image receptor?

A. The acanthiomeatal line
B. The mentomeatal line
C. The glabellomeatal line
D. The infraorbitomeatal line

88. A radiographic image is obtained in the tangential projection with the central ray is directed parallel to the glabellomeatal line and perpendicular to an occlusal film placed in the mouth. This projection is most often used for the demonstration of the:

A. Mandibular symphysis
B. Nasal bones
C. Hard palate
D. Parotid gland

89. In order to obtain an anteroposterior (AP) projection for a general survey of the cranium on an infant or young child, the head is adjusted so the orbital meatal line is perpendicular to the image receptor. In order to project the petrous ridges into the lower third of the orbital shadow, the central ray should be directed at a:

A. Cephalic angle of 15 degrees
B. Caudal angle of 20 degrees
C. Cephalic angle of 25 degrees
D. Caudal angle of 37 degrees

90. A radiographic image of the mandibular ramus is obtained in the axiolateral projection. With the head in the lateral position and the interpupillary line perpendicular to the image receptor, the central ray is directed to enter 1 centimeter anterior and 5 centimeters superior to the external auditory meatus at a:

A. Cephalic angulation of 45 degrees
B. Cephalic angulation of 25 degrees
C. Cephalic angulation of 15 degrees
D. Caudal angulation of 10 degrees

91. This radiographic image of the cranium was obtained in the AP axial projection with the orbitomeatal line perpendicular to the image receptor. The location of the petrous ridges in the lower 1/3 of the orbital shadows is most consistent with a central ray that is directed at a:

   A. Caudal angulation of 15 degrees to the image receptor
   B. Cephalic angulation of 15 degrees to the image receptor
   C. Cephalic angulation of 23 degrees to the image receptor
   D. Cephalic angulation of 37 degrees to the image receptor

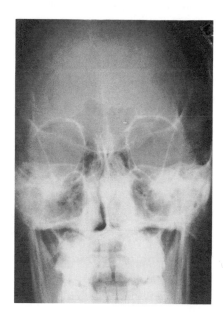

92. This projection of the cranium is often employed for the visualization of the:

   A. Mastoid process and the auditory canals
   B. Anterior and the posterior clinoid processes
   C. Crista galli and the frontal sinuses
   D. Zygomatic arches and the TMJs

Referring to the radiographic image, answer questions 93 & 94.

93. This radiographic image projection was obtained in the:

   A. Tangential projection
   B. Submentovertical projection
   C. Parieto-orbital projection
   D. Axiolateral projection

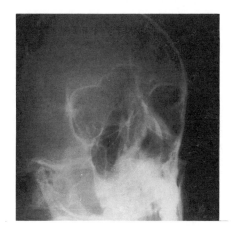

94. The principal structure of interest in this radiographic projection is the:

   A. Sella turcica
   B. Optic foramen
   C. Nasal septum
   D. Zygomatic arch

Referring to the radiographic image, answer questions 95 & 96.

95. In this parietoacanthial (Waters) projection, the bony process seen obstructing the maxillary sinuses are the:

   A. Zygomatic arches   C. Alveolar process
   B. Infraorbital ridges   D. Petrous ridges

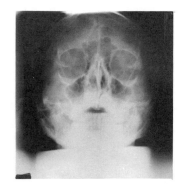

96. The desired visualization of the entire maxillary sinuses could best be accomplished by:

   A. Tilting of the head
   B. Angling the tube cranially
   C. Extending the head & chin
   D. Flexing the head & chin

97. These radiographic images were obtained in the axiolateral projection to demonstrate the:

   A. Temporomandibular joints
   B. Sella turcica
   C. Sphenoid sinuses
   D. Mastoid processes

98. Which of the radiographic images was taken with patient having their mouth open?

   A. Film A
   B. Film B

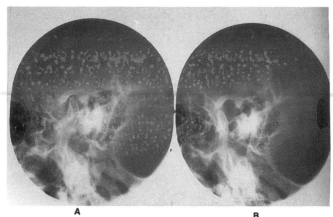

99. This radiographic image was obtained in the:

   A. True lateral projection
   B. Axiolateral oblique projection
   C. PA axial projection
   D. Tangential projection

100. Which of the following structures is NOT well demonstrated in this projection?

   A. Temporomandibular joint
   B. Body of the mandible
   C. Alveolar process
   D. Ramus of the mandible

1. D   The glabellomeatal line (GML) is the most superior of the four common radiographic lines used in positioning of the skull. The orbitomeatal line (OML) extending from the outer canthus to the external auditory meatus (EAM) forms an angle of 8 degrees with the GML. The infraorbitomeatal line (IOML) which extends from the lower orbital margin to the EAM is separated from the OML by 7 degrees. The most inferior acanthiomeatal line (ACML) extends from the point just below the nose (acanthion) to the EAM and is about 15 degrees below the IOML.

2. C   A radiographic image of the paranasal sinuses is obtained in the lateral projection. The interpupillary line is perpendicular to the image receptor. The central ray is directed perpendicular to enter at a point 2 centimeters posterior to the outer canthus. This projection is taken to demonstrate all four paranasal sinus groups separately.

3. C   A radiographic image of the nasal bones is obtained in the lateral projection. The interpupillary line is perpendicular to the image receptor. The central ray is directed perpendicular to enter at the bridge of the nose 2 centimeters distal to the nasion. This projection is taken to demonstrate the nasal bones closest to the image receptor and the anterior nasal spine.

4. B   A radiographic image of a single zygomatic arch is obtained in the tangential (May) projection. The patient is placed with their chin on the image receptor with the neck extended as far as is possible. The central ray is directed perpendicular to the infraorbitomeatal line to a point 4 centimeters posterior to the outer cantus. The head is rotated away from the affected side so the midsagittal plane forms a 15 degree angle with the plane of the image receptor. This projection demonstrates a single zygomatic arch free of superimposition with the adjacent facial and skull bones.

5. C   The "cranial type" influences the location of the petrous ridges and their relationship to the midsagittal plane (MSP). In the more broad brachycephalic cranium, an angle of 54 degrees is found. The mesocephalic or most common skull is associated with an angle of 47 degrees, the thin long skull of the dolichocephalic averages about 40 degrees. Variation of patient positioning and/or compensation of the central angulation may be required when radiographing these different cranial types.

6. B   The word vertex literally means "summit", and refers to the superior aspect of the cranium.

7. A   A radiographic image of the maxillae is obtained in the parietoacanthial (Waters) projection. The patient is placed prone with the head resting on the chin, the neck is extended so that the orbitomeatal line forms an angle of 37 degrees with the plane of the image receptor. The central ray is directed perpendicular to exit at the acanthion. This projection is used to demonstrate the orbits, the zygomatic arches and most of the other facial bones. Because the foramina rotundum are horizontally located in the anteromedial portion of the sphenoid bone, projecting through the greater wings at an angle of 37 degrees to the vertical these foramina will be opened in the parietoacanthial projection.

8. D   A radiographic projection of the temporomandibular articulations is obtained in the axiolateral projection. With the head in the lateral position and the interpupillary line perpendicular to the image receptor, the central ray is directed to enter 1 centimeter anterior and 5 centimeters superior to the external auditory meatus at a caudal angle of 25 degrees. This projection is taken to show the TMJ closest to the image receptor, the mandibular condyle at the superoposterior border of the mandible and the mandibular fossa on the inferior lateral surface of the temporal bone. Images should be obtained of both joints for comparison and in the open and closed mouth positions to compare the movement of the condyles into and out of the fossa.

9. C   The acanthiomeatal line (ACML) extends from the point just below the nose (acanthion) to the external auditory meatus. This line is about 15 degrees below the infraorbitomeatal line.

10. A   The side-to-side measurements of the cranium, are made between the prominence (eminences) found on the external surface of the parietal bones. For the average (mesocephalic) skull a value of 15 centimeters is considered normal.

11. A   A radiographic image of the sella turcica is obtained in the lateral projection. The interpupillary line is placed perpendicular to the image receptor. The central ray is directed perpendicular to enter at a point 2 centimeters anterior and 2 centimeters superior to the external auditory meatus (EAM). This projection is used to evaluate the sella turcica, a saddle-like bony depression located on the upper surface of the body of the sphenoid bone.

12. C   A radiographic image of the petromastoid region is obtained in the axiolateral oblique (Stenvers) projection. The patient is placed prone with the midsagittal plane adjusted to form an angle of 45 degrees to the plane of the image receptor. The central ray is directed at a cephalic angulation of 12 degrees to enter 9 centimeters posterior to and 1 centimeter inferior to the EAM closest to the tube. This projection shows a profile image of the petrous ridge, mastoid process and internal auditory canals closest to the image receptor. The petrous ridge is parallel to the image receptor in this position.

13. A   A radiographic image of the cranial base is obtained in the verticosubmental (Schuller) projection. The patient is placed prone with the chin fully extended. The central ray is directed through the vertex of the skull 2 centimeters anterior to the EAM and perpendicular to the infraorbital meatal line. This projection is used to demonstrate the structures of the cranial base, the sphenoid sinuses and the mastoid processes.

14. G   15. B   16. C   17. I   18. E   19. A

20. H

21. B   The orbitomeatal line (OML) extending from the outer canthus to the external auditory meatus (EAM) forms an angle of 7 degrees with the infraorbitomeatal line (IOML) which extends from the lower orbital margin to the EAM.

22. D   A radiographic image of the optic foramen is obtained in the orbitoparietal (Rhese) projection. The patient is placed supine with the acanthiomeatal line perpendicular, the head is rotated so the midsagittal plane forms an angle of 53 degrees or an angle of 37 degrees to the vertical. The central ray is directed perpendicular to enter through the orbit closest to the image receptor. This projection of the optic canal and optic foramen shows these within the inferolateral shadow of the orbit.

23. C   In order to remove overlying shadows when radiographic images are obtained of the mastoid processes in the axiolateral or tangential projections, the ears should be taped forward. Because the mastoid air cells are thin and may be difficult to visualize, the taping of the cartilaginous auricles of the ear forward helps prevent their obscuring of the air cells.

24. D   A radiographic image of the cranial base is obtained in the submentovertical (Schuller) projection. The patient is placed supine, with the head resting on the vertex of the skull. The central ray is directed at a point just above the thyroid cartiledge perpendicular to the infraorbital meatal line. This full basal projection is used to demonstrate the structures of the cranial base, the sphenoid sinuses and the mastoid processes. It can also be used to show the foramina ovale and spinosum in the base of the skull, the maxillary, the anterior ethmoid sinuses, the odontoid process of the axis and the entire atlas.

25. A   The orbitomeatal line (OML) that extends from the outer canthus to the external auditory meatus (EAM) was in the past called the radiographic baseline.

26. D   A radiographic image of the cranial base is obtained in the submentovertical (Schuller) projection. The patient is placed supine with the head resting on the vertex of the skull.. This full basal projection is used to demonstrate the structures of the cranial base, the sphenoid sinuses and the mastoid processes. The inferosuperior path of the central ray in the full basal (SMV) projection is used to provide a tangential view of the zygomatic arches as well as visualization of the foramen magnum, internal auditory canals and mandibular condyles.

27. B   The distance between the vertex and submental portion in an "average" skull is 22 centimeters. Any variation of this external measurement will result in deviation of the internal structures and must be compensated by the central ray or part angulations.

28. C   A radiographic image of the cranium is obtained in the lateral projection. With the patient in the lateral position, the interpupillary line (IPL) which bisects the pupils of the eyes, should be placed perpendicular to the image receptor and the midsagittal plan. The central ray is directed perpendicular to the enter 5 centimeters superior to the EAM. This projection is used to show the sella turcica, the dorsum sella and the posterior clinoid processes.

29. B   A radiographic image of the cranial base is obtained in the submentovertical (Schuller) projection. This projection demonstrates the axis, atlas, clivus, and foramen magnum in the middle of the skull posterior to the petrous ridges. The paranasal sinuses, and mandibular structures are located anterior to the petrous ridges.

30. B   A radiographic image of the cranium is obtained in the PA axial (Caldwell) projection. With the patient prone, the orbitomeatal line is placed perpendicular to the image receptor. The central ray is directed to exit at the nasion at a caudal angle of 15 degrees. This projection will project the petrous ridges into the lower 3rd of the orbital shadows, freeing the frontal and anterior ethmoids from their superimposition with the petrous ridges.

31. B   The glabellomeatal line (GML) is the most superior of the four common radiographic lines used in positioning of the skull. The orbitomeatal line (OML) extending from the outer canthus to the external auditory meatus (EAM) forms an angle of 8 degrees with the GML.

32. B   A radiographic image of the optic foramen is obtained in the parieto-orbital (Rhese) projection. The patient is placed prone with the acanthiomeatal line perpendicular, the head is rotated so the midsagittal plane forms a an angle of 53 degrees or an angle of 37 degrees to the vertical. When the nose, zygoma and chin in contact with the table, the midsagittal plane forms a 53 degree angle to the image receptor. This will enable a vertical central ray to exit through the dependent optic foramen and provide an oblique image of the facial bones and the dependent optic foramen.

33. B   Because of the normal overbite, occluding the incisors would likely dislocate the mandibular condyles. In the normal bite, the molars are occluded, avoiding the unnatural displacement of the mandible.

34. D   A radiographic image of the optic foramen is obtained in the parieto-orbital (Rhese) projection. The patient is placed prone with the acanthiomeatal line perpendicular, the head is rotated so the midsagittal plane forms an angle of 53 degrees with the image receptor or an angle of 37 degrees to the vertical. This projection is used to help project the optic foramen into the inferolateral quadrant of the orbit and opens the optic foramen for its optimum visualization.

35. D   The infraorbitomeatal line (IOML) extends from the lower orbital margin to the EAM. This line is separated from the OML by 7 degrees.

36. C   The long axis of the orbit is directed obliquely at an anterior angle of 37 degrees from the midsagittal plane and 53 degrees from the vertical.

37. D   A radiographic image of the mandibular rami obtained in the PA axial projection is used to demonstrate the side-to-side (mediolateral) fracture displacement of the mandible. With the patient prone, the orbitomeatal line is placed perpendicular to the image receptor. The central ray is directed perpendicular to exit at the acanthion. The axiolateral oblique projection with the patient's head in the lateral position and the central ray directed at a cephalic angle of 25 degrees, is used to demonstrate the ramus of the mandible. The axiolateral oblique projection with the patient's head rotated 45 degrees and the central ray directed at a cephalic angle of 25 degrees, is used to demonstrate the symphysis of the mandible. The axiolateral oblique projection with the patient's head rotated 30 degrees toward the image receptor and the central ray directed at a cephalic angle of 25 degrees, is used to demonstrate the body of the mandible. These projections are used to demonstrate anteroposterior displacements of the mandible.

38. C A radiographic image of the petromastoid region is obtained in the axiolateral oblique (Stenvers) projection. The patient is placed prone with the midsagittal plane adjusted to form an angle of 45 degrees to the plane of the image receptor. The central ray is directed at a cephalic angulation of 12 degrees to enter 9 centimeters posterior to and 1 centimeter inferior to the EAM closest to the tube. This projection shows a profile image of the petrous ridge, mastoid process, internal auditory canals closest to the image receptor. The petrous ridge is parallel to the image receptor in the position.

39. B A radiographic image of the cranium is obtained in the lateral projection. With the patient in the lateral position, the interpupillary line (IPL) which bisects the pupils of the eyes should be placed perpendicular to the image receptor and the midsagittal plane. In a true lateral projection of the skull, the midsagittal plane is parallel to the image receptor and coronal planes are perpendicular to the image receptor. The central ray is directed perpendicular to enter 5 centimeters superior to the EAM. This projection is used to show the sella turcica, the dorsum sella and the posterior clinoid processes.

40. D A radiographic image of the maxillae is obtained in the parietoacanthial (Waters) projection. The patient is placed prone with the head resting on the chin, the neck is extended so that the orbitomeatal line forms an angle of 37 degrees with the plane of the image receptor. In this projection, the extension of the head and chin places the petrous ridges below the maxillary sinuses. In order to determine the appropriate amount of extension, the orbitomeatal line (OML) forms an angle of 37 degrees with the image receptor. This will normally place the nose about .5-1.5 centimeters from the image receptor. The failure to extend the chin and place the nose in this position will result petrous ridges obscuring the maxillary sinuses.

41. G   42. E   43. I   44. B   45. K   46. F

47. C   48. J

49. A A radiographic image of the cranium is obtained in the lateral projection. With the patient in the lateral position, the interpupillary line (IPL) which bisects the pupils of the eyes should be placed perpendicular to the image receptor and the midsagittal plane. In a true lateral projection of the skull, the midsagittal plane is parallel to the image receptor and coronal planes are perpendicular to the image receptor.

50. C A radiographic image of the petromastoid region is obtained in the axiolateral oblique (Arcelin) projection. The patient is placed supine with the midsagittal plane adjusted to form an angle of 45 degrees to the plane of the image receptor. The central ray is directed at a caudal angulation of 10 degrees to enter 2 centimeters anterior to and 2 centimeters superior to the EAM farthest from the tube. This projection shows a profile image of the petrous ridge, mastoid process, internal auditory canals farthest from the image receptor. The petrous ridge is parallel to the image receptor in this position.

51. D A radiographic image of the skull is obtained in the AP axial (Towne) projection. With the patient supine, the orbitomeatal line is placed perpendicular to the image receptor. The central ray is directed to exit the foramen magnum at a caudal angulation of 30 degrees to the orbitomeatal line or an angle of 37 degrees to the infraorbitomeatal line. This projection of the skull is used to give a symmetrical image of the petrous pyramids and zygomatic arches as well demonstrating the dorsum sella, foramen magnum, and occipital bone.

52. C A radiographic image of the petromastoid region is obtained in the axiolateral oblique (Arcelin) projection. The patient is placed supine, with the midsagittal plane adjusted to form an angle of 45 degrees to the plane of the image receptor. The central ray is directed at a caudal angulation of 10 degrees to enter 2 centimeters anterior to and 2 centimeters superior to the EAM farthest from the tube.

53. B A radiographic image of the cranial base is obtained in the submentovertical (Schuller) projection. The patient is placed supine, with the head resting on the vertex of the skull.. This full basal projection is used to demonstrate the structures of the cranial base, the sphenoid sinuses and the mastoid processes. The inferosuperior path of the central ray in the full basal (SMV) projection is used to provide a tangential view of the zygomatic arches as well as visualization of the foramen magnum, internal auditory canals and mandibular condyles.

54. C   A radiographic image of the maxillae is obtained in the parietoacanthial (Waters) projection. The patient is placed prone with the head resting on the chin, the neck is extended so that the orbitomeatal line forms an angle of 37 degrees with the plane of the image receptor. The floor of the orbit and roof of the maxillary sinus are formed by the maxillary bones. In cases of facial trauma, pressure on the orbit can fracture (blow out) this bony plate filling the maxillary sinus with fluid.

55. B   A radiographic image of the internal auditory canals is obtained in the PA axial (Haas) projection. With the patient prone, the orbitomeatal line is placed perpendicular to the image receptor. The central ray is directed to enter a point 4 centimeters below the inion at a cephalic angulation of 25 degrees. This projection of the cranium is used to give a symmetrical image of the petrous pyramids and zygomatic arches as well demonstrating the dorsum sella, foramen magnum, and occipital bone. The internal auditory canals are located within the medial portion of the petrous ridges of the temporal bones. This projection produces a bilateral symmetrical image of these canals projected within the orbital shadows. The prone position is preferred because it may result in a dramatic reduction in the exposure to the eye.

56. C   The PA axial (Haas) projection may be used on a hypersthenic patient who, because of a large torso and short neck, is unable to drop the chin enough to place the orbitomeatal (OML) line perpendicular to the image system. By turning the patient to a prone position and angling the central ray at a cephalic angulation of 25 degrees, the problem is corrected.

57. A   A radiographic image of a single zygomatic arch is obtained in the tangential (May) projection. The patient is placed with their chin on the image receptor with the neck extended as far as is possible. The central ray is directed perpendicular to the infraorbitomeatal line to a point 4 centimeters posterior to the outer cantus. The head is rotated ("May Away") away from the affected side so the midsagittal plane forms a 15 degree angle with the plane of the image receptor. This projection demonstrates a single zygomatic arch free of superimposition with the adjacent facial and skull bones.

58. D   A radiographic image of the petromastoid region is obtained in the axiolateral oblique (Mayer) projection. The patient is placed supine with the midsagittal plane adjusted to form an angle of 45 degrees to the plane of the image receptor. The central ray is directed at a caudal angulation of 45 degrees to exit to the EAM closest to the image receptor. This projection shows a profile image of the petrous ridge, mastoid process, internal auditory canals closest to the image receptor. The 45-45 degree double angulation of the midsagittal plane and the caudal central ray projects the petrosa inferior to the mastoid air cells, allowing the visualization of the auditory ossicles.

59. C   A radiographic image of the skull is obtained in the anteroposterior axial (Towne) projection. With the patient supine, the orbitomeatal line is placed perpendicular to the image receptor. The central ray is directed to exit the foramen magnum at a caudal angulation of 30 degrees to the orbitomeatal line. Compared to the overall body size of a child or infant, the skull is disproportionally large. When laying supine, the child's neck is more flexed than in an adult and the orbitomeatal line cannot be placed perpendicular to the image receptor. To compensate, the central ray angle must be less about 5-10 degrees than the usual 30 degrees for an adult.

60. A   A radiographic image of the optic foramen is obtained in the parieto-orbital (Rhese) projection. The patient is placed prone with the acanthiomeatal line perpendicular to the image receptor. The head is rotated so the midsagittal plane forms an angle of 53 degrees or an angle of 37 degrees to the vertical. This projection will project the optic foramen into the inferolateral orbital quadrant.

61. D   A radiographic image of the petromastoid region is obtained in the axiolateral oblique (Stenvers) projection. The patient is placed prone with the midsagittal plane adjusted to form an angle of 45 degrees to the plane of the image receptor. A 45 degree rotation of the midsagittal plane will place the dependent petrosa parallel to the image receptor. The 12 degree cephalic angle projects the petrous ridges into the lower 2/3rds of the orbits and superimposes the mandibular ramus on the lateral aspect of the cervical spine.

62. C   A radiographic image of the cranium is obtained in the PA projection for a general survey examination. The patient is prone with the head adjusted so the orbitomeatal line is perpendicular to the image receptor. When the central ray is directed perpendicular to exit at the nasion, the orbits will be filled with the shadows of the petrous pyramids. This projection will demonstrate the posterior ethmoidal air cells, the crista galli, the frontal bone and the frontal sinus.

63. A   A radiographic image of the petromastoid portion is obtained in the axiolateral oblique (Modified Law) projection. The patient's head is placed in the relaxed lateral position with the central ray directed at a caudal angulation of 15 degrees. The axiolateral oblique (original Law) projection, taken with the patient in the true lateral position, requires a double angulation of 15 degrees to the face and 15 degrees to the feet. Both projections are used to demonstrate the mastoid cells, lateral portion of the petrous pyramid and the internal auditory canals on end.

64. A   A radiographic image of the frontal sinus is obtained in the PA axial (Caldwell) projection. With the patient prone, the orbitomeatal line is placed perpendicular to the image receptor. The central ray is directed to exit at the nasion at a caudal angle of 15 degrees. Since there is an 8 degree difference between the GML and OML, a central ray angulation of 23 degrees caudad is used to produce a comparable image when the GML is perpendicular to the image receptor. Other modifications involve placing the patient against an upright Bucky that has been angled 15 degrees. This enables the image to be taken with no central ray angulation. Another modification requires the patient to be placed on the table so that the orbitomeatal line forms an angle of 15 degrees with the plane of the image receptor. All of these projections will project the petrous ridges into the lower 3rd of the orbital shadows, freeing the frontal and anterior ethmoids from their superimposition with the petrous ridges.

65. C   A radiographic image of the inferior orbital fissure and orbital floor is obtained in the PA axial (Bertel) projection. If the patient's head is placed prone and the infraorbitomeatal line adjusted perpendicular to the image receptor, the central ray should be directed to a point 8 centimeters below the external occipital protuberance at a cephalic angulation of 20-25 degrees. The inferior orbital fissure is a narrow cleft extending anterolaterally to the lateral wall of the orbit at an angle of 25 degrees. This projection is used to open the fissure and enable its demonstration within shadow of the orbit.

66. C   A radiographic image of the optic foramen is obtained in the parieto-orbital (Rhese) projection. The patient is placed prone with the acanthiomeatal line perpendicular, the head is rotated so the midsagittal plane forms an angle of 53 degrees with the plane of the image receptor or an angle of 37 degrees to the vertical in order to project the optic foramen into the inferolateral orbital quadrant.

67. B   When demonstrating the paranasal sinuses, an even optical density between the upper and lower orbits is the best indication that proper exposure factors have been employed.

68. A   A radiographic image is obtained in the PA axial (Cahoon) projection. If the head is adjusted so that the orbitomeatal line is perpendicular to the image receptor, the central ray should be directed to pass through the nasion at a cephalic angle of 25 degrees. This projection is used for demonstrating a symmetrical image of the styloid processes free of superimposition of adjacent bony structures and the clinoid process of the mandible. Wigby-Taylor describes an AP oblique projection with the mouth opened to demonstrate an oblique image of the styloid free of superimposition of the clinoid process. Fuchs describes an axiolateral oblique projection with the mouth opened and double central ray angulation of 10 degrees cephalic and 10 degrees anteriorly producing a lateral image of the styloid process free of superimposition of the clinoid process.

69. A   A radiographic image of the superior orbital fissure is obtained in the PA axial projection. The patient is placed prone with the head adjusted so that the orbitomeatal line and midsagittal plane are perpendicular to the plane of the image receptor. The central ray is directed 25 degrees caudal to pass through the inferior margin of the orbit. The superior orbital fissure is a cleft in the roof of the orbit between the greater and lesser wings of the sphenoid bone. This projection projects the shadows of the superior orbital fissures on the medial side of the orbit.

70. A   A radiographic image of a single zygomatic arch is obtained in the tangential (May) projection. The patient is placed with their chin on the image receptor with the neck extended as far as is possible. The central ray is directed perpendicular to the infraorbitomeatal line to a point 4 centimeters posterior to the outer cantus. The head is rotated ("May Away") away from the affected side so the midsagittal plane forms a 15 degree angle with the plane of the image receptor. This projection demonstrates a single zygomatic arch free of superimposition with the adjacent facial and skull bones. This position is particularly useful with patients who have depressed fractures or "flat" cheekbones.

71. D   A radiographic image of the cranium is obtained in the PA projection. The patient is prone with the head adjusted so the orbitomeatal line is perpendicular to the image receptor. When the central ray is directed perpendicular to exit at the nasion, the orbits will be filled with the shadows of the petrous pyramids. This projection will demonstrate the posterior ethmoidal air cells, the crista galli, the frontal bone and the frontal sinus. This projection places the frontal bone both parallel and closest to the image receptor.

72. A   A radiographic image of the sella turcica is obtained in the PA axial (Valdini) projection. The patient is placed prone with the head adjusted to place the orbitomeatal line perpendicular to the image receptor. The central ray is directed perpendicular to enter at a level just above the external acoustic meatus. This projection is used to give a symmetrical image of the petrous pyramids and zygomatic arches as well demonstrating the dorsum sella, foramen magnum, and occipital bone. Less angular distortion of any part is accomplished by the correct positioning of the anatomic structure, rather than using a central ray angulation required in AP projections of this region.

73. D   A radiographic image is obtained in the AP axial position (Towne) projection. With the patient supine, the head is adjusted to place the infraorbitomeatal line perpendicular to the image receptor. The central ray should be directed through the foramen magnum at a cephalic angulation of 30 degrees. This projection of the skull is used to give a symmetrical image of the petrous pyramids, the zygomatic arches, the dorsum sella, foramen magnum, and occipital bone.

74. A   A radiographic image of the cranium is obtained in the lateral projection, with the patient in the dorsal decubitus position. The patient is adjusted to place the interpupillary line perpendicular to the image receptor and the midsagittal plane vertical. This cross-table lateral projection is used to determine the presence of sphenoid sinus effusion due to basal fracture. This position will best demonstrate a fluid-filled sphenoid sinus, which may be the only clue to the presence of a basal skull fracture. The central ray is directed horizontal to enter at a point 5 centimeters superior to the EAM.

75. D   A radiographic image of the facial bones is obtained in the acanthioparietal (reverse Waters) projection. With the patient supine, the head is adjusted so the mentomeatal line is perpendicular to the image receptor. The central ray should be directed perpendicular to enter at the acanthion. This projection gives a similar image to the parietoacanthial projection of the facial bones. Another modification can be used for patient with facial trauma. This acanthioparietal projection requires that the patient be supine, with the infraorbitomeatal line perpendicular to the image receptor, the central ray is directed at a cephalic angulation of 30 degrees parallel to the mentomeatal line.

76. B   A radiograph image of the optic foramen is obtained in the parieto-orbital oblique (Rhese) projection. The chin is extended to place the acanthiomeatal line perpendicular to the image receptor. This image demonstrates the optic foramen in the inferolateral portion of the orbital shadow. This projection provides the optimum visualization of the optic foramen and reduce the amount of focal spot blur in the image.

77. C   Radiographic images of the paranasal sinuses are most often taken erect in order to demonstrate the existence of fluid levels that can only be detected by having the central ray horizontal to the image receptor. Only upright or decubitus projections can be used to accomplish this effect.

78. A   A radiographic image of the petromastoid portion is obtained in the axiolateral oblique (Modified Law) projection. The patient's head is placed in the relaxed lateral position with the central ray directed at a caudal angulation of 15 degrees. The axiolateral oblique (original Law) projection, taken with the patient in the true lateral position, requires a double angulation of 15 degrees to the face and 15 degrees to the feet. Both projections are used to demonstrate the mastoid cells, lateral portion of the petrous pyramid and the internal auditory canals on end.

79. D   A radiographic image of the mandibular rami is obtained in the PA axial projection. With the patient prone, the orbitomeatal line is placed perpendicular to the image receptor. The central ray is directed to exit at the acanthion at a cephalic angulation of 20-25 degrees. This projection is used to project the mastoid process and the base of the skull above the condylar processes of the mandible.

80. B   A radiographic image is obtained in the parieto-orbital (open mouth Waters) projection. The central ray is directed perpendicular to the image receptor to exit at the acanthion. To help remove the superimposition of the teeth that would otherwise obstruct visualization of the sphenoid sinuses, this projection requires that the sinuses be projected through the open mouth. The phonation "ah" during the exposure will immobilize the tongue in the floor of the mouth producing better demonstration of the sphenoid sinuses.

81. C   A radiographic image of the cranial base is obtained in the submentovertical (Schuller) projection. The patient is placed supine with the head resting on the vertex of the skull. This full basal projection is used to demonstrate the structures of the cranial base, the sphenoid sinuses and the mastoid processes. The inferosuperior path of the central ray in the full basal (SMV) projection is used to provide a tangential view of the zygomatic arches as well as visualization of the foramen magnum, internal auditory canals and mandibular condyles. The mandibular symphysis is superimposed on the frontal bone of the skull and is therefore poorly visualized.

82. D   A radiographic image of the internal auditory canals is obtained in the PA axial (Haas) projection. With the patient prone, the orbitomeatal line is placed perpendicular to the image receptor. The central ray is directed to enter a point 4 centimeters below the inion at a cephalic angulation of 25 degrees. This projection of the cranium is used to give a symmetrical image of the petrous pyramids and zygomatic arches as well demonstrating the dorsum sella, foramen magnum, and occipital bone. This projection, also called the reverse Towne, was developed to overcome the difficulty in radiographing hypersthenic patients, for the visualization of the petrous pyramids and occipital regions of the cranium. The prone position is preferred because it may result in a dramatic reduction in the exposure to the eye.

83. A   A radiographic image of the sella turcica is obtained in the PA axial (Valdini) projection. The patient is placed prone with the head adjusted to place the orbitomeatal line perpendicular to the image receptor. The central ray is directed perpendicular to enter at a level just above the external acoustic meatus. This projection is used to give a symmetrical image of the petrous pyramids and zygomatic arches as well demonstrating the dorsum sella, foramen magnum, and occipital bone. This projection is similar to a Towne or Haas projection but does not require an angulation of the central ray. The patient is positioned so that the infraorbital meatal line forms a 50 degree angle with the plane of the image receptor. The projection shows the sella turcica and petrous ridges including the auditory canals, labyrinths and mastoid air cells with less angular distortion.

84. C   A radiographic image of the foramen rotundus can be obtained in the parietoacanthial (Waters) projection. The patient is placed prone with the head resting on the chin, the neck is extended so that the orbitomeatal line forms an angle of 37 degrees with the plane of the image receptor. This projects the dense shadows of the petrous ridges below the antral floors allowing for visualization of the maxillary sinuses, foramen rotundum and axial images of the facial bones. It is important to note that when the neck is extended too much (less than 37 degrees) the antral shadows will be foreshortened; when the neck is extended too little, the shadows of the petrous ridges are projected into the antra and obscure underlying pathology.

85. D   A series of radiographic images of the paranasal sinuses are to be obtained for preoperative measurements. A 180 centimeter (72") source-to-image receptor distance (SID) is employed to minimize image magnification and reduce the amount of focal spot blur allowing for more precise measurements.

86. C   A radiographic image of the cranium is obtained in the PA projection. The patient is prone with the head adjusted so the orbitomeatal line is perpendicular to the image receptor. When the central ray is directed perpendicular to exit at the nasion, the orbits will be filled with the shadows of the petrous pyramids. This projection will demonstrate the posterior ethmoidal air cells, the crista galli, the frontal bone and the frontal sinus.

87. D   Radiographic images for the cranial base are to be obtained in the submentovertical (SMV) or verticosubmental (VSM) projections. Both projections require that the infraorbitomeatal line be adjusted parallel to the plane of the image receptor.

88. B   A radiographic image is obtained in the tangential projection. The central ray is directed parallel to the glabellomeatal line and perpendicular to an occlusal film placed in the mouth pebbled surface up. This projection is most often used for the demonstration of the nasal bones.

89. A   In order to obtain an AP projection for a general survey of the cranium on an infant or young child, the head is adjusted so the orbital meatal line is perpendicular to the image receptor in order to project the petrous ridges into the lower third of the orbital shadow. The central ray should be directed at a cephalic angulation of 15 degrees. Compared to the size of the trunk, a young child or infant's skull is relatively large. When lying supine, the neck is somewhat flexed. To compensate, the tube must be angled cranially so the central ray is directed along the orbitomeatal line to produce a true AP projection.

90. B   The axiolateral oblique projection of the mandible, the patient's head is placed in the lateral position with the central ray is directed at a cephalic angle of 25 degrees. This projection is used to demonstrate the ramus of the mandible. The axiolateral oblique projection with the patient's head rotated 45 degrees and the central ray directed at a cephalic angle of 25 degrees is used to demonstrate the symphysis of the mandible. The axiolateral oblique projection with the patient's head rotated 30 degrees toward the image receptor and the central ray directed at a cephalic angle of 25 degrees is used to demonstrate the body of the mandible. These projections are used to demonstrate anteroposterior displacements of the mandible.

91. A   In an AP axial projection of the skull, a 15 degree caudal angle to the orbitomeatal line will project the petrous ridges into the lower 1/3 of the orbits as demonstrated in the image. A perpendicular central ray will demonstrate the petrous ridges filling the orbits. To completely free the orbits of the shadows of the petrous ridges, an angle of 23 degrees caudal should be employed.

92. C   The AP axial projection is used to demonstrate the crista galli and frontal sinuses free of bony super imposition.

93. C   The size differences seen between the orbits indicates that this is an example of a parieto-orbital projection.

94. B   The optic foramen, located at the medial end of the sphenoid bone, can be demonstrated by rotating the head at an angle of 53 degrees with the image receptor and adjusting the acanthiomeatal line perpendicular to the table. In the parieto-orbital oblique projection, the optic foramen is projected into the inferolateral portion of the orbit.

95. D   In the parietoacanthial (Waters) projection, the extension of the head and chin places the petrous ridges below the maxillary sinuses. In order to determine the appropriate amount of extension, the orbitomeatal line (OML) forms an angle of 37 degrees to the plane of the image receptor leaving the nose about .5- 1.5 centimeters from the image receptor.

96. C   In the parietoacanthial (Waters) projection, the extension of the head and chin places the petrous ridges below the maxillary sinuses.

97. A   A radiographic projection of the temporomandibular articulations is obtained in the axiolateral projection. This projection is taken to show the TMJ closest to the image receptor, the mandibular condyle at the superoposterior border of the mandible and the mandibular fossa on the inferior lateral surface of the temporal bone. Images should be obtained of both joints for comparison and in the open and closed mouth positions to compare the movement of the condyles into and out of the fossa. The lateral axiolateral positions are employed to demonstrate the temporomandibular joints closest to the image receptor.

98. B   The closed (A) and open mouth (B) are easily recognized by the relationship of the mandibular condyle to the mandibular fossa. The condyle lies in the fossa when the mouth is closed and inferior to the articular tubercle when the mouth is open. The closed mouth is employed to demonstrate fractures of the neck and condyle of the ramus, and the open mouth best demonstrates anterior and inferior excursion of the condyle from the mandibular fossa.

99. B   The axiolateral oblique projection of the mandible, the patient's head is placed in the lateral position with the central ray is directed at a cephalic angle of 25 degrees. This projection is used to demonstrate the ramus of the mandible. The axiolateral oblique projection with the patient's head rotated 45 degrees and the central ray directed at a cephalic angle of 25 degrees, is used to demonstrate the symphysis of the mandible. The axiolateral oblique projection with the patient's head rotated 30 degrees toward the image receptor and the central ray directed at a cephalic angle of 25 degrees, is used to demonstrate the body of the mandible. These projections are used to demonstrate anteroposterior displacements of the mandible.

100. A   See the explanation above.

1. A radiographic image of the chest is obtained in the lateral projection. If the patient is erect, it is important to place the midsagittal plane:

    A. Parallel to the image receptor and vertical to the floor
    B. Perpendicular to the image receptor and horizontal to the floor
    C. Parallel to the image receptor and horizontal to the floor
    D. Perpendicular to the image receptor and vertical to the floor

2. A radiographic image of the chest is obtained in the posteroanterior (PA) projection. The patient is placed erect facing the image receptor with the arms flexed and the backs of the hands resting on the hips. This position of the arms is used to:

    A. Move the diaphragms to their highest position
    B. Move the diaphragms to their lowest position
    C. Direct the clavicles above the apices of the lungs
    D. Rotate the scapulae laterally away from the chest

3. A radiographic image of the chest is obtained in the lateral projection. If the patient is erect, the central ray is directed horizontal to the floor to enter at the midcoronal plane at the level of the:

    A. Second thoracic vertebra
    B. Fourth thoracic vertebra

    C. Seventh thoracic vertebra
    D. Ninth thoracic vertebra

4. A radiographic image of the chest is obtained in the PA axial projection. In order to demonstrate the apices free of superimposition with the shadows of the clavicles, the central ray should be directed:

    A. Cephalic at an angulation of 10-15 degrees
    B. Caudal at an angulation of 10-15 degrees

    C. Cephalic at an angulation of 35-40 degrees
    D. Caudal at an angulation of 35-40 degrees

5. A radiographic image of the chest is obtained in the posteroanterior (PA) projection. A 180 centimeter source-to-image receptor distance is employed in order to:

    A. Reduce the amount of motion due to respiration
    B. Blur the shadows of the scapulae from the lung fields
    C. Reduce the amount of magnification in the heart shadow
    D. Project the clavicles above the top of the thoracic cavity

6. A radiographic image of the chest is obtained in the right lateral projection. This projection is most commonly taken to improve the visualization of lesions involving the:

    A. Mediastinal region
    B. Right lung

    C. Left ventricle of the heart
    D. Left hemidiaphragm

7. A radiographic image of the chest is obtained in the posteroanterior (PA) projection. If the central ray is directed perpendicular to the image receptor, the resulting image is most likely to show the shadows of the clavicles:

    A. At the level of the midscapula
    B. At the level of the costophrenic recesses
    C. About 2 centimeters above the apices of the lung
    D. About 5 centimeters below the apices of the lung

8. Whenever possible, radiographic images of the chest should be performed with the patient erect in order to:

    A. Prevent engorgement of the great vessels
    B. Reduce the amount of magnification of the heart
    C. Demonstrate the presence free fluids
    D. Reduce the amount of focal spot blur in the image

9. Which radiographic projection of the chest is most often used to evaluate the presence of free air in the left pleural cavity?

A. A PA projection with the patient erect
B. An AP projection with the patient supine
C. An AP projection with the patient in the right lateral decubitus position
D. An AP projection with patient in the left lateral decubitus position

10. Two radiographic images of the chest are obtained in the posteroanterior (PA) projection. Separate images are taken with the patient suspending their respiration at the end of full inspiration and one on full expiration. These projections are commonly used to evaluate:

A. The presence of vascular bleeding
B. The movement of foreign bodies of the lungs
C. The excursion of the diaphragms
D. The presence of mediastinal lesions

11. A radiographic image of the chest is obtained in the PA oblique projection in the LAO position. The patient is placed with their left shoulder in contact with the image receptor and their right hand on the hip. In order to demonstrate the heart and descending aorta, the patient should be rotated.

A. 10-20 degrees from the plane of the image receptor
B. 25-35 degrees from the plane of the image receptor
C. 45-60 degrees from the plane of the image receptor
D. 75-85 degrees from the plane of the image receptor

12. A radiographic image of the chest is obtained in the posteroanterior (PA) projection. Which body position and phase of respiration will move the diaphragms to their lowest position?

A. The erect position with inspiration
B. The erect position with expiration
C. The supine position with inspiration
D. The supine position with expiration

13. A radiographic image of the chest is obtained in the posteroanterior (PA) projection. Unwanted rotation of the chest is best detected by evaluating the:

A. Rotation of the scapulae
B. Dimensions of the costophrenic angle
C. Dimensions of the sternoclavicular joint spaces
D. Rotation of the sternum

14. A radiographic image of the chest is obtained in the posteroanterior (PA) projection. If the thoracic viscera and the heart are of primary interest, the patient should be asked to suspend respiration at the end of:

A. Full inspiration
B. Normal inspiration
C. Full expiration
D. Normal expiration

Referring to the radiograph of the chest, place the letter that corresponds to the following structures for questions 15-24.

15._____ Axillary border of the ribs
16._____ Soft tissue shadow of the breast
17._____ Apical region of the right lung
18._____ Right clavicle
19._____ Hilar region
20._____ Air-filled trachea
21._____ Left hemidiaphragm
22._____ Left ventricle of the heart
23._____ Posterior portion of the eighth rib
24._____ Right costophrenic angle

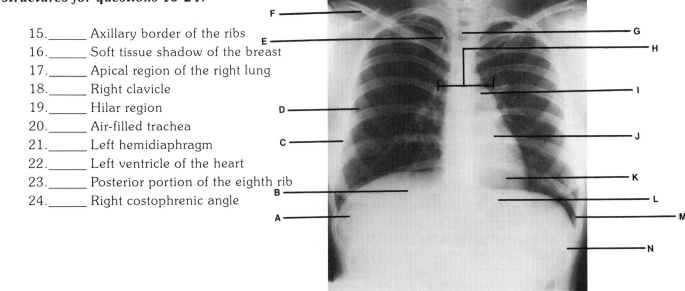

25. Which of the following radiographic projections is most often requested to provide the best demonstration of the left atrium?

A. Right PA oblique projection in the RAO position   C. Left lateral projection with the patient supine
B. Left PA oblique projection in the LAO position   D. PA projection with the patient erect

26. A radiographic image of the trachea is obtained in the AP projection. The central ray should be directed perpendicular to the midsagittal plane and adjusted to enter at the:

A. Level of the tip of the xiphoid process   C. Level of the acromioclavicular joints
B. Level of the sixth thoracic vertebra   D. Level of the manubrium

27. Which projection of the chest is most commonly used for the demonstration of interlobar effusions and to give an unobstructed view of the pulmonary apices?

A. PA oblique projection with the patient RAO
B. AP axial (Lindholm) projection
C. AP projection with the patient in the ventral decubitus position
D. Axiolateral (Twining) projection

28. A radiographic image of the chest is obtained in the PA oblique projection in the RAO position. The patient is placed with their right shoulder in contact with the image receptor and their left hand on the hip. In order to demonstrate pulmonary diseases the patient should be rotated:

A. 10-20 degrees from the plane of the image receptor
B. 25-35 degrees from the plane of the image receptor
C. 45-60 degrees from the plane of the image receptor
D. 75-85 degrees from the plane of the image receptor

29. The projection of the chest that is commonly employed for the demonstration of free fluid in the costophrenic recess is the:

A. AP projection with the patient supine   C. AP axial projection with the patient erect
B. PA projection with the patient erect   D. Lateral projection with the patient recombinant

30. A radiographic image of the chest is obtained in the posteroanterior (PA) projection. With the patient erect, an adequate inspiratory effort is indicated by the visualization of:

A. 10 pairs of ribs above the diaphragms   C. 6 pairs of ribs above the diaphragms
B. 8 pairs of ribs above the diaphragms   D. 4 pairs of ribs above the diaphragms

31. A radiographic image of the chest is obtained in the lateral projection. With the patient erect in the left lateral position, the central ray should be directed perpendicular to the midcoronal plane at the level of the:

A. Fifth thoracic vertebra   C. Eighth thoracic vertebra
B. Seventh thoracic vertebra   D. Tenth thoracic vertebra

32. A radiographic image of the chest is obtained in the posteroanterior (PA) projection. This projection is preferred over the AP projection because:

A. The diaphragms are in their lowest position   C. It places the heart closer to the image receptor
B. It opens the aortic arch   D. It places the clavicles above the apices

33. A radiographic image of the chest is obtained in the posteroanterior (PA) projection. In order to demonstrate the maximum amount of lung tissue, the patient should be instructed to suspend respiration at the:

A. End of shallow expiration   C. End of the second breath inspiration
B. End of shallow inspiration   D. End of the first breath expiration

34. A radiographic image of the chest is obtained in the PA oblique projection. With the patient in the LAO position the maximum area of the lung demonstrated is on:

    A. The right side of the thorax
    B. The left side of the thorax

35. A radiographic image of the chest is obtained in the AP projection. With the patient supine, the scapulae can be rotated away from the lung fields by:

    A. Flexing the elbows and pronating the hands   C. Supinating the hands and elbows
    B. Raising the arms over the head of the patient   D. Abducting the upper arms 90 degrees

Referring to the radiographic image, answer questions 36 & 37.

36. This radiographic image of the chest was obtained in the:

    A. AP projection with the patient in the left lateral
       decubitus position
    B. AP oblique projection with the patient supine
    C. Lateral projection with the patient erect
    D. PA axial projection with the patient prone

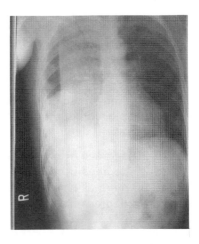

37. This radiographic image is most often used in the evaluation of:

    A. Mediastinal shifts
    B. Air or fluid levels
    C. Small pneumothoraces
    D. Megalocardia

Referring to the radiographic image, answer questions 38 & 39.

38. This radiographic image of the chest was obtained in the:

    A. PA projection with the patient supine
    B. Lateral projection with the patient recombinant
    C. PA oblique projection with the patient erect
    D. Lateral projection with the patient erect

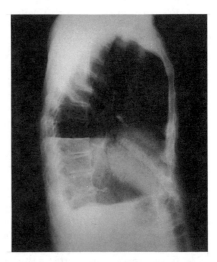

39. The radiographic image indicates that the patient has a/an:

    A. Large consolidation in the middle lobe of the lung
    B. Large abscess in the apical region of the lung
    C. Enlarged left ventricle of the heart
    D. Large amount of free fluid in the lower lobe of the lung

Referring to the radiographic image, answer question 40.

40. This radiographic image was obtained in the:

    A. AP axial projection with the patient erect
    B. PA projection with the patient supine
    C. AP oblique projection with the patient erect
    D. Lateral projection with the patient erect

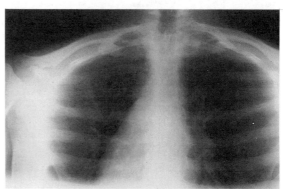

1. A A radiographic image of the chest is obtained in the lateral projection. If the patient is erect, it is important to place the midsagittal vertical to the floor and midaxillary parallel to the image receptor. The central ray is directed horizontal to enter the midcoronal plane at the level of the seventh thoracic vertebra. When possible, the patient is placed in the left lateral position to place the heart closest to the image receptor with the arms raised upwards. A 180 centimeter source-to-image receptor distance is employed to decrease the magnification of the heart and reduce the amount of focal spot blur. The patient is asked to suspend their respiration at the end of full inspiration to maximize the volume of the lung tissue and to move the diaphragms to their lowest position. This projection is used to evaluate the heart, the aorta, left side pulmonary lesions and interlobular fissures.

2. D A radiographic image of the chest is obtained in the posteroanterior (PA) projection. The patient is placed erect facing the image receptor, with the arms flexed and the backs of the hands resting on the hips to rotate the scapulae laterally. The central ray is directed horizontal to enter the midsagittal plane at the level of the seventh thoracic vertebra. A 180 centimeter source-to-image receptor distance is employed to decrease the magnification of the heart and reduce the amount of focal spot blur. The patient is asked to suspend their respiration at the end of full inspiration to maximize the volume of the lung tissue and to move the diaphragms to their lowest position. This projection shows the thoracic viscera, the air filled trachea, the heart and aortic knob and the domes of the hemidiaphragms.

3. C A radiographic image of the chest is obtained in the posteroanterior (PA) projection. The patient is placed erect facing the image receptor, with the arms flexed and the backs of the hands resting on the hips to rotate the scapulae laterally. When a chest is to be imaged on a 35 x 43 cm image receptor, the entire thoracic cavity will be included if the central ray is directed horizontal to enter the midsagittal plane at the level of the seventh thoracic vertebra.

4. A Because the clavicles are normally superimposed with the apical regions of the lungs, an unobstructed view of this region is difficult. A radiographic image of the chest obtained in the PA or AP axial projection is used to demonstrate the apices free of superimposition with the shadows of the clavicles. In the PA axial projection, the central ray should be directed at a cephalic angulation of 10-15 degrees to enter at the level of T3. In the AP axial projection, the central ray should be directed at a cephalic angulation of 15-20 degrees to enter at the midsternum. Exposures are normally made after suspending the breath on full inspiration. Either projection will show the apical portion of the lungs lying below the clavicles.

5. C The use of a long 180 centimeter (72") source-to-image distance (SID) will reduce the amount of focal spot blur and the degree of magnification of the entire chest and heart. The PA projection is preferred to further reduce the size of the heart shadow by placing the heart closest to the image receptor.

6. B A radiographic image of the chest is obtained in the lateral projection. If the patient is erect, it is important to place the midsagittal vertical to the floor and midaxillary parallel to the image receptor. The central ray is directed horizontal to enter the midcoronal plane at the level of the seventh thoracic vertebra. A patient may be placed in the right lateral position when the principal structures of interest are located in the right lung. Because of the importance of the heart, the left lateral position is normally employed for routine chest radiography.

7. D A radiographic image of the chest is obtained in the posteroanterior (PA) projection. The central ray is directed horizontal to enter the midsagittal plane at the level of the seventh thoracic vertebra. The patient is asked to depress the shoulders to direct the clavicles below the apical regions of the lungs.

8. C In order to demonstrate the presence of free fluids in the lung or free air in the abdominal cavity and reduce the engorgement of the pulmonary vessels, the upright or erect projections are preferred. Radiographic projections of the chest are normally taken at the end of inspiration to maximize the size of the lung tissue and to lower the diaphragms.

9. C   When a patient cannot be placed upright, air-fluid levels can be identified in the decubitus positions. The right lateral decubitus position is preferred for the visualization of air in the axillary region of the left lung and fluid levels in the right lung.

10. C   Two radiographic images of the chest are obtained in the posteroanterior (PA) projection. Separate images are taken with the patient suspending their respiration at the end of full inspiration and one on full expiration. Inspiration and expiration radiographs are most frequently taken to show the differences in position of the diaphragm (excursion) and small pneumothoraces that may only be visible on the expiration images.

11. C   A radiographic image of the chest is obtained in the PA oblique projection in the erect LAO position. The patient is placed with their left shoulder in contact with the image receptor and their right hand on the hip. The central ray is directed horizontal at a level of the T7. The normal degree of obliquity for the chest requires that the body be rotated 45 degrees from the PA projection. In order to demonstrate the heart and descending aorta, the patient should be rotated 45-60 degrees. A rotation of 35-45 degrees is generally recommended to evaluate the esophagus. A shallow rotation of 10-20 degrees is employed to evaluate pulmonary diseases.

12. A   The diaphragm moves to its lowest position during erect inspiration; forced inhalation should be avoided because it may cause an abnormal elongation of the heart distorting its normal appearance.

13. C   Rotation of the chest is most easily seen by comparing the spacing of the sternoclavicular (SC) joints located at the sides of the third thoracic vertebra.

14. B   The proper breathing instruction for chest radiography of the heart, lungs and mediastinum is to stop breathing at the end of full inspiration. When the heart is of primary interest, the patient is instructed to stop breathing at the end of normal inspiration.

15. D   16. C   17. E   18. F   19. H   20. G

21. L   22. K   23. J   24. A

25. A   The right anterior oblique (RAO) and left posterior oblique (LPO) positions are used to provide the optimum visualization of the left atrium when the patient is rotated 45 degrees.

26. D   The trachea extends from the cricoid cartilage at a level near the first thoracic vertebra to its bifurcation near the level of the 5th thoracic vertebra. In order to center this structure on the image, the central ray is directed through the body of the manubrium.

27. B   A radiographic image of the chest obtained in the PA or AP axial projection is used to demonstrate the apices free of superimposition with the shadows of the clavicles. In the PA axial projection, the central ray should be directed at a cephalic angulation of 10-15 degrees to enter at the level of T3. In the AP axial projection, the central ray is directed at a cephalic angulation of 15-20 degrees to enter at the midsternum. Exposures are normally obtained with the breath suspended on full inspiration. Either projection will show the apical portion of the lungs lying below the clavicles and are used in the evaluation of interlobar effusions.

28. A   A radiographic image of the chest is obtained in the PA oblique projection in the erect RAO position. The patient is placed with their right shoulder in contact with the image receptor and their left hand on the hip. The central ray is directed horizontal at a level of the T7. The normal degree of obliquity for the chest requires that the body be rotated 45 degrees from the PA projection. A shallow rotation of 10-20 degrees is employed to evaluate pulmonary diseases.

29. B   The upright AP or PA projection is the preferred projection for the demonstration of free fluid levels in the costophrenic angles.

30. A   A radiographic image of the chest is obtained in the posteroanterior (PA) projection. The central ray is directed horizontal to enter the midsagittal plane at the level of the seventh thoracic vertebra. The patient is asked to suspend their respiration at the end of full inspiration to maximize the volume of the lung tissue and to move the diaphragms to their lowest position. During full inspiration, the diaphragm is lowered to a position at or near the eleventh rib enabling the visualization of the first ten pairs of ribs.

31. B   A radiographic image of the chest is obtained in the lateral projection. If the patient is erect, it is important to place the midsagittal vertical to the floor and midaxillary parallel to the image receptor. The central ray is directed horizontal to enter the midcoronal plane at the level of the seventh thoracic vertebra.

32. C   A radiographic image of the chest is obtained in the posteroanterior (PA) projection. The central ray is directed horizontal to enter the midsagittal plane at the level of the seventh thoracic vertebra. A 180 centimeter source-to-image receptor distance is employed to decrease the magnification of the heart and reduce the amount of focal spot blur.

33. C   A radiographic image of the chest is obtained in the posteroanterior (PA) projection. The central ray is directed horizontal to enter the midsagittal plane at the level of the seventh thoracic vertebra. The patient is asked to suspend their respiration at the end of full inspiration to maximize the volume of the lung tissue and to move the diaphragms to their lowest position. Normal inspiration on the second breath will most often increase the lung volume for chest radiography, especially in those patients with a heavy or hypersthenic body habitus.

34. B   A radiographic image of the chest is obtained in the PA oblique projection in the erect LAO position. The patient is placed with their left shoulder in contact with the image receptor and their right hand on the hip. The central ray is directed horizontal at a level of the T7. The normal degree of obliquity for the chest requires that the body be rotated 45 degrees from the PA projection. This projection with the patient LAO, shows the left side of the chest foreshortened and the right side elongated, showing the maximum area of the right lung.

35. A   A radiographic image of the chest is obtained in the AP projection. In either the upright or supine position, the shoulders should be rolled forward and hands placed on the hips as they would be in the PA projection to remove the scapula from the lung fields.

36. A   The air-fluid level appearing on the right side of the chest cavity indicates that this AP projection was obtained with the patient in the left lateral decubitus position.

37. B   See the explanation above.

38. D   A radiographic image of the chest is obtained in the lateral projection. If the patient is erect, it is important to place the midsagittal vertical to the floor and midaxillary parallel to the image receptor. The central ray is directed horizontal to enter the midcoronal plane at the level of the seventh thoracic vertebra. When possible, the patient is placed in the left lateral position to place the heart closest to the image receptor with the arms raised upwards. This lateral projection in the erect position shows a large amount of fluid filling the lower lobe of the lung.

39. D   See the explanation above.

40. A   This radiographic image of the chest was obtained in the AP axial projection. This projection is used to demonstrate the apical regions of the lungs free of superimposition with the shadows of the clavicles. The projection was obtained by directing the central ray at a cephalic angulation of 15-20 degrees to enter at the midsternum.

1. On the day preceding a radiographic imaging study of the gallbladder, before the patient takes the oral iodinated contrast agent, they should be instructed to eat lunch consisting of:

   A. Clear fluids
   B. High fat content foods
   C. Low fat content foods
   D. High fibrous foods

2. What is the maximum amount of air that should normally be used to inflate a retention tip (Bardex)?

   A. 10 milliliters
   B. 30 milliliters
   C. 90 milliliters
   D. 250 milliliters

3. A radiographic examination of the upper gastrointestinal tract is to be obtained with a double contrast. The two contrast agents that are most often used for this purpose are a combination of:

   A. Air and oil based iodinated contrast agents
   B. Glucagon and nonionic iodinated contrast agents
   C. Barium sulfate and carbon dioxide producing tablets
   D. Gastrografin and barium sulfate tablets

4. Diaphragmatic or hiatal hernias are common disorders of the gastrointestinal tract that can be diagnosed during the radiographic study called gastrointestinal (GI) series. The evaluation of this condition is often aided by placing the patient in the:

   A. Erect position
   B. Trendelenburg position
   C. Sims position
   D. Billings's position

5. All of the following radiographic procedures can be used for the evaluation of the organs of the upper urinary tract following the injection of a iodinated contrast agent EXCEPT:

   A. Infusion nephrotomography
   B. Intravenous urography
   C. Retrograde urography
   D. Dacryocystography

6. In order to demonstrate intestinal fluids or free air in the abdomen, radiographic images should be obtained in AP or lateral projections with the patient in all of the following positions EXCEPT:

   A. Supine position
   B. Erect position
   C. Lateral decubitus position
   D. Ventral decubitus position

7. During the radiographic evaluation of the esophagus during a swallowing study using barium sulfate suspension, breathing instructions are NOT normally required because:

   A. Rapid exposure times can eliminate all peristaltic motion
   B. The esophagus does not move during respiration
   C. Deglutition will inhibit normal respiration
   D. Breathing motion aids in the visualization most esophageal lesions

8. In order to insure that the entire small intestines have been demonstrated during the radiographic portion of a small bowel series, delayed films are normally taken until the barium sulfate suspension passes into the:

   A. Distal jejunum
   B. Proximal colon
   C. Duodenal bulb
   D. Cardiac orifice

9. Before a radiographic image procedure of the large intestines is performed, it is important that the colon be properly cleaned. This will generally involve having the patient:

   1. Undergo a colonic lavage
   2. Follow a set of dietary restrictions
   3. Undergo a gastric lavage

   A. 1 & 2 only
   B. 1 & 3 only
   C. 2 & 3 only
   D. 1, 2 & 3

10. All of the following radiographic or fluoroscopic imaging procedures will require the patient to follow dietary restrictions preceding the examination EXCEPT:

A. A barium enema
B. A gastrointestinal series

C. An oral cholecystography
D. Diskography

11. The radiographic evaluation of the pelvocalyceal system is obtained following the intravenous administration of an aqueous nonionic iodinated contrast agent. The greatest concentration will be demonstrated at:

A. 1 minute following the injection
B. 3 minutes following the injection

C. 15 minutes following the injection
D. 90 minutes following the injection

12. A radiographic image of the abdomen is obtained in the AP projection. The normal breathing instructions given will require that the patient:

A. Suspend respiration at the end of exhalation
B. Suspend respiration at the end of inhalation

C. Breathe normally throughout the exposure

Referring to the radiographic image of the barium-filled stomach, place the letter that corresponds to the following structures for questions 13-18.

13._____Pylorus
14._____Cardiac orifice
15._____Sweep of the duodenum
16._____Fundus of the stomach
17._____Lesser curve of the stomach
18._____Duodenal bulb

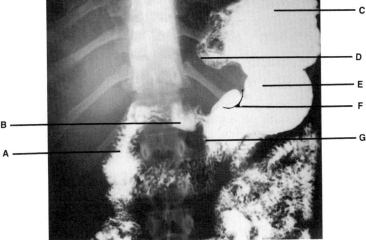

19. To insure proper preparation of the stomach prior to the radiographic evaluation of the gastrointestinal tract, the patient is instructed to avoid foods and liquids for at least:

A. 30 minutes prior to the procedure
B. 1 hour prior to the procedure

C. 8 hours prior to the procedure
D. 24 hours prior to the procedure

20. When radiolucent foreign bodies are suspected in the pharynx and upper esophagus, radiographic images may be obtained following the introduction of:

A. Nasogastric tubes
B. Mercury-filled bags

C. Barium saturated cotton balls
D. Iodinated tablets

21. A radiographic image of the abdomen is obtained in the AP projection on a male patient. In order to insure that all abdominal structures are included on a survey image, it is important to include the:

A. Greater trochanters on the image
B. Diaphragms on the image

C. Cardiac shadows on the image
D. Gonads on the image

22. A radiographic evaluation of the large intestines is obtained following the introduction of a liquid barium sulfate suspension. The lateral and axial (Chassard-Lapiné) projections are most often taken to evaluate lesions of the:

A. Rectosigmoid junction
B. Ileocecal junction
C. Splenic flexure
D. Distal ascending colon

23. A radiographic procedure of the urinary tract performed following the intravenous injection of an aqueous nonionic iodinated contrast agent is called:

A. Intravenous urography
B. Retrograde pyelography
C. Voiding urethrography
D. Intravenous cholangiography

24. A radiographic image of the abdomen is obtained in the AP projection on a female patient. The exposure and technical factors selected for this image should demonstrate:

A. A long scale contrast
B. A short scale contrast
C. A high optical density
D. A low optical density

25. A radiographic procedure of the gallbladder is obtained following the introduction of an oral iodinated contrast agent. Images that are obtained in the erect or decubitus positions are normally included for evaluation of:

A. Solitary calculi
B. Inflammatory lesions
C. Layered calculi
D. Radiopaque lesions

26. The radiographic projection that will best demonstrate the biliary ducts following the introduction of an iodinated contrast agent is:

A. Anteroposterior (AP) projection
B. Posteroanterior (PA) projection
C. Right posterior oblique projection
D. Left posterior oblique projection

27. A radiographic procedure of the upper gastrointestinal tract is performed following the introduction of a liquid barium sulfate suspension. Which of the following projections of the stomach will most likely demonstrate a gas-filled fundus and duodenum?

A. An AP projection with the patient supine
B. A PA projection with the patient prone
C. An PA oblique projection with the patient erect
D. A lateral projection with the patient erect

28. A radiographic evaluation of the urinary tract (intravenous urography) is accomplished following the introduction of an aqueous nonionic iodinated contrast agent. The AP projection with the patient placed in Trendelenburg is often employed to optimize the visualization of the:

A. Lower urinary bladder
B. Upper urinary bladder
C. Upper ureters
D. Lower ureters

29. A radiographic procedure of the large intestines is performed following the introduction of a liquid barium sulfate suspension and air. Which of the following projections is used to best visualize an open hepatic flexure?

A. A PA axial projection
B. A left AP oblique projection
C. A right AP oblique projection
D. A lateral projection

Referring to the radiographic image of the barium-filled colon, place the letter that corresponds to the following structures for questions 30-38.

30. _____ Descending colon
31. _____ Hepatic flexure
32. _____ Ileocecal value
33. _____ Rectum
34. _____ Sigmoid colon
35. _____ Transverse colon
36. _____ Cecum
37. _____ Ascending colon
38. _____ Splenic flexure

39. Which of the following procedures best describes the placement of a long hollow tube into the stomach and/or intestine to remove liquids of gasses from these structures?

A. Enteroplasty
B. Gastrointestinal intubation
C. Gastrectomy
D. Colonoscopy

40. A radiographic image of the abdomen is obtained in the AP projection with the patient supine. When possible, these images should be performed only after an attempt has been made to reduce all of the following EXCEPT:

A. Gas that is in the intestines
B. Fecal material that is in the intestines
C. Medications that are in the intestines
D. Fluids that are in the intestines

41. A radiographic procedure of the large intestines is performed following the introduction of a liquid barium sulfate suspension and air. During the fluoroscopic portion of this procedure which of the following structures is normally filled first?

A. The cecum
B. The hepatic flexure
C. The descending colon
D. The splenic flexure

42. TA radiographic evaluation of the urinary tract (intravenous urography) is accomplished following the introduction of an aqueous nonionic iodinated contrast agent. The usual amount of contrast media used for this imaging procedure for a normal adult is about:

A. 5-10 milliliters
B. 30-100 milliliters
C. 100-200 milliliters
D. 200-300 milliliters

43. A radiographic procedure of the gallbladder is obtained following the introduction of an oral iodinated contrast agent. To help insure a maximum concentration of contrast media in the gallbladder, it is important for the patient to avoid all of the following EXCEPT:

A. Laxatives
B. Liquids that contain fats
C. Solid foods
D. Oxygen therapy

44. The preliminary radiographs that should be taken on infants who may have swallowed or aspirated foreign objects should include radiographic projections of the:

A. Skull, mouth and esophagus
B. Nose, oropharynx, and mouth
C. Hands and fingers
D. Neck, chest and abdomen

45. A radiographic procedure of the gallbladder is obtained following the introduction of an oral iodinated contrast agent. Because movement of the gallbladder in relation to the phase of respiration can be as great as 15 centimeters, it is advisable to expose an image at the end of:

A. Inhalation
B. Exhalation

46. A radiographic procedure of the large intestines is to be performed. The enema tip that is used for the introduction of the contrast agent is generally inserted with the patient on his-her left side, leaning forward, and the right knee flexed. This describes the:

A. Sims position
B. Billings position
C. Lithotomy position
D. Chassard position

47. A radiographic procedure of the upper gastrointestinal tract is performed following the introduction of a liquid barium sulfate suspension. Images of the stomach commonly obtained in the right PA oblique projection, are primarily intended to visualize the:

A. Fundus and the gastroesophageal junction
B. Pyloric canal and the duodenal bulb
C. Haustral folds
D. Semicircular canals

48. A radiographic procedure of the large intestines is performed following the introduction of a liquid barium sulfate suspension and air. A PA axial projection is obtained with the central ray directed to the level of the anterior superior iliac spine at a caudal angulation of 30-40 degrees. This projection is used to improve the demonstrate of the:

A. Hepatic flexure
B. Rectosigmoid area
C. Splenic flexure
D. Ileocecal valve

49. A radiographic procedure of the gallbladder and/or biliary ducts following the introduction of an iodinated contrast agent can be accomplished through the:

1. Oral administration of the agent
2. Intravenous administration of the agent
3. Direct (T-tube) administration of the agent

A. 1 & 2 only
B. 1 & 3 only
C. 2 & 3 only
D. 1, 2 & 3

50. A radiographic procedure of the upper gastrointestinal tract is performed following the introduction of a liquid barium sulfate suspension. A PA projection taken in the Trendelenburg position is often employed to improve the visualization of:

A. Gastroesophageal reflux
B. Duodenal ulcerations
C. Pyloric stenosis
D. Hypermotility

51. A radiographic image of the abdomen is obtained in the AP projection. If it is important to determine the presence of air/fluid levels, the patient should be placed in the semi-erect position. The central ray should be directed to enter at the level of the iliac crests:

A. Perpendicular to the image receptor
B. Horizontal to the floor
C. Vertical to the floor
D. Perpendicular to the patient

52. The minor side effects that result from the use of oral cholecystographic agents are nearly all related to the action of the contrast agents on the:

A. Liver
B. Gallbladder
C. Kidney
D. Intestinal mucosa

53. A radiographic procedure of the esophagus is performed in the PA oblique projection following the introduction of a thick liquid barium sulfate suspension. The central ray should be directed perpendicular to enter at the vertebral level of:

A. C7-D1
B. T5-T6
C. T10-T11
D. D12-L1

54. A radiographic evaluation of the urinary tract (intravenous urography) is accomplished following the introduction of an aqueous nonionic iodinated contrast agent. Following the completion of the desired radiographic projections, the patient is instructed to go to the lavatory and void. A post-voiding image may be obtained to aid in the evaluation of:

A. Retained urinary calculi
B. Delayed contrast reaction
C. Prostate gland enlargements
D. Hydronephrosis

55. A radiographic procedure of the gallbladder and/or biliary ducts, following intravenous administration of an iodinated contrast agent, is called intravenous cholangiography. The maximum opacification of biliary ducts occurs about:

A. Two minutes after the injection of contrast media
B. Five minutes after the injection of contrast media.
C. Forty minutes after the injection of contrast media
D. Ninety minutes after the injection of contrast media

56. Preliminary preparation for a radiographic image of a patient with the non-acute abdomen may include all of the following, EXCEPT:

A. Administration of laxatives
B. Colonic lavage
C. Gastric lavage
D. Controlled food intake

57. In certain barium enema studies, a retention tip (Bardex) is used to help the patient retain the enema. Because of the danger of intestinal rupture, it is recommended that inflation of the tip be done in conjunction with (a):

A. Fluoroscopic control
B. Pressure injector
C. Compression paddle
D. High pressure injector

58. A patient has a suspected ruptured viscus. Which projection for the abdomen is often used to demonstrate this condition?

A. An AP projection with the patient supine
B. A lateral projection with the patient recumbent
C. An AP projection with the patient erect
D. An axial (Chassard-Lapiné) erect

59. A radiographic procedure of the large intestines is performed following the introduction of a liquid barium sulfate suspension and air. When a large amount of air is distending the colon, it is important to:

A. Increase mAs values to compensate for the greater tissue thickness
B. Decrease the kVp settings to avoid overpenetration of the image
C. Reduce the source-to-image receptor distance
D. Reduce the size of the effective focal spot

60. A radiographic evaluation of the urinary tract (intravenous urography) is accomplished following the introduction of an aqueous nonionic iodinated contrast agent. If tomographic images are to be obtained with the patient supine, the fulcrum level should be adjusted to a height of:

A. 3-6 centimeters
B. 8-10 centimeters
C. 11-14 centimeters
D. 15-22 centimeters

61. In recent years, the use of high density barium suspensions for the demonstration of the large intestines have gained greater acceptance because they have the ability to:

A. Reduce the number of adverse reactions
B. Increase the rate at which the contrast media is absorbed
C. Be retained for longer periods of time
D. Provide a more uniform coating of the intestines

62. All of the following radiographic procedures are often performed without any type of patient preparation EXCEPT:

A. Dacryocystography
B. Defecography
C. Cholecystography
D. Laryngography

63. A radiographic procedure of the small intestines is performed following the introduction of a liquid barium sulfate suspension. The most common means by which the contrast agent is administered is by way of:

A. A topical administration
B. A parenteral administration
C. An oral administration
D. A colonic administration

64. A radiographic procedure of the large intestine is performed following the introduction of a liquid barium sulfate suspension. In order to prevent possible damage to the rectum, the enema tip should be well lubricated and inserted no more than:

A. 2.5 centimeters into the rectum
B. 4 centimeters into the rectum
C. 6 centimeters into the rectum
D. 10 centimeters into the rectum

65. A radiographic procedure of the gallbladder is obtained following the introduction of an oral iodinated contrast agent. Most radiologists prefer PA projections for the visualization of the gallbladder because these serve to:

A. Reduce motility of the liver
B. Increase the passage of bile through the tract
C. Diffuse the appearance of the overlying ribs
D. Reduce part-to-image receptor distance

66. The radiographic evaluation of the urinary tract (intravenous urography) is accomplished following the introduction of an aqueous nonionic iodinated contrast agent. Which of the following can be used in order to improve the visualization of the distal ureters during the procedure?

A. A compression band
B. A Foley catheter
C. An ambu bag
D. A compression paddle

67. During a double contrast enema, a right lateral decubitus position is used with a horizontal central ray and a stationary grid. This projection is primarily intended to show lesions of the:

A. Medial ascending colon
B. Ileocecal valve
C. Transverse colon
D. Medial descending colon

68. The radiographic procedure of the upper gastrointestinal tract is performed following the introduction of a liquid barium sulfate suspension. The anteroposterior (AP) projection of the stomach, taken with the patient supine, is often obtained to show the barium-filled structures of the:

A. Cardiac orifice and esophagus
B. Fundus and duodenum
C. Greater and lesser curves
D. Pylorus and fundus

69. A radiographic image of the chest is obtained in the right PA oblique projection. Prior to taking the exposure the patient is asked to swallow a thick barium sulfate preparation. This erect projection of the chest taken with a barium filled esophagus will often aid in the demonstration of:

A. Small sliding hiatal hernias
B. An esophageal diverticulum
C. An enlargement of the left atrium of the heart
D. A bleed in the distal esophagus

70. Before a patient takes the oral iodinated contrast agent required for a routine for a gallbladder study, a scout image should be obtained with no contrast media to help evaluate:

A. The presence of radiolucent stones
B. Chronic inflammatory conditions of the gallbladder
C. The presence of radiopaque stones
D. Acute inflammatory conditions of the gallbladder

71. A radiographic evaluation of the urinary tract (intravenous urography) is accomplished following the introduction of an aqueous nonionic iodinated contrast agent. Some radiologists recommend that the first radiographic image be obtained immediately after the contrast injection is completed. This 1 minute image is taken to show the outline of the kidney and is referred to as a:

A. Tomographic image
B. Nephrogram
C. Cystogram image
D. Arteriogram

72. A radiographic procedure of the gallbladder is obtained following the introduction of an oral iodinated contrast agent. The projections which are generally preferred to image the gallbladder are the:

A. PA, the PA oblique and the right lateral projections
B. PA, the left lateral and the tangential projections
C. AP, the AP axial oblique and the axiolateral projections
D. AP, the AP axial and the left lateral projections

73. A radiographic evaluation of the urinary tract (intravenous urography) is used for the imaging of infants and children. It has been recommended that images be obtained after the child has been administrated a carbonated drink. This has been shown to improve visualization by:

A. Dispersing the fecal material in the distal colon
B. Compressing the ureters and slowing the elimination rate
C. Demonstrating the kidneys through the distended stomach
D. Increasing the rate at which the contrast agent is absorbed

74. The radiographic procedure of the upper gastrointestinal tract is performed following the introduction of a liquid barium sulfate suspension. During the fluoroscopic portion of the examination, the doctor asks the patient to perform the Valsalva maneuver by "bearing down". This will increase the pressure in the abdominal cavity which will often improve the visualization of:

A. Gastric polyps
B. Esophageal diverticula
C. Esophageal varices
D. Peptic ulcers

75. A radiographic procedure of the large intestines is to be performed following the introduction of a liquid barium sulfate suspension. It has been found that all of the following are advantages associated with cold enema preparations, EXCEPT:

A. They seem to have a mild anesthetic effect
B. They result in fewer colonic spasms
C. They have a higher retention rate
D. They increase the absorption rate of the barium

Referring to the radiographic image, answer questions 76 & 77.

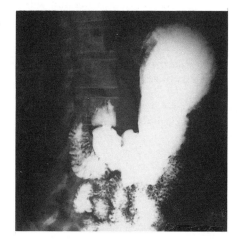

76. The radiographic image of the stomach was obtained in the:

A. AP projection with the patient erect
B. PA projection with the patent supine
C. Right PA oblique projection
D. Lateral projection in the ventral decubitus position

77. This projection of the barium-filled stomach is most often employed to provide a non-superimposed image of the:

1. Pylorus   2. Duodenal bulb   3. Retrogastric space

A. 1 only
B. 2 only
C. 3 only
D. 1 & 2 only

Referring to the radiographic image, answer questions 78 & 79.

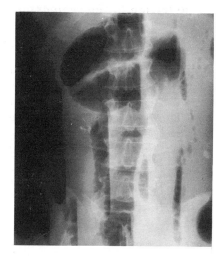

78. This radiographic image of the abdomen was obtained in the:

A. AP projection
B. PA oblique projection
C. Lateral projection
D. Axial projection

79. Based on the appearance of this image, it can be determined that the patient was in the:

A. Right lateral decubitus position
B. Left lateral decubitus position
C. Ventral decubitus position
D. Dorsal decubitus position

Referring to the radiographic image, answer questions 80 & 81.

80. This radiographic image obtained of the large intestines, using a double contrast agent, was taken in the:

    A. AP projection   C. Lateral projection
    B. AP oblique projection   D. Axial projection

81. Based on the appearance of the image, it can be determined that the patient was in the:

    A. Erect position
    B. Supine position
    C. Lateral decubitus position
    D. Dorsal decubitus position

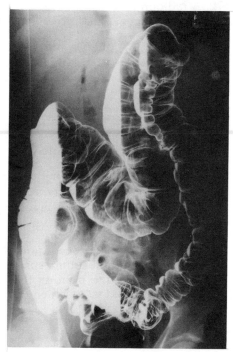

Referring to the radiographic image, answer questions 82 & 83.

82. This projection of the barium-filled colon was obtained in the:

    A. AP oblique projection with the patient in the LPO position
    B. AP oblique projection with the patient in the RPO position
    C. Lateral position
    D. AP projection with the patient erect

83. This projection is most commonly employed to improve the visualization of the:

    A. Lesser curvature   C. Hepatic flexure
    B. Greater curvature   D. Splenic flexure

Referring to the radiographic image, answer questions 84 & 85.

84. This projection of the abdomen is an example of a/an:

    A. AP projection with the patient erect
    B. Lateral projection with the patient erect
    C. AP oblique projection with the patient in the LPO position
    D. AP projection with the patient supine

85. This projection is most useful in the determination of:

    A. Diaphragm hernias   C. Colonic spasms
    B. Pyloric stenosis   D. Air/fluid levels

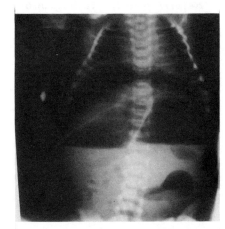

Explanations

1. B   A radiographic procedure of the gallbladder is obtained following the introduction of an oral iodinated contrast agent. The projections which are generally preferred to image the gallbladder are the PA, the PA oblique and the right lateral projections. Though the level of the gallbladder depends on the body habitus, it is nearly always found at the level of the tip of the elbow. Therefore, the central ray should be directed at this level. The GB study should include at least one image obtained with the patient erect to demonstrate the layering of cholesterol gallstones that may float on the contrast rich bile. In order to ensure proper filling of the gallbladder on the day preceding a radiographic imaging study, before the patient takes the oral iodinated contrast agent (tablets), they are often instructed to eat lunch, consisting of food containing fats, to empty the bladder. After the contrast agent is introduced, it is important for the patient to avoid any liquids or foods that contain fats and laxatives which can all trigger the release of bile. Maximum filling occurs about 12 hours after the ingestion of Cholographin or Bilopaque tablets. Bile is a secretion of the liver that aids in the digestion of fats. The main function of the gallbladder is to temporarily store and concentrate bile before it is released into the duodenum. To improve the visualization of the biliary ducts, a fatty meal; i.e., milkshake, can be given to the patient to empty the gallbladder and encourage the flow of the contrast agent into the biliary ducts.

2. C   In a radiographic evaluation of the large intestines (barium enema), the liquid barium sulfate suspension is introduced through an enema tip with an inflatable retention collar or a (Bardex) type enema tube. The inflatable collar, which sits against the anal sphincter, lessens the chance of premature evacuation of the contrast media. Because of the potential danger of intestinal rupture from over-inflation of the collar, many of these devices come with a 90 millimeter reusable squeeze-type inflation as a safety precaution. The insufflation of the collar with air should be performed under fluoroscopic control. The tip should not be inserted more than 10 centimeters or 4 inches into to the rectum.

3. C   A radiographic examination of the upper gastrointestinal tract is normally accomplished following the introduction of an oral suspension of barium sulfate. The preliminary imaging is performed by the radiologist using fluoroscopy followed by a series of static radiographic images most often including AP, PA, PA oblique, and lateral projections with the patient recumbent. The central ray is normally directed perpendicular to the image receptor to enter at a the level of L1-L2. In the AP projection, with the patient supine, the fundus of the stomach and duodenum are more posterior than the body of the stomach resulting in these structures being filled with barium. The PA projection taken with the patient prone, will normally show these structures filled with air or gas and the body filled with barium. The PA oblique projection most often provides the best image of the duodenal bulb and the pyloric canal. The lateral projection is obtained to show the anterior and posterior aspects of the stomach. Prior to the imaging of the stomach the patient is told to eat only low residue foods for at least one day and be NPO for 8 hours before the procedure. Food that may be in the stomach during the procedure may be mistaken for gastric lesions. When an improved visualization of the mucosal linings is desired, the liquid barium sulfate may be combined with a radiolucent gas to provide a "double" contrast. This is most often accomplished by having the patient ingest carbon dioxide gas-producing tablets to improve visualization of small lesions that are not normally seen with barium alone.

4. B   A radiographic examination of the upper gastrointestinal tract is normally accomplished following the introduction of an oral suspension of barium sulfate. Small sliding gastroesophageal (hiatal) hernias are found in the thoracoabdominal region passing through the esophageal hiatus of the diaphragm. The placement of the head lower than the feet as in the Trendelenburg position will increase the intra-abdominal pressure, and help to fill the hernia with barium.

5. D   The radiographic evaluation of the pelvocalyceal system and ureters is normally obtained following the intravenous administration of an aqueous nonionic iodinated contrast agent. Images are usually accomplished in the AP projection with the central ray directed perpendicular to the image receptor at the level of the iliac crests. During an intravenous urogram, the contrast media is introduced by way of an intravenous (IV) injection. The contrast is rapidly filtered by the kidneys and passed into the rest of the urinary tract. In nephrotomography, improved visualization is possible through the infusion of contrast agents and the use of tomographic sections of the kidneys at the desired levels. Retrograde pyelography involves the placement of a catheter transurethrally into the proximal ureter and the direct injection of the contrast media into the proximal end of the ureters.

6. A   A radiographic image of the abdomen is normally obtained in the AP projection with the patient supine. The central ray is directed perpendicular to enter at the level of the iliac crests. This projection is used to show the size and shape of the liver, spleen, kidneys, and demonstrate inter abdominal calcifications of solid tumor masses. In order to demonstrate intestinal fluids or free air in the abdomen, radiographic images should be obtained in AP or PA projections taken erect or in the lateral decubitus positions. Upright and/or decubitus positions employing a horizontal central ray, is used to show the levels between any free air and/or fluids that may have escaped from the viscera into the peritoneum. When possible, the patient should be placed in the lateral decubitus position for at least 10-20 minutes prior to the exposure to give the gases time to rise.

7. C   A radiographic procedure for the imaging of the esophagus is normally accomplished following the introduction of an oral suspension of barium sulfate. The preliminary imaging is performed by the radiologist using fluoroscopy followed by a series of static radiographic images most often including AP, PA, PA oblique and lateral projections with the patient recumbent. The central ray is normally directed perpendicular to the image receptor to enter at a the level of T5-T6. Images obtained during an esophagram are often made during the act of swallowing, No breathing instructions are required because deglutition will inhibit respiration for about two seconds. Patients with suspected esophageal varices may be instructed to perform the Valsalva maneuver.

8. B   A radiographic examination of the upper gastrointestinal tract, to show the small intestines, is normally accomplished following the introduction of an oral suspension of barium sulfate. The preliminary imaging is performed by the radiologist using fluoroscopy followed by a series static radiographic images that most often include AP projections with the patient recumbent taken at various times after the ingestion of the contrast agent. The central ray is normally directed perpendicular to the image receptor to enter at a the level of the iliac crests. The 22 feet of small intestine extends from the pyloric sphincter of the stomach to the ileocecal valve at the beginning of the colon. Digested food or barium takes about 1-2 hours to travel through the 3 portions of the small intestine; the duodenum, jejunum and ileum. The last image obtained should show the spillage of the barium into the cecum.

9. A   In a radiographic evaluation of the large intestines (barium enema), the liquid barium sulfate suspension is introduced through an enema tip. Before the contrast media is introduced it is important that the colon be properly cleaned. This will generally involve having the patient take laxatives, undergo a colonic lavage (cleansing enemas) and follow a set of dietary restrictions to remove fecal material, intestinal gas or previous contrast media from the colon. Any remaining fecal material in the tract may mimic the appearance of small tumor masses or polypoid lesions of the large bowel.

10. D   Dietary restrictions are necessary prior to all contrast studies of the gastrointestinal tract; i.e., GI, SI, BE, GB. Food or other liquids can dilute or simulate the appearance of small masses or defects in the mucosal linings of the organs. Prior to a gallbladder examination food and laxatives must be avoided to prevent the release of bile from the gallbladder.

11. C   The radiographic evaluation of the pelvocalyceal system and ureters is normally obtained following the intravenous administration of an aqueous nonionic iodinated contrast agent. Following the intravenous injection or infusion of about 30-100 ml of iodinated contrast media, the pelvocalyceal system begins to visualize within 2-8 minutes. Depending on a number of physiologic variables, the maximum concentration will occur in about 15-20 minutes.

12. A   A radiographic image of the abdomen is normally obtained in the AP projection with the patient supine. The central ray is directed perpendicular to enter at the level of the iliac crests. Since the exhalation phase of breathing relaxes the intra-abdominal organs and maintains the abdominal organs at a consistent location, all abdominal exposures should be taken after the patient suspends respiration at the end of expiration. A delay of 1-2 seconds after the patient stops breathing should be allowed for more complete cessation of abdominal organ movement.

13. G   14. D   15. A   16. C   17. F   18. B

19. C   A radiographic examination of the upper gastrointestinal tract is normally accomplished following the introduction of an oral suspension of barium sulfate. Prior to the imaging of the stomach, the patient is told to eat only low residue foods for at least one day and be NPO for 8 hours before the procedure. Food that may be in the stomach during the procedure may be mistaken for gastric lesions.

20. C   Foreign bodies that are lodged in the upper esophagus may not be clearly demonstrated on non-contrast images. Having the patient swallow barium-soaked cotton balls or thick barium paste will slow the contrast media's descent and may improve the visualization of obstructed regions.

21. B   A radiographic image of the abdomen is normally obtained in the AP projection with the patient supine. Because the abdomen is bounded superiorly by the diaphragm and inferiorly by the area at the level of the symphysis pubis, when possible, both the diaphragm and the pubic symphysis should be included on the radiographic image. In larger patients, two crosswise 34" x 43" centimeter image receptors may be necessary to image the entire region.

22. A   A radiographic examination of the lower gastrointestinal tract is normally accomplished following the introduction of a liquid suspension of barium sulfate through an enema tube. The preliminary imaging is performed by the radiologist using fluoroscopy followed by a series of static radiographic images most often including AP, PA, PA axial, AP oblique and lateral projections with the patient recumbent. The central ray is normally directed perpendicular to the image receptor to enter at a the level of the iliac crests. The AP projection will show the entire colon filled with barium. The AP oblique projections are used to improve the image of the splenic (RPO) and hepatic flexures (LPO). The lateral projection is obtained to show the rectosigmoid junction. The sigmoid colon and rectosigmoid junctions are also well visualized in the axial, ventral and dorsal decubitus and Chassard-Lapiné projections and are used to supplement routine projections used in a barium enema study.

23. A   The radiographic evaluation of the pelvocalyceal system and ureters is normally obtained following the intravenous administration of an aqueous nonionic iodinated contrast agent. Images are usually accomplished in the AP projection with the central ray directed perpendicular to the image receptor at the level of the iliac crests. During an intravenous urogram, the contrast media is introduced by way of an intravenous (IV) injection. The contrast is rapidly filtered by the kidneys and passed into the rest of the urinary tract.

24. A   A radiographic image of the abdomen of a female is normally obtained in the AP projection with the patient supine. In order to produce a radiograph with a greater number of gray scales or long scale contrast, a kilovoltage setting in the 80-90 range is required.

25. C   A radiographic procedure of the gallbladder is obtained following the introduction of an oral iodinated contrast agent. The projections which are generally preferred to image the gallbladder are the PA, the PA oblique and the right lateral projections. Due to the superimposition of the ribs, the vertebrae or overlying fecal material, small calculi in the gallbladder are often poorly visualized. For this reason, upright and/or decubitus positions are often included as routine images. The erect image may also be included to demonstrate the layering of low density cholesterol gallstones that may float on the contrast rich bile.

26. C   A radiographic procedure to evaluate the biliary ducts following the introduction of an iodinated contrast agent are better visualized in the right posterior oblique (RPO) position. This is because the ducts extend medially and posteriorly from the gallbladder and this position places them parallel to the image receptor.

27. B   A radiographic examination of the upper gastrointestinal tract is normally accomplished following the introduction of an oral suspension of barium sulfate. In the AP projection with the patient supine, the fundus of the stomach and duodenum are more posterior than the body of the stomach resulting in these structures being filled with barium. The PA projection taken with the patient prone, will normally show these structures filled with air or gas and the body filled with barium. The PA oblique projection most often provides the best image of the duodenal bulb and the pyloric canal. The lateral projection is obtained to show the anterior and posterior aspects of the stomach.

28. D   The radiographic evaluation of the pelvocalyceal system and ureters is normally obtained following the intravenous administration of an aqueous nonionic iodinated contrast agent. Images are usually accomplished in the AP projection with the central ray directed perpendicular to the image receptor at the level of the iliac crests. If the patient is placed in the Trendelenburg position, the upper torso is lowered allowing retrograde spillage of the contrast media from a filled bladder into the distal ureter.

29. B   A radiographic examination of the lower gastrointestinal tract is normally accomplished following the introduction of a liquid suspension of barium sulfate through an enema tube. The AP oblique projections are used to improve the image of the splenic (RPO) and hepatic flexures (LPO). The right colic or hepatic flexure formed at the junctions of the ascending and transverse colons is best visualized by rotating the patient 35-45 degrees away from the affected side. This is more commonly referred to as the left AP oblique projection. An image taken in the PA oblique projection with the patient RAO, will demonstrate this flexure closest to the image receptor.

30. G.   31. E   32. B   33. I   34. H   35. D

36. A   37. C   38. F

39. B   Gastrointestinal intubation is often used to remove gasses or fluids from the gastrointestinal organs. This is accomplished through the placement of a thin rubber tube into the stomach or small intestines. The tube is most often inserted through the nose and passed into the oropharynx, esophagus, stomach and small intestine.

40. C   A radiographic evaluation of the abdomen is to be obtained in the AP projection with the patient supine. When possible, the abdomen should be clear of gas, fluids and fecal material. This will generally involve having the patient take laxatives and follow a set of dietary restrictions to remove fecal material and intestinal gas from the abdominal cavity.

41. C   In a radiographic evaluation of the large intestines (barium enema), the liquid barium sulfate suspension is introduced through an enema tip. This will result in the flow of barium through the colon in a backward direction. The barium will pass, in order, from the rectum to the sigmoid colon, descending colon, splenic flexure, transverse colon, hepatic flexure, ascending colon and into the cecum.

42. B   The radiographic evaluation of the pelvocalyceal system and ureters is normally obtained following the intravenous administration of an aqueous nonionic iodinated contrast agent. The normal amount of iodinated contrast media that is introduced to opacity the urinary system is between 30-100 milliliters (cc) given by intravenous injection. Amounts may vary depending on the age, size and weight of the patient.

43. D   A radiographic procedure of the gallbladder is obtained following the introduction of an oral iodinated contrast agent. After the ingestion of the cholecystographic contrast media, it is important to avoid foods, liquids and laxatives which can all stimulate the emptying of the gallbladder.

44. D   Localization of foreign bodies of the respiratory or gastrointestinal tracts will normally require images of the neck, chest and abdomen. In cases where nonopaque objects are suspected, the use of contrast agents may also be indicated.

45. B   A radiographic image of the abdomen is normally obtained in the AP projection with the patient supine. Since the exhalation phase of breathing relaxes the intra-abdominal organs and maintains the abdominal organs at a consistent location, all abdominal exposures should be taken after the patient suspends respiration at the end of expiration. A delay of 1-2 seconds after the patient stops breathing should be allowed to complete cessation of abdominal organ movement.

46. A   In a radiographic evaluation of the large intestines, the liquid barium sulfate suspension is introduced through an enema tip with an inflatable retention collar or a (Bardex) type enema tube. The enema tube should be inserted into the patient in the Sims position. This position requires that the patient lay on their left side leaning forward. This position relaxes the abdominal muscles and decreases the intra-abdominal pressure making the insertion of the tip easier.

47. B   A radiographic examination of the upper gastrointestinal tract is normally accomplished following the introduction of an oral suspension of barium sulfate. The patient should be placed in the right anterior oblique position to increase the peristaltic action of the stomach and speed the emptying of the gastric contents into the duodenum.

48. B   In a radiographic evaluation of the large intestines, the liquid barium sulfate suspension is introduced through an enema tip. The PA or AP axial projections is used for the evaluation of the rectosigmoid area. In the AP axial projection, the central ray is directed at a cephalic angle of 30-40 degrees to enter at a point 5 centimeters below the ASIS. In the PA axial projection, the central ray is directed at a caudal angle of 30-40 degrees to enter at the level of the ASIS. Both serve to project the S-shaped sigmoid colon free of superimposition with itself.

49. D   A radiographic procedure of the gallbladder and or biliary ducts following the introduction of an iodinated contrast agent can be accomplished in a number of ways. Because the gallbladder has a double blood supply (arterial and venous), oral contrast agents absorbed by the organs in the gastrointestinal tract will pass into the portal circulation and are filtered by the liver. Intravenous contrast agents take the arterial pathways to the liver. A number of other direct methods involving the placement of a needle or catheter into the biliary ducts, called T-tube cholangiography and endoscopic retrograde cholangiopancreatography (ERCP), will also visualize the biliary ducts.

50. A   A radiographic examination of the upper gastrointestinal tract is normally accomplished following the introduction of an oral suspension of barium sulfate. Gastroesophageal reflux is most frequently seen radiographically by increasing pressure on the gastroesophageal junction through the use of the Trendelenburg position or the Valsalva maneuver.

51. B   Regardless of the position of the patient, air-fluid levels can only be demonstrated if the central ray is directed horizontal to the floor.

52. D   A radiographic procedure of the gallbladder is obtained following the introduction of an oral iodinated contrast agent. The ingestion of the oral cholecystographic contrast media may cause a myriad of systemic symptoms. The vast majority are related to their affects on the intestinal mucosa of the gastrointestinal tract. The most commonly seen are nausea, vomiting and diarrhea.

53. B   A radiographic procedure for the imaging of the esophagus is normally accomplished following the introduction of an oral suspension of barium sulfate. A PA projection obtained with the patient in the RAO position, with a 30-40 degree rotation, will help free the esophagus from its superimposition with the heart and vertebral column. The central ray is normally directed perpendicular to the image receptor to enter at a the level of T5-T6.

54. C   The radiographic evaluation of the pelvocalyceal system and ureters is normally obtained following the intravenous administration of an aqueous nonionic iodinated contrast agent. Post voiding images may be required to detect residual amounts of urine in the bladder. This is most often related to small tumor masses near the bladder or enlargements of the prostate gland.

55. C A radiographic procedure of the gallbladder is obtained in a non-cholecystectomized patient, following an intravenous injection of an iodinated contrast media. The biliary ducts will usually be demonstrated 15 minutes following the introduction of the contrast media. These ducts will reach their maximum concentration after about 40 minutes. The gallbladder will reach its maximum concentration after about 2 hours.

56. C A radiographic image of the abdomen is normally obtained in the AP projection with the patient supine. Like a barium enema, the visualization of a non-acute abdomen can be dramatically improved by colonic lavage, laxatives and limiting the intake of foods.

57. A In a radiographic evaluation of the large intestines, the liquid barium sulfate suspension is introduced through an enema tip with an inflatable retention collar or a (Bardex) type enema tube. The inflatable collar which sits against the anal sphincter lessens the chance of premature evacuation of the contrast media. Because of the potential danger of intestinal rupture from over-inflation of the collar, many of these devices come with a 90 cc reusable squeeze-type inflation as a safety precaution. The insufflation of the collar with air should be performed under fluoroscopic control.

58. C A radiographic image of the abdomen is normally obtained in the AP projection with the patient supine. In order to demonstrate intestinal fluids or free air in the abdomen, radiographic images should be obtained in AP or PA projections taken with the patient erect or in the lateral decubitus positions. Upright and/or decubitus positions employing a horizontal central ray, is to show levels between any free air and/or fluids that may have escaped from a ruptured viscus into the peritoneum. When possible, the patient should be placed in the lateral decubitus position for at least 10-20 minutes prior to the exposure to give the gases time to rise.

59. B A radiographic evaluation of the large intestines is performed following the introduction of the liquid barium sulfate suspension and air. Because the air will lower tissue density of the colon, a reduction of about 10 kVp is required to prevent overpenetration of these structures.

60. B The radiographic evaluation of the pelvocalyceal system and ureters is normally obtained following the intravenous administration of an aqueous nonionic iodinated contrast agent. If tomographic sections are to be obtained in the AP projection on a small patient, the fulcrum should be adjusted to a height of about 7 centimeters to demonstrate the center of the kidney. A 9 centimeter fulcrum height is used for the average patient and a height of 11 centimeters in a large patient.

61. D In recent years, the use of high density barium (Polybar) suspensions for the demonstration of the large intestines have gained great acceptance. These products are used to provide a more uniform coating of the intestinal walls.

62. C Many radiographic procedures will require the patient to follow dietary restrictions and have bowel preps including oral cholecystography. All of the other procedures listed can be performed without any patient preparation.

63. C The small bowel can be examined in 3 ways: by mouth, complete reflux or by enteroclysis. The most common and comfortable is by mouth through the ingestion of 36-60 ounces of liquid barium sulfate suspension. The complete reflux method requires the filling of the colon and the retrograde filling of the small intestine. This uncomfortable procedure is rarely done in the modern department. Enteroclysis involves introduction of barium through a nasogastric tube that is placed in the duodenum.

64. D In a radiographic evaluation of the large intestines, the liquid barium sulfate suspension is introduced through an enema tip with an inflatable retention collar or a (Bardex) type enema tube. The tip should not be inserted more than 10 centimeters or 4 inches into to the rectum.

65. D A radiographic procedure of the gallbladder is obtained following the introduction of an oral iodinated contrast agent. The projections which are generally preferred to image the gallbladder are the PA, the PA oblique and the right lateral projections. The PA projections and the lateral projections will minimize magnification and reduce the amount of focal spot blur by placing the gallbladder as close to the image receptor as possible.

66. A The radiographic evaluation of the pelvocalyceal system and ureters is normally obtained following the intravenous administration of an aqueous nonionic iodinated contrast agent. In order to improve the filling of the ureters and renal pelvis, a compression band may be applied over the distal ureters in the lower pelvis. This retards the filling of the bladder and encourages the retention of urine within the drainage apparatus.

67. A A radiographic evaluation of the large intestines is performed following the introduction of the liquid barium sulfate suspension and air. The AP or PA projection with the patient in the right lateral decubitus position is used to demonstrate the medial border of the ascending colon and lateral border of the descending colon. The left lateral decubitus position is used to demonstrate the lateral border of the ascending colon and medial border of the descending colon.

68. D A radiographic examination of the upper gastrointestinal tract is normally accomplished following the introduction of an oral suspension of barium sulfate. In the AP projection with the patient supine, the fundus of the stomach and duodenum are more posterior than the body of the stomach resulting in these structures being filled with barium. The PA projection taken with the patient prone, will normally show these structures filled with air or gas and the body filled with barium.

69. C A radiographic procedure for the imaging of the esophagus is normally accomplished following the introduction of an oral suspension of barium sulfate. A PA projection obtained with the patient in the RAO position with a 30-40 degree rotation, will help free the esophagus from its superimposition with the heart and vertebral column. Images taken in this projection, with a barium-filled esophagus, are also used to determine enlargements of left atrium of the heart.

70. C A radiographic procedure of the gallbladder is obtained following the introduction of an oral iodinated contrast agent. Prior to the administration of the contrast agent, a scout image should be obtained to visualize radiopaque stones whose appearance may be masked by the positive contrast media. Radiolucent (cholesterol) stones are not seen on the scout images, but may be demonstrated during the contrast enhanced images.

71. B The radiographic evaluation of the pelvocalyceal system and ureters is normally obtained following the intravenous administration of an aqueous nonionic iodinated contrast agent. The nephrogram is the first image taken of the kidney immediately after the completion of the bolus injection. This 1 minute image shows the blush phase associated with the uptake of contrast by the parenchyma of the kidney.

72. A A radiographic procedure of the gallbladder is obtained following the introduction of an oral iodinated contrast agent. The projections which are generally preferred to image the gallbladder are the PA, the PA oblique and the right lateral projections to avoid superimposition of the gallbladder with the lateral aspect of the spine.

73. C The radiographic evaluation of the pelvocalyceal system and ureters is obtained following the intravenous administration of an aqueous nonionic iodinated contrast agent on a young child or infant. In order to displace the overlying shadows caused by intestinal gas or fecal material, the stomach may be filled with carbonated liquids which push the intestines inferior to the renal pelvis and kidneys.

74. C A radiographic examination of the upper gastrointestinal tract is normally accomplished following the introduction of an oral suspension of barium sulfate. The increased venous pressure resulting from the Valsalva (bearing down) maneuver, places the esophagus under stress causing the varices to enlarge, thus improving their appearance.

75. D A radiographic evaluation of the large intestines is performed following the introduction of the liquid barium sulfate suspension. The use of a cold (41 degrees F, 5 degrees C) water in preparing the barium suspensions is recommended because it can produce a mild anesthetic effect and stimulate the contraction of the anal sphincter. Since the colon is unable to detect temperature this can help to make the procedure more comfortable and improve the retention of the barium.

76. C   A radiographic examination of the upper gastrointestinal tract is normally accomplished following the introduction of an oral suspension of barium sulfate. The PA oblique projection with the patient in the right anterior oblique (RAO) position, most often provides the best image of the duodenal bulb and the pyloric canal. In this projection, free gas in the stomach will rise into the fundus and demonstrate the pylorus, duodenum and body of the stomach filled with barium.

77. D   See the explanation above.

78. B   A radiographic image of the abdomen is normally obtained in the AP projection with the patient supine. The central ray is directed perpendicular to enter at the level of the iliac crests. In order to demonstrate intestinal fluids or free air in the abdomen, radiographic images should be obtained in AP or PA projections in the lateral decubitus position. This AP projection clearly shows air-fluid levels along the long axis of the body indicating that it was obtained with the patient in the left lateral decubitus position.

79. A   See the explanation above.

80. A   A radiographic evaluation of the large intestines is performed following the introduction of the liquid barium sulfate suspension and air. The AP or PA projection with the patient in the right lateral decubitus position, is used to demonstrate the medial border of the ascending colon and lateral border of the descending colon. The left lateral decubitus position, is used to demonstrate the lateral border of the ascending colon and medial border of the descending colon. The presence of barium on the lateral surface of the ascending colon and medial border of the descending colon, is characteristic of the right lateral decubitus position.

81. C   See the explanation above.

82. A   A radiographic examination of the lower gastrointestinal tract is normally accomplished following the introduction of a liquid suspension of barium sulfate through an enema tube. The AP oblique projections are used to improve the image of the splenic (RPO) and hepatic flexures (LPO). The right colic or hepatic flexure formed at the junctions of the ascending and transverse colons is best visualized by rotating the patient 35-45 degrees away from the affected side. This is more commonly referred to as the left AP oblique projection. An image taken in the PA oblique projection with the patient RAO will demonstrate this flexure closest to the image receptor. This projection shows an open hepatic flexure and the more superior splenic flexure superimposed on itself.

83. C   See the explanation above.

84. A   A radiographic image of the abdomen is normally obtained in the AP projection with the patient supine. The central ray is directed perpendicular to enter at the level of the iliac crests. In order to demonstrate intestinal fluids or free air in the abdomen, radiographic images should be obtained in AP projection with the patient upright. The air-fluid levels which are seen crossing the abdomen from side-to-side indicates that this child was upright in the AP projection.

85. D   See the explanation above.

1. The radiographic evaluation of the central nervous system and the structures located within the spinal canal is termed:

   A. Meningography
   B. Myography

   C. Ventriculography
   D. Myelography

2. The specific term applied to any radiographic opacification study of the veins is called:

   A. Angiography
   B. Venography

   C. Digital angiography
   D. Arteriography

3. The radiographic examination of the breast is termed:

   A. Sialography
   B. Mammography

   C. Salpingography
   D. Duodenography

4. Arthrography is employed for the radiographic visualization of the soft tissue structures of the:

   A. Mediastinum
   B. Blood vessels

   C. Pelvic cavity
   D. Joints

5. Radiographic images obtained following the injection of an iodinated contrast media into a T-tube is used to evaluate:

   A. The function of the gallbladder
   B. Acute cholecystitis

   C. The patency of the biliary ducts
   D. The layering of the gallstones

6. The radiographic evaluation of the uterine cavity and lumen of the fallopian tubes following the introduction of an iodinated contrast media is called:

   A. Hysterosalpingography
   B. Pelvic pneumography

   C. Pelvimetry
   D. Vaginography

7. Cephalopelvimetry is a radiographic procedure used for the evaluation and measurement of the:

   1. Fetal head size   *2. Pelvic inlet*   *3. Pelvic outlet*

   A. 1 only
   B. 2 only

   C. 3 only
   D. 1, 2 & 3

8. The evaluation of the pelvicaliceal system and ureters by a direct catheterization of the ureter is termed:

   A. Intravenous urography
   B. Infusion tomography

   C. Retrograde urography
   D. Nephrotomography

9. In order to maximize the contrast of human breast tissue, the kilovoltage range recommended for mammography is:

   A. 5-15 kVp
   B. 15-35 kVp

   C. 45-60 kVp
   D. 70-80 kVp

10. The iodinated contrast media that is used for a myelogram is introduced by way of a spinal needle that is inserted between the:

    A. First and second lumbar vertebrae
    B. Third and fourth lumbar vertebrae

    C. Tenth and eleventh thoracic vertebrae
    D. Sixth and seventh cervical vertebrae

11. During a retrograde urogram, the vesicouretral orifices are located and catheterized using a:

    A. Colonoscope
    B. Gastroscope

    C. Proctoscope
    D. Cystoscope

12. During dorsal and cervical myelography, the ascent of contrast media into the ventricles of the brain can be limited by acute:

    A. Flexion of the head
    B. Extension of the head
    C. Lateral flexion of the head
    D. Lateral rotation of the head

13. All of the following are routine radiographic projections used in the evaluation of the normal breast EXCEPT:

    A. The mediolateral projection
    B. The craniocaudal position
    C. The axillary projection
    D. The stereo projection

14. Sialography is a radiographic procedure that involves the introduction of an iodinated contract agent into the:

    A. Nasolacrimal ducts
    B. Salivary glands
    C. Mastoid air cells
    D. Paranasal sinuses

15. During a myelogram of the upper thoracic region, the site most often chosen for the introduction of the contrast agent will be the:

    A. Cisterna magna
    B. Upper dorsal region
    C. Lower dorsal region
    D. Lumbar region

16. The principal type of contrast agent most often employed for the radiographic evaluation of the renal parenchyma will be a/an:

    A. Aqueous barium suspension
    B. Aqueous iodinated solution
    C. Oil-based iodine
    D. Gaseous media

17. After the introduction of a nonionic iodinated contrast media in to the subarachnoid space for a cervical myelogram, the contrast is moved into the desired region by:

    A. Flexing and extending the spine
    B. Tilting the table to raise the head
    C. Moving the spine from side-to-side
    D. Tilting the table to lower the head

18. The term cystography is employed for the radiographic examination of the:

    A. Gallbladder
    B. Urinary bladder
    C. Splenic vein
    D. Pancreas

19. A specialized mammographic used for the radiographic evaluation of the breast system is likely to include all of the following EXCEPT:

    A. A radiographic tube with a molybdenum target
    B. A system that can safely compress the breast tissue
    C. A radiographic tube that contains a fractional focal spot
    D. A system for precise measurement of the breast volume

20. After the removal of the gallbladder, a tube may be placed in the collecting system for evaluation of the ducts. This is termed a/an:

    A. Cantor tube
    B. PICC line
    C. T-tube
    D. Ostomy tube

21. Prior to the insertion of a spinal needle into L2-L3 interspace for a myelogram, the patient is asked to flex their spine to:

    A. Increase the size of the vertebral interspace
    B. Reduce the pressure in the spinal canal
    C. Compress the spinal cord
    D. Compress the intervertebral disk spaces

22. In the postoperative (T-tube) cholangiography, the status of the sphincter of Oddi is determined by contrast media spillage into the:

    A. Peritoneum
    B. Splenic vein
    C. Duodenum
    D. Gallbladder

23. During the radiographic imaging of the breast, the use of a compression devise will serve to:

   1. Produce a more uniform thickness in the breast
   2. Reduce the amount of radiation that is needed for the exposure
   3. Decrease the amount of scattered radiation produced by the breast

   A. 1 only
   B. 2 only
   C. 3 only
   D. 1, 2 & 3

24. The principal advantage of the use of a nonionic water-soluble iodinated contrast media compared to an oil based iodinated contrast media during a myelogram is:

   A. The greater viscosity of the agent
   B. Its slower absorption rate from the spinal canal
   C. It eliminates the need to remove the contrast material at the end of the procedure
   D. It eliminates all adverse reactions to the agent

25. Which of the following anatomical areas are commonly examined by contrast arthrography?

   1. Elbow joints   2. Hip joints   3. Wrists

   A. 1 & 2 only
   B. 1 & 3 only
   C. 2 & 3 only
   D. 1, 2 & 3

26. The projection that is most commonly employed for cystourethrogram on an adult male patient is the:

   A. Anteroposterior (AP) projection with the patient supine
   B. A PA 35 degree oblique projection with the patient recumbent
   C. The lateral projection with the patient erect
   D. An AP projection with the patient in the dorsal decubitus projection

27. Expiratory and inspiratory phonation are two common techniques that are employed to improve the radiographic evaluation of an image obtained during:

   A. Dacryocystography
   B. Arthrography
   C. Laryngopharyngography
   D. Thyroid angiography

28. During lower extremity venography, the tilting of the x-ray table into the semi-erect position or the placement of tourniquets on the lower extremities, are used to help insure the filling of the:

   A. Superficial leg veins
   B. Deep leg veins
   C. Inferior vena cava
   D. Pelvic veins

29. During an endoscopic retrograde cholangiopancreatography(ERCP), an endoscope is used to help insert a small cannula into the:

   A. Uretal orifice
   B. Common bile duct
   C. Pyloric antrum
   D. Hepatic vein

30. Which of the following radiographic procedures is normally performed on a gravid patient?

   A. Cephalometry
   B. Hysterosalpingography
   C. Pelvic pneumography
   D. Vaginography

31. During a hysterosalpingogram, the contrast media is introduced into the uterine cavity and should normally spill over into the:

   A. Ovary
   B. Cervix
   C. Peritoneum
   D. Vagina

32. A stent is a metallic devise that can be used in connection with balloon angioplasty to:

   A. Remove fatty deposits from the walls of an artery
   B. Prevent the collapse of a stenotic vessel
   C. Monitor the blood pressure in the pulmonary artery
   D. Filter out blood clots before they reach the lungs

33. Which of the following injection sites could be used for a cervical myelogram, if an obstruction of the lumbar region prevents the ascent of the contrast agent within the spinal canal?

A. A ventricular injection site
B. A subdural injection site
C. A cisternal injection site
D. A cranial sinus injection site

34. During an arthrogram, both the iodinated contrast agent and air are normally introduced into the:

A. Joint capsule
B. Bursae
C. Menisci
D. Ligaments

35. During cystography, which of the following techniques is employed to evaluate urethral reflux?

A. The inclusion of a voiding image
B. The use of erect images
C. The use of exercise images
D. The use of a breathing technique

36. The most common secretory stimulant used in radiographic contrast examination of the salivary glands is (a/an):

A. Lemon wedge
B. Fatty meal
C. Chocolate bar
D. Esophotrast

37. Endoscopic retrograde cholangiopancreatography (ERCP), is an imaging procedure that is used in the evaluation of the:

A. Biliary duct network
B. Upper respiratory system
C. Cystourinary system
D. Lower gastrointestinal tract

38. Operative cholangiography is often employed during biliary tract surgery to evaluate the patency of the:

1. Intrahepatic ducts   2. Extrahepatic ducts   3. Sphincter of Oddi

A. 1 only
B. 2 only
C. 3 only
D. 1, 2 & 3

39. During a myelogram, after the needle is placed into the spinal canal, an amount of spinal fluid equal to the amount of contrast media that is to be introduced, must be removed to help prevent:

A. The mixing of the contrast agent and the spinal fluid
B. The rapid absorption of the contrast material
C. Bacterial contamination of the spinal canal
D. A large increase in the intracranial pressure

40. Which of the following devices is required for orthoroentgenographic (long bone) measurements of the extremities (scanography)?

A. The use of a metallic ruler that is placed next to the extremity
B. The use of a tourniquet to restrict the blood flow to the extremity
C. The use of an air-filled splint to prevent motion of the extremity
D. The use of a traction devise that can increase the length of the extremity

41. In recent years, interventional angiography has expanded rapidly. A procedure that can be used for the occlusion of arteriovenous malformations is termed:

A. Transluminal angioplasty
B. Elective embolism
C. Peripheral angiography
D. Spontaneous thrombolization

42. Dilatation of stenotic vessels is often possible using a Grüntzig balloon tip catheter inserted into the desired vessel with the balloon expanded. This describes a procedure termed:

A. Therapeutic anastomosis
B. Elective vascularization
C. Elective embolization
D. Transluminal angioplasty

43. At the present time, all of the following are advantages of digital subtraction angiography over conventional angiography EXCEPT:

A. The ability to obtain images in multiple planes
B. The ability to enable 3-dimensional reconstructions
C. The ability to obtain and process image more rapidly
D. The ability to obtain images without the use of contrast agents

44. Which of the following imaging procedures is associated with the least amount of radiation exposure to the fetus of a pregnant female?

A. Obstetrical ultrasonography
B. Pelvimetry
C. An abdominal computed tomographic scan
D. Radionuclide scintigraphy of the liver

45. The injection site for a lower extremity venogram is normally the/a:

A. Femoral artery
B. Femoral vein
C. Abdominal aorta
D. Superficial foot vein

46. An abnormal passageway between two or more internal organs is termed a:

A. Boil
B. Ductus
C. Fistula
D. Abscess

47. A radiographic procedure that is used for the evaluation of the urethra is termed:

A. Defecography
B. Sinography
C. Retrograde hysterosalpingography
D. Voiding cystourethrography

48. Cerebral pneumography was, until recently, used for the examination of the air-filled ventricles of the brain. This procedure has been replaced by use of:

A. Computed tomography
B. Digital radiography
C. Ultrasonography
D. Polytomography

49. The general term employed for the visualization of the vessels of the body following the introduction of a water based iodinated contrast agent, is termed:

A. Arteriography
B. Venography
C. Angiography
D. Lymphangiography

50. The percutaneous introduction of catheters in most angiographic studies involves a multi-set process using compound needles and guidewire, which is referred to as the:

A. Valsalva technique
B. Haschek technique
C. Linden Than technique
D. Seldinger technique

51. When performing a venogram of the lower extremities, the leg should be positioned in the:

A. True anteroposterior (AP) projection
B. AP 30° medial oblique projection
C. Lateral projection
D. PA 15° lateral oblique projection

52. The demonstration of the lymphatic vessels and nodes by a contrast media is termed:

A. Sialography
B. Lymphangiography
C. Splenography
D. Portography

53. During fetography, fetal movement can often be prevented by:

1. Hyperaeration of the mother   *2. Use of a compression band*   *3. Long exposure times*

A. 1 only
B. 2 only
C. 3 only
D. 1, 2 & 3

54. An abnormal channel leading to a site of infection such an abscess is called a:

A. Viscus
B. Fistula

C. Canicule
D. Sinus tract

55. A radiographic procedure that is used for the evaluation of the anorectal angle and the anal canal and the rectum is termed:

A. Defecography
B. Haustrography

C. Enterostomy
D. Bronchography

In the following, match the angiographic procedures to the principal structures demonstrated.

56. ____ Pulmonary arteriography
57. ____ Cerebral angiography
58. ____ Celiac arteriography
59. ____ Renal arteriography
60. ____ Superior mesenteric arteriography

A. Vessels of the stomach
B. Vessels of the lungs
C. Vessels of the kidneys
D. Vessels of the small intestines
E. Vessels of the brain

Referring to the radiographic images, answer questions 61 & 62.

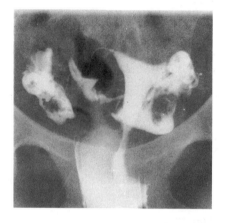

A

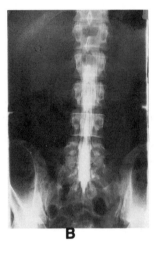

B

61. The radiographic image of Film A represents a special contrast study used in the evaluation of the:

A. Urinary bladder
B. Rectum

C. Uterus and oviducts
D. Ovaries

62. The radiographic image of Film B represents a special contrast study used in the evaluation of the:

A. Intervertebral discs
B. Ventricular system

C. Spinal canal
D. Venous system

Referring to the radiographic image, answer question 63.

63. This radiographic image is an example of an examination used in the evaluation of (the):

1. Hairline fracture of a bone
2. Soft tissue structure of a joint
3. Joint calcifications

A. 1 only
B. 2 only

C. 3 only
D. 1, 2 & 3

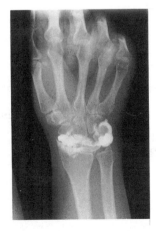

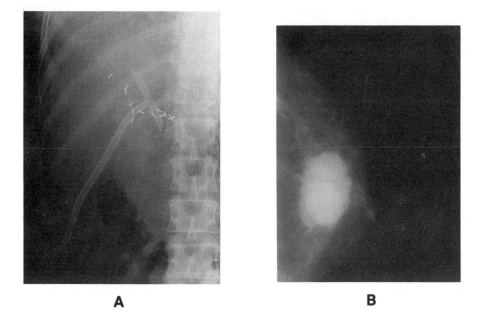

A   **B**

64. The radiographic image of Film A represents a scout film prior to a contrast study called:

   A. Endoscopic retrograde cholangiopancreatography
   B. Retrograde urography
   C. T-Tube cholangiography
   D. Abdominal angiography

65. The radiographic image of Film B represents a non-contrast study of the:

   A. Salivary ducts   C. Breast
   B. Ventricular system   D. Paranasal sinuses

Explanations

1. **D** A myelogram is a radiographic procedure that is used to examine the fluid-filled space around the spinal cord (subarachnoid space). During the procedure a spinal needle is inserted into the L3-L4 interspace and passed into the space around the spinal cord. A spinal tap or a myelographic injection is nearly always done at the L3-4 interspace to avoid damage to the spinal cord, which ends at the level of L1 or L2. The vast majority of the studies performed today use a nonionic iodinated contrast agent. This media must be non-irritating and easy to eliminate. The older oil-based iodinated agents used in the past had to be removed from the subarachnoid at the end of the procedure. This required that the spinal needle remain in place for the procedure. The newer water-based contrast agents are absorbed into the blood stream in only a few hours. The greater density of the myelographic contrast media allows it to be moved by gravity. As the head of a table is tilted down, the contrast flows toward the brain. Acute extension of the head and neck closes the opening to the fourth ventricle, and prevents the spillage of the media into the ventricles.

2. **B** There are three major techniques used for the radiographic evaluation of the vessels of the body. All involve the introduction of an iodinated contrast agent into the vessel and the subsequent imaging of the contrast enhanced blood. Arteriography describes an opacification study involving the arteries; venography involving the opacification of the veins; and angiography involving the opacification of any vessels including capillaries and lymphatic vessels.

3. **B** A mammogram is the non-contrast radiographic examination of the soft tissue structures of the breast. Because the breast only contains soft and fatty tissues, it will posses only a small amount of radiographic contrast. The use of a low kVp in the 15-35 kVp range is employed to maximize these small tissue differences. In order to maintain the highest possible contrast, the amount of scattered radiation must be limited. This can be accomplished by the use of a low ratio grid (2:1) and the use of breast compression. A breast compression device serves to reduce the thickness of the breast, reduce the amount of scattered radiation produced in the tissues and provide higher contrast. The most common projections used to evaluate the breast are the craniocaudal (cc) projection, the mediolateral oblique (MLO) projection and the axilla projection.

4. **D** Arthrography is a procedure involving the injection of a positive water-based iodinated contrast agent and a negative contrast agent such as air into the fluid-filled joint cavity. These contrast agents will help to outline the major support structures of the joint, the ligaments and cartilages.

5. **C** The radiographic evaluation of the biliary ducts after the removal of the gallbladder and cystic duct can be accomplished by the direct injection of a water-based iodinated contrast media. This is accomplished by an injection into a T-shaped tube that is placed into the common bile duct during surgery to facilitate the drainage of fluids. This procedure, called operative T-tube cholangiography, is used to demonstrate undetected stones in the biliary tree.

6. **A** The radiographic evaluation of the uterus and the fallopian tubes can be accomplished through the injection of a water-based contrast agent into the uterine canal. The contrast media is introduced by placing a cannula through the cervix into the uterus. The media will pass from the uterus into the oviducts. When the ducts is patent (open) the contrast media will spill over into the peritoneum.

7. **D** Cephalopelvimetry is a rarely performed radiographic procedure that has been largely replaced by advances in obstetrical ultrasound. This procedure was performed to make measurements of the fetal head and pelvic inlet and outlet diameters. These measurements are useful trying to determine if a vaginal birth should be attempted.

8. **C** The demonstration of the ureter and kidney by direct means, requires the passage of a catheter through the urethra and bladder into the proximal ureter. This is accomplished by using a cystoscope to locate the opening on the posterolateral surface of the urinary bladder. Since the iodinated contrast media is introduced against the normal urine flow, the procedure called a retrograde urogram. This procedure is used for the evaluation for the pelvocaly system and ureters.

9. B   The soft and fatty tissue of the human breast are best demonstrated by the use of low kVp in the 15-35 kVp range.

10. B   A spinal tap or a myelographic injection is nearly always done at the L3-4 interspace to avoid damage to the spinal cord, which ends at the level of L1 or L2.

11. D   Prior to the placement of a catheter into the ureter, the opening is located on the posterolateral surface of the urinary bladder, by way of a cystoscope.

12. B   The greater density of the myelographic contrast media allows it to be moved by gravity. As the head of a table is tilted down, the contrast flows toward the brain. Acute extension of the head and neck closes the opening to the fourth ventricle, and prevents the spillage of the media into the ventricles.

13. D   The most common projections used to evaluate the breast are the craniocaudal (cc) projection, the mediolateral oblique (MLO) projection and the axilla projection.

14. B   The radiographic evaluation of the salivary glands and their ducts by the canalization of one of the salivary ducts and the introduction of an iodinated contrast media is called sialography. The emptying of the salivary gland prior to and after a contrast injection to the gland is readily accomplished by sucking on a wedge of fresh lemon.

15. D   Because the spinal cord doesn't end until about L2, the injection sight for nearly all myelograms is at L3-4.

16. B   The kidney or renal parenchyma will be demonstrated by newly formed urine which contains the water-based iodinated contrast media.

17. D   After the puncture of the subarachnoid space around the spinal cord, the contrast media is introduced. Since this agent has a higher mass density than the cerebral spinal fluid, it can be moved by tilting the table to let gravity move the agent to its desired location. The movement toward the cervical region is accomplished by tilting the table to place the patient's head lower than the feet.

18. B   The root "cysto" refers to the urinary bladder: the suffix "graph" is from the word picture.

19. D   A dedicated mammographic system will often possess molybdenum targets for the low energy x-rays, micro or fractional focal spots to provide for high recorded detail, and compression devices to improve contrast and control motion.

20. C   The radiographic evaluation of the biliary ducts after the removal of the gallbladder and cystic duct can be accomplished by the direct injection of a water-based iodinated contrast media into a T- tube. This tube is placed into common bile duct during surgery to facilitate the drainage of fluids. This procedure called operative T-tube cholangiography and is used to demonstrate undetected stones in the biliary tree.

21. A   During a spinal tap or myelogram, acute flexion of the spine will help open the L3-L4 interspace and facilitate access to the meningeal coverings.

22. C   During a T-tube injection, contrast media flows through the common bile duct and passes through the sphincter of Oddi on its way to the duodenum.

23. D   The use of compression during a mammogram serves three main functions; namely to reduce breast tissue thickness differences, reducing the amount of scattered radiation, and provide a higher contrast.

24. C   During myelography the use of water-soluble compounds such as amnipaque or metrizamide for myelography have the advantage of a lower viscosity, improved nerve root visualization, and eliminate the need for post-procedural contrast removal.

25. D   Arthrography can currently be performed on all joints or the extremities including the TMJs. In recent years the use of the less invasive magnetic resonance imaging (MRI) has become the method of choice for the examination of most soft tissue joint injuries.

26. B Radiographic images obtained for male cystourethrography are most often taken in the AP oblique projection with the patient adjusted so the midsagittal plane forms a 30-40 degree angle with the plane of the image receptor. This projection is used to reduce the bony superimposition of the urethra, the neck of the bladder, and the bony pelvis.

27. C Laryngopharyngography is a radiographic procedure that is used to evaluate the opening of the vocal cords during inspiration, or the closing of the vocal cords with the "eee" phonation. This procedure is helpful in the evaluation of the normal function of the larynx and pharynx.

28. B During lower leg venography, the use of tourniquets and/or a 45 degree-semi-erect position of the contrast-filled leg, will aid in the filling of the deep veins.

29. B Endoscopic retrograde cholangiopancreatography (ERCP) involves the scoping of the duodenum and the passage of a cannula into the hepatopancreatic ampulla and common bile ducts system.

30. A A gravid or pregnant patient may undergo one of four types of radiographic imaging procedures, namely; fetography, placentography, cephalometry, or pelvimetry.

31. C During a hysterosalpingogram, because the ovary is not directly connected to the fallopian tube, any contrast within these tubes will spill over into the peritoneum.

32. B A stent is a metallic devise that is placed into an artery by way of a balloon type catheter. This devise is used to maintain the diameter of the vessel following angioplasty.

33. C During a myelogram, a blockage above the normal injection sight in the lumbar region will prevent the ascent of the contrast agent. In these situations, the media may be introduced into an area near the back of the head called the cisterna magna.

34. A In order to demonstrate the soft tissue structures of the joints (arthrogram), a needle is passed into the joint capsule for introduction of an iodinated contrast media.

35. A The radiographic evaluation of the urinary tract following the introduction of a water-based iodinated contrast agent, is termed intravenous urography. In order to evaluate the reflux of urine into the distal ureters, the patient is asked to void. This increases the pressure in the bladder and may cause the backward flow of urine.

36. A The emptying of the salivary gland prior to and after a contrast injection to the gland is readily accomplished by sucking on a wedge of fresh lemon.

37. A Endoscopic retrograde cholangiopancreatography (ERCP) involves the scoping of the duodenum and the passage of a cannula into the hepatopancreatic ampulla and common bile ducts system.

38. D During an operative-tube cholangiogram, a direct injection into the biliary ducts will enable the evaluation of the intrahepatic and extrahepatic ducts as well as the sphincter of Oddi.

39. D In all myelographic procedures, it is important to remove an amount of spinal fluid equal to the amount of contrast that is to be administered. This is done to avoid an increase in the intracranial pressure.

40. A Orthoroentgenography is a radiographic procedure used for the measurements of the long bones in the lower extremities. The measurement of the lower leg length is accomplished most readily through the placement of a specially designed metallic ruler which is radiographed between the desired extremities. Scanography is another term used for this study.

41. B Interventional angiography is the general term that is used to describe a radiographic procedure involving the repair or evaluation of the vessels of the circulatory system. A procedure that can be used for the occlusion of arteriovenous malformations is termed elective embolization. This catheterization technique is used to repair many disorders that previously required surgery such as arteriovenous malformations (AVM's) and some vascular bleeds. The procedure involved the injection of materials into a vessel.

42. D At the present time, dilatation of constricted vessels by the manipulation of a balloon tip catheter and/or the placement of a permanent stent to keep the vessel open are two of the more commonly used types of transluminal angioplasty.

43. D Digital angiography has all but replaced many types of conventional angiographic techniques. This technique can be used to obtain and process images more rapidly, obtained images in multiple planes and enable the 3-dimensional reconstruction of the images.

44. A Because obstetrical ultrasonography is not associated with the production of ionizing radiation, it is associated with the least amount of radiation exposure to the fetus of a pregnant female.

45. D Because the blood flow in a vein is toward the heart, the injection site for a venogram will be one of the superficial veins of the foot or hand.

46. C An abnormal passageway between two or more internal organs is termed a fistula. Fistulography is the term applied to the radiographic evaluation of this connection following the introduction of a contrast agent.

47. D A radiographic procedure that is used for the evaluation of the urethra is termed voiding cystourethrography.

48. A The advent of computed tomography (CT), and more recently, magnetic resonance imaging (MRI), both enable the improved visualization of the soft structures of the central nervous system and many joint cavities.

49. C There are three major techniques used for the radiographic evaluation of the vessels of the body. All involve the introduction of a iodinated contrast agent into the vessel and the subsequent imaging. Angiography is the term applied to the opacification of any vessels including capillaries and lymphatic vessels.

50. D The placement of a blunt cannula or flexible catheter guidewire device into an artery will often require the use of a multiple step process called the Seldinger technique.

51. B To improve the visualization of the deep veins of the legs, the position most often selected is the AP oblique projection with the leg medially rotated 30 degrees to the plane of the image receptor.

52. B The demonstration of the lymphatic vessels and nodes may be essential in determining the type and spread of many types of lymphatic cancers and is termed lymphangiography. The visualization of these small vessels requires special injection techniques, and slow introduction of an oil-based iodinated contrast agent into the system.

53. A During fetography, as well as obstetrical ultrasonography, the use of hyperaeration is often beneficial in reduction of fetal movement.

54. D An abnormal channel leading to a site of infection, such as an abscess, is called a sinus tract. Sinography is the term applied to the radiographic evaluation of this connection following the introduction of a contrast agent.

55. A Defecography is a radiographic procedure that is used for the evaluation of the anorectal angle and the anal canal and the rectum. This procedure is used to obtain measurements of the long axis of the anal canal and the anorectal angle.

56. B   57. E   58. A   59. C   60. D

61. C This image is one of a series obtained during a hysterosalpingogram.

62. C This image is one of series obtained during a lumbar myelogram.

63. B This image is one of series obtained during wrist arthrogram.

64. C This image is one of series obtained during T-tube cholangiogram.

65. C This image is one of series obtained during mammogram.

Section VI

Patient Care and Management

1. The use of inappropriate arm or leg restraints on a patient is considered a form of:

   A. Nonverbal assault
   B. False imprisonment
   C. Unintentional malpractice
   D. Malicious prosecution

2. Which of the following is considered one of the seven C's of malpractice prevention?

   1. Confidentiality   *2. Courtesy*   *3. Charting*

   A. 1 & 2 only
   B. 1 & 3 only
   C. 2 & 3 only
   D. 1, 2 & 3

3. A document which enables a trusted person to act on the behalf of an incapacitated individual is called a:

   A. No code document
   B. Self-care release
   C. Prior directives (healthcare power-of-attorney)
   D. Good Samaritan document

4. A patient shows signs of respiratory distress. This can often be helped by placing the patient supine in a semi-fowlers position requires the:

   A. Head to be placed level with the body
   B. Head to be lowered below the level of the heart
   C. Head to be elevated about 30 degrees
   D. Patient to be placed erect

5. A tube and optical system for observing the inside of a hollow organ or capacity is called a(an):

   A. Endoscope
   B. Cytometer
   C. Microscope
   D. Organometer

6. During oxygen therapy, the passage of the oxygen through distilled water serves to:

   A. Increase the concentration of the oxygen in the blood to improve uptake
   B. Improve purity of the oxygen to prevent infections
   C. Increase the humidity to prevent drying of the mucous membranes
   D. Decrease the possibility of combustion in oxygen therapies

7. The failure to keep an IV bottle above the level of the patient during an IV infusion is most likely to result in:

   A. Infiltration of the IV solutions into the surrounding tissues
   B. Bacterial contamination of the IV fluids
   C. Blood clot formation in the needle or line
   D. An increase in the infusion rate of the contrast agent

8. The normal resting pulse rate for a child up to the age of 10 years old is about:

   A. 40 degrees
   B. 55 degrees
   C. 75 degrees
   D. 95 degrees

9. The most common type of catheter placed in the pulmonary artery to monitor pulmonary disorders and heart function is a:

   A. Miler-Abbott tube
   B. Greenfield tube
   C. Fowler's catheter
   D. Swan-Ganz catheter

10. After a severe trauma, a patient has returned to the emergency room following a radiographic study of the ribs, a small pool of blood is found on the floor. The proper action of the radiographer includes all of the following EXCEPT:

    A. Putting on protective gloves
    B. Testing the blood for HIV
    C. Cleaning up the spill immediately
    D. Washing the area with a disinfectant

11. The abnormal fear of death or of dying is termed:

   A. Narcolepsy
   B. Noctosis
   C. Narcophobia
   D. Necrophobia

12. A form which must be signed by the patient indicating their understanding of the potential risks of a particular treatment or diagnostic procedure is called a:

   A. Living Will
   B. Prior Directives
   C. Informed Consent
   D. Power-of-Attorney

13. A medication which is given to an in-patient by the physician during a radiographic procedure must be recorded:

   A. On the patient prescription
   B. On the physician's report
   C. On the patient's chart
   D. Incident report

14. Kaposi's sarcoma and pneumocystis carinii pneumonia are among the most common opportunistic infections associated with:

   A. AIDS
   B. Nosocomial infections
   C. Urinary tract infections
   D. Hepatitis B

15. The loss of muscle coordination which may be seen with a cerebral vascular accident or other neurologic pathologies is termed:

   A. Paralysis
   B. Anemia
   C. Ataxia
   D. Dysphasia

16. Gastrointestinal intubation can be used for:

   1. Visual inspection   *2. Aspiration of contents*   *3. Direct feeding*

   A. 1 & 2 only
   B. 1 & 3 only
   C. 2 & 3 only
   D. 1, 2 & 3

17. When a patient is connected to the chest tube drainage equipment, it is important to _____ when moving the patient.

   1. Clamp off the chest tube
   2. Raise the bottle above the chest
   3. Avoid kinks in the drainage tube

   A. 1 only
   B. 2 only
   C. 3 only
   D. 1, 2 & 3

18. Which of the following equipment is considered a part of the high-flow oxygen system?

   1. Respirator   *2. Oxygen tent*   *3. O^2 mask*

   A. 1 only
   B. 2 only
   C. 3 only
   D. 1, 2 & 3

19. The normal range for diastolic blood pressure is about:

   A. 30 – 50 mmHg
   B. 60 – 90 mmHg
   C. 90 – 110 mmHg
   D. 110 – 140 mmHg

20. The common sites for the injection of subcutaneous medications are the:

   A. Elbow and wrist
   B. Arm and thigh
   C. 3rd and 4th fingers
   D. Webs of fingers and toes

21. In children under five years of age, to avoid damage to the sciatic nerve the preferred site for intramuscular injections is the:

A. Deltoid muscle
B. Gluteus muscle

C. Vastus lateralis muscle
D. Quadriceps femoris

22. The lumen of a hypodermic needle is measured by the unit called a gauge. Which of the following possesses the largest diameter opening?

A. 10 gauge
B. 15 gauge

C. 20 gauge
D. 25 gauge

23. A patient has a pulse rate of 35. This may be an indication of:

A. Shock
B. Fever

C. Hypothermia
D. Hemorrhage

24. Chest tube drainage systems are introduced to _____ to/in the pleural cavity.

   1. Create a negative pressure   2. Remove air   3. Create a positive pressure

A. 1 & 2 only
B. 1 & 3 only

C. 2 & 3 only
D. 1, 2 & 3

25. An increase in body metabolism, usually in response to an infectious process is called (a):

A. Fever
B. Shock

C. Defibrillation
D. Sepsis

26. The escape of fluids into the surrounding tissues, often seen when intravenous needles infiltrate, is termed:

A. Extravasation
B. Emasculation

C. Embolization
D. Exacerbation

27. The normal adult resting respiration is about:

A. 3 – 5 times/min.
B. 7 –10 times/min.

C. 16 – 20 times/min.
D. 25 – 30 times/min.

28. If pressure is NOT maintained on a vein immediately after the removal of the needle, a collection of blood may result called a:

A. Hemolysis
B. Hemorrhage

C. Hematoma
D. Diaphoresis

29. Which of the following special precautions should be followed in a patient's room for a patient receiving oxygen therapy?

   1. Avoid open flames   2. Keep doors closed   3. Wear masks

A. 1 only
B. 2 only

C. 3 only
D. 1, 2 & 3

30. The flow rate of an oxygen delivery system is normally measured in units of:

A. Cubic feet per hour
B. Liters per minute

C. Cubic centimeter per hour
D. Respirations per minute

31. In general, oral temperature readings possess a high degree of accuracy. A common factor that may cause erroneous readings is:

A. Oxygen therapy
B. Tachycardia

C. Dyspnea
D. Chest pain

32. A normal body temperature reading of 38° C (99.6° F) can be expected for:

A. Oral temperatures
B. Axillary temperatures

C. Rectal temperatures
D. Tympanic temperatures

33. Cardiac and respiratory sounds and apical pulse determinations are made using a device called a:

A. Sphygmomanometer
B. Stethoscope

C. Cardiac monitor
D. Spirometer

34. In order to reduce the need for oxygen during surgery and to reduce high fever, a technique called _____ is sometimes employed.

A. Hypoalimentation
B. Hypothermia

C. Hyperpnea
D. Hypochondria

35. The determination of the pulse can be made at various sites. Among these are:

1. Radial arteries   *2. Carotid arteries*   *3. Abdominal aorta*

A. 1 & 2 only
B. 1 & 3 only

C. 2 & 3 only
D. 1, 2 & 3

36. During the evaluation of a patient's pulse rate it is important to:

1. Determine the pulse using the thumb
2. Record both strength and regularity
3. Compare readings on opposite extremities

A. 1 only
B. 2 only

C. 3 only
D. 1, 2 & 3

37. Blood or other secretions, which may be accidentally aspirated, are often removed by using:

A. A chest drainage tube
B. T-tube drainage

C. Suction equipment
D. Respirator

38. The unlawful touching of a person without their consent is called:

A. Assault
B. Battery

C. Invasion of privacy
D. Negligence

39. In order to obtain the appropriate vital signs for a patient will require the use of a:

1. Watch   *2. Stethoscope*   *3. Sphygmomanometer*

A. 1 & 2 only
B. 1 & 3 only

C. 2 & 3 only
D. 1, 2 & 3

40. The normal retention period for which radiographs must legally be kept on file for an adult patient is:

A. 6 months
B. 1 – 2 years

C. 3 – 4 years
D. 5 – 7 years

41. The five phases of the grieving process were identified by:

A. Elizabeth Kubler Ross
B. Sigmund Freud

C. Abraham Maslow
D. Miller Abbott

42. In many emergency situations, a rapid systemic response to the medication is necessary. The route that will in most cases provide this is:

A. Subcutaneous
B. Intravenous

C. Oral
D. Intramuscular

43. In order to properly identify a radiograph, the following items should be on the film:

   1. Patient's name   2. Institutional identity   3. The date of the exam

   A. 1 only   C. 3 only
   B. 2 only   D. 1, 2 & 3

44. Which of the following is a common type of nasogastric or nasoenteric tube?

   1. Miller-Abbott   2. Levin   3. Cantor

   A. 1 & 2 only   C. 2 & 3 only
   B. 1 & 3 only   D. 1, 2 & 3

45. Which of the following will be noted if an injury has effected an individual's motor control functions?

   1. Muscle coordination   2. Pain sensitivity   3. Visual reflex

   A. 1 only   C. 3 only
   B. 2 only   D. 1, 2 & 3

46. When I.V. administration is taking place, the solution should always remain _____ above the level of the vein.

   A. 6" – 8"   C. 18" – 20"
   B. 8" – 12"   D. 36" – 40"

47. A sphygmomanometer is a device that is employed in the evaluation of:

   A. Respiration   C. Pulse rate
   B. Cardiac sounds   D. Blood pressure

48. Chest drainage apparatus should always be placed:

   A. 12" – 20" above the level of chest   C. Below the level of the chest
   B. At the level of the chest

49. To enlarge the vein to make penetration easier, a tourniquet is often placed on the patient's arm:

   A. Proximal to the injection site   C. Over the injection site
   B. Distal to the injection site

50. The pulse rate for individuals is dependent upon many factors. Among these is:

   1. Age of the patient   2. Sex of the patient   3. Patient's anxiety

   A. 1 only   C. 3 only
   B. 2 only   D. 1, 2 & 3

51. A persistently high arterial blood pressure is called:

   A. Hyperthermia   C. Hypotension
   B. Hyperkinesia   D. Hypertension

52. A rapid heart rate is one of the common indicators of cardiac insufficiency or heart damage and is termed:

   A. Bradycardia   C. Tachycardia
   B. Hypertension   D. Hypoxia

53. You are asked by a physician to draw up 20 milligrams of epinephrine. The vial reads 5 milligrams per cc. How many cc's should be drawn for the proper dosage?

   A. 100 cc   C. 4 cc
   B. 20 cc   D. 1½ cc

54. When IV solutions are employed, to insure the proper flow of the solution, the fill chamber below the IV bottle should be:

A. Completely empty
B. About half full
C. Completely full
D. Three-quarters full

55. A tube that is placed into the upper respiratory tract to improve access to the lower respiratory tract is called a:

A. Nasogastric tube
B. Chest tube
C. Tracheostomy tube
D. T-tube

56. A patient has a blood pressure reading of 140 mm Hg systolic and 70 mm diastolic. The reading for this person indicated he/she has:

A. Above normal systolic pressure
B. Below normal systolic pressure
C. Below normal diastolic pressure
D. Normal range of pressures

57. During an intravenous infusion, a swelling develops around the injection site from a dislodged needle or catheter. The immediate action to be taken by the radiographer is:

A. Remove the needle
B. Clamp off the tubing
C. Lower the IV bottle
D. Apply warm compresses

58. Medications that cannot be absorbed by the intestinal mucosa are most often administered through:

A. Topical administrations
B. Oral administrations
C. Intrathecal administrations
D. Parenteral administrations

59. Which of the following can be classified as a physical sign?

1. Cyanosis   *2. Diaphoresis*   *3. Skin Pallor*

A. 1 & 2 only
B. 1 & 3 only
C. 2 & 3 only
D. 1, 2 & 3

60. Which of the following types of medical apparatus should normally remain at a level above the patient at all times?

A. Chest drainage tubes
B. Suction apparatus
C. Intravenous infusion equipment
D. Urinary drainage equipment

Explanations

1. B. The use of inappropriate restraints on patients is legally classified as a form of intentional malpractice termed false imprisonment.

2. D. In order to avoid lawsuits all radiographers should know and follow the seven C's of malpractice prevention. These include 1. Competence 2. Compliance 3. Charting 4. Communication 5. Confidentiality 6. Courtesy 7. Carefulness.

3. C. When an individual become incapacitated, they may wish to designate a trusted person to make health care decisions for themselves. This is called prior directives or the power-of- attorney for healthcare decisions. A living will is a second document that the individual can develop that describes the level of care they wish to be used in case of serious illness or injury. The no code or do not resuscitate (DNR) orders often stem from this document.

4. C. The semi-fowlers position where the head is raised about 30 degree above the body is helpful in lessening respiratory distress.

5. A. Any type of flexible tube and the associated optics for observing the inside of an organ or hollow cavity is called an endoscope.

6. C. The passage of oxygen through a water bath helps increase the humidity of the oxygen and help reduce the dehydration this gas can cause to the mucus membranes.

7. C. All IV infusions should be accomplished with the IV apparatus about 12"-20" above the level of the vein. This will help to insure that blood does not collect and clot in the needle of the IV line.

8. D. The resting pulse rate for a child is considerably higher than the 75 beats per minute of an adult. For a 10- year old the normal heart rate is about 95 beats per minute.

9. D. Direct measurement of the pressure in the pulmonary arteries can be made by the placement of a special balloon catheter called a Swan-Ganz.

10. B. Whenever blood or other potentially dangerous body fluids are found, the radiographer should put on protective gloves and attempt to cleanup the spill and wash the area with disinfectants.

11. D. The abnormal fear of dying is called necrophobia.

12. C. A form signed by the patient after being informed of the potential risks and benefits of a specific procedure is called an informed consent form.

13. C. Any drugs or medications or medical treatments that have been given to a patient during your care should be noted in the patient's chart.

14. A. AIDS is a disease that destroys the body's ability to fight off opportunistic infections. Two of the more common infections seen with Aids patients are Kaposi's sarcoma and pneumocystis carinii pneumonia.

15. A. The loss of the ability to move or control the movement of muscles due to neurologic injury is termed paralysis.

16. C Many tubes are available to give direct access to the stomach. Among the more common are the Levin, Nutriflex, Moss and sump tubes. These are used to aid the healing process by removing fluids and gas from the stomach. Some of these tubes also can be used to introduce foods, medications, or contrast agents into the stomach. Visual inspection of the gastric mucosa can be accomplished through the use of an endoscope.

17. C In order to maintain the suction of a chest drainage tube as a patient is moved, the tubing must be free of kinks and tightly sealed in addition to keeping the unit below chest level.

18. A   A respirator or ambu bag which is involved with the forcible introduction of respiratory gases into the body is classified as a part high flow oxygen system. Passive or low flow systems include masks, cannulas and oxygen tents.

19. B   The normal blood pressure for an adult is between 110-140 mm Hg systolic and 60-80 mm Hg diastolic pressure.

20. B   The injection sites for most subcutaneous medications are selected for their relatively low pain sensitivity and easy access. Among the more common sites are the upper arm, posterior and lateral thigh.

21. C   In adults with fully developed muscles, intermuscular injections are performed into the deltoid and gluteus muscles with little complication. In children, a more lateral sight over the vastus lateralis muscle is preferred to avoid possible damage to the sciatic nerve.

22. A   The normal range for a hypodermic needle is from about 14 gauge for the largest and 28 gauge for the smallest needle diameters.

23. C   A pulse of 35 which is well below the normal 60-90 beats/min range is called bradycardia and may be associated with hypothermia shock. A fever or hemorrhaging will normally cause tachycardia or an increased heart rate.

24. A   A chest drainage device is used to extract fluids or air from the pleural cavity.

25. D   The natural response to an invading organism is frequently the development of a fever. The increased blood flow and higher body temperature creates a better environment for the elimination of the organism.

26. A   During an injection into a vessel, leakage at the injection site is termed extravasation. This frequently painful condition is treated with warm compresses which helps to speed the dissipation of the medication into the surrounding tissues.

27. C   A number of conditions can effect the normal respiration rate of 12-16 breaths/minute. Blood loss and shock will normally increase this rate while the administration of oxygen will often slow this rate.

28. C   A collection of blood in the soft tissue called a hematoma may occur following the removal of an intravenous needle. This can usually be avoided by the removal of the tourniquet prior to the administration of fluids and by applying 1 minute of direct pressure over the injection site after the needle is removed.

29. A   Because oxygen supports combustion, it can increase the potential for a fire. When oxygen is in use, smoking, open flames, and the use of ungrounded electrical equipment must be avoided.

30. B   The normal flow rate of oxygen delivery is about 4-5 liters per minute. When a patient is transferred to a new oxygen source a radiographer should always maintain the rate that was previously set. Patients with emphysema normally receive about half of the normal rate.

31. A   Inaccurate oral temperatures may result if a patient has recently ingested hot or cold beverages. Patients under going oxygen therapies will also often have lower readings for oral temperatures.

32. C   Normal body temperature varies depending upon the area from which it is taken. Rectal readings are highest at 99.6 F. (38 C.), oral readings of 98.6 F. (37 C.) are considered normal with a 97.6 F. (36.4 C.) being average for an axillary temperature.

33. B   The distal end of a stethoscope usually consists of a dual side bladder or bell which transmits sound vibrations through flexible tubing to the bilateral ear pieces.

34. B   A technique by which the body temperature is lowered to between 80-90 degrees F. called hyperthermia, is used to reduce persistent fever, or reduce the body's need for oxygen.

35. C   The common sites for the determination of the pulse include the radial, carotid, femoral, popliteal, temporal, dorsalis pedis and brachial arteries.

36. B   During the recording of the pulse, it is important to note both the rate and quality. The thumb, which has its own pulse, is not used because it may give erroneous readings. A comparison of the opposite extremity is only required in cases of injury or an interventional procedure which might compromise the blood flow to that area.

37. C   The collection of fluids or mucus in the upper respiratory passages can be removed by negative pressure equipment. For most adult patients the usual pressure setting is 120 mm Hg.

38. B   Personal injury law suits may involve 1) assault - the threat of touching a person in an injurious way, 2) battery - the unlawful touching of a person, 3) false imprisonment - detaining a patient against their will, 4) invasion of privacy - failure to maintain the patient's right to privacy, 5) negligence - failure to use reasonable care or caution.

39. D   Obtaining the vital signs of a patient will require a sphygmomanometer and stethoscope for blood pressure reading, a watch for taking a pulse, and a thermometer for determination of the body temperature.

40. D   Most radiology departments will retain films for 5-7 years based on the statute of limitations for most law suits. Radiographs of children are normally maintained until the child is 21 years of age. Mammographic films are also retained indefinitely so that comparisons with previous images are possible.

41. A   Dr. Elizabeth Kubler-Ross identified the five stages that a terminally ill patient will undergo. They are denial, anger, bargaining, depression and acceptance.

42. B   When an immediate response to medication is desired, the intravenous route is most commonly employed.

43. D   In order to ensure the proper identification and access to a patient's radiographs the following data is required: patient name and/or ID number, the date of the exam, appropriate side marker, and the institution's identity.

44. D   The three most common nasoenteric tubes are the Cantor, Harris and Miller Abbott. Nasogastric tubes include the Levin, sump, and Sengstaken-Blakemore tubes. The Swan-Ganz catheter is a balloon- tipped catheter used in pressure measurements of the pulmonary artery.

45. A   The motor control functions are related to the neuromuscular system; i.e., movement and reflexes. Pain sensitivity, sight, sound, taste, and touch are classified as sensory in nature.

46. C   During the intravenous administration of fluids, a bag or bottle which is too low will slow the delivery rate. Raising the bag too much can increase the hydrostatic pressure and cause infiltrations into the tissues near the injection site.

47. D   A blood pressure cuff or sphygmomanometer consists of an inflatable cuff, a mercury or aneroid manometer and a pressure bulb for the inflation of the cuff. A stethoscope is used to listen for pulsations in the brachial artery.

48. C   In order to maintain the negative pressure required with a chest tube, it must remain below the level of the chest. Raising the unit will cause a positive pressure that can cause fluids to flow back into the chest cavity.

49. A   Because venous blood flows toward the heart a tourniquet is placed proximal to the injection site to increase the venous pressure and make the vein more accessible. The tourniquet is released after the penetration of the vein but prior to the injection of the medication to avoid the formation of a hematoma at the injection sight.

50. D   The number of heart beats (pulse rate) is most easily determined by detecting the pulsations felt in a superficial artery. The average adult rate of between 60-90 is dependent upon age, sex, anxiety, physical exertion, and many other physiological conditions.

51. D   Blood pressure is a measurement of the pressures associated with the contraction (systole) and relaxation (diastole) phases of the heart. Blood pressures above the normal range are referred to as hypertensive, while low blood pressures are called hypotensive.

52. C A rapid heart rate called tachycardia may be indicative of low blood volume, anxiety, cardiac problems or any number of other physiological states.

53. C The proper dosage is determined by dividing the unit dose per cc into the total dose desired.

54. B Failure to partially fill the drip chamber below the IV bottle may result in an insufficient flow and the development of air bubbles in the tubing.

55. C Severe injuries to the nose, mouth, pharynx, or larynx which obstruct the respiratory passages may require an artificial opening into the trachea just below the cricoid cartilage of the larynx called a tracheostomy.

56. D These pressures fall within the normal adult range of 110-140 mm Hg systolic and 60-80 mm Hg diastolic pressures.

57. B The failure of a needle to remain within a vein will cause the release or infiltration of the medication into the soft tissues. If this occurs, the infusion should be stopped immediately and a physician notified to administer additional treatment.

58. D Though most patients prefer oral medications, not all drugs can be absorbed by the intestinal organs. Parental administrations, including subcutaneous, intramuscular, intravenous, and intradermal injections are used to introduce substances directly into the body.

59. D Physical signs such as skin color, sweating, and tremors can all be determined by careful observation of the patient.

60. C All devices which operate by maintaining a negative pressure (suction) or gravity, such as chest tubes and urinary catheters, should be placed below the level of the patient.

1. Under system of standard precautions, healthcare workers are encouraged to:

   A. Determine the nature of possible infectious disease for all patients
   B. Use every means of protection available on all patients and at all times
   C. Use protective barriers to prevent blood and body fluid transfer between individuals
   D. Disinfect all surfaces prior to touching them

2. The most effective treatment for the control of bacterial infections are a class of drugs called:

   A. Cathartics   C. Diuretics
   B. Antihistamines   D. Antibiotics

3. The final stage of an infectious process during which symptoms are diminished is termed:

   A. Terminal acute stage   C. Convalescent stage
   B. Full term stage   D. Prodromal stage

4. An infectious process in which the blood serves as the medium for the growth of microorganisms is called:

   A. Aseptic necrosis   C. Hematuria
   B. Septicemia   D. Pyrexia

5. A sterile field is considered contaminated in all of the following situations EXCEPT:

   A. If the field is left unattended
   B. If it is touched by a non-sterile object
   C. If a sterile object is thrown on the sterile field
   D. If the sterile field is touched by the side of a sterile gown

6. Hepatitis B virus is frequently transmitted from a patient to a healthcare worker through:

   A. The improper handling of urine
   B. Direct contact with the patient's skin
   C. Droplets caused by sneezing or coughing
   D. Accidental needle stick injuries

7. Surgical asepsis is a procedure which attempts to remove all organisms and spores:

   A. From the patient's skin prior to an invasive procedure
   B. In the patient through the use of antibiotics and antiseptic agents
   C. From all surfaces the patient is likely to touch
   D. From all the equipment that will be used to perform an invasive procedure

8. The only type of disease defense which provides a specific resistance or immunity is the:

   A. Phagocytic response   C. Complement response
   B. Epithelial response   D. Antigen/antibody response

9. The one type of protective barrier that is required in all types of isolation situations is the use of:

   A. Protective gowns   C. Protective gloves
   B. Protective masks   D. Protective eye shields

10. Bacterium which release pyrotoxins are responsible for the development of:

    A. Fever   C. Autoimmune responses
    B. Allergic response   D. Tremors

11. The main route by which the human immunodeficiency virus (HIV) enters the body is through the:

    A. Respiratory tract   C. Urinary tract
    B. Circulatory system   D. Digestive system

12. When preparing for an isolation procedure, the proper sequence for getting on protective clothing is:

   1. Gown   2. Mask   3. Cap   4. Gloves

   A. 1, 2, 3, 4   C. 2, 3, 4, 1
   B. 3, 2, 1, 4   D. 4, 1, 2, 3

13. A contaminated needle or a scalpel blade which is responsible for an infectious condition in a medical worker, could be referred to as a type of a/an:

   A. Nosocomial transmission   C. Fomite transmission
   B. Vector transmission   D. Endogenous transmission

14. The incubation period of an infection is the time between the entrance of the organism and:

   A. Disappearance of the symptoms   C. Appearance of the symptoms
   B. Death of the individual

15. Interferon is a chemical produced by the immune system to help control the spread of:

   A. Viruses   B. Bacteria
   C. Fungi   D. Yeasts

16. A radiographer is accidentally punctured by a used syringe needle. This is classified as a/an _____ transmission.

   A. Endogenous   C. Fomite
   B. Nosocomial   D. Vector

17. The spread of infection may occur from:

   1. Personal contact   2. Aerial transmission   3. Fomites

   A. 1 only   C. 3 only
   B. 2 only   D. 1, 2 & 3

18. The smallest disease-producing organism yet discovered is (the):

   A. Yeast   C. Fungus
   B. Virus   D. Bacteria

19. The two most common and effective methods of sterilization for use in a hospital are:

   A. Pressurized steam and dry heat   C. Gas and pressurized stem
   B. Radiation and dry heat   D. Chemical and dry heat

20. When a sterile gown is worn during an operative procedure, the portion of the gown considered to be sterile is the:

   1. Front above the waist   2. Back above the waist   3. Front below the waist

   A. 1 only   C. 3 only
   B. 2 only   D. 1, 2 & 3

21. The muscle groups that should be employed for lifting heavy objects are located in the:

   1. Arms   2. Legs   3. Back

   A. 1 & 2 only   C. 2 & 3 only
   B. 1 & 3 only   D. 1, 2 & 3

22. A patient has a serious complication during a myelographic procedure, and a malpractice suit is lodged. The claimant is able to sue the:

   1. Physician   2. Hospital   3. Radiographer

   A. 1 only   C. 3 only
   B. 2 only   D. 1, 2 & 3

23. Which of the following diseases is likely to require some type of strict isolation procedure?

1. Small pox   **2. Diphtheria**   **3. Staphylococcus aureus**

A. 1 only
B. 2 only

C. 3 only
D. 1, 2 & 3

24. The tissue of the spinal cord is protected by the canal through the vertebrae. Due to the mobility of the _____ and _____ spines, injuries in these areas are most common.

A. Cervical/dorsal
B. Dorsal/lumbar

C. Cervical/lumbar
D. Sacral/lumbar

25. The term that best describes the complete removal of all micro-organisms is called:

A. Medical asepsis
B. Disinfection

C. Antisepsis
D. Sterilization

26. The two most common methods for the control of the spreading of a contagious disease are:

A. Vaccination and inoculation
B. Decompression and contamination

C. Disinfection and isolation
D. Inoculation and asepsis

27. The most susceptible to infection ("at risk") patients are usually:

1. The very young   **2. The very old**   **3. Those with poor hygiene**

A. 1 only
B. 2 only

C. 3 only
D. 1, 2 & 3

28. When opening a sterile pack, the first fold of cloth or paper should be folded:

A. Away from the individual
B. To the right of the individual

C. To the left of the individual
D. Towards the individual

29. With patients getting in and out of wheelchairs and on and off x-ray tables, the radiographer should give:

A. Verbal assistance only
B. Help only when necessary

C. Help only when asked
D. Help at all times

30. A patient with a communicable disease should be separated from other patients and placed in a/an:

A. Protective isolation unit
B. Strict isolation unit

C. Intensive care unit
D. Burn unit

31. The proper use of body mechanics is primarily intended to prevent damage to the muscles and structures of the:

A. Upper extremities
B. Lower extremities

C. Synarthritic joints
D. Spinal column

32. After a portable machine has been used in a strict isolation unit, care must be taken to disinfect the _____ after leaving the room.

1. Cassette   **2. Mobile unit**   **3. Radiographer**

A. 1 only
B. 2 only

C. 3 only
D. 1, 2 & 3

33. Which of the following items in the O.R. are usually considered free of all microorganisms?

1. Surgical instruments   **2. O.R. table**   **3. Suction device**

A. 1 only
B. 2 only

C. 3 only
D. 1, 2 & 3

34. The best method to prevent the spread of droplet infections is by:

   1. Hand washing   2. Wearing a mask   3. Using gloves

   A. 1 only   C. 3 only
   B. 2 only   D. 1, 2 & 3

35. According to the Center for Disease Control, the Acquired Immune Deficiency Syndrome (AIDS) requires that _____ isolation technique be employed.

   A. Strict   C. Enteric
   B. Respiratory   D. Blood/body fluid

36. In a system of protective precautions (reverse isolation), two persons are required. The "clean" member is responsible for:

   1. Patient positioning   2. Positioning of the tube head   3. Making the exposure

   A. 1 only   C. 3 only
   B. 2 only   D. 1, 2 & 3

37. The contaminated linens from an isolation patient should be placed in a plastic bag and:

   A. Disposed of in the usual manner   C. Maintained in the isolation unit for disposal
   B. Marked and sent to the laundry

38. Because of the availability of anti-infective agents and good medical asepsis, hospital infections are on the decline.

   A. True
   B. False

39. A common unit used for high pressure steam sterilization is called the:

   A. Thermal ventilator   C. Vector chamber
   B. Autoclave   D. Inhalator

40. The cycle of infection can be broken by:

   1. Destruction of the infectious organism
   2. Removal of the source of infection
   3. Preventing the means by which an infection is spread

   A. 1 & 2 only   C. 2 & 3 only
   B. 1 & 3 only   D. 1, 2 & 3

41. Which of the following is normally classified as a type of vehicle that spreads infection?

   1. Mosquito bite   2. Contaminated blood transfusion   3. Coughing

   A. 1 only   C. 3 only
   B. 2 only   D. 1, 2 & 3

42. A common disease caused by a vector is:

   1. Rabies   2. Bubonic plaque   3. Lyme disease

   A. 1 & 2 only   C. 2 & 3 only
   B. 1 & 3 only   D. 1, 2 & 3

43. Micro-organisms that cause infections can be classified as:

   1. Fungi   2. Bacteria   3. Vectors

   A. 1 & 2 only   C. 2 & 3 only
   B. 1 & 3 only   D. 1, 2 & 3

44. Pertussis, influenza and tuberculosis are three common disorders that require a patient to be placed in:

A. Enteric isolation
B. Protective isolation
C. Respiratory isolation
D. No isolation is required

45. Most strict isolation techniques are designed to:

A. Control the spread of infection
B. Speed the healing process
C. Protect the patient from disease
D. Control secondary infection

46. A chemical substance that is used to kill pathogenic bacteria on instruments and equipment is called a/an:

1. Germicide (disinfectant)   **2. Antiseptic solution**   **3. Antibiotic**

A. 1 only
B. 2 only
C. 3 only
D. 1, 2 & 3

47. The viral agent that is responsible for Acquired Immune Deficiency Syndrome (AIDS) is:

A. Human immune virus
B. Escherichia coli
C. Candida albicans
D. Herpes simplex 2

48. In order to prevent needle stick injuries to hospital staff, used needled should be:

1. Recapped after use   **2. Bent or broken after use**   **3. Placed in puncture-resistant containers**

A. 1 only
B. 2 only
C. 3 only
D. 1, 2 & 3

49. A patient with salmonella is placed in an enteric isolation room. All linens used should be:

A. Disposed of in the usual manner
B. Incinerated immediately
C. Double bagged and marked contaminated
D. Sterilized in an autoclave

50. The log-roll method of patient transfer should be used to turn patients with suspected or known injuries to the:

1. Spinal column   **2. Pelvis**   **3. Upper extremities**

A. 1 only
B. 2 only
C. 3 only
D. 1, 2 & 3

51. Which of the following areas will generally require the use of protective isolation procedures?

A. Neonatal nursery
B. Tuberculosis ward
C. Recovery room
D. Pediatric ward

52. The flu, measles, herpes and hepatitis are some of the common types of infections caused by:

A. Bacterial agents
B. Yeast organisms
C. Viral agents
D. Fungal agents

53. During an intravenous infusion, a swelling develops near the injection site from a dislodged needle. The first action that should be taken is:

A. To apply a warm compress
B. Increase the rate of the infusion
C. Clamp off the tubing
D. Squeeze the IV bag

54. Povidone-iodine and isopropyl alcohol are two common agents used to:

A. Sterilize surgical instruments
B. Lubricate enema tips
C. Disinfect the skin
D. Reduce skin temperature

55. A disease that is usually transmitted by accidental blood or needle stick injury is:

   1. Acquired Immune Deficiency Syndrome (AIDS)
   2. Hepatitis B
   3. Creutzfeldt-Jakob disease

   A. 1 only
   B. 2 only
   C. 3 only
   D. 1, 2 & 3

56. The practice that helps reduce the spread of micro-organisms is termed:

   A. Lethargy
   B. Contracture
   C. Asepsis
   D. Isolation

57. In general, the best method for a radiographer to maintain medical asepsis and avoid the spread of infection is by frequent and proper:

   A. Mask usage
   B. Hand washing
   C. Isolation usage
   D. Personal hygiene

58. The most common means by which infections are spread is by:

   A. Airborne particulates
   B. Indirect (fomite) contact
   C. Direct (touching) contact
   D. Endogenous contact

59. Nosocomial infections are those that are contracted by and individual:

   A. From contaminated water
   B. From occupational environment
   C. Within a hospital
   D. With poor hygiene

60. Any medical procedure that involves the penetration of the skin requires the use of:

   A. Medical asepsis
   B. Surgical asepsis
   C. Isolation techniques
   D. Protective isolation techniques

61. A sterile pack is considered to be unsterile or contaminated if it has been exposed to:

   1. Excessive heat   2. Excessive cold   3. Water or moisture

   A. 1 only
   B. 2 only
   C. 3 only
   D. 1, 2 & 3

62. A surgical procedure involving the repair of a fractured bone is termed a/an:

   A. Frame reduction
   B. Compound reduction
   C. Open reduction
   D. Closed reduction

63. Which is a common method for the direct spread of infectious agent?

   1. Sexual transmission   2. Fomite transmission   3. Vector transmission

   A. 1 only
   B. 2 only
   C. 3 only
   D. 1, 2 & 3

64. Surgical aseptic technique should be employed during which of the following procedures?

   1. Myelography   2. Intravenous pyelography   3. Cystography

   A. 1 only
   B. 2 only
   C. 3 only
   D. 1, 2 & 3

65. Which is a common form of indirect contact method of spreading contagious diseases?

   1. Droplet transmission   2. Touching   3. Sexual contact

   A. 1 only
   B. 2 only
   C. 3 only
   D. 1, 2 & 3

66. A person is bitten by a tick or mosquito and develops an infection. This is classified as a type of :

    A. Vector transmission   C. Endogenous transmission
    B. Fomite transmission   D. Nosocomial transmission

67. The center of gravity for a person standing in the erect position is at the level of the:

    A. Mid-abdomen   C. Mid-pelvis
    B. Mid-thorax   D. Mid-thigh

68. A burn patient is generally very susceptible to infection and is often protected by the use of a/an:

    A. Enteric isolation unit   C. Strict isolation unit
    B. Respiratory isolation unit   D. Protective isolation unit

69. To preserve sterile integrity when a person's path is crossed, contact should be made:

    A. Face to face   C. Back to back
    B. Side to side   D. Front to back

70. A barium enema is a type of procedure that requires the use of:

    A. Surgical asepsis   C. No asepsis
    B. Medical asepsis

1. B Universal or standard precautions refers to a set of procedures that should be used to help the medical worker avoid contact with blood and body fluids. The impetus for these guidelines was the threat associated with the spread of hepatitis B and HIV infections. The polices assume that all persons are potential carries of disease and requires the workers to use whatever barrier devices (mask, gowns, gloves) that are necessary to prevent contact with the patient body fluids. These fluids include blood, semen, vaginal secretions, cerebrospinal fluid, pleural fluid, saliva, pus, urine, feces, amniotic fluid, tears, mucus, perspiration and peritoneal fluids.

2. D Antibiotics are a large class of medications that have been formulated to assist in the destruction of bacteria after they have entered into the body. Antibiotics are ineffective against viral or fungal infections.

3. C The process of infection follows 4 distinct stages:

   1. Latent period- stage where the infection is not apparent

   2. Incubation- stage where the microorganism reproduces and disease progresses

   3. Manifest disease- the stage where the infection takes hold and symptoms are found

   4. Convalescence-stage where the symptoms diminish and recovery is complete

4. B An infection of the blood that involves the growth of microorganisms in this medium is termed septicemia.

5. C The basic rules for maintaining a sterile field of surgical asepsis include never leaving a field unattended, never having contact with non sterile objects, Only the front of a sterile gown above the waist is considered sterile. The sides, back, and bottom of the gown are not sterile. To avoid contamination persons should pass back-to-back in a sterile room.

6. D It has been reported that as many as 250 health care workers died from hepatitis B in the years prior to the hepatitis B vaccine. The blood borne pathogen is most commonly spread to health care workers through needle stick or sharps injuries.

7. D Surgical asepsis is the procedure that attempts to remove all microorganisms and their spores from all equipment and surfaces.

8. D There are two types of immunity, specific and nonspecific. Non-specific immunity involves barriers such as the skin, and mucus membranes, the phagocytic cells of the blood stream and antimicrobial agents. Specific immunity involves the identification of specific organisms by the T-cells and the development of specific antigens and antibodies to complete the immune response.

9. C. The one barrier device that is required in all forms of precaution and isolation procedures are latex or non-latex protective gloves.

10. A Pyrotoxins are chemical agents released by some bacteria that will trigger an increase in the body's temperature, called a fever.

11. B HIV is carried mainly by blood or semen. It commonly enters the circulatory system by way of needle stick injuries, sharing of used needles (IV drug users), or through the direct contact with the semen, with the broken skin or mucus membranes during unprotected vaginal or anal sex.

12. B The proper sequence for putting on barrier protective devices is in order: the cap, the mask, the gown and the gloves.

13. C Infections can be transmitted through a number of sources. Fomites are inanimate objects such instruments, needles, linens. Vectors are animals or insects that carry diseases.

14. C The incubation time is defined as the time between the entrance of the organism and the appearance of the symptoms.

15. A Interferon is one of the body's agents that is used to prevent spread of viral agents.

16. C Any inanimate object that can spread a pathogenic organism is termed a fomite. X-ray tables, cassettes, sponges, calipers, needles, and even pillow cases are potential carriers of infection.

17. D The spread of infection may be caused by direct contact, by touching or sex; or indirectly by air, fomites, vectors, droplets, or food and water.

18. B Diseases are caused by a number of organisms. The smallest of these is a virus which injects its genetic instructions (DNA) into a host cell turning it into a virus factory.

19. C The destruction of all micro-organisms can be accomplished by exposing pathogens to pressurized steam (autoclave) or gasses such as ethylene oxide.

20. A Only a small portion above the waist in the front of the gown is considered sterile.

21. A Due to their strength and lower potential for serious injury, the muscles of the extremities are preferred for lifting heavy objects.

22. D Law suits may be brought against any caregiver or organization responsible for the patient. The most serious is medical malpractice which can include both inappropriate action or lack of action by the medical personnel.

23. D Serious infections which can be easily transmitted by air or indirect contact will require the strict isolation of the patient in which masks, gowns, and gloves are worn. These diseases include diphtheria, pneumonic plague, chicken pox, smallpox, and staph infections.

24. C The greater flexibility of the cervical and lumbar spines make them the more common sites for spinal injuries.

25. D The destruction of all microorganisms and their spores, sterilization, can be accomplished by steam, dry heat, ionizing radiation, gasses, or chemical agents.

26. C The most effective means for the control of a contagious disease is by a combination of an isolation technique and disinfection.

27. D Three classes of individuals most likely to contract infections; the very young, the very old, and those with poor hygiene.

28. A The first fold of a sterile pack is always unfolded away from the body so that all other folds can be made without leaning or reaching over the open sterile pack.

29. D Wheel chairs and stretchers are inherently dangerous if used improperly. To avoid problems, direct physical contact and assistance with the patient is required at all times.

30. B Strict isolation units are category related. Specific isolation categories include respiratory, enteric, blood and body fluids and are all employed to prevent specific types of infections.

31. D During lifting or moving patients, the structures at greatest risk to injury are those of the spinal column. Good body mechanics are essential to help minimize potential injury to the medical workers.

32. D After completing a study in a strict isolation unit, all equipment and members of the care team must be disinfected prior to leaving the unit to avoid the spread of infection.

33. A Sterilization is only possible on relatively small items which can tolerate heat and chemical agents. Suction devices, OR tables, mobile x-ray units, and the patients themselves can only be disinfected, not sterilized.

34. B The spread of droplets or air borne contaminates is most easily prevented by the use of a mask. As with all patient contact, hand washing is also essential for all types of asepsis.

35. D   Blood and body fluid precautions should be observed for AIDS, tick born infections, hepatitis B, hepatitis non A, non B, and malaria.

36. A   During strict isolation techniques for a compromised patient, the clean member of the team is responsible for all contact with the patient.

37. C   The disposal of contaminated linens or materials requires that a double wrapped clearly marked (red) bag be used for all materials.

38. B   Hospital acquired (nosocomial) infections are unfortunately on the rise. This is due to a combination of a more aged population, the existence of more resistant bacterial strains, and an increase in the use of immunosuppressive drugs in organ transplant patients.

39. B   The most convenient type of device for the sterilization of most types of reusable instruments is the pressurized steam device called an autoclave.

40. D   The four factors included in the cycle of infection are the pathogenic organism, reservoir of infection, means of transmission and susceptible host. The infection is controllable when any link in the cycle is eliminated or controlled.

41. B   Vehicle routes of transmission spread include water, food, medications, and contaminated blood.

42. D   Animals or insects which carry and transmit infections are called vectors. Malaria, bubonic plague, Lyme disease, and rabies are all spread in this manner.

43. A   The four major classes of microorganisms are viruses, bacteria, fungi, and protozoa.

44. C   Because pertussis, influenza and TB are all spread by airborne means, the use of a respiratory isolation technique is required.

45. A   Most isolation procedures are designed to protect the caregiver and reduce the spread of infection.

46. A   The destruction of all pathogenic bacteria can be accomplished by the use of a disinfectant or germicide.

47. A   AIDS is caused by the human immune virus, which unlike most infections, does not attack a specific organ, but instead infects the T-cells which are an essential component of the immune system.

48. C   The recapping and breakage of needles, practiced up until the mid 80's, has been replaced with the less dangerous policy requiring their disposal into a puncture resistant container.

49. C   All linens from a patient in enteric isolation should be double bagged using the appropriate tag or color coded bag.

50. A   The log roll, requiring a five person team, is used to prevent further injury to a patient with neck or back trauma.

51. A   Special techniques to prevent infecting compromised patients are routinely used for burn patients and for neonates.

52. C   The smallest microorganism. the virus, is responsible for a number of disorders, including AIDS, hepatitis A & B, herpes, mumps, measles, the flu, and the common cold.

53. C   The first action that should be taken when a needle infiltrates is the clamping of the IV tubing. A doctor or nurse should then be notified immediately.

54. C   The disinfection of the skin can be accomplished by an antiseptic solution such as alcohol or povidone-iodine (Betadine).

55. D   Accidental needle-sharp (needle stick) injuries are related to the transmission of AIDS, Hepatitis A & B, and Creutzfeldt-Jakob disease.

56. C   Reduction in the probability of infectious organism transmission to susceptible individuals is accomplished through proper medical asepsis which involves keeping the room clean and washing of the hands between patients.

57. B   Microbes are most commonly spread from one individual to another by the direct physical contact of two persons.

58. C   Touching, a form of direct contact, is the most common method by which infections are transmitted from one person to another.

59. C   Patients entering the hospital for one condition sometimes acquire secondary infections, due to the hospital environment. These infections are called nosocomial infections.

60. B   The increased potential for the introduction of microorganisms when the skin is broken requires the use of the more strict surgical asepsis.

61. C   The sterility of a pack may be compromised by its exposure to moisture which can penetrate through the outer layers of the pack and bring bacteria or viruses toward the interior of the pack.

62. C   A serious fracture which requires surgical intervention to accomplish proper alignment is termed an open reduction. The more usual non-surgical repair is called a simple or closed reduction.

63. A   Direct contact with infectious agents may occur from touching or sexual contact. Indirect agents for the spread of infection include fomites, vectors, air borne, food, and water contaminates.

64. D   In addition to any breakage of the skin, surgical asepsis should be practiced with any procedures which penetrate the urinary tract; i.e., catheterization and tracheostomy tubes.

65. A   Droplet or airborne contaminates associated with respiratory infections are substantially reduced by the use of surgical type masks and negative air flow systems.

66. A   When an animal serves as an intermediary in an infectious process, it is termed a vector. Mosquitoes that transmit malaria, fleas that carry bubonic plague, ticks that spread yellow fever or Lyme disease and many types of animal bites are types of vector-spread infections.

67. C   The point around which the weight of the body is balanced, the center of gravity, is found in the mid-pelvis region. The weight distribution in females results in a lower center of gravity compared to the males.

68. D   An isolation procedure that is designed to prevent a compromised patient's exposure to infectious agents is a form of strict isolation often called protective or reverse isolation.

69. C   In an operating room or other sterile area, the accidental contamination of the gown can be avoided by passing back to back to keep the sterile surfaces separated.

70. B   In all routine radiographic and patient care procedures the main component for medical asepsis is proper hand cleanliness.

1. A patient is brought into the radiology department after the placement of a fiberglass cast. During the radiograph, the radiographer notices the fingers are cold. This should be reported because it may indicate:

   A. The patient is having an allergic reaction
   B. Warm, sweaty palms
   C. The cast may be too loose
   D. The circulation may be compromised

2. A patient comes to the radiology department two days after a laparotomy wearing an abdominal binder. After the patient is moved onto the table he complains the binder is wet. The radiographer should suspect:

   A. Postural hypotension
   B. Wound dehiscence
   C. Septicemia
   D. Epistaxis

3. The basic steps in cardiopulmonary resuscitation (CPR) include all the following EXCEPT:

   A. Maintaining the open airway
   B. Monitoring cardiac response
   C. Providing the appropriate breathing rate
   D. Compressing the heart 60 times per minute

4. The signs and symptoms of a severe headache, aphasia, ataxia, and facial paralysis are most often associated with:

   A. Hypoglycemia
   B. Anaphylaxis
   C. Cerebral vascular accident
   D. Hypertension

5. Premature infants and newborn babies are more vulnerable to hyperthermia than older children because they:

   1. Have a relatively large surface area for their body mass
   2. Have a slow heart rate and high skin temperature
   3. Lack the fatty tissue which helps to maintain the internal heat

   A. 1 & 2 only
   B. 1 & 3 only
   C. 2 & 3 only
   D. 1, 2 & 3

6. The shrinkage or reduction in the size of a tissue or organ caused by the death of cells or lack of normal use is called:

   A. Dermatitis
   B. Anemia
   C. Atrophy
   D. Desquamation

7. A tracheostomy is a surgical procedure which is performed to:

   A. Reduce the pressure which can buildup in the heart
   B. Help relieve a severe blockage of the upper respiratory tract
   C. Drain off excessive fluids which have collected in the pleura
   D. Sample or portion a lesion for laboratory testing

8. Syncope or fainting is a mild form of neurogenic shock that may occur:

   A. After a highly unpleasant or painful event
   B. Following a sudden change of posture which increases the blood pressure
   C. After the exposure to an air-borne contamination
   D. Following exposure to a pathogen

9. The constriction of the bronchial tree caused by the swelling of the mucous membranes during an asthma attack is most easily eliminated by the use of a/an:

   A. Nasopharyngeal suction
   B. Hypoglossal administration of vasodilator
   C. Inhalation of bronchodilating agent
   D. Topical application of an aseptic agent

10. The signs of cardiac arrest will normally include:

 1. Apnea   2. Absence of the pulse   3. Falling blood pressure readings

A. 1 only
B. 2 only

C. 3 only
D. 1, 2 & 3

11. A diabetic patient that begins to complain of extreme hunger prior to a gastrointestinal series should be:

A. Given an ice cube to suck on
B. Encouraged to give themselves their insulin shot
C. Given fruit juice or candy orally
D. Given spirits of ammonia nasally

12. A serious hereditary disease in which certain blood clotting factors are absent and leads to excessively long coagulation time is called:

A. Hemophilia
B. Leukemia

C. Erythroblastosis
D. Thrombophlebitis

13. The main clinical symptoms of a partial respiratory obstruction include:

 1. Wheezing   2. Rales   3. Tachypnea

A. 1 & 2 only
B. 1 & 3 only

C. 2 & 3 only
D. 1, 2 & 3

14. The ABC's of cardiopulmonary resuscitation are:

A. Airway, body fluid, coma
B. Airway, breathing, circulation
C. Asepsis, bleeding, control
D. Agitation, breathing, compression

15. Croup and haemophilus influenza are two of the agents which are associated with:

A. Lower respiratory tract infections
B. Upper alimentary tract infections

C. Lower urinary tract infections
D. Biliary tract infections

16. A young patient is radiographed and shows an incomplete fracture of the radius. This is often classified as a/an:

A. Epiphyseal fracture
B. Greenstick fracture

C. Compound fracture
D. Comminuted fracture

17. Cardiopulmonary resuscitation should be started immediately on patients with:

A. Hypovolemic shock
B. Grand mal seizures

C. Cardiac failure
D. Neurogenic shock

18. During a convulsive seizure, the most important action to be taken by a radiographer is:

 1. Prevention of patient injury from falling
 2. Inserting the fingers into the mouth
 3. Administer some type of sugar immediately

A. 1 only
B. 2 only

C. 3 only
D. 1, 2 & 3

19. Fainting (syncope) is caused by an insufficient blood supply to the brain. A patient experiencing this sensation should be instructed to:

A. Stand quietly
B. Stand and walk

C. Sit in a chair
D. Lay supine

20. During cardiopulmonary resuscitation, the rate of cardiac compressions to assisted respirations should normally be about:

A. 60:1
B. 30:1
C. 315:1
D. 4:1

21. A medication that increases urine output is called a/an:

A. Narcotic
B. Diuretic
C. Vasodilator
D. Antipyretic

22. A fracture that causes an external wound because of protruding fragments is classified as a/an:

A. Simple fracture
B. Compound fracture
C. Incomplete fracture
D. Comminuted fracture

23. Anaphylactic shock is usually preceded by:

A. Profuse bleeding
B. Systemic infection
C. Drug administration
D. Cardiac failure

24. During cardiopulmonary resuscitation, the external cardiac massage is started by placing the heel of the hand 1 ½ inches above the:

A. Manubrium
B. Gladiolus
C. Xiphoid
D. Solar plexus

25. Rigid muscles and jerky body movements are often seen in a patient experiencing (a/an):

A. Heart attack
B. Shock
C. Convulsive seizure
D. Airway obstruction

26. The human brain can survive without oxygen for about:

A. 2 - 4 minutes
B. 7 - 9 minutes
C. 11- 14 minutes
D. 15 - 20 minutes

27. When a patient with a fractured extremity is moved, special care must be taken to support the site:

A. Distal to the injury
B. Proximal to the injury
C. Both of the above
D. Neither of the above

28. Constriction of the blood vessels going to the heart often leads to a painful sensation of the chest called angina pectoris. This is often relieved by administration of:

A. Heparin
B. Spirits of ammonia
C. Benadryl
D. Nitroglycerin

29. When a patient is in shock the Trendelenburg position will:

A. Increase pulmonary blood flow
B. Relieve colon spasms
C. Increase circulating blood volume
D. Promote venous return to the heart

30. Uremia is caused by excessive buildup of the by-products of protein metabolism that is often associated with:

A. Gastric disorders
B. Kidney disorders
C. Respiratory disorders
D. Circulatory disorders

31. A male infant is brought to the department with a diagnosis of pyloric stenosis. The examination that will most likely confirm this condition is a/an:

A. Gastrointestinal series
B. Cholangiography
C. Barium enema
D. Intravenous urogram

32. Convulsive seizures may be associated with:

    1. Hyperthermia   **2. Epilepsy**   **3. High intercranial pressure**

    A. 1 & 2 only   C. 2 & 3 only
    B. 1 & 3 only   D. 1, 2 & 3

33. A diabetic has taken an insulin injection but is NPO for a GI series examination. You note excessive sweating and dizziness. This person my be experiencing:

    A. A hypoglycemic reaction   C. Ketoacidosis
    B. A hyperglycemic reaction   D. Cardiac failure

34. A common allergic reaction to iodine contrast media is bronchial constriction. This is often treated with a/an:

    A. Vasodilator   C. Antihistamine
    B. Vasoconstrictor   D. Antidiuretic

35. A patient comes into the x-ray department with a fractured rib which has punctured the lung. This is an example of a:

    A. Compound fracture   C. Comminuted fracture
    B. Complicated fracture   D. Greenstick fracture

36. The dizziness associated with sitting or standing after prolonged bed rest is called:

    A. Vertigo   C. Postural hypotension
    B. Decubitus seizure   D. Dehiscence

37. A patient has been in a car accident and is being radiographed. Without warning, he begins to projectile vomit. This should be reported to a nurse or physician because it is often associated with:

    A. High intercranial pressure   C. Respiratory distress
    B. Internal bleeding   D. Cardiac tamponade

38. Severe respiratory dysfunction often results from:

    1. Grand mal seizure   **2. Airway obstruction**   **3. Ketoacidosis**

    A. 1 only   C. 3 only
    B. 2 only   D. 1, 2 & 3

39. A patient is brought to the Radiography Department from a car accident with a weak, rapid pulse, low blood pressure, and feels cold and lacks color. The patient is showing signs of:

    A. Tetanus   C. Cardiovascular disease
    B. Shock   D. Epilepsy

40. A patient with a pre-existing disease, such as osteomalacia, develops non-trauma fracture. The term for this is a/an:

    A. Greenstick fracture   C. Pathologic fracture
    B. Incomplete fracture   D. Colles fracture

41. A patient is said to be experiencing dyspnea has:

    A. Breathing difficulties   C. A fever
    B. High blood pressure   D. A urinary problem

42. A car accident victim has lost a large amount of blood, and while performing the radiographs, you see his condition worsen. The patient is most likely lapsing into:

    A. Septic shock   C. Grand mal seizure
    B. Hypovolemic shock   D. Anaphylactic shock

43. Angina pectoris is a common indication of:

A. Heart disease
B. Lung disease

C. Thyroid disease
D. Kidney disease

44. The most important consideration in the care of unconscious patients is to insure that:

1. Fluids are maintained   2. Airways remain clear   3. No medications are given

A. 1 only
B. 2 only

C. 3 only
D. 1, 2 & 3

45. A sharp turning of the ankle often results in a fracture of the distal fibula called a/an:

A. Evulsion fracture
B. Potts fracture

C. Slipped epiphysis
D. Depressed fracture

Match the following medications with their principal action for questions 46-50.

46. _____ Penicillin, tetracycline
47. _____ Aminophylline, epinephrine
48. _____ Aspirin, codeine
49. _____ Phenobarbital, ethanol
50. _____ Valium, Thorazine

A. Analgesics
B. Sedatives
C. Bronchodilators
D. Antianxiety
E. Antibiotic

51. The process of identifying and assigning priorities to a patient for emergency care is termed:

A. Sequencing
B. Scheduling

C. Triage
D. Prophylaxis

52. A patient's request contains contradictory information; i.e., right hand is injured, but the request is for the left hand. The radiographer should:

A. Do the ordered exam without question
B. Do the ordered exam, and include the opposite hand
C. Call the ordering physician or station and get the proper information
D. Perform the exam you feel is appropriate

53. The patient has the right to:

1. Refuse medical treatment
2. Confidential treatment of their record
3. Know the qualifications of the medical personnel

A. 1 only
B. 2 only

C. 3 only
D. 1, 2 & 3

54. Prognosis is defined as:

A. The science of pathology
B. Forecast of the probable result of disease

C. Method of examination
D. Detailed program of treatment

55. A simple technique used to dislodge foreign bodies from the respiratory system is called the:

A. Heimlich maneuver
B. Syncope technique

C. Cardiopulmonary resuscitation
D. Cardiac arrest

56. A type of shock occurring in response to a massive buildup of bacteria and their toxic by-products is termed:

A. Septic shock
B. Neurogenic shock

C. Hypovolemic shock
D. Allergic shock

57. The major sign or symptom associated with cardiac failure is:

1. Apnea   2. Unconsciousness   3. Weak or absence of a pulse

A. 1 only   C. 3 only
B. 2 only   D. 1, 2 & 3

58. The common indicators of shock generally include:

1. Hypotension   2. Tachycardia   3. Convulsive seizures

A. 1 & 2 only   C. 2 & 3 only
B. 1 & 3 only   D. 1, 2 & 3

59. A comminuted fracture is best described as a:

A. Displaced fracture   C. Fragmented fracture
B. Compound fracture   D. Simple fracture

60. The back, chest and abdominal thrust maneuvers can be of assistance in the relief of:

A. Convulsive seizures   C. Airway obstructions
B. Anaphylactic shock   D. Cardiac tamponade

61. During cardiopulmonary resuscitation, the cardiac compression rate should be about:

A. 10 – 20 compressions/min.   C. 30 – 40 compressions/min.
B. 20 – 30 compressions/min.   D. 40 – 60 compressions/min.

62. When a patient is vomiting, the head is lifted or turned to the side to prevent:

A. Abdominal cramps   C. Drowsiness
B. Dehydration syndrome   D. Aspiration

63. A strong indication of a low oxygen content in the tissues of the body is:

A. High body temperature   C. Eclampsia
B. Cyanosis   D. Diaphoresis

64. The failure to metabolize glucose may result in a condition termed:

A. Antisepsis   C. Ketoacidosis
B. Septicemia   D. Anaphylaxis

65. The control of bleeding in the extremities is often helped by:

1. Direct pressure over the wound   2. Raising the extremity   3. Cold compresses

A. 1 only   C. 3 only
B. 2 only   D. 1, 2 & 3

Explanations

1. D   The placement of a cast on an extremity can compromise the circulation of the body part. The main symptom of this is a reduced blood supply to the region and the reduction in the temperature of the digits.

2. B   Following abdominal surgery a binder is often used to help the sutures from opening up. The appearance of fluids on this binder may be a sign of wound dehiscence.

3. B   The three main steps in CPR are maintaining an open airway, breathing for the patient and providing cardiac compressions. These are sometimes called the ABC'S of CPR. that stands for airway, breathing and circulation.

4. C   The loss of blood supply to the brain as in a cerebral vascular accident (CVA) or stroke may lead to a number of symptoms including headaches, aphasia, ataxia and paralysis of the face or extremities.

5. D   A new-born baby has a large surface area relative to their mass, and a low percentage of body fat. Both of these factors tend to make the baby vulnerable to hypothermia.

6. C   The inability to use a tissue or organ will result in a wasting away of the tissues, called disuse atrophy.

7. B   A blockage of upper respiratory tract can be extremely dangerous. The Heimlich maneuver is used to clear the passages of the upper respiratory tract of trapped food. A simple surgical procedure called a tracheostomy is used to bypass these passages when the blockage is caused by injury or disease.

8. A   A mild form of neurogenic shock called syncope may follow a painful injury, unexpected trauma or other unpleasant event.

9. C   The restriction of the bronchial tree caused by the swelling of the walls and linings of the respiratory tract can often be reduced by the introduction of a bronchodilating agent

10. D   The signs of cardiac arrest include; the absence of a pulse, a dramatic decrease in blood pressure, and most often, apnea.

11. C   In a patient with diabetes, extreme hunger is often a symptom of hypoglycemia. The ingestion of a small amount of sugar (juice or candy) can help to lessen these symptoms.

12. A   One of the most common diseases associated with an excessive bleeding is called hemophilia. This hereditary condition can now be treated through the use of a special chemical derived from whole blood called the "clotting factor".

13. A   Rales and wheezing are two of the common symptoms associated with partial obstructions of the respiratory tract.

14. B   The three main steps in CPR are maintaining an open airway, breathing for the patient and providing cardiac compressions. These are sometimes called the ABC'S of CPR, that stands for airway, breathing and circulation.

15. A   Croup and haemophilus influenza are among the most common agents involved with lower respiratory tract infections.

16. B   An incomplete fracture of a bone most often involving the more flexible bones of a developing child is called a greenstick fracture, due to its similarity to the appearance of a bent live tree branch.

17. C   Cardiopulmonary resuscitation is essential in cases of cardiac arrest and respiratory failure with or without airway obstructions.

18. A   During a seizure, little medical intervention is necessary or helpful. The radiographer's main responsibility is that of preventing secondary injury from falling. Trying to restrain the patient or taking other actions may cause more harm than letting the seizure run its few minute course.

19. D   Fainting has many possible causes, such as heart irregularities, hunger, poor ventilation, fatigue, and emotional shock. Assisting the patient to a recumbent position or lowering the head to a level below that of the heart increases the blood flow to the brain and may prevent fainting in some cases. After fainting, if the patient does not rouse immediately, spirits of ammonia held under the nose will bring a rapid return to consciousness.

20. D   The ratio between the normal heart rate of 72 beats a minute and normal respiration rate of 18 breaths per minute is called the vital sign ratio. This ratio is useful in determining the correct compression to breathing rate that should be employed during cardiopulmonary resuscitation.

21. B   The abnormal buildup of fluids in the body's tissues can be reduced by the use of a diuretic such as furosemide (lasix).

22. B   When external bone fragments are present, the fracture is described as compound. This type of injury is more prone to secondary infections such as osteomyelitis.

23. C   The severe allergic reaction to a chemical agent such as a contrast media is classified as a type of anaphylactic reaction.

24. C   External cardiac compression during CPR is started by placing the heel of the hand at a level of the midsternum thus avoiding possible injury to the xiphoid process located at the lower end of the sternum.

25. C   Seizure disorders and their associated involuntary muscle contractions can be a result of epilepsy, internal cranial pressure, poisoning, uremia, and high fevers.

26. A   During respiratory or cardiac failure it is critical to supply the brain with adequate oxygen in order to avoid brain damage which can occur within 2-5 minutes.

27. C   All non-splinted fractures require support on both ends of the injured segment.

28. D   A severe pain or constriction in the area of the heart usually radiating down the left arm, angina pectoris, is caused by a low cardiac blood supply and can often be relieved by the administration of a vasodilator such as nitroglycerin or amylnitrite.

29. D   The Trendelenburg position is used to promote the return of veinous blood to the heart. It is often used for patients experiencing shock.

30. B   A severe toxic condition associated with renal failure or renal insufficiency will result in a dangerous buildup of nitrogenous by-products in the blood stream. Renal or hemodialysis is normally required to clean these toxic by-products from the blood.

31. A   A gastrointestinal series is most often used to aid in the diagnosis of pyloric stenosis.

32. D   Though epilepsy is by far the most common cause of seizures, high intracranial pressure, toxins, infection, fever, tetanus, uremia, and low oxygen levels can also trigger seizures.

33. A   An inadequate diet following an insulin injection will result in a low blood sugar or hypoglycemic reaction. This serious condition requires medical treatment which should include the immediate administration of sugar to prevent the lapse of the patient into a coma.

34. C   Bronchial constriction due to an allergic reaction is triggered by histamine production. The treatment most often prescribed is an antihistamine such as diphenhydramine hydrochloride (Benadryl).

35. B   Any fracture which is associated with an injury to another organ or structure is termed a complicated fracture.

36. C   Sometimes confused with fainting, postural hypotension results from blood pooling in the extremities while recumbent. Upon sitting, a transient cerebral anoxia may occur.

37. A Head trauma, depending on its severity, may result in minor symptoms to a full blown coma. Careful observation of a head injury patient is important. Symptoms of lethargy, irritability, lowered vital signs, and projectile vomiting may indicate the potential for high intracranial pressure or brain injury.

38. B Among the most common causes of respiratory dysfunctions are mechanical airway obstructions, chronic lung disease, and severe cardiovascular disorders.

39. B After any type of mental or physical trauma, a thready rapid pulse, low blood pressure, shallow respirations, and a reduced peripheral circulation are the classic signs associated with shock.

40. C Undiagnosed bone tumors, osteoporosis, and any number of other conditions which weaken the bone and make a fracture more likely are called pathological pre-disposers.

41. A The word root, pnea, for breathing is often combined with the common prefixes: dys - difficult, brady -slow, tachy - fast, and u - normal.

42. B The type of shock most frequently observed in individuals experiencing a severe blood or fluid loss (burns) is termed low volume or hypovolemic shock.

43. A The constriction of blood vessels around the heart can lead to a sharp pain over the chest radiating down the left arm called angina pectoris.

44. B All caregivers must be aware of the importance of breathing and the special care that must be taken to maintain clear airways on unconscious patients.

45. B A fracture of the lower end of the fibula and medial malleolus of the tibia with outward dislocation of the foot is termed a Pott's fracture.

46. E   47. C   48. A   49. B   50. D

51. C When medical services must be provided to a large number of individuals, patients are assigned according to the severity of their illness in a process called triage.

52. C Prior to performing a requested examination, a radiographer must review and confirm any questionable information contained on the requisition.

53. D The American Hospital Association affirms the following patient rights:

1) considerate care 2) complete information concerning their diagnosis and prognosis 3) informed consent 4) refusal to be treated 5) privacy 6) confidential treatment of their records 7) receive appropriate services 8) know the relationship which may exist between the caregiver and the institution 9) when and if experimental treatment is being employed 10) have any bills explained 11) expect continuity of care 12) know appropriate conduct expected of a patient.

54. B Diagnosis is the use of scientific methods to establish the cause of an illness. Prognosis is the forecast of the likely outcome of an illness.

55. A Improper ventilation caused by a foreign blockage of the throat can often be alleviated by the abdominal thrust or Heimlich maneuver in which arms are placed around the patient and rapidly squeezed to force expelled air from the compression of the diaphragm.

56. A Some bacteria produce harmful compounds called endotoxins which may be released into the blood stream. This causes symptoms which may include fever, chills, and in severe cases, septic shock.

57. D Cardiac arrest or heart failure is associated with a complete loss of blood circulation in the body resulting in unconsciousness and the absence of the pulse and respiratory functions.

58. A The physiological response to a trauma or injury is often a condition called shock. The symptoms of this serious disturbance which effects the blood flow to the vital organs ,include a weak, rapid pulse, shallow breathing, low blood pressure, and a decreased perfusion to the vital organs and skin.

59. C   A comminuted fracture, commonly associated with a severe blunt trauma, appears radiographically as a splintered or crushed area of a bone.

60. C   Thrust maneuvers, such as the Heimlich procedure, are used in the relief of blocked respiratory passages.

61. D   During cardiopulmonary resuscitation, the cardiac compression rates should be about one compression per second. The breathing rate is about five times slower to maintain 12-15 respirations per minute.

62. D   A serious complication associated with an unconscious patient is the aspiration of vomitus which may result in an airway obstruction. This can be prevented by laying the patient on his/her side or stomach.

63. B   A low oxygen content blood obtains a dark red tint when seen through the skin of the extremities, this results in a characteristic blue color (cyanosis) of the skin and in the nail beds of the fingers and toes.

64. C   When insufficient insulin is present in the blood stream, glucose metabolism is reduced causing the mobilization of fatty acids resulting in an acidic state termed ketoacidosis.

65. D   The control of bleeding in an extremity is assisted by lowering its blood pressure. This is most easily accomplished by raising the effected area above the level of the heart. The use of direct pressure over the bleeding sight and application of cold compresses to constrict the blood vessels are also useful.

1. A patient for a urographic examination comes into the radiology department with a running IV at a rate of 16 drops per minute. It is most likely that:

   A. The needle has been inadvertently dislodged from the vessel
   B. This rate is appropriate for the patient
   C. This rate is too slow to prevent blood clotting in the needle
   D. The rate is too high and may cause pulmonary edema

2. A medication that decreases the amount of fluid retained by the urinary system is called a/an:

   A. Antipyretic
   B. Antihistamine
   C. Vasodilator
   D. Diuretic

3. Blood urea nitrogen (BUN) and creatinine are two of the more common tests used for the evaluation of:

   A. Kidney function
   B. Liver function
   C. Thyroid function
   D. Reproductive function

4. Which of the following medication administrations is likely to effect the state of consciousness of a patient?

   1. Diazepam (Valium)   *2. Midazolam (Versed)*   *3. Lidocaine*

   A. 1 & 2 only
   B. 1 & 3 only
   C. 2 & 3 only
   D. 1, 2 & 3

5. In order to visualize the gall bladder through the use if a contrast media, the agent must first be removed from the blood stream by the:

   A. Liver
   B. Spleen
   C. Kidney
   D. Pancreas

6. A 25-gauge hypothermic needle containing 1 cc of medication is injected at a 45 degree angle into the tissue beneath the skin. This represents a/an:

   A. Intravenous administration
   B. Intrathecal administration
   C. Intramuscular administration
   D. Subcutaneous administration

7. In the elderly, cathartics may be ordered to prevent _____ following a gastrointestinal series or barium enema.

   A. The loss of electrolytes due to diarrhea
   B. Constipation and possible fecal impaction
   C. Excessive loss of fluids and dehydration
   D. Excessive retention of fluid and pulmonary edema

8. Which of the following patient factors may influence the pharmacological effects of a given drug?

   1. Weight   *2. Sex*   *3. Anxiety level*

   A. 1 only
   B. 2 only
   C. 3 only
   D. 1, 2 & 3

9. Which of the following are included with the five rights of medication administration?

   1. The right dose   *2. The right time*   *3. The right patient*

   A. 1 & 2 only
   B. 1 & 3 only
   C. 2 & 3 only
   D. 1, 2 & 3

10. The vasovagal or vasomotor response is most often triggered by:

    A. Any chemical allergen
    B. Presence of adrenal gland hormones
    C. Fear or anxiety
    D. Histamine

11. The amount of time a contrast agent remains in the body is termed:

    A. Persistence   C. Viscosity
    B. Miscibility   D. Opacity

12. Which of the following is classified as an isotonic solution?

    A. Hypaque 50   C. .9% NaCl
    B. 5% glucose   D. Ringer's lactate

13. Any type of drug administration which involves the delivery of medications or contrast agents into the tissue by way of an injection through the skin is termed:

    A. Direct   C. Epidural
    B. Parenteral   D. Subarachnoid

14. Prior to the intravenous administration of an iodinated contrast media, it is important for a healthcare worker to determine if the patient has a history of:

    1. Transient ischemic attacks   *2. Previous contrast reactions*   *3. Severe renal disease*

    A. 1 & 2 only   C. 2 & 3 only
    B. 1 & 3 only   D. 1, 2 & 3

15. Anaphylaxis is a complex allergic type reaction that is mediated by the release of histamine by the:

    A. Immune system   C. Exocrine system
    B. Biliary system   D. Urinary system

16. The use of sodium diatrizoate (Hypaque) is NOT considered safe for _____ injections.

    A. Intravenous   C. Intrathecal
    B. Intramuscular   D. Intra-arterial

17. A contrast media that absorbs more radiation than the organ in which it is placed, is termed a:

    A. Negative agent   C. Suspension agent
    B. Positive agent   D. Miscible agent

18. An allergic reaction to iodinated contrast media is more frequent in patients known to have:

    1. Diabetes mellitus   *2 Multiple myeloma*   *3. Asthma*

    A. 1 & 2 only   C. 2 & 3 only
    B. 1 & 3 only   D. 1, 2 & 3

19. The signs of flushing, sneezing and urticaria from an iodine contrast media injection are some of the warning signs of:

    A. Prophylaxis   C. Extravasation
    B. Iodine coma   D. Anaphylaxis

20. An intravenous urogram patient has an anaphylactic shock reaction The drug that is most likely to be administered first from the emergency tray is:

    A. Heparin   C. Antibiotics
    B. Epinephrine   D. Atropine

21. The introduction of a small dose of contrast media (sensitivity testing) before full dosage is given, has proven to be _____ in predicting reactions.

    A. Totally reliable   C. Somewhat reliable
    B. Very reliable   D. Very unreliable

22. The most commonly employed contrast media for intravenous cholangiography is:

A. Iopanoic acid
B. Iodipamide meglumine

C. Iophendylate
D. Diatrizoate meglumine

23. Which of the following contrast media is classified as a negative contrast media?

A. Room air
B. Iopanoic acid

C. Sodium diatrizoate
D. Iodipamide meglumine

24. Barium sulfate suspensions are nearly totally safe if they remain within the G.I. tract. Their toxicity increases dramatically if spillage occurs into the:

1. Duodenum   2. Peritoneum   3. Circulatory system

A. 1 & 2 only
B. 1 & 3 only

C. 2 & 3 only
D. 1, 2 & 3

25. In infants with swallowing disorders and adults with suspected bronchoesophageal fistulas, it is often prudent to select a/an:

A. Water-soluble iodine contrast media
B. Water-based barium contrast media

C. Oil-based iodine contrast media
D. Oil-based barium sulfate contrast media

26. Iopanoic acid, sodium tyropanoate, and sodium ipodate are among the most common agents used for:

A. Oral cholecystography
B. Intravenous urography

C. Angiography
D. Myelography

Match the radiographic procedure with the class of contrast media most often selected to improve visualization for questions 27-35.

27. _____ Arthrography
28. _____ Sialography
29. _____ Esophagraphy
30. _____ Cholangiography
31. _____ Urography
32. _____ Mammography
33. _____ Hysterosalpingography
34. _____ Cholecystography
35. _____ Venography

A. Ionic water-based iodine contrast media
B. Barium sulfate contrast media
C. Oil-based iodine contrast media
D. No contrast media

36. The principal advantage of a low osmolality non-ionic contrast agent is its decreased:

A. Opacity
B. Toxicity

C. Miscibility
D. Viscosity

37. A patient is to be scheduled for a barium enema, intravenous urogram and a gastrointestinal series. The sequence for these exams is:

A. IVP, BE, GI
B. BE, IVP, GI

C. GI, BE, IVP
D. IVP, GI, BE

38. As the concentration of an iodine contrast media is increased, an increase in _____ should be anticipated.

1. Viscosity   2. Toxicity   3. Opacity

A. 1 & 2 only
B. 1 & 3 only

C. 2 & 3 only
D. 1, 2 & 3

39. A patient is to be scheduled for the following series of radiographic examinations. The proper sequence for these exams from first to last is:

1. GI series   **2. Gallbladder**   **3. Lumbar spine**

A. 3, 2, 1   C. 2, 3, 1
B. 1, 2, 3   D. 1, 3, 2

40. Because of the possibility of residual contrast media, which of the following studies is normally performed last in a sequence of exams?

A. Esophagram   C. Intravenous urography
B. Barium enema   D. Gallbladder series

41. A 4-year old is brought down to the Radiology Department for an intravenous urogram. The radiologist orders 40% of the normal 50 cc adult dosage be used. The amount to be prepared is:

A. 45 cc   C. 30 cc
B. 38 cc   D. 20 cc

42. A barium enema is to be performed on a patient suspected of having a perforated bowel. The contrast media preferred for this procedure would be a/an:

A. High-density barium sulfate preparation   C. Water-based iodine preparation
B. Low density barium sulfate preparation   D. Oil-based iodine preparation

43. A low osmolality contrast agent is able to deliver a relatively _____ concentration of iodine with _____ particles in solution than conventional ionic contrast agents.

A. High, fewer   C. High, more
B. Low, more   D. Low, fewer

44. During radiographic exams requiring long intravenous infusions, a/an _____ is often preferred over typical intravenous needles.

A. Angiographic catheter   C. Seldinger technique
B. Butterfly set   D. Transdermal injection

45. Most of the new low osmolality contrast agents which are classified as multipurpose can be employed for:

1. Arthrography   **2. Angiography**   **3. Urography**

A. 1 only   C. 3 only
B. 2 only   D. 1, 2 & 3

46. A commonly employed category of drugs that are used to promote defecation in patients are termed:

A. Emetics   C. Diuretics
B. Cathartics   D. Stimulants

47. The principal reason that iopamidol and iohexol are now beginning to replace meglumine salts as the primary agents for intravenous urography is their:

1. Lowered iodide content   **2. Lower toxicity**   **3. Lower cost**

A. 1 only   C. 3 only
B. 2 only   D. 1, 2 & 3

48. In order to prevent possible damage to the rectum, the enema tip should never be inserted more than:

A. 1" (2.5 cm) into the rectum   C. 4" (10 cm) into the rectum
B. 2" (5 cm) into the rectum

49. Which of the following is a major compound used in the formulation of a low osmolality (non-ionic) contrast agent?

1. Iohexol   2. Iopamidol   3. Sodium meglumine

A. 1 & 2 only
B. 1 & 3 only

C. 2 & 3 only
D. 1, 2 & 3

50. A serious complication following the administration of an iodine contrast agent is development of a moving blood clot that obstructs the blood supply to the lung. This is called a _____.

A. Phlebolith
B. Pulmonary edema

C. Pulmonary embolus
D. Pericardial effusion

51. The contrast agent that does NOT dissociate into charged particles when placed in a solution is designated as a/an:

A. High osmolar contrast agent
B. Ionic contrast agent

C. Anaphylactic contrast agent
D. Non-ionic contrast agent

52. The cleansing of the bowel after a barium enema to avoid barium impactions can be accomplished by the use of:

1. Milk of Magnesia   2. Benadryl   3. Castor oil

A. 1 & 2 only
B. 1 & 3 only

C. 2 & 3 only
D. 1, 2 & 3

53. A common medication given for relief of bronchial spasm which occasionally accompanies iodine contrast injections is:

A. Nitroglycerin
B. Heparin

C. Benadryl
D. Spirits of ammonia

54. A male infant is brought to the department with a diagnosis of pyloric stenosis. The examination that will most likely confirm this condition is a/an:

A. Intravenous urogram
B. Cholangiography

C. Gastrointestinal series
D. Barium enema

55. The types of contrast media that can be employed for use in the subarachnoid space of the spinal canal are:

1. Oil-based iodides   2. Ionic water-based iodides   3. Non-ionic water-based iodides

A. 1 & 2 only
B. 1 & 3 only

C. 2 & 3 only
D. 1, 2 & 3

56. Before a contrast media is injected, the patient should be questioned about his/her:

1. History of allergies   2. Alcohol consumption   3. Smoking habits

A. 1 only
B. 2 only

C. 3 only
D. 1, 2 & 3

57. One of the major danger signs after administration of iodine contrast media is:

A. Nausea
B. Urticaria

C. Dysphasia
D. Sneezing

58. When iodinated contrast media is employed, the maximum kilovoltage that should be used is:

A. 60 kVp
B. 80 kVp

C. 95 kVp
D. 110 kVp

59. The principal route of elimination of most aqueous iodine contrast agents is the:

A. G.I. tract
B. Respiratory system
C. Biliary tract
D. Urinary tract

60. The major property that determines the x-ray attenuation properties of the contrast media is its:

A. Atomic number
B. Valence
C. Viscosity
D. Concentration

61. The resistance of a fluid to change its form or position due to the molecular cohesion is termed:

A. Viscosity
B. Miscibility
C. Toxicity
D. Osmolality

62. A severe allergic reaction to a chemical substance (contrast media) may result in:

A. Homeostatic shock
B. Anaphylactic shock
C. Histotoxemia
D. Toxonosis

63. A serious drug reaction when water-based iodine contrast is administered intravenously occurs in:

A. Less than 1% of the patients
B. About 2% of the patients
C. About 30% of the patients
D. About 50% of the patients

64. Which of the following contrast media is classified as a type of colloidal suspension?

A. Sodium diatrizoate
B. Iopanoic acid
C. Nitrous oxide
D. Barium sulfate

65. When a low atomic number material is used to demonstrate the density difference between two structures, it is termed a _____ contrast media.

A. Negative
B. Positive
C. Saline
D. Hypotonic

66. The type of contrast media most commonly employed for intravenous administration is:

A. Water-based barium solutions
B. Water-based iodine solutions
C. Oil-based iodine solutions
D. Gaseous solutions

67. The type and severity of complications associated with contrast media is related to:

1. Route of administration   *2. Amount used*   *3. Patient's sensitivity*

A. 1 only
B. 2 only
C. 3 only
D. 1, 2 & 3

68. The extent to which a material is found to be poisonous is termed:

A. Toxicity
B. Miscibility
C. Contraindication
D. Persistence

69. A serious complication often associated with intravenous administration of iodinated contrast media is:

1. Renal Failure   *2. Hypotension*   *3. Impactions*

A. 1 & 2 only
B. 1 & 3 only
C. 2 & 3 only
D. 1, 2 & 3

70. Iodized oil, such as Pantopaque, is most often employed in structures that are extremely sensitive to:

A. Blood clotting
B. Dehydration
C. Irritation
D. Infection

Explanations

1. B The common rate for the delivery of nearly all medications during an IV infusion is about 16-20 drops per minute. A rate above this amount has the potential to infuse the fluids to fast resulting in the excessive spillage of fluids into the lungs called pulmonary edema.

2. D A diuretic either increases the rate at which blood components are filtered across the glomerulus or decreases the reabsorption by the renal tubules. Both will result in an increase in the amount of urine passing into the urinary bladder.

3. A Two common tests are performed to indicate the status of the kidney. A blood urea nitrogen (BUN) above 25 mg/dl or a creatinine level above 1.5 mg/dl are indicators that the kidney is not functioning within normal limits. A patient with kidneys that are not functioning within normal limits is at a much greater risk for nephrotoxicity and or other types of adverse reactions.

4. A Diazepam (Valium) and Midazolam (Versed) are among the most common agents used to relax patients prior to invasive procedures. Since both of these can alter the state of consciousness of the patient, these patients will need to be observed until the medications have worn off.

5. A In order for the gall bladder to fill up with a contrast media it must first be extracted from the blood stream by the liver and be combined with other the secretions of the liver. After this the bile and contrast agents can pass into the biliary tract where it can be directed into the gall bladder.

6. D A small gauge needle that is used for subcutaneous injections should be directed at the skin at an angle of about 45 degrees.

7. B One of the possible adverse reactions to barium contrast agents are impactions in the intestinal tract. These are much more common in the elderly. This can be prevented by the use of laxatives or cathartics following a GI series or barium enema.

8. D The sensitivity of an individual to any drug or contrast agent is dependent upon the weight, age, sex and general state of health of the individual. Other factors such as a previous history of allergies, anxiety and pervious reactions to a drug; all of these can influence the potential effect of a medication.

9. D The five rights of drugs administrations include: 1. Right drug 2. Time of the administration 3. Right patient 4. Right dosage 5. Route of administration.

10. C The vasovagal and/or vasomotor response that leads to hypertension and severe bradycardia are most often triggered by fear and anxiety and not the direct effects of the contrast agent.

11. A The amount of time a medication remains in the body is called its persistence. Since this will be effected by the body's ability to eliminate substance any failure of the kidneys or liver can significantly alter the rate at which a drug is eliminated.

12. A The normal salt (NaCl) concentration of the blood is about the same as that of seawater or about .9% NaCl. Any solution that contains this much salt is referred to as being isotonic.

13. B There are two general classifications for getting materials into the blood stream. Enteral administrations involve the absorption of medications through the mucus membranes that line the mouth, stomach, small, and large intestines. Parenteral administrations are those that involve the placement of the agent into the tissues or into the blood stream (injections).

14. C Both previous reactions to a medication and severe renal disease can increase the risk of a patient for a some kind of adverse reaction

15. A Anaphylaxis is a life-threatening reaction that is normally triggered by the release of histamines by the tissues of the immune system. This can lead to respiratory distress bronchospasms, laryngeal edema and a host of other symptoms.

16. C One of the first water-based ionic contrast agents called sodium diatrizoate (Hypaque) was found to be unsuitable for intrathecal or subarachnoid space injections.

17. B Iodine and barium sulfate, which both have a high atomic number, attenuate large amounts of x-radiation and are the most commonly used materials in positive contrast agents.

18. C A number of pre-existing conditions may predispose an individual to contrast reactions. Among the most notable are multiple myeloma, sickle cell disease, asthma, and thyroid disorders.

19. D An allergic reaction to contrast agents is called anaphylaxis and is classified as mild or severe; the most common mild reactions are nausea, vomiting, hives, and sneezing.

20. B The most important medication for the relief of the symptoms associated with a severe contrast reaction is epinephrine. Also known as adrenalin, this drug relieves bronchospasms, raises the blood pressure and increases the heart rate.

21. D The use of a sensitivity test prior to contrast administration has proven to be very unreliable and is rarely used in a modern radiology department. Tests included ½ cc subdermal injections (skin pop), ½ cc IV, or a single drop placed in the eye. All tests were subject to a high false positive response rate.

22. B In order to demonstrate the biliary tract, the contrast media must be eliminated by the liver. Iodipamide (Cholografin) is carried to the liver in the blood stream and is rapidly secreted by the liver. Meglumine serves as the carrier salt.

23. A The most commonly used negative contrast agent is room air.

24. C Though barium sulfate is extremely safe, when it is introduced into the gastrointestinal tract it becomes highly toxic if a leakage occurs into the peritoneum (intestinal perforations) or if injected into the blood stream.

25. C An abnormal connection between the upper GI tract and the respiratory system, such as a fistula, may allow swallowed barium to collect in the lungs. In these situations, an oil-based iodinated agent may be the contrast media of choice.

26. A Oral cholecystography requires a type of water-based iodide which is absorbed by the GI tract and eliminated by the liver. Two most common compounds for this purpose are iopanoic acid (Telepaque) or sodium iodate (Oragrafin).

27. A   28. C   29. B   30. A   31. A

32. D   33. A   34. A   35. A

36. B Osmolality is the degree to which fluid will be moved in and out of a cell. Contrast agents possessing a low osmolality tend to cause fewer contrast reactions and therefore have a lower toxicity.

37. A Proper exam sequencing is beneficial in performing a series of tests in the least amount of time and without compromising subsequent exams due to retained media. General guidelines recommend that non-contrast exams be performed first, followed by iodide studies, then lower GI tract barium studies, and lastly upper GI tract barium studies.

38. D As the concentration or weight per unit volume of a contrast media is increased, the viscosity and opacity of the media also increases.

39. A See explanation number 37 of this section.

40. A The esophagram is performed last in a series of exams because it may take up to 2 days for the barium to be completely eliminated from the GI tract.

41. D The new dose is obtained by multiplying the adult dose by 40%.

42. C Because of the high toxicity of barium entering the peritoneum, a water-based iodinated media such as a diatrizoate meglumine solution (Gastrografin) is strongly recommended.

43. A   Since there is little movement of fluids when a low osmolar agent is used, a higher concentration with more particles in the solution can be introduced.

44. B   In order to make a patient more comfortable during long intravenous infusions, a needle with flat wings (butterfly) or a flexible angio-catheter may be placed into the vein.

45. D   A multi-purpose contrast agent, as its name implies, can be used for numerous exams including angiography, urography, arthrography, and myelography.

46. B   One of the common ways to eliminate unwanted substances from the lower gastrointestinal tract is by the use of cathartics or laxatives.

47. B   The anal canal, in which the enema tip is placed, extends into the rectum. To avoid injury, the maximum safe distance for the tube insertion is about 4 inches.

48. C   The new general purpose non-ionic contrast agents, though about 10 times more expensive than the ionics they replace, have much lower toxicity.

49. A   The vast majority of the water-based non-ionic contrast media employ iohexol or iopamidol as their active components.

50. C   One of the most serious complications associated with a contrast media reaction is the formation of a pulmonary embolus blood clot in the lung which can result in death in just a few minutes.

51. D   The principal difference between ionic and non-ionic contrast media is related to the ability to form charged particles or ions. Non-ionics are chemically less reactive and therefore cause fewer harmful effects.

52. B   Geriatric patients may be less active and are often given laxatives such as castor oil or milk of magnesia to help eliminate barium following examination of the gastrointestinal tract.

53. C   Diphenhydramine (Benadryl) is commonly administered to block the action of histamine and relieve the upper respiratory constriction it may cause.

54. C   The constriction of the end of the stomach, pyloric stenosis, is best evaluated during an upper gastrointestinal series.

55. B   Until the advent of non-ionic contrast agents, subarachnoid or intrathecal injections were made with oil-based iodinated contrast agents such as Pantopaque. Ionic water-based iodides should never be employed in this region because they may cause meningitis.

56. A   Contrast sensitivity is dependent upon a number of patient factors including general health, age, weight, a previous history of allergies, and prior contrast reactions which may predispose an individual to an allergic reaction.

57. C   Among the more serious warning signs preceding a severe contrast reaction are dysphasia, laryngeal edema, and breathing difficulties.

58. B   The dilution of an iodide contrast media by the blood and the relatively low concentrations achieved in the target organs will require the use of a 60-75 kVp range to avoid overpenetration.

59. D   Nearly all of the currently employed aqueous ionic and non-ionic contrast agents are eliminated by way of the kidneys.

60. A   The degree of absorption or attenuation of a contrast agent is related to its atomic number. As the atomic number of a substance is increased its ability to absorb radiation also increases.

61. A   The rate of an injection is greatly effected by the ability of a liquid to flow, which is a measure of its viscosity. Highly concentrated contrast agents, as is often indicated by a higher number on the label of the media, are more viscous or thicker and less easily injected.

62. B   The most common type of allergic reactions seen in the radiology department are in response to iodinated contrast agents. Recently a number of cases of Latex reactions from enema retention tips have been reported.

63. A   Contrast reactions are reported in about 5% of all patients in which these agents are used. Serious or life threatening side effects occur in less than 1% of patients, with one death occurring in about every 40,000 patients.

64. D   The type of positive (radiopaque) contrast media used most frequently in the gastrointestinal tract is a white insoluble powdered compound called barium sulfate. When mixed with water it forms a type of colloidal suspension.

65. A   A number of gasses such as air, carbon dioxide, and nitrous oxide are employed as negative or radiolucent contrast agents.

66. B   The only type of contrast media suitable for intravenous uses are the water-based (aqueous) iodinated contrast agents. Barium sulfate and oil-based iodinated compounds are not readily absorbed by the tissues and will cause embolic formations in the lungs if introduced into the blood stream.

67. D   The safety of a contrast media is related to the patient's physical condition, type and amount of the agent used, as well as the route by which the agent is introduced into the body.

68. A   The toxicity or extent to which media is poisonous is principally related to the concentration it achieves in the body.

69. A   The administration of iodinated contrast agents may cause a number of serious side effects including shock, coma, renal failure, respiratory failure, and cardiac irregularities.

70. C   Before the development of the less reactive non-ionic contrast agents, a major drawback of ionic media was the high degree of irritation they caused. When imaging of sensitive organs or cavities was required, such as the lungs or subarachnoid space, oil-based mixtures were employed to overcome this difficulty.

1. Legal drugs that can be obtained by prescription but have a high potential for abuse such as morphine are classified as:

   A. Schedule C-I drugs
   B. Schedule C-II drugs
   C. Schedule C-III drugs
   D. Schedule C-IV drugs

2. The most important legal medical record which contains a history of the events occurring under the supervision of medical professions is called a:

   A. Medical chart
   B. Medical requisition
   C. Living will
   D. Informed consent form

3. The dose of a pharmaceutical that is found to be lethal to 50% of the animals tested is known as the median lethal dose. This value is used in the determination of the drugs:

   A. Affinity
   B. Toxicity
   C. Efficacy
   D. Abuse potential

4. The ratio of the median lethal dose to the median effective dose is termed the:

   A. Abuse index
   B. Toxicity index
   C. Safety index
   D. Therapeutic index

5. The physiochemical process that is required for a drug to reach the site of action is termed:

   A. Transportation
   B. Metabolism
   C. Dissolution
   D. Biopharmaceutics

6. Any substance that is mixed with the active agent to assist the drug in reaching its site of action is termed a/an:

   A. Infuser
   B. Suspender
   C. Vehicle
   D. Compressor

7. Which of the following dosage forms is normally associated with highest speed of absorption into the body?

   A. Solutions
   B. Compressed tablets
   C. Capsules
   D. Enteric coated tablets

8. The most commonly employed gas propellents used to help disperse liquified medications in aerosols is/are:

   A. Calcium alginate
   B. Fluorinated hydrocarbons
   C. Oxygen gas
   D. Methylcellulose

9. In order for a solid or semisolid drug to be absorbed across a cellular membrane it must be converted into a/an:

   A. Compressed form
   B. Emulsion form
   C. Effervescent form
   D. Solution form

10. All of the following are portions of the pharmacokinetic process EXCEPT:

    A. Absorption
    B. Metabolism
    C. Distribution
    D. Antagonism

11. Which is NOT one of the principal systems or organs that can be used for the systemic absorption of pharmaceuticals?

    A. Gastrointestinal tract
    B. Central nervous system
    C. Mucus membranes
    D. Subcutaneous tissues

12. The absorption properties of a pharmaceutical are dependent upon all the following EXCEPT:

A. The amount of blood flow to the region
B. The lipophilicity of the drug
C. The concentration of the agent at the site
D. The amount of fat at the site

13. Which of the following conditions is often associated with a reduction in the blood flow and the alteration of the absorption profile of a drug?

A. Dehydration of the tissues
B. Hyperthermia
C. Cardiovascular shock
D. Anemia

14. The most common process by which drugs traverse the cellular membranes is called:

A. Passive diffusion
B. Reverse absorption
C. Selective transport
D. Dissolution

15. A drug that is classified as a weak acid is more rapidly absorbed in an environment which is:

A. Neutrally charged
B. Highly acidic
C. Highly alkaline
D. Nonionized

16. The cell membrane is normally composed of a:

A. Single layer of columnar cells
B. Single layer of mucoids
C. Multiple layers of squamous cells
D. Double layer of lipids

17. The process by which a pharmaceutical is transported from the blood stream to the various tissues of the body is termed:

A. Compatibility
B. Elimination
C. Diffusion
D. Distribution

18. The distribution of a pharmaceutical is related to all the following EXCEPT:

A. The permeability of the liver
B. The regional blood flow
C. The cardiac output
D. The number of drug reservoirs

19. Oxidation and conjugation are the two most important chemical reactions that take place in the liver to help in the:

A. Excretion of pharmaceuticals from the body
B. Absorption of pharmaceuticals into the blood stream
C. Passage of pharmaceuticals across the blood brain barrier
D. Accumulation of pharmaceuticals in the body fluids

20. The prolonged action of a pharmaceutical due to insufficient metabolism is related to all the following EXCEPT:

A. Liver disease
B. Renal failure
C. Cardiovascular disease
D. Excessive hydration

21. The symptoms of a drug overdose even though a normal dose has been given is most often the result of:

A. Delayed metabolism
B. Excessive cardiac output
C. Excessive renal function
D. Insufficient tissue concentration

22. The study of the chemical or physical effects that a drug will have on a living system is called:

A. Physiology
B. Pharmacology
C. Toxicology
D. Etiology

23. The physical distribution of a pharmaceutical to the site of its primary action is primarily accomplished by the:

A. Lymphatic system
B. Nervous system
C. Circulatory system
D. Alimentary tract

24. The name given by the manufacturer to readily identify their particular formulation of a drug is called the:

A. Brand name
B. Chemical name
C. Generic name
D. Family name

25. According to present laws, the label on a prescription drug must contain all of the following EXCEPT:

A. Prescribing physician name
B. Inactive ingredients in the preparation
C. Brand name of the drug
D. Usual dosage of the drug

26. According to the controlled substance act, the classification of a drug is based on its:

1. Abuse potential   2. Established medical uses   3. Possibility of dependence

A. 1 only
B. 2 only
C. 3 only
D. 1, 2 & 3

27. A metabolite is best defined as a:

A. Large protein molecule that develops into a toxic agent
B. Chemical fragment that has no pharmacologic activity
C. Large chemical fragment that speeds up a chemical reaction
D. Chemically altered drug that can be eliminated from the body

28. Drugs that alter sensory or psychological processes will most often act upon the:

A. Neural synapse or neuromuscular junction
B. Myelin sheath
C. Plasma membrane
D. Blood-brain barrier

29. A medication such as Rolaids or Tums that reduces the acidity of the stomach alters a chemical process in which of the following ways?

A. It chemically alters the cellular membrane
B. It chemically alters the binding sites on the tissue receptors
C. It chemically alters the fluids in the body
D. It physically alters the cell membrane

30. The ability of a receptor to form a chemical bond that results in a chemical reaction is called:

A. Affinity
B. Efficacy
C. Compatibility
D. Synergy

31. The principal organ involved with the elimination of most metabolites and drugs is the:

A. Gallbladder
B. Colon
C. Kidney
D. Spleen

32. The majority of pharmaceuticals used today are classified as:

A. Natural drugs
B. Synthetic drugs
C. Inert drugs
D. Herbal drugs

33. Aspirin is an example of a:

A. Brand name
B. Chemical name
C. Family name
D. Generic name

34. A drug that has undergone clinical testing to prove its safety and effectiveness and requires a written order for its use is called a:

A. Practice drug
B. Stable drug
C. Prescription drug
D. Over the counter drug

35. Many types of tablets have a special enteric coating to enable the:

A. Rapid absorption of the drug by the stomach
B. Drug's passage through the highly acidic stomach
C. Drug to breakdown in the mouth
D. Drug to be absorbed by the colon

36. A suppository may be administered in all of the following ways EXCEPT:

A. Inhalation
B. Rectally

C. Urethrally
D. Vaginally

37. Which of the following drugs is often administered transdermally?

A. Demerol
B. Nicotine

C. Butyl nitrate
D. Nitroglycerine

38. The process by which a drug moves from the outside of the body into the central compartment or vascular bed is termed:

A. Biotransformation
B. Elimination

C. Absorption
D. Metabolism

39. When a drug needs a carrier molecule to assist with its movement in the cell across a concentration barrier it is classified as a form of:

A. Passive transport
B. Active transport

C. Entry transport
D. Rapid transport

40. Drugs whose actions are intended to be limited to a small region are referred to as:

A. Local administrations
B. Systemic administrations

C. Parenteral administrations
D. Enteral administrations

41. Two or more drugs that act together to produce an effect that is greater than either drug alone is termed a:

A. Synchronous effect
B. Synergistic effect

C. Antagonistic effect
D. Inhibitory effect

42. Which of the following is NOT one of the reasons for using smaller dosages for a newborn infant?

A. A newborn has a small body mass
B. A newborn's liver is not fully functional

C. A newborn has a lower respiration rate
D. A newborn's kidneys are not fully functional

43. Which of the following organs are associated with the elimination of the waste products from the body:?

1. The skin   **2. The liver**   **3. The kidney**   **4. The lungs**

A. 1 & 2 only
B. 2 & 3 only

C. 1, 2 & 3 only
D. 1, 2, 3 & 4

44-47. Place the number that corresponds to the following regions on the dose response curve.

44. ____ Biologic response axis
45. ____ Plateau region
46. ____ Relative dose axis
47. ____ Threshold region

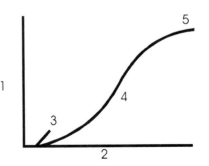

48. Pharmaceuticals that are metabolized by the liver are passed into the bile and eliminated:

A. In the feces
B. By the lymphatic system
C. In the urine
D. By the saliva

49. The pharmaceutical half-life is most often used in the evaluation of a drug's:

A. Toxicity
B. Elimination rate
C. Threshold dosage
D. Persistence

Match the following curves to the type of administration.

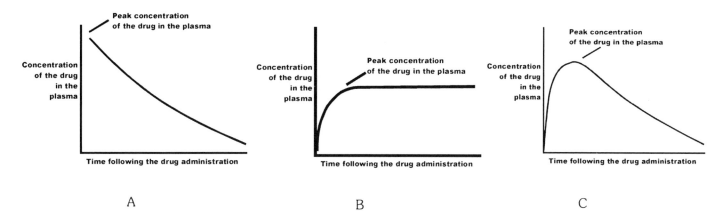

A   B   C

50. ___ Bolus injection

51. ___ Oral administration

52. ___ Intravenous infusion

53. The study of the effect of a drug and its manifestations is termed:

A. Pharmacology
B. Biopharmaceutics
C. Pharmacodynamics
D. Pharmacokinetics

54. Pharmacodynamics includes the evaluation of the following aspects of the drug EXCEPT:

A. Its side effects
B. Its adverse effects
C. Its toxicity
D. Its composition

55. The active biologic sites located on the surface of the cell to which a pharmaceutical can attach are termed:

A. Antagonists
B. Receptors
C. Predictors
D. Indices

56. The ability of a pharmaceutical to attach to a specific receptor on the cell membrane is termed its:

A. Affinity
B. Potency
C. Flexibility
D. Susceptibility

57. A drug or substance that has an affinity for a specific receptor and is able to enhance the body's natural response is termed:

A. A stimulant
B. A dimorphic
C. An enzyme
D. An agonist

58. The bronchodilation resulting from the action of the drug epinephrine at the beta receptors in the lung is an example of a/an:

A. Antagonist effect
B. Allergic effect
C. Synergistic effect
D. Agonistic effect

59. A drug or substance that blocks the receptor sites on the cell membrane and prevents the normal biologic function is termed a/an:

A. Antagonist
B. Facilitator
C. Stimulator
D. Diffuser

60. The degree to which a pharmaceutical is able to produce the desired effect in the target tissue is known as the drug's:

A. Opacity
B. Efficacy
C. Persistence
D. Miscibility

61. The time it takes for a drug or substance to be reduced to 50% of its peak serum concentration in the blood, is called its:

A. Half-dose level
B. Semiluxation level
C. Half-action level
D. Biologic half-life

62. Any action of a pharmaceutical other than that for which the drug was intended is termed a/an:

A. Step effect
B. Adverse effect
C. Side effect
D. Toxic effect

63. The amount of time a pharmaceutical preparation remains active within the body is termed its:

A. Osmolality
B. Persistence
C. Viscosity
D. Miscibility

64. Which of the following routes of administration allows for the easiest retrieval of a drug in case of an overdose?

A. Intrathecal route
B. Intradermal route
C. Parenteral route
D. Oral route

65. Following the oral administration of a drug, all of the following methods can be used to retrieve the drug during an overdose EXCEPT:

A. Gastric lavage
B. Induced vomiting
C. Hemolysis
D. Introduction of charcoal

Explanations

1. C In order to help prevent the distribution of medications that have high potential for abuse, the government has developed five schedules for controlled substances. Drugs listed in the C-I schedule have a high abuse potential and no currently accepted medical use. Examples of schedule C-I drugs include heroin, opium, LSD, crack cocaine and marijuana. Schedule C-II drugs have a high abuse potential and accepted medical uses when given under the supervision of a physician. Examples of schedule C-II medications include narcotics and sedatives such as morphine, codeine, and other opiate analogs. Drugs listed in schedule C-III have an abuse potential and accepted medical uses but cause a moderate to low psychological or physical dependence. Examples of C-III drugs include acetaminophen with codeine. Schedule C-IV medications have a low abuse potential and a are not likely to cause dependence. Examples of C-IV medications include diazepam and chloral hydrate. Schedule C-V medications have a low abuse potential and accepted medical uses and are associated with a limited psychological or physical dependance. Examples of schedule C-V drugs include codeine cough syrups.

2. A The most important medical record maintained for an individual is the patient's medical chart. This document should contain the patient's medical history and a complete history of all of the treatments and patient responses observed under the supervision of the medical personnel.

3. B The toxicity of a pharmaceutical is most often determined by the lethal dose of the drug in 50% of the animals given a comparative dose to 70 kilogram adult. This value is termed the median lethal dose.

4. D The effective dose of a medication relative to its toxicity is given by a value called the therapeutic index. This is based on the ratio of the median lethal dose and the median effective dose of the medication. Pharmaceuticals which have a high therapeutic ratio such as aspirin, are generally considered both safe and effective. Special care must be given during the administration of chemotherapeutic drugs which have an extremely low therapeutic index.

5. D The area of pharmacology that deals with the physiochemical processes involved with moving a drug to the site of its action is called biopharmaceutics.

6. C In order to move most pharmaceuticals to the site of their action will require the compound to be formulated to a carrier substance or vehicle to assist with its delivery into the body.

7. A A drug that has been combined with the appropriate vehicle is called the dosage form. Common dosages forms include tablets, capsules, troches and solutions. Since solutions are most often given via injections these are normally associated with the high rate of absorption into the body.

8. B Though many types of gas dosage forms such as oxygen and anesthetic agents require no vehicle, most aerosol inhalers require a propellant such as fluorinated hydrocarbon gas.

9. D Since the body is capable of absorbing only drugs which are in either liquid or gaseous state, all solids and semisolids must be converted into a solution in order to be absorbed across a cellular membrane.

10. D Pharmacokinetics describes the various processes by which a drug is absorbed, distributed, metabolized and eliminated from the body.

11. B The vast majority of medications that are defined as systemic are absorbed through the skin or the mucosal membranes of the gastrointestinal and respiratory tracts. Medications that are administered directly into the blood stream via intravenous injection bypass the need to be absorbed.

12. D The absorption of a systemic medication through the skin or mucus membranes is related to a number of physiochemical properties. This includes the surface of the absorbing tissue, the amount of blood flow to the absorbing tissues, the lipophilicity of the agent, the concentration of the agent at the surface of the absorbing tissues and the acid-base properties of the tissues.

13. C   One of the most important factors in the absorption profile of a medication is the supply of blood to the tissues involved with the absorption of the agent. Patients that have an altered blood flow due to cardiovascular shock will not be able to absorb medications at the normal rate. The use of intravenous route of administration is normally preferred when the absorption process may be compromised.

14. A   In order for a solution containing a medication to pass through the membrane of a cell it must either be encouraged to enter the cell by having a higher concentration outside the cell membrane or be actively transported into the cell by carrier molecules. The movement of substances from a higher to a lower concentration across a cellular membrane is termed passive diffusion. Medications that are assisted in their passage through a cellular membrane by carrier molecules is defined as a form of active transport.

15. B   In general, medications which have no charge or are nonionic will more easily pass through a cellular membrane than a charged or ionic compound. Since most pharmaceuticals are either weak acids or bases, their entry across the cellular membrane may be encouraged by the environment in which they are placed. Most drugs which are weak acids; i.e. aspirin, are more easily transported through the gastric mucosa because the environment of the stomach is highly acidic. A weak acid in an acidic solution appears to be nonionic and therefore will more readily pass through the membrane than a weak alkaline compound. The higher alkalinity of the duodenum makes this area more suitable for the absorption of medications that are weak bases.

16. D   The membranes of the cells are composed of a double layer of lipids. In order for a medication to enter the cell it must be able to cross through these layers. For this reason, drugs that have a good lipophilicity and are soluble in lipids (fats) will pass though the cell membrane more readily than pure water soluble compounds.

17. D   The process by which a compound is transported through the blood stream to the various tissues of the body is termed distribution.

18. A   The distribution of a pharmaceutical is related to the volume of blood that is pumped through the heart (cardiac output), the blood flow to the individual tissues (regional blood flow), and the amount of the drug that becomes bound to plasma or other tissues prior to being delivered to the cellular membrane (concentration). In order to protect the sensitive tissues of the brain and the fetus, the body has developed special barriers. The blood-brain barrier and the placental barrier are two sets of tissues that have a selective permeability to many compounds that enter the blood stream.

19. A   In theory, the time a pharmaceutical remains active is virtually unlimited unless it is chemically altered by the body. Since many compounds are not safe when present for long periods of time, it is important that the body has a means for converting medications into a form that will allow for their elimination from the body. The principal organ involved with the chemical alteration of a pharmaceutical is the liver. Metabolism or biotransformation is the general term that is applied to the process by which drugs are altered to facilitate their removal from the body. A drug reaching the liver will normally undergo oxidation and/or conjugation to reduce the drug to a less active form (metabolites) that can be readily eliminated from the system. In addition to the liver, the kidneys, blood plasma and enzymes of the intestinal mucosa, all play a role in the alteration of pharmaceuticals.

20. D   Because there are a number of organs or components that play a role in the alteration of a medication, a number of physical factors can alter the body's ability to eliminate a compound. Among these are severe liver and kidney disease, severe cardiovascular disease, and immature or degenerating enzyme functions.

21. A   The delayed metabolism due to the inability of the liver and kidneys to eliminate medications from the body in the usual time will often result in an accidental overdose of a medication even when the usual dose levels are observed.

22. B   The study of the chemical or physical effects a drug will have on living systems is termed pharmacology.

23. C   Once a pharmaceutical is absorbed into the blood stream its physical distribution to the site of its action is accomplished by the circulatory system.

24. A   Pharmaceuticals are identified using three different names. The most precise but least meaningful for a medical professional is the chemical name. This long name provides the chemical formula for the active compound in the formulation. The most familiar of the names for a drug is the brand or trade name. To assist in the marketing of a drug the manufacturer will give the pharmaceutical a name under which it will be marketed.

Valium, Prozac, and Tylenol are three readily recognizable brand names. After the patent protection expires and the manufacturer no longer has exclusive rights to a particular drug it can be marketed under a generic name. For example Tylenol is the brand name for a compound which has the generic name, acetaminophen. Though most generic medications are virtually identical to their brand name formulations they can often be purchased at a lower cost.

25. B   Pharmaceuticals that can only be obtained by way of a prescription are termed legend drugs. In order to be valid a prescription must contain seven components. These include the name, address and/or the identification number for the patient, the brand or generic name of the drug, the proper dosage of the medication, the dosage form, the route of administration, the date of the order and the signature of the prescriber.

26. D   In order to help prevent the distribution of medications that have high potential for abuse, the government has developed five schedules for controlled substances. Drugs listed in the C-I schedule have a high abuse potential and no currently accepted medical use. Examples of schedule C-I drugs include heroin, opium, LSD, crack cocaine and marijuana. Schedule C-II drugs have a high abuse potential and accepted medical uses when given under the supervision of a physician. Examples of schedule C-II medications include narcotics and sedatives such as morphine, codeine, and other opiate analogs. Drugs listed in schedule C-III have an abuse potential and accepted medical uses but cause a moderate to low psychological or physical dependence. Examples of C-III drugs include acetaminophen with codeine. Schedule C-IV medications have a low abuse potential and a are not likely to cause dependence. Examples of C-IV medications include diazepam and chloral hydrate. Schedule C-V medications have a low abuse potential and accepted medical uses and are associated with a limited psychological or physical dependance. Examples of schedule C-V drugs include codeine cough syrups.

27. D   A drug reaching the liver will normally undergo oxidation and/or conjugation to reduce the drug to a less active form(metabolites) that can be readily eliminated from the system. In addition to the liver, the kidneys, blood plasma and enzymes of the intestinal mucosa all play a role in the alteration of pharmaceuticals.

28. A   The vast majority of pharmaceuticals that alter sensory or psychological processes of the body cause the chemical alteration of the neural synapses and/or the neuromuscular junctions. Depending on the chemical nature of the drug, the effects at these junctions can result in facilitation of the nerve impulses, the summation of the nerve impulses, or the partial or complete inhibition of the nerve impulses.

29. C   Pharmaceuticals are employed for one of six major reasons. These include the treatment of disease (antibiotics), the relief of pain (analgesics), the prevention of disease (vaccinations), the alteration of pain responses (anesthesia), maintenance of homeostasis (vitamins), and the diagnosis of a disease (contrast agents and radiopharmaceuticals)

30. A   The most common way for a pharmaceutical to exert its action on the cell is through the chemical bonding of the drug with the receptors on the cell membrane. When there is a compatibility of the drug with a receptor there is said to be an affinity. Regardless of the affinity for a drug, the agent may be able to cause a profound effect on the cell. This is termed its efficacy. Drugs that attach to specific bonding sites that elicit a strong biologic response in a cell are called synergists. Drugs that block the receptor sites without eliciting a biologic response are termed antagonists.

31. C   The principal organ involved with the chemical alteration of a pharmaceutical is the liver. Metabolism or biotransformation is the general term that is applied to the process by which drugs are altered to facilitate their removal from the body. A drug reaching the liver will normally undergo oxidation and/or conjugation to reduce the drug to a less active form(metabolites) that can be readily eliminated from the system.

32. B Pharmaceuticals come from a number of different sources. All of the first pharmaceuticals were derived from natural sources such as botanical (plant), animal extracts, minerals, inorganic and organic chemicals. In recent years the majority of drugs that were derived from natural sources have been replaced with drugs that are synthetic or have been manufactured in the laboratory.

33. D After the patent protection expires and the manufacturer no longer has exclusive rights to a particular drug it can be marketed under a generic name. For example, Tylenol is the brand name for a compound which has the generic name, acetaminophen. Though most generic medications are virtually identical to their brand name formulations they can often be purchased at a lower cost. Aspirin is an example of a generic compound.

34. C The Pure Food, Drug and Cosmetic Act of 1906 was the country's first attempt to protect the public from adulterated or mislabeled drugs. In 1952, this Act was amended to designate certain medications as prescription drugs that required a written order from a physician. These drugs must also undergo a series of clinical trials that have shown the drug to be both safe and effective when given for its intended purpose.

35. B Many tablets are coated with a special layer of material to prevent the dissolution of the capsule in the stomach. These so-called enteric coated tablets are designed to pass though the stomach and absorbed in the small intestines to lower the potential of gastric upset.

36. A Suppositories or compressed inserts, are solid dosage forms which are placed in the rectum, vagina, nasal passages or urethra to deliver the medication directly to the mucus membranes. The solid medications are normally compounded with cocoa butter or some other compound that melts at the normal temperature of the body.

37. B One of the newer ways to deliver medications is through the placement of a patch on the skin surface (transdermal patch) that releases the pharmaceutical slowly over a long period of time. Nicotine, scopolamine (sea sickness) and more recently birth control hormones have been approved for delivery through the use of these patches.

38. C The passage of a pharmaceutical from outside the body into the vascular compartment or the blood stream is termed absorption.

39. B In order for a solution containing a medication to pass through the membrane of a cell, it must either be encouraged to enter the cell by having a higher concentration outside the cell membrane or be actively transported into the cell by carrier molecules. The movement of substances from a higher to a lower concentration across a cellular membrane is termed passive diffusion. Medications that are assisted in their passage through a cellular membrane by carrier molecules is defined as a form of active transport.

40. A A pharmaceutical whose action is limited to a small area such as a topical cream is classified as a form of local administration. Medications that enter the central compartment are termed systemic medications.

41. B Whenever two or more drugs are given during the period of their activity, a drug-to-drug interaction is possible. When this drug interaction results in a greater effect than is expected by either drug alone, the effect is termed a synergistic or additive agonistic effect. The drinking of alcohol in combination with many types of sedative drugs will often result in a potentially dangerous synergistic response. A drug interaction that involves the reduction of the action of either or both medications is termed an antagonistic response.

42. C There are a number of physical factors that affect a person's response to a pharmaceutical. These include the body mass, the age of the patient, status of the immune system, pathologies effecting the liver, kidneys or the heart, as well as some psychological factors. In addition to the lower body mass and smaller blood volume of a newborn infant, they may also have a reduced kidney and liver function which can prolong the elimination time of a medication.

43. D There are four main systems that are capable of the elimination of waste products from the body. These include the liver, the kidneys, the lungs and the skin.

44. 2   The action of a pharmaceutical can be graphically demonstrated by the use of a dose response curve. This curve is useful in the determination of the optimal dose for a medication. The dose axis labeled number 1 is plotted with the response axis labeled number 2. The portion of the resulting curve where the response is first detected is termed the threshold region and is labeled number 3. The relative dose region labeled number 4 shows the range during which the medication is considered effective. The plateau labeled number 5 represents the region in which a higher dose is unlikely to cause any beneficial effect. Dosages given above the plateau region are ineffective at increasing a response and may be toxic to the patient.

45. 5   See explanation 44 of this section.

46. 1   See explanation 44 of this section.

47. 3   See explanation 44 of this section.

48. A   After a pharmaceutical is metabolized by the liver, any of the reduced compounds (metabolites) may reenter the blood stream to be eliminated by the kidneys, lung, or skin. The metabolites that pass into the bile are passed into the gastrointestinal tract and are eliminated in the feces.

49. A.   The amount of time it takes to eliminate a pharmaceutical from the body is found by plotting the serum concentration of a drug with the time it takes to pass out of the body. By convention, the time it takes to reduce the serum concentration by 50% is termed the pharmacological or biologic half-life. This value is useful for the determination of the administration frequency and the toxicity of the medication.

50. A   The dose response curve for a pharmaceutical is closely related to the route by which the medication is administered. A bolus injection will result in a almost immediate response to the medication and a relatively short duration of its activity and is shown in curve A. An intravenous infusion is associated with a moderate response rate and a prolonged duration of the response and it shown in curve B. Oral administrations are characterized by a slow dose buildup and a relatively long duration of drug activity an is represented in curve C.

51. C   See explanation 50 of this section.

52. B   See explanation 50 of this section.

53. C   The study of the effects of a pharmaceutical in the body is termed pharmacodynamics. This area of study includes the mechanism and duration of the response, therapeutic effects, adverse and side effects, and drug toxicity.

54. D   See explanation 53 of this section.

55. B   The most common way for a pharmaceutical to exert its action on the cell is through the chemical bonding of the drug with the receptors on the cell membrane. When there is compatibility of the drug with a receptor there is said to be an affinity. Regardless of the affinity for a drug, the agent may be able to cause a profound effect on the cell. This is termed its efficacy. Drugs that attach to specific bonding sites that elicit a strong biologic response in a cell are called synergists. Drugs that block the receptor sites without eliciting a biologic response are termed antagonists.

56. A   When there is a compatibility of the drug with a specific receptor on the cell membrane there is said to be an affinity with the cell.

57. D   Drugs that have a strong affinity for a receptor and are able to stimulate or enhance the body's natural response to the pharmaceutical are termed agonists. The administration of a synthetic hormone such as insulin, is an example of an agonist compound.

58. D   Epinephrine is a pharmaceutical that can stimulate the beta receptors in the cells that line the lungs. This causes the dilation of the bronchial passages (bronchodilation). Since this action is identical to the effects of endogenous epinephrine, this is classified as an agonistic effect.

59. A   An agonist is a compound that is able to block the receptor sites on the cell membrane, thereby preventing the biologic effect that would have occurred if the compound were not present. The administration of Narcan to an individual who has received an overdose of opiates is used to block the opiate receptors and reverse the respiratory distress and hypotension that can be caused by these compounds.

60. B   Regardless of the affinity to a drug, the agent may be able to cause a profound effect on the cell. This is a termed its efficacy.

61. D   By convention, the time it takes to reduce the concentration by 50% is termed the pharmacological or biologic half-life. This value is useful for the determination of the administration frequency and the toxicity of the medication.

62. C   All prescription drugs must show that they are effective in providing the action for which they are given. Unfortunately, some of these compounds will also cause other actions which may be unintended called side effects. An unwanted side effect of a pharmaceutical is termed an adverse effect.

63. B   The amount of time during which a drug remains active following its administration is termed its persistence.

64. D   One of the main advantages of oral medications is the ability to retrieve the drug from the stomach or intestinal tract when a severe adverse effect is seen or is anticipated.

65. C   A patient that has taken an overdose of aspirin can be given a medication to induce vomiting or undergo gastric lavage to prevent the complete absorption of the compound that has been ingested. The administration of active charcoal and cathartics may also be helpful in absorbing or speeding the passage of toxic compounds from the body.

Millisecond Conversion

1.-20. State the decimal equivalent in seconds for the following millisecond values.

Example: 100 ms = .10 seconds

Solution: The decimal is moved three places to the left.

1. 800 ms = _____ s	8. 88 ms = _____ s	15. 7 ms = _____ s
2. 130 ms = _____ s	9. 24 ms = _____ s	16. 350 ms = _____ s
3. 80 ms = _____ s	10. 3 ms = _____ s	17. 175 ms = _____ s
4. 16 ms = _____ s	11. 540 ms = _____ s	18. 35 ms = _____ s
5. 5 ms = _____ s	12. 125 ms = _____ s	19. 45 ms = _____ s
6. 750 ms = _____ s	13. 18 ms = _____ s	20. 1 ms = _____ s
7. 200 ms = _____ s	14. 30 ms = _____ s	

21.-40. State the millisecond equivalent for the following decimal values.

Example: .020 seconds = 20 ms

Solution: The decimal is moved three places to the right.

21. .07 s = _____ ms	28. .095 s = _____ ms	35. .5 s = _____ ms
22. .001 s = _____ ms	29. .02 s = _____ ms	36. .61 s = _____ ms
23. .80 s = _____ ms	30. .06 s = _____ ms	37. .16 s = _____ ms
24. .03 s = _____ ms	31. .005 s = _____ ms	38. .345 s = _____ ms
25. .004 s = _____ ms	32. .09 s = _____ ms	39. .040 s = _____ ms
26. .65 s = _____ ms	33. .90 s = _____ ms	40. .021 s = _____ ms
27. .31 s = _____ ms	34. .17 s = _____ ms	

41.-50. State the decimal equivalent for the following fractional values.

Example: 1/2 = .5 seconds

Solution: $1 \div 2$

41. $\frac{1}{10}$ second = _____ seconds	46. $\frac{1}{120}$ second = _____ seconds
42. $\frac{1}{20}$ second = _____ seconds	47. $\frac{1}{240}$ second = _____ seconds
43. $\frac{1}{40}$ second = _____ seconds	48. $\frac{1}{480}$ second = _____ seconds
44. $\frac{1}{60}$ second = _____ seconds	49. $\frac{3}{10}$ second = _____ seconds
45. $\frac{1}{15}$ second = _____ seconds	50. $\frac{3}{4}$ second = _____ seconds

Milisecond Conversion

1.-20. State the decimal equivalent in seconds for the following millisecond values.

Example: 300 ms = .30 seconds

Solution: Decimal place moved three places to the left.

1. 600 ms = _____ s	8. 84 ms = _____ s	15. 8 ms = _____ s
2. 230 ms = _____ s	9. 26 ms = _____ s	16. 130 ms = _____ s
3. 180 ms = _____ s	10. 3 ms = _____ s	17. 275 ms = _____ s
4. 36 ms = _____ s	11. 340 ms = _____ s	18. 75 ms = _____ s
5. 7 ms = _____ s	12. 225 ms = _____ s	19. 35 ms = _____ s
6. 950 ms = _____ s	13. 28 ms = _____ s	20. 5 ms = _____ s
7. 150 ms = _____ s	14. 40 ms = _____ s	

21.-40. State the millisecond equivalent for the followind deicmal values.

Example: .400 seconds = 400ms

Solution: Decimal place moved three places to the right.

21. .09 s = _____ ms	28. .075 s = _____ ms	35. .6 s = _____ ms
22. .003 s = _____ ms	29. .04 s = _____ ms	36. .11 s = _____ ms
23. .50 s = _____ ms	30. .02 s = _____ ms	37. .12 s = _____ ms
24. .08 s = _____ ms	31. .001 s = _____ ms	38. .315 s = _____ ms
25. .007 s = _____ ms	32. .06 s = _____ ms	39. .048 s = _____ ms
26. .25 s = _____ ms	33. .10 s = _____ ms	40. .031 s = _____ ms
27. .53 s = _____ ms	34. .27 s = _____ ms	

41.-50. State the decimal equivalent for the following fractional values.

Example: $\frac{1}{5}$ = .2 seconds

Solution: $1 \div 5$

41. $\frac{1}{30}$ second = _____ seconds	46. $\frac{3}{50}$ second = _____ seconds
42. $\frac{1}{50}$ second = _____ seconds	47. $\frac{1}{250}$ second = _____ seconds
43. $\frac{1}{2}$ second = _____ seconds	48. $\frac{1}{4}$ second = _____ seconds
44. $\frac{3}{10}$ second = _____ seconds	49. $\frac{3}{8}$ second = _____ seconds
45. $\frac{2}{5}$ second = _____ seconds	50. $\frac{2}{3}$ second = _____ seconds

Decimal Multiplication

1.-20. In the following, state the decimal that is two times the stated value.

Example: .05 x 2 = .10

1. .07 _____	8. .009 _____	15. .08 _____
2. .15 _____	9. .60 _____	16. .33 _____
3. .012 _____	10. .032 _____	17. .45 _____
4. .004 _____	11. .125 _____	18. .062 _____
5. .05 _____	12. .005 _____	19. .035 _____
6. .25 _____	13. .03 _____	20. .002 _____
7. .75 _____	14. .23 _____	

21.-40. In the following, state the decimal that is four times the stated value.

Example: .05 x 4 = .20

21. .06 _____	28. .005 _____	35. .85 _____
22. .15 _____	29. .40 _____	36. .092 _____
23. .013 _____	30. .022 _____	37. .008 _____
24. .002 _____	31. .140 _____	38. .104 _____
25. .035 _____	32. .001 _____	39. .017 _____
26. .33 _____	33. .075 _____	40. .46 _____
27. .125 _____	34. .62 _____	

41.-50. In the following, state the squared value for the stated value.

Example: $3^2 = 9$ (3 x 3 = 9)

41. 2 = _____	45. 180 = _____	49. 6 = _____
42. 36 = _____	46. 40 = _____	50. 100 = _____
43. 150 = _____	47. 30 = _____	
44. 72 = _____	48. 200 = _____	

Decimal Multiplication

1.-20. In the following, state the decimal that is two times the stated value.

Example: .05 x 2 = .10

1.   .09 _____
2.   .16 _____
3. .012 _____
4. .007 _____
5.   .07 _____
6.   .35 _____
7.   .85 _____

8. .006 _____
9. .80 _____
10. .052 _____
11. .125 _____
12. .003 _____
13.   .06 _____
14.   .44 _____

15. .038 _____
16.   .66 _____
17.   .48 _____
18. .032 _____
19. .075 _____
20. .002 _____

21.-40. In the following, state the decimal that is four times the stated value.

Example: .05 x 4 = .20

21.   .04 _____
22.   .10 _____
23. .023 _____
24. .004 _____
25. .025 _____
26.   .16 _____
27. .125 _____

28. .003 _____
29.   .80 _____
30. .033 _____
31. .240 _____
32. .007 _____
33. .095 _____
34.   .82 _____

35.   .65 _____
36. .072 _____
37. .006 _____
38. .204 _____
39. .037 _____
40.   .03 _____

41.-50. In the following, state the squared value for the stated value.

Example: $4^2 = 16$ (4 x 4 = 16)

41.   4 = _____
42.   32 = _____
43. 140 = _____
44.   92 = _____

45. 180 = _____
46.   60 = _____
47.   40 = _____
48. 400 = _____

49.   63 = _____
50. 200 = _____

Decimal Division

1.-20. State the decimal that is one-half of the stated value.

Example: .05 x 1/2 = .025   or   .05 x .5 = .025

1. .10 _____
2. .46 _____
3. .72 _____
4. .004 _____
5. .026 _____
6. 1.5 _____
7. .006 _____

8. .010 _____
9. .20 _____
10. .07 _____
11. .38 _____
12. .042 _____
13. .18 _____
14. .012 _____

15. 2.0 _____
16. .066 _____
17. .036 _____
18. .002 _____
19. .016 _____
20. .84 _____

21.-40. State the decimal that is one-quarter of the stated value.

Example: .05 x 1/4 = .0125   or   .05 x .25 = .0125

21. .06 _____
22. .004 _____
23. .82 _____
24. 1.6 _____
25. .120 _____
26. .036 _____
27. .76 _ _____

28. .008 _____
29. .32 _____
30. 2.4 _____
31. .60 _____
32. .08 _____
33. .40 _____
34. .90 _____

35. .010 _____
36. .30 _____
37. .44 _____
38. 3.6 _____
39. .68 _____
40. .88 _____

41.-50. State the square root for the stated values.

Example: $\sqrt{400} = 20$

41. 49 _____
42. 1600 _____
43. 36 _____
44. 5184 _____

45. 10,000 _____
46. 22,500 _____
47. 100 _____
48. 14,400 _____

49. 1296 _____
50. 25 _____

Decimal Division

1.-20. State the decimal that is one-half of the stated value.

Example: .06 x 1/2 = .030 or .06 x .5 = .030

1. .30 _____	8. .040 _____	15. 1.0 _____
2. .86 _____	9. .70 _____	16. .016 _____
3. .92 _____	10. .09 _____	17. .076 _____
4. .006 _____	11. .88 _____	18. .004 _____
5. .046 _____	12. .052 _____	19. .026 _____
6. 1.6 _____	13. .78 _____	20. .64 _____
7. .14 _____	14. .012 _____	

21.-40. State the decimal that is one-quarter of the stated value.

Example: .06 x 1/4 = .015 or .06 x .25 = .015

21. .09 _____	28. .004 _____	35. .030 _____
22. .008 _____	29. .12 _____	36. .28 _____
23. .84 _____	30. 4.8 _____	37. .52 _____
24. 2.4 _____	31. .30 _____	38. 1.2 _____
25. .220 _____	32. .02 _____	39. .64 _____
26. .018 _____	33. .60 _____	40. .16 _____
27. .94 _____	34. .76 _____	

41.-50. State the square root for the stated values.

Example: $\sqrt{196}$ = 14

41. 64 _____	45. 3600 _____	49. 4225 _____
42. 1521 _____	46. 19,600 _____	50. 81 _____
43. 441 _____	47. 5184 _____	
44. 1936 _____	48. 14,400 _____	

Scientific Notation

1.-20. Place the following ordinary numbers into the scientific notation system.

Example 1: $8600 = 8.6 \times 10^3$ Example 2: $.0065 = 6.5 \times 10^3$

1.	26 _____	8.	7100 _____	15.	.000605 _____
2.	340 _____	9.	98,000 _____	16.	.0000083 _____
3.	6700 _____	10.	1,200,000 _____	17.	.0000500 _____
4.	82,000 _____	11.	.003 _____	18.	.00307 _____
5.	163,000 _____	12.	.00045 _____	19.	.49 _____
6.	790 _____	13.	.01 _____	20.	.0095 _____
7.	84 _____	14.	.000007 _____		

21.-40. State the ordinary number equivalent for those in scientific notation.

Example 1: $3.6 \times 10^3 = 3600$ Example 2: $7.2 \times 10^3 = .0072$

21.	9.3×10^3 _____	28.	8.7×10^5 _____	35.	3.3×10^{-1} _____
22.	3.2×10^6 _____	29.	4.66×10^2 _____	36.	5.9×10^{-5} _____
23.	8.1×10^4 _____	30.	7.3×10^7 _____	37.	1.7×10^{-7} _____
24.	6.02×10^4 _____	31.	1.2×10^{-3} _____	38.	6.9×10^{-3} _____
25.	3.1×10^2 _____	32.	9.5×10^{-6} _____	39.	7.4×10^{-4} _____
26.	5.5×10^1 _____	33.	5.3×10^4 _____	40.	8.2×10^2 _____
27.	6.8×10^3 _____	34.	2.6×10^{-2} _____		

41.-50. Give the English unit equivalents for the following metric system values.

41.	200 cm = _____ in.	45.	180 cm = _____ in.	49.	90 cm = _____ in.
42.	10Kg = _____ lb.	46.	4 litres = _____ gal.	50.	65Kg = _____ lb.
43.	30 litres = _____ qt.	47.	100 Kg = _____ lb.		
44.	50 Km = _____ miles	48.	100 cm _____ in.		

The following conversions have been used to calculate these problems.

2.5 cm = 1 inch

1 Kg = 2.2 pounds

1 Km = .6 miles

1 litre = 1 quart

Scientific Notation

1.-20. Place the following ordinary numbers into the scientific notation system.

Example 1: 32000 = 3.2 x 10^4 Example 2: .00037 = 3.7 x 10^4

1.	17 _____	8.	6200 _____	15.	.00008 _____
2.	250 _____	9.	89,000 _____	16.	.000009 _____
3.	5800 _____	10.	2,300,000 _____	17.	.61 _____
4.	73,000 _____	11.	.004 _____	18.	.009 _____
5.	172,000 _____	12.	.0021 _____	19.	.00089 _____
6.	970 _____	13.	.018 _____	20.	.00025 _____
7.	48 _____	14.	.00038 _____		

21.-40. State the ordinary number equivalent for those in scientific notation.

Example 1: 7.2 x 10^4 = 72,000   Example 2: 3.3 x 10^4 = .00033

21.	1.3 x 10^2 _____	28.	3.4 x 10^2 _____	35.	1.6 x 10^{-1} _____
22.	2.5 x 10^4 _____	29.	9.9 x 10^4 _____	36.	3.3 x 10^4 _____
23.	2.8 x 10^5 _____	30.	8.4 x 10^5 _____	37.	8.9 x 10^7 _____
24.	1.28 x 10^3 _____	31.	4.1 x 10^3 _____	38.	2.2 x 10^{-3} _____
25.	7.6 x 10^7 _____	32.	5.6 x 10^4 _____	39.	5.1 x 10^{-2} _____
26.	6.5 x 10^6 _____	33.	7.3 x 10^{-6} _____	40.	1.3 x 10^{-3} _____
27.	5.6 x 10^1 _____	34.	4.7 x 10^2 _____		

41.-50. Give the metric unit equivalents for the following English system values.

41.	36 in. = _____ cm	45.	60 in. = _____ cm	49.	5 qts. = _____ cc
42.	110 lbs. = _____ Kg	46.	121 lbs. = _____ Kg	50.	18 ins. = _____ cm
43.	60 qts. = _____ litres	47.	10 gals. = _____ litres		
44.	87 miles = _____ Km	48.	12 ins. _____ cm		

The following conversions have been used to calculate these problems.

2.5 cm = 1 inch

1 Kg = 2.2 pounds

1 Km = .6 miles

1 litre = 1 quart

mAs Conversion

1. A radiograph is taken using 200 mA setting and .05 second exposure time. What mAs was used for this technique?

State the mAs that will result for the following mA and time value

mA	Time	mAs	mA	Time	mAs
2. 100	.10	_____	11. 300	.01	_____
3. 100	.05	_____	12. 300	.33	_____
4. 100	.25	_____	13. 300	.75	_____
5. 100	.50	_____	14. 400	.03	_____
6. 200	.2	_____	15. 400-	.40	_____
7. 200	.03	_____	16. 400	.05	_____
8. 200	.10	_____	17. 600	.50	_____
9. 200	.30	_____	18. 600	.15	_____
10. 200	.25	_____	19. 600	.30	_____

20. A radiograph is taken using 40 mAs at 400 mA setting exposure. What time factor was used for this technique?

State the time factor that will result for the following mAs and mA values.

mAs	mA	Time	mAs	mA	Time
21. 50	100	_____	31. 15	150	_____
22. 100	200	_____	32. 25	400	_____
23. 60	300	_____	33. 50	1000	_____
24. 20	600	_____	34. 80	800	_____
25. 20	800	_____	35. 30	600	_____
26. 5	400	_____	36. 2	200	_____
27. 50	500	_____	37. 3	300	_____
28. 30	300	_____	38. 75	100	_____
29. 25	100	_____	39. 150	100	_____
30. 40	200	_____	40. 600	200	_____

41. A radiograph is taken using 20 mAs at .05 second exposure time setting. What mA setting was used for this technique?

State the mA setting that will result for the following mAs and time values.

mAs	Time	mA	mAs	Time	mA
42. 50	.10	_____	47. 15	.01	_____
43. 40	.05	_____	48. 75	.75	_____
44. 25	.20	_____	49. 35	.35	_____
45. 30	.05	_____	50. 40	.25	_____
46. 100	.08	_____			

The following formulas have been used to solve these problems.

$$mAs = mA \times Time \qquad mA = \frac{mAs}{Time} \qquad Time = \frac{mAs}{mA}$$

mAs Conversion

1. A radiograph is taken using 100 mA setting and .05 second exposure time. What mAs was used for this technique?

State the mAs that will result for the following mA and time values.

mA	Time	mAs	mA	Time	mAs
2. 100	.15	_____	11. 300	.04	_____
3. 100	.35	_____	12. 300	.012	_____
4. 100	.005	_____	13. 300	.008	_____
5. 100	.75	_____	14. 400	.024	_____
6. 200	.30	_____	15. 400	.46	_____
7. 200	.02	_____	16. 400	.018	_____
8. 200	.20	_____	17. 600	.60	_____
9. 200	.60	_____	18. 600	.44	_____
10. 200	.15	_____	19. 600	.10	_____

20. A radiograph is taken using 40 mAs at 400 mA setting. What time exposure factor was used for this technique?

State the time factor that will result for the following mAs and mA values.

mAs	mA	Time	mAs	mA	Time
21. 30	100	_____	31. 100	100	_____
22. 50	200	_____	32. 25	200	_____
23. 150	300	_____	33. 250	1000	_____
24. 30	600	_____	34. 40	800	_____
25. 20	600	_____	35. 10	300	_____
26. 15	300	_____	36. 75	200	_____
27. 60	600	_____	37. 45	300	_____
28. 30	900	_____	38. 75	600	_____
29. 75	300	_____	39. 250	100	_____
30. 80	400	_____	40. 400	200	_____

41. A radiograph is taken using 20 mAs at .05 second exposure setting. What mA setting was used for this technique?

State the mA setting that will result for the following mAs and time values.

mAs	Time	mA	mAs	Time	mA
42. 100	.20	_____	47. 225	.75	_____
43. 20	.025	_____	48. 70	.35	_____
44. 60	.20	_____	49. 48	.16	_____
45. 96	.24	_____	50. 20	.04	_____
46. 200	.25	_____			

The following formulas have been used to solve these problems.

$$mAs = mA \times Time \qquad mA = \frac{mAs}{Time} \qquad Time = \frac{mAs}{mA}$$

Algebraic Principles

1.-20. In the following equations, solve for the unknown quantity.

Example 1: $2a = 4$ $a = \dfrac{4}{2}$ $a = 4 \div 2$ $a = 2$ Example 2: $4a = 1$ $a = \dfrac{1}{4}$ $a = 1 \div 4$ $a = .25$

Example 3: $2a = 5 \times 4$ $a = \dfrac{5 \times 4}{2}$ $a = \dfrac{20}{2}$ $a = 20$ $a = 10$

1. $4a = 20$ $a =$ _____
2. $6a = 24$ $a =$ _____
3. $8y = 64$ $y =$ _____
4. $3x = 36$ $x =$ _____
5. $5x = 80$ $x =$ _____
6. $2a = 12$ $a =$ _____
7. $4a = 100$ $a =$ _____

8. $8a = 2$ $a =$ _____
9. $32a = 4$ $a =$ _____
10. $16y = 4$ $y =$ _____
11. $5x = 1$ $x =$ _____
12. $25a = 10$ $a =$ _____
13. $40a = 4$ $a =$ _____
14. $50a = 25$ $a =$ _____

15. $3a = 5 \times 3$ $a =$ _____
16. $4y = 16 \times 4$ $y =$ _____
17. $7y = 49 \times 7$ $y =$ _____
18. $ay = b$ $y =$ _____
19. $xy = b$ $x =$ _____
20. $ab = cd$ $b =$ _____

21.-40. In the following equations, solve for the unknown quantity.

Example 1: $\dfrac{a}{4} = 8$ $a = 8 \times 4$ $a = 32$ Example 2: $\dfrac{a}{3} = 12 \times 2$ $a = 12 \times 2 \times 3$ $a = 72$

Example 3: $\dfrac{a}{6} = \dfrac{9}{3}$ $a = \dfrac{9 \times 6}{3}$ $a = \dfrac{54}{3}$ $a = 54 \div 3$ $a = 18$

Tip: The expression $\dfrac{a}{2}$ is mathematically the same as $\dfrac{1}{2}a$

21. $\dfrac{a}{6} = 3$ $a =$ _____
22. $\dfrac{a}{4} = 2$ $a =$ _____
23. $\dfrac{x}{9} = 4$ $x =$ _____
24. $\dfrac{y}{3} = 7$ $y =$ _____
25. $\dfrac{3}{4}a = 3$ $a =$ _____
26. $\dfrac{2}{5}y = 40$ $y =$ _____
27. $\dfrac{3}{7}x = 21$ $x =$ _____

28. $\dfrac{a}{3} = 10 \times 2$ $a =$ _____
29. $\dfrac{a}{6} = 4 \times 3$ $a =$ _____
30. $\dfrac{x}{4} = 5 \times 7$ $x =$ _____
31. $\dfrac{y}{5} = 10 \times 12$ $y =$ _____
32. $\dfrac{2}{5}y = 6 \times 3$ $y =$ _____
33. $\dfrac{2}{3}b = 12 \times 3$ $b =$ _____
34. $\dfrac{3}{20}a = 7 \times 6$ $a =$ _____

35. $\dfrac{a}{2} = \dfrac{10}{4}$ $a =$ _____
36. $\dfrac{a}{6} = \dfrac{24}{8}$ $a =$ _____
37. $\dfrac{y}{5} = \dfrac{40}{8}$ $a =$ _____
38. $\dfrac{x}{b} = c$ $x =$ _____
39. $\dfrac{y}{x} = cb$ $y =$ _____
40. $\dfrac{a}{y} = x$ $a =$ _____

41.-50. In the following equations, solve for the unknown quantity.

41. $\dfrac{mAs}{Time} = mA$ $mAs =$ _____
42. $D \times It = Log\ Io$
43. $mf \times SOD = SID$ $D =$ _____
44. $I \times R = V$ $mf =$ _____
45. $\dfrac{E}{v} = h$ $I =$ _____
   $E =$ _____

46. $D \times F = W$ $F =$ _____
47. $\dfrac{SOD \times P}{FSS} = OID$ $P =$ _____
48. $kV \times = 12.4$ $\lambda =$ _____
49. $\dfrac{Is\ Ns}{Ip} = Np$ $Is =$ _____
50. $Rad \times QF = Rem$ $Rad =$ _____

Algebraic Principles

1.-20. In the following equations, solve for the unknown quantity.

Example 1: $3a = 6$ $a = \dfrac{6}{3}$ $a = 6 \div 3$ $a = 2$ Example 2: $a = \dfrac{2}{8}$ $a = 2 \div 8$ $a = .25$

Example 3: $3a = 6 \times 5$ $a = \dfrac{6 \times 5}{3}$ $a = \dfrac{30}{3}$ $a = 30 \div 3$ $a = 10$

1. $2a = 44$	$a =$ ___	8. $7a = 28$	$a =$ ___	15. $6a = 9 \times 8$	$a =$ ___
2. $4x = 28$	$x =$ ___	9. $6b = 3$	$b =$ ___	16. $5y = 25 \times 8$	$y =$ ___
3. $6y = 42$	$y =$ ___	10. $3a = 1$	$a =$ ___	17. $4b = 32 \times 6$	$b =$ ___
4. $3b = 81$	$b =$ ___	11. $4x = 2$	$x =$ ___	18. $3x = 27 \times 9$	$x =$ ___
5. $7y = 42$	$y =$ ___	12. $30y = 5$	$y =$ ___	19. $xyb = ac$	$y =$ ___
6. $8x = 72$	$x =$ ___	13. $40x = 8$	$x =$ ___	20. $ax = cd$	$x =$ ___
7. $3a = 75$	$a =$ ___	14. $80a = 10$	$a =$ ___		

21.-40. In the following equations, solve for the unknown quantity.

Example 1: $\dfrac{a}{5} = 10$ $a = 10 \times 5$ $a = 50$ Example 2: $\dfrac{a}{6} = 3 \times 4$ $a = 3 \times 4 \times 6$ $a = 72$

Example 3: $\dfrac{a}{3} = \dfrac{18}{6}$ $a = \dfrac{18 \times 3}{6}$ $a = \dfrac{54}{6}$ $a = 9$

Tip: The expression $\dfrac{a}{2}$ is mathematically the same as $\dfrac{1}{2}a$

21. $\dfrac{a}{7} = 5$	$a =$ ___	28. $\dfrac{a}{3} = 6 \times 4$	$a =$ ___	35. $\dfrac{a}{3} = \dfrac{10}{6}$	$a =$ ___
22. $\dfrac{b}{4} = 9$	$b =$ ___	29. $\dfrac{x}{7} = 3 \times 7$	$x =$ ___	36. $\dfrac{x}{4} = \dfrac{20}{2}$	$x =$ ___
23. $\dfrac{x}{6} = 12$	$x =$ ___	30. $\dfrac{b}{6} = 4 \times 8$	$b =$ ___	37. $\dfrac{y}{5} = \dfrac{50}{25}$	$y =$ ___
24. $\dfrac{b}{3} = 42$	$b =$ ___	31. $\dfrac{y}{6} = 9 \times 11$	$y =$ ___	38. $\dfrac{x}{a} = \dfrac{c}{b}$	$x =$ ___
25. $\dfrac{2}{3}a = 30$	$a =$ ___	32. $\dfrac{3}{4}a = 8 \times 6$	$a =$ ___	39. $\dfrac{xy}{a} = cb$	$y =$ ___
26. $\dfrac{3}{4}y = 12$	$y =$ ___	33. $\dfrac{4}{9}y = 6 \times 2$	$y =$ ___	40. $\dfrac{xa}{b} = \dfrac{c}{d}$	$x =$ ___
27. $\dfrac{4}{5}x = 20$	$x =$ ___	34. $\dfrac{2}{3}b = 5 \times 6$	$b =$ ___		

41.-50. Solve the following equations for the stated quantity.

41. $\dfrac{PE}{D} = F$	$PE =$ ___	46. $\dfrac{Vs}{Vp} = \dfrac{Ns}{Np}$	$Vs =$ ___
42. $\dfrac{C}{2r} = \pi$	$C =$ ___	47. $\dfrac{mAs}{mA} = Time$	$mAs =$ ___
43. $\dfrac{rem}{QF} = Rad$	$rem =$ ___	48. $MF \times SOD = SID$	$MF =$ ___
44. $\dfrac{2A}{b} = h$	$A =$ ___	49. $Is \times Vs = IpVp$	$Is =$ ___
45. $\dfrac{V}{R} = I$	$V =$ ___	50. $\dfrac{5}{9}C = (F° 32)$	$C° =$ ___

Algebraic Principles (Squares)

1.-20. In the following equations solve for the unknown quantity.

Example 1: $a^2 = 36$ $a = \sqrt{36}$ $a = 6$

Example 2: $2a^2 = 8$ $a^2 = \dfrac{8}{2}$ $a = \sqrt{4}$ $a = 2$

Example 3: $a^2 = 8$ $a^2 = 8 \times 2$ $a = \sqrt{16}$ $a = 4$

1. $a^2 = 64$ $a =$ _____
2. $x^2 = 81$ $x =$ _____
3. $b^2 = 25$ $b =$ _____
4. $y^2 = 25$ $y =$ _____
5. $a^2 = 121$ $a =$ _____
6. $x^2 = 169$ $x =$ _____
7. $y^2 = 361$ $y =$ _____

8. $2a^2 = 72$ $a =$ _____
9. $3a^2 = 243$ $a =$ _____
10. $2x^2 = 338$ $x =$ _____
11. $5y^2 = 245$ $y =$ _____
12. $6a^2 = 54$ $a =$ _____
13. $3b^2 = 48$ $b =$ _____
14. $7x^2 = 847$ $x =$ _____

15. $\dfrac{a^2}{2} = 200$ $a =$ _____
16. $\dfrac{a^2}{3} = 27$ $a =$ _____
17. $\dfrac{y^2}{3} = 147$ $y =$ _____
18. $\dfrac{1}{4}x^2 = 81$ $x =$ _____
19. $\dfrac{1}{8}x^2 = 50$ $x =$ _____
20. $\dfrac{1}{4}y^2 = 64$ $y =$ _____

21.-40. In the following equations solve for the unknown quantity.

Example 1: $a^2 = 4 \times 16$ $a = \sqrt{4 \times 16}$ $a = \sqrt{64}$ $a = 8$

Example 2: $3a^2 = 4 \times 27$ $a = \sqrt{4 \times 27 \div 3}$ $a = \sqrt{36}$ $a = 6$

Example 3: $\dfrac{a^2}{3} = 2 \times 6$ $a = \sqrt{2 \times 6 \times 3}$ $a\sqrt{36}$ $a = 6$

21. $a^2 = 10 \times 10$ $a =$ _____
22. $x^2 = 40 \times 10$ $x =$ _____
23. $a^2 = 8 \times 32$ $a =$ _____
24. $y^2 = 5 \times 45$ $y =$ _____
25. $x^2 = 4 \times 49$ $x =$ _____
26. $b^2 = 2 \times 72$ $b =$ _____
27. $x^2 = 12 \times 48$ $x =$ _____

28. $2a^2 = 121 \times 2$ $a =$ _____
29. $3a^2 = 96 \times 18$ $a =$ _____
30. $2y^2 = 100 \times 8$ $y =$ _____
31. $5y^2 = 25 \times 125$ $y =$ _____
32. $4b^2 = 27 \times 12$ $b =$ _____
33. $6x^2 = 225 \times 24$ $x =$ _____
34. $4a^2 = 882 \times 2$ $a =$ _____

35. $a^2/2 = 648$ $a =$ _____
36. $a^2/3 = 60/5$ $a =$ _____
37. $a^2/8 = 224/7$ $a =$ _____
38. $1/2x^2 = 1014/3$ $x =$ _____
39. $1/4x^2 = 288/8$ $x =$ _____
40. $1/3x^2 = 735/5$ $x =$ _____

Algebraic Principles (Squares)

41.-50. Solve the following equations for the stated quanity.

41. $\dfrac{E}{C^2} = m$ E = _____

42. $\dfrac{2E}{V^2} = m$ E = _____

43. $Is \times Vs = Ip \times Vp$ Vs = _____

44. $\dfrac{New\ SID^2}{Original\ SID^2} = \dfrac{Original\ Dose}{New\ Dose}$ New SID² = _____

45. $\dfrac{New\ SID^2}{Original\ SID^2} = \dfrac{Original\ Dose}{New\ Dose}$ New SID = _____

46. $D^2 F = h\ Q1 \times Q^2$ F = _____

47. $\dfrac{New\ mAs}{New\ SID^2} = \dfrac{Original\ mAs}{Original\ SID^2}$ New mAs = _____

48. $\dfrac{New\ SID^2}{Original\ SID^2} = \dfrac{New\ mA}{Original\ mA}$ New SID² = _____

49. $\dfrac{New\ SID^2}{Original\ SID^2} = \dfrac{New\ Time}{Original\ Time}$ New SID² = _____

50. $Original\ Time \times \dfrac{New\ SID^2}{Original\ SID^2} = New\ Time$ New SID = _____

Algebraic Principles (Squares)

1.-20. In the following equations, solve for the unknown quantity.

Example 1: $a^2 = 16$ $a = \sqrt{16}$ $a = 4$

Example 2: $2a^2 = 18$ $a^2 = \dfrac{2}{18}$ $a^2 = 9$ $a = \sqrt{9}$ $a = 3$

Example 3: $\dfrac{a^2}{2} = 8$ $a^2 = 8 \times 2$ $a^2 = 16$ $a = \sqrt{16}$ $a = 4$

1. $a^2 = 49$ $a =$ _____

2. $x^2 = 100$ $x =$ _____

3. $b^2 = 144$ $b =$ _____

4. $y^2 = 256$ $y =$ _____

5. $a^2 = 324$ $a =$ _____

6. $x^2 = 16$ $x =$ _____

7. $y^2 = 441$ $y =$ _____

8. $2y^2 = 162$ $y =$ _____

9. $6a^2 = 150$ $a =$ _____

10. $4z^2 = 64$ $a =$ _____

11. $5x^2 = 45$ $x =$ _____

12. $7x^2 = 700$ $x =$ _____

13. $3a^2 = 363$ $a =$ _____

14. $9y^2 = 1521$ $y =$ _____

15. $\dfrac{x^2}{2} = 288$ $x =$ _____

16. $\dfrac{a^2}{3} = 12$ $a =$ _____

17. $\dfrac{y^2}{4} = 49$ $y =$ _____

18. $\dfrac{1}{3} z^2 = 192$ $a =$ _____

19. $\dfrac{1}{6} x^2 = 54$ $x =$ _____

20. $\dfrac{1}{4} y^2 = 100$ $y =$ _____

21.-40. In the following equations, solve for the unkonwn quanity.

Example 1: $a^2 = 3 \times 12$ $a = \sqrt{3 \times 12}$ $a\sqrt{36}$ $a = 6$

Example 2: $2a^2 = \dfrac{2 \times 9}{2}$ $a = \dfrac{\sqrt{2 \times 9}}{2}$ $a = \sqrt{9}$ $a = 3$

Example 3: $\dfrac{a^2}{2} = 3 \times 6$ $a = \sqrt{3 \times 6 \times 2}$ $a = \sqrt{36}$ $a = 6$

21. $a^2 = 4 \times 81$ $a =$ _____

22. $a^2\, 8 \times 32$ $a =$ _____

23. $y^2 = 4 \times 49$ $y =$ _____

24. $x^2 = 12 \times 27$ $x =$ _____

25. $b^2 = 2 \times 32$ $b =$ _____

26. $a^2 = 3 \times 27$ $a =$ _____

27. $z^2 = 12 \times 48$ $z =$ _____

28. $2a^2 = 24 \times 3$ $a =$ _____

29. $3x^2 = 48 \times 36$ $x =$ _____

30. $2y^2 = 81 \times 8$ $y =$ _____

31. $5b^2 = 20 \times 4$ $b =$ _____

32. $2y^2 = 72 \times 4$ $y =$ _____

33. $6b^2 = 450 \times 3$ $b =$ _____

34. $4a^2 = 882 \times 2$ $a =$ _____

35. $\dfrac{a^2}{2} = \dfrac{1568}{4}$ $a =$ _____

36. $\dfrac{a^2}{3} = \dfrac{200}{6}$ $a =$ _____

37. $\dfrac{b^2}{6} = \dfrac{162}{12}$ $b =$ _____

38. $\dfrac{1}{4}x^2 = \dfrac{338}{8}$ $x =$ _____

39. $\dfrac{1}{2}b^2 = \dfrac{968}{4}$ $b =$ _____

40. $\dfrac{1}{3}y^2 = \dfrac{128}{6}$ $y =$ _____

41.-50. Solve the following equations for the state quantity.

41. $\dfrac{P}{I^2} = R$ $P =$ ————

42. $\dfrac{A}{r^2} = \pi$ $A =$ ————

43. $\dfrac{E}{M} = C^2$ $E =$ ————

44. $\dfrac{A}{b} = h$ $A =$ ————

45. $\dfrac{\text{New mA}}{\text{Original mA}} = \dfrac{\text{New SID}^2}{\text{Original SID}^2}$ New mA = ————

46. $\dfrac{\text{New SID}^2}{\text{Original SID}^2} = \dfrac{\text{New mA}}{\text{Original mA}}$ New SID^2 = ————

47. $\dfrac{\text{Original Dose}}{\text{New Dose}} = \dfrac{\text{New SSD}^2}{\text{Original SSD}^2}$ New Dose = ————

48. $\dfrac{\text{New SSD}^2}{\text{Original SSD}^2} = \dfrac{\text{Original Dose}}{\text{New Dose}}$ New SSD^2 = ————

49. $\dfrac{\text{New SSD}^2}{\text{Original SSD}^2} = \dfrac{\text{Original Dose}}{\text{New Dose}}$ New SSD = ————

50. $\dfrac{\text{Original mAs} \times \text{New SID}^2}{\text{Original SID}^2} = \text{New mAs}$ New SID = ————

Answers

Millisecond Conversion

Exercise No. 1

1. .8.	16. .350	31. 5	46. .008
2. .13.	17. .175	32. 90	47. .004
3. .08.	18. .035	33. 900	48. .002
4. .016.	19. .045	34. 170	49. .3
5. .005.	20. .001	35. 500	50. .75
6. .75.	21. 70.	36. 610	
7. .2.	22. 1.	37. 160	
8. .088	23. 800	38. 345	
9. .024	24. 30	39. 40	
10. .003	25. 4	40. 21	
11. .54	26. 650	41. .1	
12. .125	27. 310	42. .05	
13. .018	28. 95	43. .025	
14. .030	29. 20	44. .016	
15. .007	30. 60	45. .066	

Decimal Multiplication

Exercise No. 2

1. .18	16. 1.32	31. .96	46. 3600
2. .32	17. .98	32. .028	47. 1600
3. .024	18. .064	33. .38	48. 160,000
4. .014	19. .15	34. 3.28	49. 3969
5. .14	20. .004	35. 2.6	50. 40,000
6. .70	21. .16	36. .288	
7. 1.7	22. .40	37. .024	
8. .012	23. .092	38. .816	
9. 1.6	24. .016	39. .148	
10. .104	25. .1	40. .12	
11. .250	26. .48	41. 16	
12. .006	27. .5	42. 1024	
13. .12	28. .012	43. 19600	
14. .88	29. 3.2	44. 8464	
15. .072	30. .132	45. 32,400	

Millisecond Conversion

Exercise No. 2

1. .6	16. .13	31. 1	46. .06
2. .23	17. .275	32. 60	47. .004
3. .18	18. .075	33. 100	48. .25
4. .036	19. .035	34. 270	49. .375
5. .007	20. .005	35. 600	50. .666
6. .95	21. 90	36. 110	
7. .15	22. 3	37. 120	
8. .084	23. 500	38. 315	
9. .026	24. 80	39. 48	
10. .003	25. 7	40. 31	
11. .34	26. 250	41. .033	
12. .225	27. 530	42. .02	
13. .028	28. 75	43. .5	
14. .04	29. 40	44. .3	
15. .008	30. 20	45. .4	

Decimal Division

Exercise No. 1

1. .05	16. .033	31. .15	46. 150
2. .23	17. .018	32. .02	47. 10
3. .36	18. .001	33. .10	48. 120
4. .002	19. .008	34. .225	49. 36
5. .013	20. .42	35. .0025	50. 5
6. .75	21. .015	36. .075	
7. .003	22. .001	37. .110	
8. .005	23. .205	38. .9	
9. .10	24. .4	39. .17	
10. .035	25. .03	40. .22	
11. .19	26. .009	41. 7	
12. .021	27. .19	42. 40	
13. .09	28. .002	43. 6	
14. .006	29. .08	44. 72	
15. 1.0	30. .6	45. 100	

Decimal Multiplication

Exercise No. 1

1. .14	16. .66	31. .56	46. 1600
2. .30	17. .90	32. .004	47. 900
3. .024	18. .124	33. .30	48. 40,000
4. .008	19. .07	34. 2.48	49. 36
5. .10	20. .004	35. 3.40	50. 10,000
6. .50	21. .24	36. .368	
7. 1.5	22. .60	37. .032	
8. .018	23. .052	38. .416	
9. 1.2	24. .008	39. .068	
10. .064	25. .14	40. 1.84	
11. .250	26. 1.32	41. 4	
12. .010	27. .50	42. 1296	
13. .06	28. .02	43. 22,500	
14. .46	29. 1.6	44. 5184	
15. .16	30. .088	45. 32,400	

Decimal Division

Exercise No. 2

1. .15	16. .008	31. .075	46. 140
2. .43	17. .038	32. .005	47. 72
3. .46	18. .002	33. .15	48. 120
4. .003	19. .013	34. .19	49. 65
5. .023	20. .32	35. .0075	50. 9
6. .8	21. .0225	36. .007	
7. .07	22. .002	37. .13	
8. .020	23. .21	38. .3	
9. .35	24. .6	39. .16	
10. .045	25. .055	40. .04	
11. .44	26. .0045	41. 8	
12. .026	27. .235	42. 39	
13. .39	28. .001	43. 21	
14. .006	29. .03	44. 44	
15. .5	30. 1.2	45. 60	

Scientific Notation

1. 2.6×10^1
2. 3.4×10^2
3. 6.7×10^3
4. 8.2×10^4
5. 1.63×10^5
6. 7.9×10^2
7. 8.4×10^1
8. 7.1×10^3
9. 9.8×10^4
10. 1.2×10^6
11. 3.0×10^{-3}
12. 4.5×10^{-4}
13. 1.0×10^{-2}
14. 7.0×10^{-6}
15. 6.05×10^{-4}
16. 8.3×10^{-6}
17. 5.0×10^{-5}
18. 3.07×10^{-3}
19. 4.9×10^{-1}
20. 9.5×10^{-3}
21. 9300
22. 3,200,000
23. 81,000
24. 60,200
25. 310
26. 55
27. 6800
28. 870,000
29. 466
30. 73,000,000
31. .0012
32. .0000095
33. .00053
34. .026
35. .33
36. .000059
37. .00000017
38. .0069
39. .00074
40. .082
41. 80 inches
42. 22 lbs.
43. 30 qts.
44. 30 miles
45. 72 inches
46. 1 gallon
47. 220 lbs.
48. 40 inches
49. 36 inches
50. 143 lbs.

Scientific Notation

1. 1.7×10^1
2. 2.5×10^2
3. 5.8×10^3
4. 7.3×10^4
5. 1.72×10^5
6. 9.7×10^2
7. 48×10^1
8. 6.2×10^3
9. 8.9×10^4
10. 2.3×10^6
11. 4.0×10^{-3}
12. 2.1×10^{-3}
13. 1.8×10^{-2}
14. 3.8×10^{-4}
15. 8.0×10^{-5}
16. 9.0×10^{-6}
17. 6.1×10^{-1}
18. 9.0×10^{-3}
19. 8.9×10^{-4}
20. 2.5×10^{-4}
21. 130
22. 25,000
23. 280,000
24. 1280
25. 76,000,000
26. 6,500,000
27. 56
28. 340
29. 99000
30. 840,000
31. .0041
32. .00056
33. .0000073
34. .047
35. .16
36. .00033
37. .00000089
38. .0022
39. .0051
40. .0013
41. 90 cm
42. 50 kg
43. 60 liters
44. 145 km
45. 150 cm
46. 55 kg
47. 40 liters
48. 30 cm
49. 5000 cc
50. 45 cm

mAs Conversion

1. 10 mAs
2. 10 mAs
3. 5 mAs
4. 25 mAs
5. 50 mAs
6. 40 mAs
7. 6 mAs
8. 20 mAs
9. 60 mAs
10. 50 mAs
11. 3 mAs
12. 99 mAs
13. 225 mAs
14. 12 mAs
15. 160 mAs
16. 20 mAs
17. 300 mAs
18. 90 mAs
19. 180 mAs
20. .1 sec.
21. .5 sec.
22. .5 sec.
23. .2 sec.
24. .033 sec.
25. .025 sec.
26. .125 sec.
27. .1 sec.
28. .1 sec.
29. .25 sec.
30. .2 sec.
31. .1 sec.
32. .0625 sec.
33. .05 sec.
34. .1 sec.
35. .05 sec.
36. .01 sec.
37. .01 sec.
38. .75 sec.
39. 1.5 sec.
40. 3 sec
41. 400 mA
42. 500 mA
43. 800 mA
44. 125 mA
45. 600 mAs
46. 1250 mA
47. 1500 mA
48. 100 mA
49. 100 mA
50. 160 mA

mAs Conversion

1. 5 mAs
2. 15 mAs
3. 35 mAs
4. .5 mAs
5. 75 mAs
6. 60 mAs
7. 4 mAs
8. 40 mAs
9. 120 mAs
10. 30 mAs
11. 12 mAs
12. 3.6 mAs
13. 2.4 mAs
14. 9.6 mAs
15. 184 mAs
16. 7.2 mAs
17. 360 mAs
18. 264 mAs
19. 60 mAs
20. .1 sec.
21. .3 sec.
22. .25 sec.
23. .5 sec.
24. .05 sec.
25. .033 sec.
26. .05 sec.
27. .1 sec.
28. .033 sec.
29. .25 sec.
30. .2 sec.
31. 1 sec.
32. .125 sec.
33. .25 sec.
34. .05 sec.
35. .033 sec.
36. .375 sec.
37. .15 sec.
38. .125 sec.
39. 2.5 sec.
40. 2 sec.
41. 400 mA
42. 500 mA
43. 800 mA
44. 300 mA
45. 400 mA
46. 800 mA
47. 300 mA
48. 200 mA
49. 300 mA
50. 500 mA

Algebraic Principles

1. 5
2. 4
3. 8
4. 12
5. 16
6. 6
7. 25
8. .25
9. .125
10. .25
11. .2
12. .4
13. .1
14. .5
15. 5
16. 16
17. 49
18. $\dfrac{b}{a}$
19. $\dfrac{b}{y}$
20. $\dfrac{cd}{a}$
21. 18
22. 8
23. 36
24. 21
25. 4
26. 100
27. 49
28. 60
29. 72
30. 140
31. 600
32. 45
33. $\sqrt{54}$
34. 280
35. 5
36. 18
37. 25
38. cb
39. cbx
40. xy
41. mA x Time
42. $\text{Log} \dfrac{IO}{IT}$
43. $\dfrac{SID}{SOD}$
44. $\dfrac{V}{R}$
45. hv
46. $\dfrac{W}{D}$
47. $\dfrac{FSS \times OID}{SOD}$
48. $\dfrac{12.4}{Kv}$
49. $\dfrac{1pNp}{Ns}$
50. $\dfrac{REM}{QF}$

Algebraic Principles

1. 22
2. 7
3. 7
4. 27
5. 6
6. 9
7. 25
8. 4
9. .5
10. .333
11. .5
12. .166
13. .2
14. .125
15. 12
16. 40
17. 48
18. 81
19. $\dfrac{ac}{xb}$
20. $\dfrac{cd}{a}$
21. 35
22. 36
23. 72
24. 126
25. 45
26. 16
27. 25
28. 72
29. 147
30. 192
31. 594
32. 64
33. 27
34. $\sqrt{45}$
35. 5
36. 40
37. 10
38. $\dfrac{ac}{b}$
39. $\dfrac{acb}{x}$
40. $\dfrac{cb}{da}$
41. FD
42. $\pi 2r$
43. Rad x QF
44. 1/2 bh
45. 1R
46. $\dfrac{Ns \times Vp}{Np}$
47. mA x Time
48. $\dfrac{SID}{SOD}$
49. $\dfrac{1pVp}{Vs}$
50. 9/5 (F°-32)

Algebraic Principles (Squares)
Exercise No. 1

1. 8	21. 10	41. mc^2
2. 9	22. 20	42. $1/2mv^2$
3. 5	23. 16	43. $\dfrac{Ip \times Vp}{Is}$
4. 5	24. 15	
5. 11	25. 14	44. $\dfrac{\text{Original Dose} \times \text{Original SID}^2}{\text{New Dose}}$
6. 13	26. 12	
7. 19	27. 24	45. $\sqrt{\dfrac{\text{Original Dose} \times \text{Original SID}^2}{\text{New Dose}}}$
8. 6	28. 11	
9. 9	29. 24	
10. 13	30. 20	46. $\dfrac{h \times Q1 \times Q2}{D^2}$
11. 7	31. 25	
12. 3	32. 9	47. $\dfrac{\text{Original mAs} \times \text{New SID}^2}{\text{Original SID}^2}$
13. 4	33. 30	
14. 11	34. 21	48. $\dfrac{\text{New mA} \times \text{Original SID}^2}{\text{Original MA}}$
15. 20	35. 36	
16. 9	36. 6	
17. 21	37. 16	49. $\dfrac{\text{New Time} \times \text{Original SID}^2}{\text{Original Time}}$
18. 18	38. 26	
19. 20	39. 12	50. $\sqrt{\dfrac{\text{New Time} \times \text{Original SID}^2}{\text{Original Time}}}$
20. 16	40. 21	

Algebraic Principles (Squares)
Exercise No. 2

1. 7	21. 18	41. I^2R
2. 10	22. 16	42. πr^2
3. 12	23. 14	43. mc^2
4. 16	24. 18	44. bh
5. 18	25. 8	45. $\dfrac{\text{New SID}^2 \times \text{Original mA}}{\text{Original SID}^2}$
6. 4	26. 9	
7. 21	27. 24	46. $\dfrac{\text{New mA} \times \text{Original SID}^2}{\text{Original mA}}$
8. 9	28. 6	
9. 5	29. 24	
10. 4	30. 18	47. $\dfrac{\text{Original SSD}^2 \times \text{Original Dose}}{\text{New SSD}^2}$
11. 3	31. 4	
12. 10	32. 12	48. $\dfrac{\text{Original Dose} \times \text{Original SSD}^2}{\text{New Dose}}$
13. 11	33. 15	
14. 13	34. 21	49. $\sqrt{\dfrac{\text{Original Dose} \times \text{Original SSD}^2}{\text{New Dose}}}$
15. 24	35. 28	
16. 6	36. 10	50. $\sqrt{\dfrac{\text{New mAs} \times \text{Original SID}^2}{\text{Original mAs}}}$
17. 14	37. 9	
18. 24	38. 13	
19. 18	39. 22	
20. 20	40. 8	